T0320388

This comprehensive text focuses on the homotopical technology in use at the forefront of modern algebraic topology. Following on from a standard introductory algebraic topology sequence, it will provide students with a comprehensive background in spectra and structured ring spectra. Each chapter is an extended tutorial by a leader in the field, offering the first really accessible treatment of the modern construction of the stable category in terms of both model categories of point-set diagram spectra and infinity-categories. It is one of the only textbook sources for operadic algebras, structured ring spectra, and Bousfield localization, which are now basic techniques in the field, and the book provides a rare expository treatment of spectral algebraic geometry. Together the contributors Emily Riehl, Daniel Dugger, Clark Barwick, Michael A. Mandell, Birgit Richter, Tyler Lawson, and Charles Rezk offer a complete, authoritative source to learn the foundations of this vibrant area.

Mathematical Sciences Research Institute
Publications

# 69

Stable Categories and
Structured Ring Spectra

# Mathematical Sciences Research Institute Publications

**Volumes 1–4, 6–8, and 10–27 are published by Springer-Verlag**

# Stable Categories
# and Structured Ring Spectra

*Edited by*

**Andrew J. Blumberg**
*Columbia University*

**Teena Gerhardt**
*Michigan State University*

**Michael A. Hill**
*University of California, Los Angeles*

Andrew J. Blumberg
Department of Mathematics
Columbia University
andrew.blumberg@columbia.edu

Teena Gerhardt
Department of Mathematics
Michigan State University
teena@math.msu.edu

Michael A. Hill
Department of Mathematics
University of California, Los Angeles
mikehill@math.ucla.edu

Silvio Levy (*Series Editor*)
Mathematical Sciences Research Institute
Berkeley, CA 94720
levy@msri.org

Frontispiece image: Jing Hu 胡菁

The Mathematical Sciences Research Institute wishes to acknowledge support by the National Science Foundation and the *Pacific Journal of Mathematics* for the publication of this series.

CAMBRIDGE UNIVERSITY PRESS
Cambridge, New York, Melbourne, Madrid, Cape Town, Singapore, São Paulo, Delhi, Dubai, Tokyo

Cambridge University Press
One Liberty Plaza, 20th Floor, New York, NY 10006, USA

www.cambridge.org
Information on this title: www.cambridge.org/9781009123297

© Mathematical Sciences Research Institute 2022

First published 2022

*A catalog record for this publication is available from the British Library.*

ISBN 978-1-009-12329-7 hardback

# Contents

# Contributors

**Emily Riehl**
Johns Hopkins University, Baltimore, MD

**Daniel Dugger**
University of Oregon, Eugene, OR

**Clark Barwick**
University of Edinburgh, Edinburgh, UK

**Michael A. Mandell**
Indiana University, Bloomington, IN

**Birgit Richter**
University of Hamburg, Hamburg, Germany

**Tyler Lawson**
University of Minnesota, Minneapolis, MN

**Charles Rezk**
University of Illinois, Urbana-Champaign, Illinois

# 1    Introduction

by Andrew J. Blumberg, Teena Gerhardt, Michael A. Hill

## 1.1    Goals of this book

The modern era in homotopy theory began in the 1960s with the profound realization, first codified by Boardman in his construction of *the stable category*, that the category of spaces up to stable homotopy equivalence is equipped with a rich algebraic structure, formally similar to the derived category of a commutative ring $R$. For example, for pointed spaces the natural map from the categorical coproduct to the categorical product becomes more and more connected as the pieces themselves become more and more connected. In the limit, this map becomes a stable equivalence, just as finitely indexed direct sums and direct products coincide for $R$-modules.

From this perspective, the objects of the stable category are modules over an initial commutative ring object that replaces the integers: the sphere spectrum. However, technical difficulties immediately arose. Whereas the tensor product of $R$-modules is an easy and familiar construction, the analogous construction of a symmetric monoidal smash product on spectra seemed to involve a huge number of ad hoc choices [1]. As a consequence, the smash product was associative and commutative only up to homotopy. The lack of a good point-set symmetric monoidal product on spectra precluded making full use of the constructions from commutative algebra in this setting — even just defining good categories of modules over a commutative ring spectrum was difficult. In many ways, finding ways to rectify this and to make the guiding metaphor provided by "modules over the sphere spectrum" precise has shaped the last 60 years of homotopy theory.

This book arose from a desire by the editors to have a reference to give to their students who have taken a standard algebraic topology sequence and who want to learn about spectra and structured ring spectra. While there are many excellent texts which introduce students to the basic ideas of homotopy theory and to spectra, there has not been a place for students to engage directly with the ideas needed to connect with commutative ring spectra and work with these objects. This book strives to provide an introduction to this whole circle of ideas, describing the tools that homotopy theorists have developed to build, explore, and use symmetric monoidal categories of spectra that refine the stable homotopy category:

1. model category structures on symmetric monoidal categories of spectra,
2. stable ∞-categories, and
3. operads and operadic algebras.

These three concepts are closely intertwined, and they all engage deeply with a fundamental principle: if the choices for some construction or map are parameterized by a space, then recording that space as part of the data makes the construction more natural.

To make this maxim precise in practice, we must keep track of the spaces of maps between objects in our categories, not just sets of maps, describing the homotopies by which two equivalent maps are seen to be equivalent. A first example of this is given by the cup product on ordinary cohomology. Students in a first algebraic topology class learn that while the cochains on a space with coefficients in a commutative ring are not a commutative ring, the cohomology of a space is canonically a graded commutative ring. Steenrod observed that over $\mathbb{F}_p$, we can keep track of cochains that enforce the symmetry between $a \smile b$ and $b \smile a$, and out of these, we can build a hierarchy of cochains and, when $a = b$, cocycles in increasingly high degree: the Steenrod reduced powers [283]. May recast this via operads: mathematical objects which exactly record spaces parametrizing particular kinds of multiplications [198, 194].

Again returning to our maxim, we want to be sure that our constructions, including of the mapping spaces, are homotopically meaningful in the sense that the resulting homotopy type of any output depends only on the homotopy types of the inputs. Model categories provide one way to ensure this, giving us not only checkable conditions to facilitate computation but also a language and explanation for fundamental constructions in homological algebra like resolutions and derived functors. More recently, ∞-categories have given another way to ensure homotopically meaningful information by recording this data from the very beginning.

Homotopy theory is at an inflection point, with much of the older literature written in the language of model categories and with newer results and machinery expressed using ∞-categories. Both approaches have distinct benefits, and we provide an introduction to both: our aim is to give people learning about stable categories and structured ring spectra a way to connect with both "neoclassical" tools and newer ones.

The book closes with applications of the tools so developed, showing how the machinery of ∞-categories allows us to fully realize Boardman's observation and "do algebraic geometry" with commutative ring spectra. Transformative work of Goerss–Hopkins–Miller in the last 1990s ushered in the era of spectral algebraic geometry, showing first that the Lubin–Tate deformation theory of formal groups naturally lifts to a diagram of commutative ring spectra and then that the structure sheaf of the moduli stack of formal groups has an essentially unique lift to a sheaf of commutative ring spectra [106, 126]. This produced a host of new cohomology theories which are naturally tied to universal constructions in algebraic geometry and moduli problems. Additionally, it refined classical invariants of rings like modular forms to invariants of ring spectra: topological modular forms. Lurie has created a vast generalization of this, showing how one can lift algebraic geometry whole-cloth to commutative ring spectra, creating spectral algebraic geometry. The final chapter of this book provides an introduction to this new area.

## 1.2 Summaries of the chapters

### Chapter 2 (Riehl)

The chapter begins by framing a foundational question: What do we mean by the homotopy category of a category and by derived functors? It proceeds through a historical arc: describing first categories of fractions, then moving on to Quillen's theory of model categories and their simplicial enrichments, and finally describing the newer, $(\infty, 1)$-categories. The goal here is to introduce the reader to the basic tools that will be used, fitting them into a broader narrative, demonstrating how they can be used, and connecting everything clearly to the literature for further study.

### Chapter 3 (Dugger)

This chapter gives a comprehensive overview of the modern symmetric monoidal categories of spectra that were invented in the 90s: symmetric spectra, orthogonal spectra, and EKMM spectra. The technical foundations are carried out in the setting of model categories, and there is an emphasis on concrete formulas for the smash product and related constructions. The goal is for the reader to become comfortable with working in these categories of spectra.

### Chapter 4 (Barwick)

This chapter returns to the construction of the category of spectra and explains the approach to spectra and stable categories more generally in the framework of $(\infty, 1)$-categories. We hope that comparing and contrasting the treatment in this chapter and the preceding one will give a flavor of the similarities and differences between the two technical approaches for abstract homotopy theory. Of necessity, many details about the underlying foundations are left to the references, but enough detail is provided to indicate how the theory works.

### Chapter 5 (Mandell)

This chapter is a thorough treatment of the theory of operadic algebras in modern homotopy theory. It gives a streamlined view of the foundations, collecting in one place results that are scattered throughout the literature, with a unifying viewpoint on techniques for understanding the homotopy theory of operadic algebras and modules.

### Chapter 6 (Richter)

This chapter gives a broad sampling of applications of commutative ring spectra in modern stable homotopy theory. Beginning with a treatment of the foundations, it then surveys applications in topological Hochschild homology, obstruction theory and topological André–Quillen homology, and the Picard and Brauer groups.

### Chapter 7 (Lawson)

This chapter gives a detailed introduction to the theory of Bousfield localization, starting from the basic constructions and studying the multiplicative properties of the localization in the context of structured ring spectra. Bousfield localization is one of the most important techniques in the modern arsenal, and the goal of this chapter is to prepare the reader to understand how to use it.

### Chapter 8 (Rezk)

This chapter draws on all of the earlier sections, showing how the machinery developed allows us to "do algebraic geometry" in a very general context. The chapter begins discussing ∞-topoi and sheaves on them, providing along the way useful tools and ways to reinterpret results to show how these constructions can be used. It then moves into more algebraic geometry notions, exploring how classical notions like étale morphism, affine and projective spaces, and stacks lift to commutative ring spectra. This culminates in a treatment of Lurie's refinement of the Goerss–Hopkins–Miller theorem that the structure sheaf of the moduli stack of elliptic curves lifts to commutative ring spectra.

## 1.3    Acknowledgements

We would like to thank first and foremost the authors of the chapters for their incredible work. Hélène Barcelo at MSRI and our editors at Cambridge University Press have been both enthusiastic about this project and patient. We would particularly like to thank Series Editor Silvio Levy for his extensive work on this volume. We were supported throughout the course of this work by the NSF: DMS-1812064, DMS-1810575, and DMS-1811189. This book is an outgrowth of the transformative MSRI semester on algebraic topology that took place in 2014, and we are grateful to everyone who participated in that program and honored to be able to continue the conversation.

We would also like to thank our advisors and mentors, Gunnar Carlsson, Ralph Cohen, Lars Hesselholt, Mike Hopkins, Mike Mandell, Peter May, and Haynes Miller, who taught us much of what we understand about this material and so many other things. We would also like to thank Doug Ravenel for encouraging comments and Christian Carrick, Robert Housden, Jason Schuchardt, and Andrew Smith for their comments on some of the chapters.

# 2     Homotopical categories: from model categories to $(\infty, 1)$-categories

by Emily Riehl

## 2.1     The history of homotopical categories

A *homotopical category* is a category equipped with some collection of morphisms traditionally called "weak equivalences" that somewhat resemble isomorphisms but fail to be invertible in any reasonable sense, and might in fact not even be reversible: that is, the presence of a weak equivalence $X \xrightarrow{\sim} Y$ need not imply the presence of a weak equivalence $Y \xrightarrow{\sim} X$. Frequently, the weak equivalences are defined as the class of morphisms in a category K that are "inverted by a functor" $F \colon \mathsf{K} \to \mathsf{L}$, in the sense of being precisely those morphisms in K that are sent to isomorphisms in L. For instance:

- Weak homotopy equivalences of spaces or spectra are those maps inverted by the homotopy group functors $\pi_* \colon \mathsf{Top} \to \mathsf{GrSet}$ or $\pi_* \colon \mathsf{Spectra} \to \mathsf{GrAb}$.
- Quasi-isomorphisms of chain complexes are those maps inverted by the homology functor $H_* \colon \mathsf{Ch} \to \mathsf{GrAb}$.
- Equivariant weak homotopy equivalences of $G$-spaces are those maps inverted by the homotopy functors on the fixed point subspaces for each compact subgroup of $G$.

The term used to describe the equivalence class represented by a topological space up to weak homotopy equivalence is a *homotopy type*. Since the weak homotopy equivalence relation is created by the functor $\pi_*$, a homotopy type can loosely be thought of as a collection of algebraic invariants of the space $X$, as encoded by the homotopy groups $\pi_* X$. Homotopy types live in a category called the *homotopy category of spaces*, which is related to the classical category of spaces as follows: a genuine continuous function $X \to Y$ certainly represents a map (graded homomorphism) between homotopy types. But a weak homotopy equivalence of spaces, defining an isomorphism of homotopy types, should now be regarded as formally invertible.

In their 1967 manuscript *Calculus of fractions and homotopy theory*, Gabriel and Zisman [100] formalized the construction of what they call the *category of fractions* associated to any class of morphisms in any category together with an associated localization functor $\pi \colon \mathsf{K} \to \mathsf{K}[\mathcal{W}^{-1}]$ that is universal among functors with domain K that invert the class $\mathcal{W}$ of weak equivalences. This construction and its universal property are presented in §2.2. For instance, the homotopy category of spaces arises as the category of fractions associated to the weak homotopy equivalences of spaces.

There is another classical model of the homotopy category of spaces that defines an equivalent category. The objects in this category are the *CW-complexes*, spaces built by gluing disks along their boundary spheres, and the morphisms are now taken to be homotopy classes of maps. By construction the isomorphisms in this category are the homotopy equivalences of CW-complexes. Because any space is weak homotopy equivalent to a CW-complex and because Whitehead's theorem proves that the weak homotopy equivalences between CW-complexes are precisely the homotopy equivalences, it can be shown that this new homotopy category is equivalent to the Gabriel–Zisman category of fractions.

Quillen introduced a formal framework which draws attention to the essential features of these equivalent constructions. His axiomatization of an abstract "homotopy theory" was motivated by the following question: When does it make sense to invert a class of morphisms in a category and call the result a homotopy category, rather than simply a localization? In the introduction to his 1967 manuscript *Homotopical Algebra* [229], Quillen reports that Kan's theorem that the homotopy theory of simplicial groups is equivalent to the homotopy theory of connected pointed spaces [143] suggested to Quillen that simplicial objects over a suitable category A might form a homotopy theory analogous to classical homotopy theory in algebraic topology. In pursuing this analogy he observed that

there were a large number of arguments which were formally similar to well-known ones in algebraic topology, so it was decided to define the notion of a homotopy theory in sufficient generality to cover in a uniform way the different homotopy theories encountered. [229, pp. 1–2]

Quillen named these homotopy theories *model categories*, meaning "categories of models for a homotopy theory." He entitled his explorations "homotopical algebra," as they describe both a generalization of and a close analogy to homological algebra — in which the relationship between an abelian category and its derived category parallels the relationship between a model category and its homotopy category. We introduce Quillen's model categories and his construction of their homotopy categories as a category of "homotopy" classes of maps between sufficiently "fat" objects in §2.3. A theorem of Quillen proven as Theorem 2.3.29 below shows that the weak equivalences in any model category are precisely those morphisms inverted by the Gabriel–Zisman localization functor to the homotopy category. In particular, in the homotopical categories that we will most frequently encounter, the weak equivalences satisfy a number of closure properties, to be introduced in Definition 2.3.1.

To a large extent, homological algebra is motivated by the problem of constructing derived versions of functors between categories of chain complexes that fail to preserve weak equivalences. A similar question arises in Quillen's model categories. Because natural transformations can point either to or from a given functor, derived functors come with a "handedness": either left or right. In §2.4, we introduce dual notions of left and right Quillen functors between model categories and construct their derived functors via a slightly unusual route that demands a stricter (but in our view improved) definition of derived functors than the conventional one. In parallel, we study the additional properties borne by Quillen's original model structure on simplicial sets, later axiomatized by Hovey [130] in the notion of a monoidal or enriched model category, which derives to define monoidal structures or enrichments on the homotopy category.

These considerations also permit us to describe when two "homotopy theories" are equivalent. For instance, the analogy between homological and homotopical algebra is solidified by a homotopical reinterpretation of the Dold–Kan theorem as an equivalence between the homotopy theory of simplicial objects of modules and chain complexes of modules presented in Theorem 2.4.33.

As an application of the theory of derived functors, in §2.5 we study homotopy limits and colimits, which correct for the defect that classically defined limit and colimit constructions frequently fail to be weak equivalence invariant. We begin by observing that the homotopy category admits few strict limits. It does admit weak ones, as we shall see in Theorem 2.5.3, but their construction requires higher homotopical information which will soon become a primary focus.

By convention, a full Quillen model structure can only be borne by a category possessing all limits and colimits, and hence the homotopy limits and homotopy colimits introduced in §2.5 are also guaranteed to exist. This supports the point of view that a model category is a presentation of a homotopy theory with all homotopy limits and homotopy colimits. In a series of papers from 1980 [89, 87, 88], Dwyer and Kan describe more general "homotopy theories" as *simplicial localizations* of categories with weak equivalences, which augment the Gabriel–Zisman category of fractions with homotopy types of the mapping spaces between any pair of objects. The *hammock localization* construction described in §2.6 is very intuitive, allowing us to re-conceptualize the construction of the category of fractions not by imposing relations in the same dimension, but by adding maps in the next dimension — "imposing homotopy relations" if you will.

The hammock localization defines a simplicially enriched category associated to any homotopical category. A simplicially enriched category is a non-prototypical exemplification of the notion of an $(\infty, 1)$-*category*, that is, a category weakly enriched over $\infty$-groupoids or homotopy types. Model categories also equip each pair of their objects with a well-defined homotopy type of maps, and hence also present $(\infty, 1)$-categories. Before exploring $(\infty, 1)$-*categories* in a systematic way, in §2.7 we introduce the most popular model, the *quasi-categories* first defined in 1973 by Boardman and Vogt [48] and further developed by Joyal [140, 141] and Lurie [169].

In §2.8 we turn our attention to other models of $(\infty, 1)$-categories, studying six in total: quasi-categories, Segal categories, complete Segal spaces, naturally marked quasi-categories, simplicial categories, and relative categories. The last two models are strictly-defined objects, which are quite easy to define, but the model categories in which they live are poorly behaved. By contrast, the first four of these models live in model categories that have many pleasant properties, which are collected together in a new axiomatic notion of an $\infty$-*cosmos*.

After introducing this abstract definition, we see in §2.9 how the $\infty$-cosmos axiomatization allows us to develop the basic theory of these four models of $(\infty, 1)$-categories model-independently, that is, simultaneously and uniformly across these models. Specifically, we study adjunctions and equivalences between $(\infty, 1)$-categories and limits and colimits in an $(\infty, 1)$-category to provide points of comparison for the corresponding notions of Quillen adjunction, Quillen equivalence, and homotopy limits and colimits developed for model categories in §2.4 and §2.5. A brief epilogue, §2.10, contains a few closing thoughts and anticipates future chapters in this volume.

### 2.1.1   Acknowledgments

The author wishes to thank Andrew Blumberg, Teena Gerhardt, and Mike Hill for putting together this volume and inviting her to contribute. Daniel Fuentes-Keuthan gave detailed comments on a draft version of this chapter, while Chris Kapulkin, Martin Szyld, and Yu Zhang pointed out key eleventh hour typos. She was supported by the National Science Foundation via the grants DMS-1551129 and DMS-1652600.

## 2.2   Categories of fractions and localization

In one of the first textbook accounts of abstract homotopy theory [100], Gabriel and Zisman construct the universal category that inverts a collection of morphisms together with accompanying "calculi-of-fractions" techniques for calculating this categorical "localization." Gabriel and Zisman prove that a class of morphisms in a category with finite colimits admits a "calculus of left fractions" if and only if the corresponding localization preserves them, which then implies that the category of fractions also admits finite colimits [100, §1.3]; dual results relate finite limits to their "calculus of right fractions." For this reason, their calculi of fractions fail to exist in the examples of greatest interest to modern homotopy theorists, and so we will not introduce them here, focusing instead in §2.2.1 on the general construction of the category of fractions.

### 2.2.1   The Gabriel–Zisman category of fractions

For any class of morphisms $\mathcal{W}$ in a category $\mathsf{K}$, the category of fractions $\mathsf{K}[\mathcal{W}^{-1}]$ is the universal category equipped with a functor $\iota\colon \mathsf{K} \to \mathsf{K}[\mathcal{W}^{-1}]$ that inverts $\mathcal{W}$, in the sense of sending each morphism to an isomorphism. Its objects are the same as the objects of $\mathsf{K}$ and its morphisms are finite zigzags of morphisms in $\mathsf{K}$, with all "backwards" arrows finite composites of arrows belonging to $\mathcal{W}$, modulo a few relations which convert the canonical graph morphism $\iota\colon \mathsf{K} \to \mathsf{K}[\mathcal{W}^{-1}]$ into a functor and stipulate that the backwards copies of each arrow in $\mathcal{W}$ define two-sided inverses to the morphisms in $\mathcal{W}$.

Definition 2.2.1 (category of fractions [100, 1.1]). For any class of morphisms $\mathcal{W}$ in a category $\mathsf{K}$, the **category of fractions** $\mathsf{K}[\mathcal{W}^{-1}]$ is a quotient of the free category on the directed graph obtained by adding backwards copies of the morphisms in $\mathcal{W}$ to the underlying graph of the category $\mathsf{K}$ modulo certain relations:

- Adjacent arrows pointing forwards can be composed.
- Forward-pointing identities may be removed.
- Adjacent pairs of zigzags

$$x \xrightarrow{\ s\ } y \xleftarrow{\ s\ } x \quad \text{or} \quad y \xleftarrow{\ s\ } x \xrightarrow{\ s\ } y$$

indexed by any $s \in \mathcal{W}$ can be removed.[1]

The image of the functor $\iota\colon \mathsf{K} \to \mathsf{K}[\mathcal{W}^{-1}]$ is comprised of those morphisms that can be represented by unary zigzags pointing forwards.

The following proposition expresses the 2-categorical universal property of the category of fractions construction in terms of categories $\mathsf{Fun}(\mathsf{K}, \mathsf{M})$ of functors and natural transformations:

**Proposition 2.2.2** (the universal property of localization [100, 1.2]). *For any category* $\mathsf{M}$*, restriction along* $\iota$ *defines a fully faithful embedding*

$$\mathsf{Fun}(\mathsf{K}[\mathcal{W}^{-1}], \mathsf{M}) \xhookrightarrow{\ \ -\circ\iota\ \ } \mathsf{Fun}(\mathsf{K}, \mathsf{M})$$
$$\underset{\cong}{\searrow} \quad \underset{\mathcal{W}\mapsto\cong}{\mathsf{Fun}}(\mathsf{K}, \mathsf{M}) \quad \nearrow$$

*defining an isomorphism*

$$\mathsf{Fun}(\mathsf{K}[\mathcal{W}^{-1}], \mathsf{M}) \cong \underset{\mathcal{W}\mapsto\cong}{\mathsf{Fun}}(\mathsf{K}, \mathsf{M})$$

*of categories onto its essential image, the full subcategory spanned by those functors that invert* $\mathcal{W}$*.*

*Proof.* As in the analogous case of rings, the functor $\iota\colon \mathsf{K} \to \mathsf{K}[\mathcal{W}^{-1}]$ is an epimorphism and so any functor $F\colon \mathsf{K} \to \mathsf{M}$ admits at most one extension along $\iota$. To show that any functor $F\colon \mathsf{K} \to \mathsf{M}$ that inverts $\mathcal{W}$ does extend to $\mathsf{K}[\mathcal{W}^{-1}]$, we define a graph morphism from the graph described in Definition 2.2.1 to $\mathsf{M}$ by sending the backwards copy of $s$ to the isomorphism $(Fs)^{-1}$ and thus a functor from the free category generated by this graph to $\mathsf{M}$. Functoriality of $F$ ensures that the enumerated relations are respected by this functor, which therefore defines an extension $\hat{F}\colon \mathsf{K}[\mathcal{W}^{-1}] \to \mathsf{M}$ as claimed.

The 2-dimensional aspect of this universal property follows from the 1-dimensional one by considering functors valued in arrow categories [146, §3]. ☐

**Example 2.2.3** (groupoid reflection). When all the morphisms in $\mathsf{K}$ are inverted, the universal property of Proposition 2.2.2 establishes an isomorphism $\mathsf{Fun}(\mathsf{K}[\mathsf{K}^{-1}], \mathsf{M}) \cong \mathsf{Fun}(\mathsf{K}, \mathrm{core}\,\mathsf{M})$ between functors from the category of fractions of $\mathsf{K}$ to functors valued in the **groupoid core**, which is the maximal subgroupoid contained in $\mathsf{M}$. In this

---

[1] It follows that adjacent arrows in $\mathcal{W}$ pointing backwards can also be composed whenever their composite in $\mathsf{K}$ also lies in $\mathcal{W}$.

way, the category of fractions construction specializes to define a left adjoint[2] to the inclusion of groupoids into categories:

$$\mathsf{Cat} \underset{\underset{\text{core}}{\overset{\bot}{\longrightarrow}}}{\overset{\overset{\text{fractions}}{\overset{\bot}{\longrightarrow}}}{\longleftarrow}} \mathsf{Gpd}$$

The universal property of Proposition 2.2.2 applies to the class of morphisms inverted by any functor admitting a fully faithful right adjoint [100, 1.3]. In this case, the category of fractions defines a *reflective subcategory* of K, which admits a variety of useful characterizations, one being as the *local* objects orthogonal to the class of morphisms being inverted [238, 4.5.12, 4.5.vii, 5.3.3, 5.3.i]. For instance, if $R \to R[S^{-1}]$ is the localization of a commutative ring at a multiplicatively closed set, then the category of $R[S^{-1}]$-modules defines a reflective subcategory of the category of $R$-modules [238, 4.5.14], and hence the extension of scalars functor $R[S^{-1}] \otimes_R -$ can be understood as a Gabriel–Zisman localization.

However, reflective subcategories inherit all limits and colimits present in the larger category [238, 4.5.15], which is not typical behavior for categories of fractions that are "homotopy categories" in a sense to be discussed in §5.8. With the question of when a category of fractions is a homotopy category in mind, we now turn our attention to Quillen's homotopical algebra.

## 2.3    Model category presentations of homotopical categories

A question that motivated Quillen's introduction of model categories [229] and also Dwyer, Kan, Hirschhorn, and Smith's later generalization [92] is: When is a category of fractions a homotopy category? Certainly, the localization functor must invert some class of morphisms that are suitably thought of as "weak equivalences." Perhaps these weak equivalences coincide with a more structured class of "homotopy equivalences" on a suitable subcategory of "fat" objects that spans each weak equivalence class — such as given in the classical case by Whitehead's theorem that any weak homotopy equivalence between CW complexes admits a homotopy inverse — in such a way that the homotopy category is equivalent to the category of homotopy classes of maps in this full subcategory. Finally, one might ask that the homotopy category admit certain derived constructions, such as the loop and suspension functors definable on the homotopy category of based spaces. On account of this final desideratum, we will impose the blanket requirement that a category that bears a model structure must be complete and cocomplete.

---

[2] More precisely, this left adjoint takes values in a larger universe of groupoids, since the category of fractions $K[K^{-1}]$ associated to a locally small category K need not be locally small. Toy examples illustrating this phenomenon are easy to describe. For instance, let K be a category with a proper class of objects whose morphisms define a "double asterisk": each non-identity morphism has a common domain object and for each other object there are precisely two non-identity morphisms with that codomain.

Consider a class of morphisms $\mathcal{W}$, denoted by "$\xrightarrow{\sim}$", in a category M. Such morphisms might reasonably be referred to as "weak equivalences" if they somewhat resemble isomorphisms, aside from failing to be invertible in any reasonable sense. The meaning of "somewhat resembling isomorphisms" may be made precise via any of the following axioms, all of which are satisfied by the isomorphisms in any category.

Definition 2.3.1. The following hypotheses are commonly applied to a class of "weak equivalences" $\mathcal{W}$ in a category M:

- The **two-of-three property**: for any composable pair of morphisms if any two of $f$, $g$, and $gf$ is in $\mathcal{W}$ then so is the third.
- The **two-of-six property**: for any composable triple of morphisms

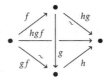

  if $gf, hg \in \mathcal{W}$ then $f, g, h, hgf \in \mathcal{W}$.
- The class $\mathcal{W}$ is **closed under retracts** in the arrow category: given a commutative diagram

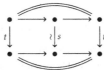

  if $s$ is in $\mathcal{W}$ then so is its retract $t$.
- The class $\mathcal{W}$ might define a **wide subcategory**, meaning that $\mathcal{W}$ is closed under composition and contains all identity morphisms.
- More prosaically, it is reasonable to suppose that $\mathcal{W}$ contains the isomorphisms.
- At a bare minimum, one might insist that $\mathcal{W}$ contains all of the identities.

Lemma 2.3.2. *Let $\mathcal{W}$ be the class of morphisms in M inverted by a functor $F \colon \mathsf{M} \to \mathsf{K}$. Then $\mathcal{W}$ satisfies each of the closure properties just enumerated.*

*Proof.* This follows immediately from the axioms of functoriality. $\square$

In practice, most classes of weak equivalences arise as in Lemma 2.3.2. For instance, the quasi-isomorphisms are those chain maps inverted by the homology functor $H_\bullet$ from chain complexes to graded modules, while the weak homotopy equivalences are those continuous functions inverted by the homotopy group functors $\pi_\bullet$. Rather than adopt a universal set of axioms that may or may not fit the specific situation at hand, we will use the term **homotopical category** to refer to any pair $(\mathsf{M}, \mathcal{W})$ comprised of a category and a class of morphisms and enumerate the specific properties we need for each result or construction. When the homotopical category $(\mathsf{M}, \mathcal{W})$ underlies a model category structure, to be defined, Theorem 2.3.29 below proves $\mathcal{W}$ is precisely the class of morphisms inverted by the Gabriel–Zisman localization functor and hence satisfies all of the enumerated closure properties.

The data of a *model structure* borne by a homotopical category is given by two additional classes of morphisms — the *cofibrations* $\mathcal{C}$ denoted "$\rightarrowtail$", and the *fibrations* $\mathcal{F}$ denoted "$\twoheadrightarrow$" — satisfying axioms to be enumerated. In §2.3.1, we present a modern reformulation of Quillen's axioms that more clearly highlights the central features of a model structure borne by a complete and cocomplete category. In §2.3.2, we discuss the delicate question of the functoriality of the factorizations in a model category with the aim of justifying our view that this condition is harmless to assume in practice.

In §2.3.3, we explain what it means for a parallel pair of morphisms in a model category to be *homotopic*; more precisely, we introduce distinct *left homotopy* and *right homotopy* relations that define a common equivalence relation when the domain is *cofibrant* and the codomain is *fibrant*. The homotopy relation is used in §2.3.4 to construct and compare three equivalent models for the homotopy category of a model category: the Gabriel–Zisman category of fractions $\mathsf{M}[\mathcal{W}^{-1}]$ defined by formally inverting the weak equivalences, the category $\mathsf{hM_{cf}}$ of fibrant-cofibrant objects in M and homotopy classes of maps, and an intermediary $\mathsf{HoM}$ which has the objects of the former and hom-sets of the later, designed to facilitate the comparison. Finally, §2.3.5 presents a fundamental example: Quillen's model structure on the category of simplicial sets.

### 2.3.1    Model category structures via weak factorization systems

When Quillen first introduces the definition of a model category in the introduction to "Chapter I. Axiomatic Homotopy Theory" [229], he highlights the factorization and lifting axioms as being the most important. These axioms are most clearly encapsulated in the categorical notion of a weak factorization system, a concept which was codified later.

**Definition 2.3.3.** A **weak factorization system** $(\mathcal{L}, \mathcal{R})$ on a category M is comprised of two classes of morphisms $\mathcal{L}$ and $\mathcal{R}$ such that:

(i) Every morphism in M may be factored as a morphism in $\mathcal{L}$ followed by a morphism in $\mathcal{R}$.

(ii) The maps in $\mathcal{L}$ have the **left lifting property** with respect to each map in $\mathcal{R}$ and the maps in $\mathcal{R}$ have the **right lifting property** with respect to each map in $\mathcal{L}$: that is, any commutative square

$$
\begin{array}{ccc}
\bullet & \longrightarrow & \bullet \\
{\scriptstyle \mathcal{L} \ni \ell} \big\downarrow & \nearrow & \big\downarrow {\scriptstyle r \in \mathcal{R}} \\
\bullet & \longrightarrow & \bullet
\end{array}
$$

admits a diagonal filler as indicated, making both triangles commute.

(iii) The classes $\mathcal{L}$ and $\mathcal{R}$ are each closed under retracts in the arrow category: given a commutative diagram

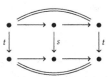

if $s$ is in that class then so is its retract $t$.

The following reformulation of Quillen's definition [229, I.5.1] was given by Joyal and Tierney [142, 7.7], who prove that a homotopical category $(M, \mathcal{W})$, with the weak equivalences satisfying the two-of-three property, admits a model structure just when there exist classes $\mathcal{C}$ and $\mathcal{F}$ that define a pair of weak factorization systems:

**Definition 2.3.4** (model category). A **model structure** on a homotopical category $(M, \mathcal{W})$ consists of three classes of maps — the **weak equivalences** $\mathcal{W}$ denoted "$\xrightarrow{\sim}$", which must satisfy the two-of-three property,[3] the **cofibrations** $\mathcal{C}$ denoted "$\rightarrowtail$", and the **fibrations** $\mathcal{F}$ denoted "$\twoheadrightarrow$" — so that $(\mathcal{C}, \mathcal{F} \cap \mathcal{W})$ and $(\mathcal{C} \cap \mathcal{W}, \mathcal{F})$ each define weak factorization systems on M.

*Remark 2.3.5* (on self-duality). Definitions 2.3.3 and 2.3.4 are self-dual: if $(\mathcal{L}, \mathcal{R})$ defines a weak factorization system on M then $(\mathcal{R}, \mathcal{L})$ defines a weak factorization system on $M^{\mathrm{op}}$. Thus the statements we prove about the left classes $\mathcal{C}$ of cofibrations and $\mathcal{C} \cap \mathcal{W}$ of **trivial cofibrations** "$\overset{\sim}{\rightarrowtail}$" will have dual statements involving the right classes $\mathcal{F}$ of fibrations and $\mathcal{F} \cap \mathcal{W}$ of **trivial fibrations** "$\overset{\sim}{\twoheadrightarrow}$".

Axiom 3 of Definition 2.3.3 was missing from Quillen's original definition of a model category; he referred to those model categories that have the retract closure property as "closed model categories." The importance of this closure property is that it implies that the left class of a weak factorization system is comprised of all maps that have the left lifting property with respect to the right class and dually, that the right class is comprised of all of those maps that have the right lifting property with respect to the left class. These results follow as a direct corollary of the famous "retract argument":

**Lemma 2.3.6** (retract argument). *Suppose $f = r \circ \ell$ and $f$ has the left lifting property with respect to its right factor $r$. Then $f$ is a retract of its left factor $\ell$.*

*Proof.* The solution to the lifting problem displayed on the left

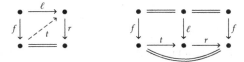

defines the retract diagram on the right.    □

---

[3] The standard definition of a model category also requires the weak equivalences to be closed under retracts, but this is a consequence of the axioms given here [142, 7.8].

Corollary 2.3.7. *Either class of a weak factorization system determines the other: the left class consists of those morphisms that have the left lifting property with respect to the right class, and the right class consists of those morphisms that have the right lifting property with respect to the left class.*

*Proof.* Any map with the left lifting property with respect to the right class of a weak factorization system lifts against its right factor of the factorization guaranteed by axiom 1 of Definition 2.3.3 and so belongs to the left class by axiom 3.     □

It follows that the trivial cofibrations can be defined without reference to either the cofibrations or weak equivalences as those maps that have the left lifting property with respect to the fibrations, and dually the trivial fibrations are precisely those maps that have the right lifting property with respect to the cofibrations.

*Exercise 2.3.8.* Verify that a model structure on M, if it exists, is uniquely determined by any of the following data:

(i) The cofibrations and weak equivalences.
(ii) The fibrations and weak equivalences.
(iii) The cofibrations and fibrations.

By a more delicate observation of Joyal [141, E.1.10] using terminology to be introduced in Definition 2.3.14, a model structure is also uniquely determined by

(iv) The cofibrations and fibrant objects.
 (v) The fibrations and cofibrant objects.

As a further consequence of the characterizations of the cofibrations, trivial cofibrations, fibrations, and trivial fibrations by lifting properties, each class automatically enjoys certain closure properties.

Lemma 2.3.9. *Let $\mathcal{L}$ be any class of maps characterized by a left lifting property with respect to a fixed class of maps $\mathcal{R}$. Then $\mathcal{L}$ contains the isomorphisms and is closed under coproduct, pushout, retract, and (transfinite) composition.*

*Proof.* We prove the cases of pushout and transfinite composition to clarify the meaning of these terms, the other arguments being similar. Let $k$ be a pushout of a morphism $\ell \in \mathcal{L}$ as in the left square below, and consider a lifting problem against a morphism $r \in \mathcal{R}$ as presented by the right square:

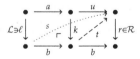

Then there exists a lift $s$ in the composite rectangle and this lift and $u$ together define a cone under the pushout diagram, inducing the desired lift $t$.

Now let $\alpha$ denote any ordinal category. The **transfinite composite** of a diagram $\alpha \to \mathsf{M}$ is the leg $\ell_\alpha$ of the colimit cone from the initial object in this diagram to its colimit. To see that this morphism lies in $\mathcal{L}$ under the hypothesis that the generating morphisms $\ell_i$ in the diagram do, it suffices to construct the solution to any lifting problem against a map $r \in \mathcal{R}$.

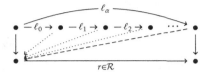

By the universal property of the colimit object, this dashed morphism exists once the commutative cone of dotted lifts do, and these may be constructed sequentially starting by lifting $\ell_0$ against $r$.                                                      □

*Exercise 2.3.10.* Verify that the class of morphisms $\mathcal{L}$ characterized by the left lifting property against a fixed class of morphisms $\mathcal{R}$ is closed under coproducts, closed under retracts, and contains the isomorphisms.

Definition 2.3.11. Let $\mathcal{J}$ be any class of maps. A $\mathcal{J}$-**cell complex** is a map built as a transfinite composite of pushouts of coproducts of maps in $\mathcal{J}$, which may then be referred to in this context as the basic **cells**.

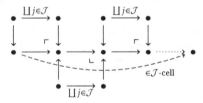

Lemma 2.3.9 implies that the left class of a weak factorization is closed under the formation of cell complexes.

*Exercise 2.3.12.* Explore the reason why the class of morphisms $\mathcal{L}$ characterized by the left lifting property against a fixed class of morphisms $\mathcal{R}$ may fail to be closed under coequalizers, formed in the arrow category.[4]

## 2.3.2    On functoriality of factorizations

The weak factorization systems that arise in practice, such as those that define the components of a model category, tend to admit *functorial* factorizations in the following sense.

Definition 2.3.13. A **functorial factorization** on a category M is given by a functor $M^2 \to M^3$ from the category of arrows in M to the category of composable pairs of arrows in M that defines a section to the composition functor $\circ \colon M^3 \to M^2$. The

---

[4] Note, however, that if the maps in $\mathcal{L}$ are equipped with specified solutions to every lifting problem posed by $\mathcal{R}$ and if the squares in the coequalizer diagram commute with these specified lifts, then the coequalizer inherits canonically defined solutions to every lifting problem posed by $\mathcal{R}$ and is consequently in the class $\mathcal{L}$.

action of this functor on objects in $\mathsf{M}^2$ (which are arrows, displayed vertically) and morphisms in $\mathsf{M}^2$ (which are commutative squares) is displayed below:

$$
\begin{array}{ccc}
X & \xrightarrow{u} & Z \\
f \downarrow & & \downarrow g \\
Y & \xrightarrow[v]{} & W
\end{array}
\qquad \mapsto \qquad
f\left(
\begin{array}{ccc}
X & \xrightarrow{u} & Z \\
\downarrow Lf & & Lg \downarrow \\
Ef & \xrightarrow{E(u,v)} & Eg \\
\downarrow Rf & & Rg \downarrow \\
Y & \xrightarrow[v]{} & W
\end{array}
\right) g
$$

This data is equivalently presented by a pair of endofunctors $L, R \colon \mathsf{M}^2 \rightrightarrows \mathsf{M}^2$ satisfying compatibility conditions relative to the domain and codomain projections $\mathrm{dom}, \mathrm{cod} \colon \mathsf{M}^2 \rightrightarrows \mathsf{M}$, namely that

$$\mathrm{dom} L = \mathrm{dom}, \quad \mathrm{cod} R = \mathrm{cod}, \quad \text{and} \quad E := \mathrm{cod} L = \mathrm{dom} R$$

as functors $\mathsf{M}^2 \to \mathsf{M}$.

The functoriality of Definition 2.3.13 is with respect to (horizontal) composition of squares and is encapsulated most clearly by the functor $E$ which carries a square $(u, v)$ to the morphism $E(u, v)$ between the objects through which $f$ and $g$ factor. Even without assuming the existence of functorial factorizations, in any category with a weak factorization system $(\mathcal{L}, \mathcal{R})$, commutative squares may be factored into a square between morphisms in $\mathcal{L}$ on top of a square between morphisms in $\mathcal{R}$

$$
\begin{array}{ccc}
X & \xrightarrow{u} & Z \\
f \downarrow & & \downarrow g \\
Y & \xrightarrow[v]{} & W
\end{array}
\qquad \mapsto \qquad
f\left(
\begin{array}{ccc}
X & \xrightarrow{u} & Z \\
\downarrow \ell\in\mathcal{L} & & \mathcal{L}\ni\ell' \downarrow \\
E & \dashrightarrow{e} & F \\
\downarrow r\in\mathcal{R} & & \mathcal{R}\ni r' \downarrow \\
Y & \xrightarrow[v]{} & W
\end{array}
\right) g
$$

with the dotted horizontal morphism defined by lifting $\ell$ against $r'$. These factorizations will not be strictly functorial because the solutions to the lifting problems postulated by axiom 2 of 2.3.3 are not unique. However, for either of the weak factorizations systems in a model category, any two solutions to a lifting problem are *homotopic* in a sense defined by Quillen, appearing below as Definition 2.3.19. As homotopic maps are identified in the homotopy category, this means that any model category has functorial factorizations up to homotopy, which suffices for most purposes.[5] Despite the moral sufficiency of the standard axioms, for economy of language we henceforth tacitly assume that our model categories have functorial factorizations and take comfort in the fact that it seems to be exceedingly difficult to find model categories that fail to satisfy this condition.

---

[5] While the derived functors constructed in Corollary 2.4.10 make use of explicit point-set level functorial factorizations, their total derived functors in the sense of Definition 2.4.4 are well-defined without strict functoriality.

### 2.3.3    The homotopy relation on arrows

Our aim now is to define Quillen's homotopy relation, which will be used to construct a relatively concrete model $hM_{cf}$ for the homotopy category of the model category M, which is equivalent to the Gabriel–Zisman category of fractions $M[\mathcal{W}^{-1}]$ but provides better control over the sets of morphisms between each pair of objects. Quillen's key observation appears as Proposition 2.3.23, which shows that the weak equivalences between objects of M that are both *fibrant* and *cofibrant,* in a sense to be defined momentarily, are more structured, always admitting a homotopy inverse for a suitable notion of homotopy. The homotopy relation is respected by pre- and post-composition, which means that $hM_{cf}$ may be defined simply to be the category of fibrant-cofibrant objects and homotopy classes of maps. In this section, we give all of these definitions. In §2.3.4, we construct the category $hM_{cf}$ sketched above and prove its equivalence with the category of fractions $M[\mathcal{W}^{-1}]$.

**Definition 2.3.14.**   An object $X$ in a model category M is **fibrant** when the unique map $X \to *$ to the terminal object is a fibration and **cofibrant** when the unique map $\emptyset \to X$ from the initial object is a cofibration.

Objects that are not fibrant or cofibrant can always be replaced by weakly equivalent objects that are, by factoring the maps to the terminal object or from the initial object, as appropriate.

*Exercise 2.3.15* (fibrant and cofibrant replacement).   Assuming the functorial factorizations of §2.3.2, define a **fibrant replacement** functor $R\colon M \to M$ and a **cofibrant replacement** functor $Q\colon M \to M$ equipped with natural weak equivalences

$$\eta\colon \mathrm{id}_M \overset{\sim}{\Rightarrow} R \qquad \text{and} \qquad \epsilon\colon Q \overset{\sim}{\Rightarrow} \mathrm{id}_M.$$

Applying both constructions, one obtains a **fibrant-cofibrant** replacement of any object $X$ as either $RQX$ or $QRX$. In the diagram

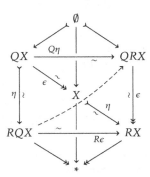

the middle square commutes because its two subdivided triangles do, by naturality of the maps $\eta$ and $\epsilon$ of Exercise 2.3.15. This induces a direct comparison weak equivalence $RQX \overset{\sim}{\to} QRX$ by lifting $\eta_{QX}$ against $\epsilon_{RX}$.

*Exercise 2.3.16.* Show that any map in a model category may be replaced, up to a zigzag of weak equivalences, by one between fibrant-cofibrant objects that moreover may be taken to be either a fibration or a cofibration, as desired.[6]

The reason for our particular interest in the subcategory of fibrant-cofibrant objects in a model category is that between such objects the weak equivalences become more structured, coinciding with a class of "homotopy equivalences" in a sense we now define.

**Definition 2.3.17.** Let $A$ be an object in a model category. A **cylinder object** for $A$ is given by a factorization of the fold map

$$A \sqcup A \xrightarrow{\ (1_A, 1_A)\ } A$$
$$\underset{(i_0, i_1)}{\searrow} \quad \underset{\mathrm{cyl}(A)}{} \quad \overset{\sim}{\underset{q}{\nearrow}}$$

into a cofibration followed by a trivial fibration. Dually, a **path object** for $A$ is given by any factorization of the diagonal map

$$\mathrm{path}(A)$$
$$\overset{j}{\underset{\sim}{\nearrow}} \qquad \overset{(p_0, p_1)}{\searrow}$$
$$A \xrightarrow{\ (1_A, 1_A)\ } A \times A$$

into a trivial cofibration followed by a fibration.

*Remark 2.3.18.* For many purposes it suffices to drop the hypotheses that the maps in the cylinder and path object factorizations are cofibrations and fibrations, and retain only the hypothesis that the second and first factors, respectively, are weak equivalences. The standard terminology for the cylinder and path objects defined here adds the accolade "very good." But since "very good" cylinder and path objects always exist, we eschew the usual convention and adopt these as the default notions.

**Definition 2.3.19.** Consider a parallel pair of maps $f, g \colon A \rightrightarrows B$ in a model category. A **left homotopy** $H$ from $f$ to $g$ is given by a map from a cylinder object of $A$ to $B$ extending $(f, g) \colon A \sqcup A \to B$:

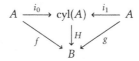

in which case one writes $f \sim_\ell g$ and says that $f$ and $g$ are **left homotopic**.

A **right homotopy** $K$ from $f$ to $g$ is given by a map from $A$ to a path object for $B$

---

[6] Exercise 2.3.16 reveals that the notions of "cofibration" and "fibration" are not homotopically meaningful: up to isomorphism in $\mathrm{M}[\mathcal{W}^{-1}]$, any map in a model category can be taken to be either a fibration or a cofibration.

extending $(f,g)\colon A \to B \times B$:

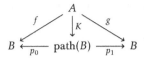

in which case one writes $f \sim_r g$ and says that $f$ and $g$ are **right homotopic**.

*Exercise 2.3.20.* Prove that the endpoint inclusions $i_0, i_1 \colon A \rightrightarrows \mathrm{cyl}(A)$ into a cylinder object are weak equivalences always and also cofibrations if $A$ is cofibrant. Conclude that if $f \sim_\ell g$ then $f$ is a weak equivalence if and only if $g$ is. Dually, the projections $p_0, p_1 \colon \mathrm{path}(B) \rightrightarrows B$ are weak equivalences always and also fibrations if $B$ is fibrant, and if $f \sim_r g$ then $f$ is a weak equivalence if and only if $g$ is.

A much more fine-grained analysis of the left and right homotopy relations is presented in the classic expository paper "Homotopy theories and model categories" of Dwyer and Spalinski [91]. Here we focus on the essential facts for understanding the homotopy relation on maps between cofibrant and fibrant objects.

**Proposition 2.3.21.** *If $A$ is cofibrant and $B$ is fibrant then left and right homotopy define equivalence relations on the set $\mathrm{Hom}(A, B)$ of arrows and moreover these relations coincide.*

In light of Proposition 2.3.21, we say that maps $f, g \colon A \rightrightarrows B$ from a cofibrant object to a fibrant one are **homotopic** and write $f \sim g$ to mean that they are left or equivalently right homotopic.

*Proof.* The left homotopy relation is reflexive and symmetric without any cofibrancy or fibrancy hypotheses on the domains or codomains. To prove transitivity, consider a pair of left homotopies $H \colon \mathrm{cyl}(A) \to B$ from $f$ to $g$ and $K \colon \mathrm{cyl}'(A) \to B$ from $g$ to $h$, possibly constructed using different cylinder objects for $A$. By cofibrancy of $A$ and Exercise 2.3.20, a new cylinder object $\mathrm{cyl}''(A)$ for $A$ may be constructed by factoring the map from the following pushout $C$ to $A$:

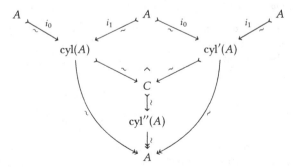

The homotopies $H$ and $K$ define a cone under the pushout diagram inducing a map $H \cup_A K \colon C \to B$. By fibrancy of $B$, this map may be extended along the trivial cofibration $C \rightarrowtail \mathrm{cyl}''(A)$ to define a homotopy $\mathrm{cyl}''(A) \to B$ from $f$ to $h$. This proves that left homotopy is an equivalence relation.

Finally we argue that if $H$: $\mathrm{cyl}(A) \to B$ defines a left homotopy from $f$ to $g$ then $f \sim_r g$. The desired right homotopy from $f$ to $g$ is constructed as the restriction of the displayed lift

$$
\begin{array}{ccc}
A & \xrightarrow{\ f\ } B \rightarrowtail\xrightarrow{\sim} \mathrm{path}(B) \\
i_0 \downarrow \wr & \qquad\qquad \downarrow (p_0, p_1) \\
A \underset{i_1}{\overset{\sim}{\rightarrowtail}} \mathrm{cyl}(A) \xrightarrow[(fq, H)]{} B \times B
\end{array}
$$

along the endpoint inclusion $i_1$. The remaining assertions are dual to ones already proven.                                                                     □

Moreover, the homotopy relation is respected by pre- and post-composition:

**Proposition 2.3.22.** *Suppose $f, g \colon A \rightrightarrows B$ are left or right homotopic maps and consider any maps $h \colon A' \to A$ and $k \colon B \to B'$. Then $kfh, kgh \colon A' \rightrightarrows B'$ are again left or right homotopic, respectively.*

*Proof.* By lifting the endpoint inclusion $(i_0, i_1) \colon A' \sqcup A' \rightarrowtail \mathrm{cyl}(A')$ against the projection $\mathrm{cyl}(A) \overset{\sim}{\twoheadrightarrow} A$ — or by functoriality of the cylinder construction in the sense discussed in §2.3.2 — there is a map $\mathrm{cyl}(h) \colon \mathrm{cyl}(A') \to \mathrm{cyl}(A)$. Then, for any left homotopy $H \colon \mathrm{cyl}(A) \to B$ from $f$ to $g$, the horizontal composite defines a left homotopy $kfh \sim_\ell kgh$:

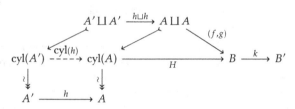

                                                                    □

**Proposition 2.3.23.** *Let $f \colon A \to B$ be a map between objects that are both fibrant and cofibrant. Then $f$ is a weak equivalence if and only if it has a homotopy inverse.*

*Proof.* For both implications we make use of the fact that any map between fibrant-cofibrant objects may be factored as a trivial cofibration followed by a fibration through an object that is again fibrant-cofibrant:

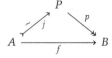

If $f$ is a weak equivalence then $p$ is a trivial fibration. We argue that any trivial fibration $p$ between fibrant-cofibrant objects extends to a deformation retraction: admitting a right inverse that is also a left homotopy inverse. A dual argument proves that the trivial cofibration $j$ admits a left inverse that is also a right homotopy inverse. These homotopy equivalences compose in the sense of Proposition 2.3.22 to define a homotopy inverse for $f$.

If $p$ is a trivial fibration, then cofibrancy of $B$ implies that it admits a right inverse $i$. The homotopy constructed in the lifting problem

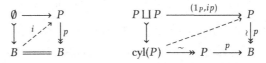

proves that $ip \sim 1_P$ as desired.

For the converse we suppose that $f$ admits a homotopy inverse $g$. To prove that $f$ is a weak equivalence it suffices to prove that $p$ is a weak equivalence. A right inverse $i$ to $p$ may be found by lifting the endpoint of the homotopy $H \colon fg \sim 1_B$:

$$
\begin{array}{ccc}
B & \xrightarrow{\;g\;} A \xrightarrow{\;j\;} & P \\
{\scriptstyle\wr}\downarrow{\scriptstyle i_0} & \nearrow & \downarrow p \\
B \; \xrightarrow[\;\sim\;]{i_1} \; \mathrm{cyl}(B) & \xrightarrow{\hspace{2cm}H\hspace{2cm}} & B
\end{array}
$$

and then restricting this lift along $i_1$. By construction this section $i$ is homotopic to $jg$. The argument of the previous paragraph applies to the trivial cofibration $j$ to prove that it has a left inverse and right homotopy inverse $q$. Composing the homotopies $1_P \sim jq$, $i \sim jg$, and $gf \sim 1_A$ we see that

$$
ip \sim ipjq = ifq \sim jgfq \sim jq \sim 1_P
$$

By Exercise 2.3.20 we conclude that $ip$ is a weak equivalence. But by construction $p$ is a retract of $ip$:

$$
\begin{array}{ccccc}
P & = & P & = & P \\
p\downarrow & & {\scriptstyle\wr}\downarrow{\scriptstyle ip} & & \downarrow p \\
B & \xrightarrow{\;i\;} & P & \xrightarrow{\;p\;} & B
\end{array}
$$

so it follows from the retract stability of the weak equivalences [142, 7.8] that $p$ is a weak equivalence, as desired. $\qquad\square$

### 2.3.4   The homotopy category of a model category

In this section, we prove that the category of fractions $\mathsf{M}[\mathcal{W}^{-1}]$, defined by formally inverting the weak equivalences, is equivalent to the category $\mathsf{hM}_{\mathrm{cf}}$ of fibrant-cofibrant objects and homotopy classes of maps. Our proof appeals to the universal property of Proposition 2.2.2, which characterizes those categories that are *isomorphic* to the category of fractions. For categories to be isomorphic, they must have the same object sets, so we define a larger version of the homotopy category $\mathsf{HoM}$, which has the same objects as $\mathsf{M}[\mathcal{W}^{-1}]$ and is equivalent to its full subcategory $\mathsf{hM}_{\mathrm{cf}}$.

**Definition 2.3.24.**   For any model category $\mathsf{M}$, there is a category $\mathsf{hM}_{\mathrm{cf}}$

– whose objects are the fibrant-cofibrant objects in $\mathsf{M}$ and

– in which the set of morphisms from $A$ to $B$ is taken to be the set of homotopy classes of maps

$$[A, B] := \mathsf{Hom}(A, B)_{/\sim}.$$

Proposition 2.3.22 ensures that composition in $\mathsf{hM}_{cf}$ is well-defined.

**Definition 2.3.25.** The **homotopy category** $\mathsf{HoM}$ of a model category $\mathsf{M}$ is defined by applying the (bijective-on-objects, fully faithful) factorization to the composite functor

$$\mathsf{M} \xrightarrow{RQ} \mathsf{M}_{cf} \xrightarrow{\pi} \mathsf{hM}_{cf} \tag{2.1}$$

That is, the objects in $\mathsf{HoM}$ are the objects in $\mathsf{M}$ and

$$\mathsf{HoM}(A, B) := \mathsf{M}(RQA, RQB)_{/\sim}.$$

*Exercise 2.3.26* ($\mathsf{HoM} \simeq \mathsf{hM}_{cf}$).

 (i) Verify that the category $\mathsf{hM}_{cf}$ defined by Definition 2.3.24 is equivalent to the full subcategory of $\mathsf{HoM}$ spanned by the fibrant-cofibrant objects of $\mathsf{M}$.
 (ii) Show that every object in $\mathsf{M}$ is isomorphic in $\mathsf{HoM}$ to a fibrant-cofibrant object.
 (iii) Conclude that the categories $\mathsf{HoM}$ and $\mathsf{hM}_{cf}$ are equivalent.

**Theorem 2.3.27** (Quillen). *For any model category* $\mathsf{M}$, *the category of fractions* $\mathsf{M}[\mathcal{W}^{-1}]$ *obtained by formally inverting the weak equivalences is isomorphic to the homotopy category* $\mathsf{HoM}$.

*Proof.* We will prove that $\gamma \colon \mathsf{M} \to \mathsf{HoM}$ satisfies the universal property of Proposition 2.2.2 that characterizes the category of fractions $\mathsf{M}[\mathcal{W}^{-1}]$. First we must verify that $\gamma$ inverts the weak equivalences. The functor $RQ$ carries weak equivalences in $\mathsf{M}$ to weak equivalences between fibrant-cofibrant objects. Proposition 2.3.23 then implies that these admit homotopy inverses and thus become isomorphisms in $\mathsf{hM}_{cf}$. This proves that the composite horizontal functor of (2.1) inverts the weak equivalences. By fully-faithfulness of $\nu$, the functor $\gamma \colon \mathsf{M} \to \mathsf{HoM}$ also inverts the weak equivalences.

It remains to verify that any functor $F \colon \mathsf{M} \to \mathsf{E}$ that inverts the weak equivalences factors uniquely through $\gamma$:

Since $\gamma$ is identity-on-objects, we must define $\bar{F}$ to agree with $F$ on objects. Recall that the fibrant and cofibrant replacement functors come with natural weak equivalences $\epsilon_X \colon QX \xrightarrow{\sim} X$ and $\eta_X \colon X \xrightarrow{\sim} RX$. Because $F$ inverts weak equivalences, these natural transformations define a natural isomorphism $\alpha \colon F \Rightarrow FRQ$ of functors from $\mathsf{M}$ to $\mathsf{E}$. By the definition $\mathsf{HoM}(X, Y) := \mathsf{M}(RQX, RQY)_{/\sim}$, the morphisms from $X$ to $Y$ in $\mathsf{HoM}$ correspond to homotopy classes of morphisms from $RQX$ to $RQY$ in $\mathsf{M}$. Choose any representative $h\colon RQX \to RQY$ for the corresponding homotopy class of maps and define its image to be the composite

$$\bar{F}h\colon FX \xrightarrow{\;\alpha_X\;} FRQX \xrightarrow{\;Fh\;} FRQY \xrightarrow{\;\alpha_Y^{-1}\;} FY.$$

This is well-defined because if $h \sim h'$ then there exists a left homotopy such that $Hi_0 = h$ and $Hi_1 = h'$, where $i_0$ and $i_1$ are both sections to a common weak equivalence (the projection from the cylinder). Since $F$ inverts weak equivalences, $Fi_0$ and $Fi_1$ are both right inverses to a common isomorphism, so it follows that $Fi_0 = Fi_1$ and hence $Fh = Fh'$.

Functoriality of $\bar{F}$ follows immediately from naturality of $\alpha$ and functoriality of $FRQ$. To see that $\bar{F}\gamma = F$, recall that for any $f\colon X \to Y$ in M, $\gamma(f)$ is defined to be the map in $\mathrm{HoM}(X, Y)$ represented by the homotopy class $RQf\colon RQX \to RQY$. By naturality of $\alpha$, $\bar{F}\gamma(f) = Ff$, so that the triangle of functors commutes.

Finally, to verify that $\bar{F}$ is unique observe that from the following commutative diagram in M any map $h \in \mathrm{HoM}(X, Y)$, the leftmost vertical arrow, is isomorphic in HoM to a map in the image of $\gamma$, the vertical arrow on the right:

$$
\begin{array}{ccccc}
RQX & \xleftarrow{\;\epsilon_{RQX}\;} & QRQX & \xrightarrow{\;\eta_{QRQX}\;} & RQRQX \\
{\scriptstyle h}\downarrow & & \downarrow{\scriptstyle Qh} & & \downarrow{\scriptstyle RQh=\gamma(h)} \\
RQY & \xleftarrow{\;\epsilon_{RQY}\;} & QRQY & \xrightarrow{\;\eta_{QRQY}\;} & RQRQY
\end{array}
$$

Since the image of $\bar{F}$ on the right vertical morphism is uniquely determined and the top and bottom morphisms are isomorphisms, the image of $\bar{F}$ on the left vertical morphism is also uniquely determined.                                □

*Remark 2.3.28.* The universal property of $\mathrm{hM}_{\mathrm{cf}}$ is slightly weaker than the universal property described in Proposition 2.2.2 for the category of fractions $\mathrm{M}[\mathcal{W}^{-1}]$. For any category E, restriction along $\gamma\colon \mathrm{M} \to \mathrm{hM}_{\mathrm{cf}}$ defines a fully faithful embedding $\mathrm{Fun}(\mathrm{hM}_{\mathrm{cf}}, \mathrm{E}) \hookrightarrow \mathrm{Fun}(\mathrm{M}, \mathrm{E})$ and equivalence onto the full subcategory of functors from M to E that carry weak equivalences to isomorphisms. The difference is that a given homotopical functor on M may not factor strictly through $\mathrm{hM}_{\mathrm{cf}}$ but may only factor up to natural isomorphism. In practice, this presents no serious difficulty.

As a corollary, it is now easy to see that the only maps inverted by the localization functor are weak equivalences. By Lemma 2.3.2, this proves that the weak equivalences in a model category have all of the closure properties enumerated at the outset of this section.

Theorem 2.3.29 ([229, 5.1]). *A morphism in a model category* M *is inverted by the localization functor*

$$\mathrm{M} \to \mathrm{M}[\mathcal{W}^{-1}]$$

*if and only if it is a weak equivalence.*

*Proof.* Cofibrantly and then fibrantly replacing the map it suffices to consider a map between fibrant-cofibrant objects. By Theorem 2.3.27 we may prove this result for

$M_{cf} \to hM_{cf}$ instead. But now this is clear by construction: since morphisms in $hM_{cf}$ are homotopy classes of maps, the isomorphisms are the homotopy equivalences, which coincide exactly with the weak equivalences between fibrant-cofibrant objects by Proposition 2.3.23. $\qquad\square$

### 2.3.5     Quillen's model structure on simplicial sets

We conclude this section with a prototypical example. Quillen's original model structure is borne by the category of **simplicial sets**, presheaves on the category $\Delta$ of finite non-empty ordinals $[n] = \{0 < 1 < \cdots < n\}$ and order-preserving maps. A **simplicial set** $X \colon \Delta^{\mathrm{op}} \to \mathsf{Set}$ is a graded set $\{X_n\}_{n \geq 0}$ — where elements of $X_n$ are called "$n$-simplices" — equipped with dimension-decreasing "face" maps $X_n \to X_m$ arising from monomorphisms $[m] \rightarrowtail [n] \in \Delta$ and dimension-increasing "degeneracy" maps $X_m \to X_n$ arising from epimorphisms $[n] \twoheadrightarrow [m] \in \Delta$. An $n$-simplex has $n + 1$ codimension-one faces, each of which avoids one of its $n + 1$ vertices.

There is a **geometric realization** functor $|-| \colon \mathsf{sSet} \to \mathsf{Top}$ that produces a topological space $|X|$ from a simplicial set $X$ by gluing together topological $n$-simplices for each non-degenerate $n$-simplex along its lower-dimensional faces. The simplicial set represented by $[n]$ defines the standard $n$-simplex $\Delta^n$. Its boundary $\partial\Delta^n$ is the union of its codimension-one faces, while a horn $\Lambda_k^n$ is the further subspace formed by omitting the face opposite the vertex $k \in [n]$.

Theorem 2.3.30 (Quillen).  *The category* $\mathsf{sSet}$ *admits a model structure whose*

- *weak equivalences are those maps* $f \colon X \to Y$ *that induce a weak homotopy equivalence* $f \colon |X| \to |Y|$ *on geometric realizations,*
- *cofibrations are monomorphisms, and*
- *fibrations are the **Kan fibration**, which are characterized by the left lifting property with respect to the set of all horn inclusions:*

$$
\begin{array}{ccc}
\Lambda_k^n & \longrightarrow & X \\
{\scriptstyle\wr}\big\downarrow & \nearrow & \big\downarrow \\
\Delta^n & \longrightarrow & Y
\end{array}
$$

All objects are cofibrant. The fibrant objects are the **Kan complexes**, those simplicial sets in which all horns can be filled. The fibrant objects are those simplicial sets that most closely resemble topological spaces. In particular, two vertices in a Kan complex lie in the same path component if and only if they are connected by a single 1-simplex, with may be chosen to point in either direction. By Proposition 2.3.23 a weak equivalence between Kan complexes is a homotopy equivalence where the notion of homotopy is defined with respect to the interval $\Delta^1$ using $\Delta^1 \times X$ as a cylinder object or $X^{\Delta^1}$ as a path object.

Quillen's model category of simplicial sets is of interest because, on the one hand, the category of simplicial sets is very well behaved and, on the other hand, the geometric realization functor defines an "equivalence of homotopy theories": in particular, the homotopy category of simplicial sets gives another model for the homotopy category of spaces. To explain this, we turn our focus to derived functors and derived equivalences between model categories, the subject of §2.4.

## 2.4 Derived functors between model categories

Quillen's model category axioms allow us to conjure a homotopy relation between parallel maps in any model category, whatever the objects of that category might be. For this reason, model categories are often regarded as "abstract homotopy theories." We will now zoom out to consider functors comparing such homotopy theories.

More generally, we might consider functors between homotopical categories equipped with weak equivalences that at least satisfy the two-of-three property. A great deal of the subtlety in "category theory up to weak equivalence" has to do with the fact that functors between homotopical categories need not necessarily preserve weak equivalences. In the case where a functor fails to preserve weak equivalence the next best hope is that it admits a universal approximation by a functor that does, where the approximation is either "from the left" or "from the right." Such approximations are referred as *left* or *right derived functors*.

The universal properties of left or right derived functors exist at the level of homotopy categories though the derived functors of greatest utility, and the ones that are most easily constructed in practice, can be constructed at the "point-set level." One of the selling points of Quillen's theory of model categories is that they highlight classes of functors — the left or right Quillen functors — whose left or right derived functors can be constructed in a uniform way making the passage to total derived functors pseudofunctorial. However, it turns out a full model structure is not necessary for this construction; morally speaking, all that matters for the specification of derived functors is the weak equivalences.

In §2.4.1, we give a non-standard and in our view greatly improved presentation of the theory of derived functors guided by a recent axiomatization of Dwyer–Hirschhorn–Kan–Smith [92] paired with a result of Maltsiniotis [176]. The key point of difference is that we give a much stronger definition of what constitutes a derived functor than the usual one. In §2.4.2 we introduce left and right Quillen functors between model categories and show that such functors have a left or right derived functor satisfying this stronger property. Then, in §2.4.3, we see that the abstract theory of this stronger class of derived functors is considerably better than the theory of the weaker ones. A highlight is an efficient expression of the properties of composite or adjoint derived functors proven by Shulman [277] and reproduced as Theorem 2.4.15.

In §2.4.4, we extend the theory of derived functors to allow functors of two variables, with the aim of proving that the homotopy category of spaces is cartesian closed, inheriting an internal hom defined as the derived functor of the point-set level mapping spaces. Implicit in our approach to the proof of this statement is a result promised at the end of §2.3.5. In §2.4.5, we define a precise notion of equivalence between abstract homotopy theories encoded by model categories, which specializes to establish an equivalence between the homotopy theory of spaces and the homotopy theory of simplicial sets. Finally, in §2.4.6 we briefly sketch the connection between *homotopical algebra* and *homological algebra* by considering suitable model structures appropriate for a theory of derived functors between chain complexes.

### 2.4.1    Derived functors and equivalence of homotopy theories

As a warning to the reader, this definition of a derived functor is stronger than the usual one in two ways:

- We explicitly require our derived functors to be defined "at the point-set level" rather than simply as functors between homotopy categories.
- We require the universal property of the corresponding "total derived functors" between homotopy categories to define *absolute* Kan extensions.

Before defining our derived functors we should explain the general meaning of absolute Kan extensions.

**Definition 2.4.1.**   A **left Kan extension** of $F\colon \mathsf{C} \to \mathsf{E}$ along $K\colon \mathsf{C} \to \mathsf{D}$ is a functor $\mathrm{Lan}_K F\colon \mathsf{D} \to \mathsf{E}$ together with a natural transformation $\eta\colon F \Rightarrow \mathrm{Lan}_K F \cdot K$ such that for any other such pair $(G\colon \mathsf{D} \to \mathsf{E}, \gamma\colon F \Rightarrow GK)$, $\gamma$ factors uniquely through $\eta$:[7]

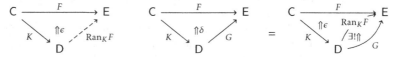

Dually, a **right Kan extension** of $F\colon \mathsf{C} \to \mathsf{E}$ along $K\colon \mathsf{C} \to \mathsf{D}$ is a functor $\mathrm{Ran}_K F\colon \mathsf{D} \to \mathsf{E}$ together with a natural transformation $\epsilon\colon \mathrm{Ran}_K F \cdot K \Rightarrow F$ such that for any $(G\colon \mathsf{D} \to \mathsf{E}, \delta\colon GK \Rightarrow F)$, $\delta$ factors uniquely through $\epsilon$:

A left or right Kan extension is **absolute** if for any functor $H\colon \mathsf{E} \to \mathsf{F}$, the whiskered composite $(H\mathrm{Lan}_K F\colon \mathsf{D} \to \mathsf{E}, H\eta)$ or $(H\mathrm{Ran}_K F\colon \mathsf{D} \to \mathsf{E}, H\epsilon)$ defines the left or right Kan extension of $HF$ along $K$.

A functor between homotopical categories is a **homotopical functor** if it preserves the classes of weak equivalences, or carries the weak equivalences in the domain to isomorphisms in the codomain in the case where no class of weak equivalences is specified. Derived functors can be understood as universal homotopical approximations to a given functor in a sense we now define.

**Definition 2.4.2** (derived functors).   Let $\mathsf{M}$ and $\mathsf{K}$ be homotopical categories with weak equivalences satisfying the two-of-three property and with localization functors $\gamma\colon \mathsf{M} \to \mathsf{HoM}$ and $\delta\colon \mathsf{K} \to \mathsf{HoK}$.

---

[7] Writing $\alpha$ for the natural transformation $\mathrm{Lan}_K F \Rightarrow G$, the right-hand pasting diagrams express the equality $\gamma = \alpha K \cdot \eta$, i.e., that $\gamma$ factors as $F \xRightarrow{\eta} \mathrm{Lan}_K F \cdot K \xRightarrow{\alpha K} GK$.

- A **left derived functor** of $F\colon \mathsf{M} \to \mathsf{K}$ is a homotopical functor $\mathbb{L}F\colon \mathsf{M} \to \mathsf{K}$ equipped with a natural transformation $\lambda\colon \mathbb{L}F \Rightarrow F$ such that $\delta\mathbb{L}F$ and $\delta\lambda\colon$ $\delta\mathbb{L}F \Rightarrow \delta F$ define an absolute *right* Kan extension of $\delta F$ along $\gamma$:

$$\mathsf{M} \underset{\mathbb{L}F}{\overset{F}{\rightrightarrows}}^{\Uparrow\lambda} \mathsf{K} \qquad \longleftrightarrow \qquad
\begin{array}{ccc}
\mathsf{M} & \xrightarrow{\ F\ } & \mathsf{K} \\
{\scriptstyle\gamma}\downarrow & {\scriptstyle\Uparrow\delta\lambda} & \downarrow{\scriptstyle\delta} \\
\mathsf{HoM} & \xrightarrow[\delta\mathbb{L}F]{} & \mathsf{HoK}
\end{array}$$

- A **right derived functor** of $F\colon \mathsf{M} \to \mathsf{K}$ is a homotopical functor $\mathbb{R}F\colon \mathsf{M} \to \mathsf{K}$ equipped with a natural transformation $\rho\colon F \Rightarrow \mathbb{R}F$ such that $\delta\mathbb{R}F$ and $\delta\rho\colon$ $\delta F \Rightarrow \delta\mathbb{R}F$ define an absolute *left* Kan extension of $\delta F$ along $\gamma$:

$$\mathsf{M} \underset{\mathbb{R}F}{\overset{F}{\rightrightarrows}}^{\Downarrow\rho} \mathsf{K} \qquad \longleftrightarrow \qquad
\begin{array}{ccc}
\mathsf{M} & \xrightarrow{\ F\ } & \mathsf{K} \\
{\scriptstyle\gamma}\downarrow & {\scriptstyle\Downarrow\delta\rho} & \downarrow{\scriptstyle\delta} \\
\mathsf{HoM} & \xrightarrow[\delta\mathbb{R}F]{} & \mathsf{HoK}
\end{array}$$

*Remark 2.4.3.* Absolute Kan extensions are in particular "pointwise" Kan extensions, these being the left or right Kan extensions that are preserved by representable functors. The pointwise left or right Kan extensions are those definable as colimits or limits in the target category [238, §6.3], so it is somewhat surprising that these conditions are appropriate to require for functors valued in homotopy categories, which have few limits and colimits.[8]

As a consequence of Proposition 2.2.2, the homotopical functors

$$\delta\mathbb{L}F, \delta\mathbb{R}F\colon \mathsf{M} \rightrightarrows \mathsf{HoK}$$

factor uniquely through $\gamma$ and so may be equally regarded as functors

$$\delta\mathbb{L}F, \delta\mathbb{R}F\colon \mathsf{HoM} \rightrightarrows \mathsf{HoK},$$

as appearing in the displayed diagrams of Definition 2.4.2.

Definition 2.4.4 (total derived functors). The **total left** or **right derived functors** of $F$ are the functors

$$\delta\mathbb{L}F, \delta\mathbb{R}F\colon \mathsf{HoM} \rightrightarrows \mathsf{HoK},$$

defined as absolute Kan extensions in Definition 2.4.2 and henceforth denoted by

$$\mathbf{L}F, \mathbf{R}F\colon \mathsf{HoM} \rightrightarrows \mathsf{HoK}.$$

There is a common setting in which derived functors exist and admit a simple construction. Such categories have a subcategory of "good" objects on which the functor of interest becomes homotopical and a functorial reflection into this full subcategory. The details are encoded in the following axiomatization introduced in [92] and exposed in [276], though we diverge from their terminology to more thoroughly ground our intuition in the model categorical case.

---

[8] With the exception of products and coproducts, the so-called "homotopy limits" and "homotopy colimits" introduced in §2.5 do not define limits and colimits in the homotopy category.

**Definition 2.4.5.** A **left deformation** on a homotopical category M consists of an endofunctor $Q$ together with a natural weak equivalence $q: Q \overset{\sim}{\Rightarrow} 1$.

The functor $Q$ is necessarily homotopical. Let $M_c$ be any full subcategory of M containing the image of $Q$. The inclusion $M_c \rightarrow M$ and the left deformation $Q: M \rightarrow M_c$ induce an equivalence between $HoM$ and $HoM_c$. As our notation suggests, any model category M admits a left deformation defined by cofibrant replacement. Accordingly, we refer to $M_c$ as the subcategory of **cofibrant objects**, trusting the reader to understand that when we have not specified any model structures, Quillen's technical definition is not what we require.

**Definition 2.4.6.** A functor $F: M \rightarrow K$ between homotopical categories is **left deformable** if there exists a left deformation on M such that $F$ is homotopical on an associated subcategory of cofibrant objects.

Our first main result proves that left deformations can be used to construct left derived functors. The basic framework of left deformations was set up in [92] while the fact that such derived functors are absolute Kan extensions was observed in [176].

**Theorem 2.4.7** ([92, 41.2-5], [176]). *If $F: M \rightarrow K$ has a left deformation $q: Q \overset{\sim}{\Rightarrow} 1$, then $\mathbb{L}F = FQ$ is a left derived functor of $F$.*

*Proof.* Write $\delta: K \rightarrow HoK$ for the localization. To show that $(FQ, Fq)$ is a point-set left derived functor, we must show that the functor $\delta FQ$ and natural transformation $\delta Fq: \delta FQ \Rightarrow \delta F$ define a right Kan extension. The verification makes use of Proposition 2.2.2, which identifies the functor category $Fun(HoM, HoK)$ with the full subcategory of $Fun(M, HoK)$ spanned by the homotopical functors. Suppose $G: M \rightarrow HoK$ is homotopical and consider $\alpha: G \Rightarrow \delta F$. Because $G$ is homotopical and $q: Q \Rightarrow 1_M$ is a natural weak equivalence, $Gq: GQ \Rightarrow G$ is a natural isomorphism. Using naturality of $\alpha$, it follows that $\alpha$ factors through $\delta Fq$ as

$$G \overset{(Gq)^{-1}}{\Longrightarrow} GQ \overset{\alpha_Q}{\Longrightarrow} \delta FQ \overset{\delta Fq}{\Longrightarrow} \delta F.$$

To prove uniqueness, suppose $\alpha$ factors as

$$G \overset{\beta}{\Longrightarrow} \delta FQ \overset{\delta Fq}{\Longrightarrow} \delta F.$$

Naturality of $\beta$ provides a commutative square of natural transformations:

$$\begin{array}{ccc} GQ & \overset{\beta_Q}{\Longrightarrow} & \delta FQ^2 \\ {\scriptstyle Gq}\big\Downarrow & & \big\Downarrow{\scriptstyle \delta FQq} \\ G & \underset{\beta}{\Longrightarrow} & \delta FQ \end{array}$$

Because $q$ is a natural weak equivalence and the functors $G$ and $\delta FQ$ are homotopical, the vertical arrows are natural isomorphisms, so $\beta$ is determined by $\beta_Q$. This restricted natural transformation is uniquely determined: $q_Q$ is a natural weak equivalence between objects in the image of $Q$. Since $F$ is homotopical on this subcategory, this means that $Fq_Q$ is a natural weak equivalence and thus $\delta Fq_Q$ is an isomorphism, so $\beta_Q$ must equal the composite of the inverse of this natural isomorphism with $\alpha_Q$.

Finally, to show that this right Kan extension is absolute, our task is to show that for any functor $H \colon \mathsf{HoK} \to \mathsf{E}$, the pair $(H\delta FQ, H\delta Fq)$ again defines a right Kan extension. Note that $(Q, q)$ also defines a left deformation for $H\delta F$, simply because the functor $H \colon \mathsf{HoK} \to \mathsf{E}$ preserves isomorphisms. The argument just given now demonstrates that $(H\delta FQ, H\delta Fq)$ is a right Kan extension, as claimed.    □

### 2.4.2    Quillen functors

We'll now introduce important classes of functors between model categories that will admit derived functors.

**Definition 2.4.8.** A functor between model categories is

- **left Quillen** if it preserves cofibrations, trivial cofibrations, and cofibrant objects, and
- **right Quillen** if it preserves fibrations, trivial fibrations, and fibrant objects.

Most left Quillen functors are "cocontinuous," preserving all colimits, while most right Quillen functors are "continuous," preserving all limits; when this is the case there is no need to separately assume that cofibrant or fibrant objects are preserved. Importantly, cofibrant replacement defines a left deformation for any left Quillen functor, while fibrant replacement defines a right deformation for any right Quillen functor, as we now demonstrate:

**Lemma 2.4.9 (Ken Brown's lemma).**

(i) *Any map between fibrant objects in a model category can be factored as a right inverse to a trivial fibration followed by a fibration:*

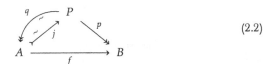

$$(2.2)$$

(ii) *Let $F \colon \mathsf{M} \to \mathsf{K}$ be a functor from a model category to a category with a class of weak equivalences satisfying the two-of-three property. If $F$ carries trivial fibrations in $\mathsf{M}$ to weak equivalences in $\mathsf{K}$, then $F$ carries all weak equivalences between fibrant objects in $\mathsf{M}$ to weak equivalences in $\mathsf{K}$.*

*Proof.* For (i), given any map $f \colon A \to B$ factor its graph $(1_A, f) \colon A \to A \times B$ as a trivial cofibration $j$ followed by a fibration $r$:

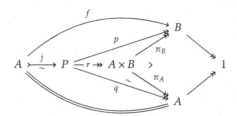

Since $A$ and $B$ are fibrant, the dual of Lemma 2.3.9 implies that the product projections are fibrations, and thus the composite maps $p$ and $q$ are fibrations. By the two-of-three property, $q$ is also a weak equivalence.

To prove (ii) assume that $f : A \to B$ is a weak equivalence in M and construct the factorization (2.2). It follows from the two-of-three property that $p$ is also a trivial fibration, so by hypothesis both $Fp$ and $Fq$ are weak equivalences in K. Since $Fj$ is right inverse to $Fq$, it must also be a weak equivalence, and thus the closure of weak equivalences under composition implies that $Ff$ is a weak equivalence as desired.  □

Specializing Theorem 2.4.7 we then have:

**Corollary 2.4.10.** *The left derived functor of any left Quillen functor F exists and is given by* $\mathbb{L}F := FQ$, *while the right derived functor of any right Quillen functor G exists and is given by* $\mathbb{R}G := GR$, *where Q and R denote any cofibrant and fibrant replacement functors, respectively.*

### 2.4.3     Derived composites and derived adjunctions

Left and right Quillen functors frequently occur in adjoint pairs, in which case the left adjoint is left Quillen if and only if the right adjoint is right Quillen:

**Definition 2.4.11.** Consider an adjunction between a pair of model categories.

$$M \xrightarrow[\;\;\;G\;\;\;]{\;\;\;F\;\;\;} \perp \; N \tag{2.3}$$

Then the following are equivalent, defining a **Quillen adjunction**.

(i) The left adjoint $F$ is left Quillen.

(ii) The right adjoint $G$ is right Quillen.

(iii) The left adjoint preserves cofibrations and the right adjoint preserves fibrations.

(iv) The left adjoint preserves trivial cofibrations and the right adjoint preserves trivial fibrations.

*Exercise 2.4.12.* Justify the equivalence of the properties in the definition by proving:

(i) In the presence of any adjunction (2.3) the lifting problem displayed below left in N admits a solution if and only if the transposed lifting problem displayed below right admits a solution in M.

$$
\begin{array}{ccc}
FA & \xrightarrow{f^{\sharp}} & X \\
{\scriptstyle F\ell}\downarrow & {\scriptstyle k^{\sharp}}\nearrow & \downarrow{\scriptstyle r} \\
FB & \xrightarrow{g^{\sharp}} & Y
\end{array}
\qquad\qquad
\begin{array}{ccc}
A & \xrightarrow{f^{\flat}} & GX \\
{\scriptstyle \ell}\downarrow & {\scriptstyle k^{\flat}}\nearrow & \downarrow{\scriptstyle Gr} \\
B & \xrightarrow{g^{\flat}} & GY
\end{array}
$$

(ii) Conclude that if M has a weak factorization system $(\mathcal{L}, \mathcal{R})$ and N has a weak factorization system $(\mathcal{L}', \mathcal{R}')$ then $F$ preserves the left classes if and only if $G$ preserves the right classes.

Importantly, the total left and right derived functors of a Quillen pair form an adjunction between the appropriate homotopy categories.

Theorem 2.4.13 (Quillen [229, I.3]). *If*

$$
\mathsf{M} \underset{G}{\overset{F}{\underset{\perp}{\rightleftarrows}}} \mathsf{N}
$$

*is a Quillen adjunction, then the total derived functors form an adjunction*

$$
\mathsf{HoM} \underset{\mathbf{R}G}{\overset{\mathbf{L}F}{\underset{\perp}{\rightleftarrows}}} \mathsf{HoN}
$$

A particularly elegant proof of Theorem 2.4.13 is due to Maltsiniotis. Once the strategy is known, the details are elementary enough to be left as an exercise:

*Exercise 2.4.14* ([176]). Use the fact that the total derived functors of a Quillen pair $F \dashv G$ define *absolute* Kan extensions to prove that $\mathbf{L}F \dashv \mathbf{R}G$. Conclude that Theorem 2.4.13 applies more generally to any pair of adjoint functors that are deformable in the sense of Definition 2.4.6 [92, 44.2].

A double categorical theorem of Shulman [277] consolidates into a single statement the adjointness of the total derived functors of a Quillen adjunction, the pseudo-functoriality of the construction of total derived functors of Quillen functors, and one further result about functors that are simultaneously left and right Quillen. A **double category** is a category internal to Cat: it has a set of objects, a category of horizontal morphisms, a category of vertical morphisms, and a set of squares that are composable in both vertical and horizontal directions, defining the arrows in a pair of categories with the horizontal and vertical morphisms as objects, respectively [147].

For instance, $\mathbb{C}$at is the double category of categories, functors, functors, and natural transformations inhabiting squares and pointing southwest. There is another double category $\mathbb{M}$odel whose objects are model categories, whose vertical morphisms are left Quillen functors, whose horizontal morphisms are right Quillen functors, and whose squares are natural transformations pointing southwest. The following theorem and a generalization, with deformable functors in place of Quillen functors [277, 8.10], is due to Shulman.

**Theorem 2.4.15** ([277, 7.6]). *The map that sends a model category to its homotopy category and a left or right Quillen functor to its total left or right derived functor defines a double pseudofunctor* Ho: $\mathbb{M}$odel $\to \mathbb{C}$at.

The essential content of the pseudofunctoriality statement is that the composite of the left derived functors of a pair of left Quillen functors is coherently naturally weakly equivalent to the left derived functor of their composite. Explicitly, given a composable pair of left Quillen functors M $\xrightarrow{F}$ L $\xrightarrow{G}$ K , the map

$$\mathbb{L}G \circ \mathbb{L}F := GQ \circ FQ \xrightarrow{G\epsilon_{FQ}} GFQ =: \mathbb{L}(GF)$$

defines a comparison natural transformation. Since $Q \colon$ M $\to$ M$_c$ lands in the sub-category of cofibrant objects and $F$ preserves cofibrant objects, $\epsilon_{FQ} \colon QFQ \Rightarrow FQ$ is a weak equivalence between cofibrant objects. Lemma 2.4.9(ii) then implies that $G\epsilon_{FQ} \colon GQFQ \to GFQ$ defines a natural weak equivalence $\mathbb{L}G \circ \mathbb{L}F \to \mathbb{L}GF$. Given a composable triple of left Quillen functors, there is a commutative square of natural weak equivalences $\mathbb{L}H \circ \mathbb{L}G \circ \mathbb{L}F \to \mathbb{L}(H \circ G \circ F)$. If we compose with the Gabriel–Zisman localizations to pass to homotopy categories and total left derived functors, these coherent natural weak equivalences become coherent natural isomorphisms, defining the claimed pseudofunctor.

Quillen adjunctions are encoded in the double category $\mathbb{M}$odel as "conjoint" relationships between vertical and horizontal 1-cells; in this way Theorem 2.4.15 subsumes Theorem 2.4.13. Similarly, functors that are simultaneously left and right Quillen are presented as vertical and horizontal "companion" pairs. The double pseudofunctoriality of Theorem 2.4.15 contains a further result: if a functor is both left and right Quillen, then its total left and right derived functors are isomorphic.

### 2.4.4    Monoidal and enriched model categories

If M has a model structure and a monoidal structure it is natural to ask that these be compatible in some way, but what sort of compatibility should be required? In the most common examples, the monoidal product is *closed* — that is, the functors $A \otimes -$ and $- \otimes A$ admit right adjoints[9] and consequently preserve colimits in each variable separately. This situation is summarized and generalized by the notion of a two-variable adjunction, which we introduce using notation that will suggest the most common examples.

**Definition 2.4.16.** A triple of bifunctors

$$K \times L \xrightarrow{\otimes} M, \quad K^{op} \times M \xrightarrow{\{,\}} L, \quad L^{op} \times M \xrightarrow{\text{Map}} K$$

---

[9] Very frequently a monoidal structure is symmetric, in which case these functors are naturally isomorphic, and a single right adjoint suffices.

equipped with a natural isomorphism

$$\mathsf{M}(K \otimes L, M) \cong \mathsf{L}(L, \{K, M\}) \cong \mathsf{K}(K, \mathrm{Map}(L, M))$$

defines a **two-variable adjunction**.

Example 2.4.17. A symmetric monoidal category is **closed** just when its monoidal product $- \otimes -\colon \mathsf{V} \times \mathsf{V} \to \mathsf{V}$ defines the left adjoint of a two-variable adjunction

$$\mathsf{V}(A \otimes B, C) \cong \mathsf{V}(B, \mathrm{Map}(A, C)), \mathsf{V}(A, \mathrm{Map}(B, C)),$$

the right adjoint $\mathrm{Map}\colon \mathsf{V}^{\mathrm{op}} \times \mathsf{V} \to \mathsf{V}$ defining an **internal hom**.

Example 2.4.18. A category $\mathsf{M}$ that is enriched over a monoidal category is **tensored** and **cotensored** just when the enriched hom functor $\mathrm{Map}\colon \mathsf{M}^{\mathrm{op}} \times \mathsf{M} \to \mathsf{V}$ is one of the right adjoints of a two-variable adjunction

$$\mathsf{M}(V \otimes M, N) \cong \mathsf{M}(M, \{V, N\}) \cong \mathsf{V}(V, \mathrm{Map}(M, N)),$$

the other two adjoints defining the **tensor** $V \otimes M$ and **cotensor** $\{V, N\}$ of an object $V \in \mathsf{V}$ with objects $M, N \in \mathsf{M}$.[10]

A Quillen two-variable adjunction is a two-variable adjunction in which the left adjoint is a left Quillen bifunctor while the right adjoints are both right Quillen bifunctors, any one of these conditions implying the other two. To state these definitions, we must introduce the following construction. The "pushout-product" of a bifunctor $- \otimes -\colon \mathsf{K} \times \mathsf{L} \to \mathsf{M}$ defines a bifunctor $- \hat{\otimes} -\colon \mathsf{K}^2 \times \mathsf{L}^2 \to \mathsf{M}^2$ that we refer to as the "Leibniz tensor" (when the bifunctor $\otimes$ is called a "tensor"). The "Leibniz cotensor" and "Leibniz hom"

$$\widehat{\{-, -\}}\colon (\mathsf{K}^2)^{\mathrm{op}} \times \mathsf{M}^2 \to \mathsf{L}^2 \qquad \text{and} \qquad \widehat{\mathrm{Map}}(-, -)\colon (\mathsf{L}^2)^{\mathrm{op}} \times \mathsf{M}^2 \to \mathsf{K}^2$$

are defined dually, using pullbacks in $\mathsf{L}$ and $\mathsf{K}$ respectively.

Definition 2.4.19 (the Leibniz construction). Given a bifunctor $- \otimes -\colon \mathsf{K} \times \mathsf{L} \to \mathsf{M}$ valued in a category with pushouts, the **Leibniz tensor** of a map $k\colon I \to J$ in $\mathsf{K}$ and a map $\ell\colon A \to B$ in $\mathsf{L}$ is the map $k \hat{\otimes} \ell$ in $\mathsf{M}$ induced by the pushout diagram on the left:

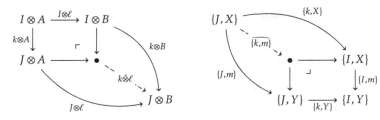

---

[10] As stated, this definition is a little too weak: one needs to ask in addition that (i) the tensors are associative relative to the monoidal product in $\mathsf{V}$, (ii) dually that the cotensors are associative relative to the monoidal product in $\mathsf{V}$, and (iii) that the two-variable adjunction is enriched in $\mathsf{V}$. Any of these three conditions implies the other two.

In the case of a bifunctor $\{-,-\} \colon \mathsf{K}^{\mathrm{op}} \times \mathsf{M} \to \mathsf{L}$ contravariant in one of its variables valued in a category with pullbacks, the **Leibniz cotensor** of a map $k \colon I \to J$ in $\mathsf{K}$ and a map $m \colon X \to Y$ in $\mathsf{M}$ is the map $\widehat{\{k, m\}}$ induced by the pullback diagram above right.

**Proposition 2.4.20.** *The Leibniz construction preserves:*

(i) *structural isomorphisms: a natural isomorphism*
$$X * (Y \otimes Z) \cong (X \times Y) \square Z$$
*between suitably composable bifunctors extends to a natural isomorphism*
$$f \,\hat{*}\, (g \,\hat{\otimes}\, h) \cong (f \,\hat{\times}\, g) \,\hat{\square}\, h$$
*between the corresponding Leibniz products;*

(ii) *adjointness: if $(\otimes, \{,\}, \mathsf{Map})$ define a two-variable adjunction, then the Leibniz bifunctors $(\hat{\otimes}, \widehat{\{,\}}, \widehat{\mathsf{Map}})$ define a two-variable adjunction between the corresponding arrow categories;*

(iii) *colimits in the arrow category: if $\otimes \colon \mathsf{K} \times \mathsf{L} \to \mathsf{M}$ is cocontinuous in either variable, then so is $\hat{\otimes} \colon \mathsf{K}^2 \times \mathsf{L}^2 \to \mathsf{M}^2$;*

(iv) *pushouts: if $\otimes \colon \mathsf{K} \times \mathsf{L} \to \mathsf{M}$ is cocontinuous in its second variable, and if $g'$ is a pushout of $g$, then $f \,\hat{\otimes}\, g'$ is a pushout of $f \,\hat{\otimes}\, g$;*

(v) *composition, in a sense: the Leibniz tensor $f \,\hat{\otimes}\, (h \cdot g)$ factors as a composite of a pushout of $f \,\hat{\otimes}\, g$ followed by $f \,\hat{\otimes}\, h$:*

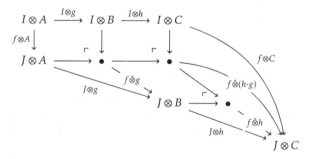

(vi) *cell complex structures: if $f$ and $g$ may be presented as cell complexes with cells $f_\alpha$ and $g_\beta$, respectively, and if $\otimes$ is cocontinuous in both variables, then $f \,\hat{\otimes}\, g$ may be presented as a cell complex with cells $f_\alpha \,\hat{\otimes}\, g_\beta$.*

Proofs of these assertions and considerably more details are given in [245, §4-5].

*Exercise 2.4.21.* Given a two-variable adjunction as in Definition 2.4.16 and classes of maps $\mathcal{A}$ in $\mathsf{K}$, $\mathcal{B}$ in $\mathsf{L}$, and $\mathcal{C}$ in $\mathsf{M}$, prove equivalences between the lifting properties:
$$\mathcal{A} \,\hat{\otimes}\, \mathcal{B} \,\square\, \mathcal{C} \quad \Leftrightarrow \quad \mathcal{B} \,\square\, \widehat{\{\mathcal{A}, \mathcal{C}\}} \quad \Leftrightarrow \quad \mathcal{A} \,\square\, \widehat{\mathsf{Map}}(\mathcal{B}, \mathcal{C}).$$

Here $\mathcal{A} \,\hat{\otimes}\, \mathcal{B} \,\square\, \mathcal{C}$, for instance, asserts that maps in $\mathcal{C}$ have the right lifting property with respect to each map in $\mathcal{A} \,\hat{\otimes}\, \mathcal{B}$.

Exercise 2.4.21 explains the equivalence between the following three definitions of a Quillen two-variable adjunction.

**Definition 2.4.22.** A two-variable adjunction

$$V \times M \xrightarrow{\otimes} N, \quad V^{\mathrm{op}} \times N \xrightarrow{\{-,-\}} M, \quad M^{\mathrm{op}} \times N \xrightarrow{\mathrm{Map}} V$$

between model categories V, M, and N defines a **Quillen two-variable adjunction** if any, and hence all, of the following equivalent conditions are satisfied:

(i) The functor $\hat{\otimes}\colon V^2 \times M^2 \to N^2$ carries any pair comprised of a cofibration in V and a cofibration in M to a cofibration in N, and this cofibration is a weak equivalence if either of the domain maps are.

(ii) The functor $\widehat{\{-,-\}}\colon (V^2)^{\mathrm{op}} \times N^2 \to M^2$ carries any pair comprised of a cofibration in V and a fibration in N to a fibration in M, and this fibration is a weak equivalence if either of the domain maps are.

(iii) The functor $\widehat{\mathrm{Map}}\colon (M^2)^{\mathrm{op}} \times N^2 \to V^2$ carries any pair comprised of a cofibration in M and a fibration in N to a fibration in V, and this fibration is a weak equivalence if either of the domain maps are.

*Exercise 2.4.23.* Prove that if $-\otimes-\colon V \times M \to N$ is a left Quillen bifunctor and $V \in V$ is cofibrant then $V \otimes -\colon M \to N$ is a left Quillen functor.

Quillen's axiomatization of the additional properties enjoyed by his model structure on the category of simplicial sets has been generalized by Hovey [130, §4.2].

**Definition 2.4.24.** A (**closed symmetric**) **monoidal model category** is a (closed symmetric) monoidal category $(V, \otimes, I)$ with a model structure so that the monoidal product and hom define a Quillen two-variable adjunction and furthermore so that the maps

$$QI \otimes v \to I \otimes v \cong v \qquad \text{and} \qquad v \otimes QI \to v \otimes I \cong v \qquad (2.4)$$

are weak equivalences if $v$ is cofibrant.[11]

**Definition 2.4.25.** If V is a monoidal model category a V-**model category** is a model category M that is tensored, cotensored, and V-enriched in such a way that $(\otimes, \{,\}, \mathrm{Map})$ is a Quillen two-variable adjunction and the maps

$$QI \otimes m \to I \otimes m \cong m$$

are weak equivalences if $m$ is cofibrant.

*Exercise 2.4.26.* In a locally small category M with products and coproducts the hom bifunctor is part of a two-variable adjunction:

$$-*-\colon \mathrm{Set} \times M \to M, \quad \{-,-\}\colon \mathrm{Set}^{\mathrm{op}} \times M \to M, \quad \mathrm{Hom}\colon M^{\mathrm{op}} \times M \to \mathrm{Set}.$$

Equipping Set with the model structure whose weak equivalences are all maps, whose cofibrations are monomorphisms, and whose fibrations are epimorphisms, prove that

---

[11] If the monoidal product is symmetric then of course these two conditions are equivalent and if it is closed then they are also equivalent to a dual one involving the internal hom [130, 4.2.7].

(i) Set is a cartesian monoidal model category.

(ii) Any model category M is a Set-model category.

Example 2.4.27. Quillen's model structure of Theorem 2.3.30 is a closed symmetric monoidal model category. The term **simplicial model category** refers to a model category enriched over this model structure.

*Exercise 2.4.28.* Show that if M is a simplicial model category then the full simplicial subcategory $M_{cf}$ is Kan-complex enriched.

The conditions (2.4) on the cofibrant replacement of the monoidal unit are implied by the Quillen two-variable adjunction if the monoidal unit is cofibrant and are necessary for the proof of Theorem 2.4.29, which shows that the homotopy categories are again closed monoidal and enriched.

Theorem 2.4.29 ([130, 4.3.2,4]).

(i) *The homotopy category of a closed symmetric monoidal model category is a closed monoidal category with tensor and hom given by the derived adjunction*

$$(\mathbb{L}\otimes, \mathbb{R}\mathsf{Map}, \mathbb{R}\mathsf{Map}) \colon \mathsf{HoV} \times \mathsf{HoV} \to \mathsf{HoV}$$

*and monoidal unit* $QI$.

(ii) *If* M *is a* V*-model category, then* HoM *is the underlying category of a* HoV*-enriched, tensored, and cotensored category with enrichment given by the total derived two-variable adjunction*

$$(\mathbb{L}\otimes, \mathbb{R}\{,\}, \mathbb{R}\mathsf{Map}) \colon \mathsf{HoV} \times \mathsf{HoM} \to \mathsf{HoM}.$$

In particular:

Corollary 2.4.30. *The homotopy category of spaces is cartesian closed. If* M *is a simplicial model category, then* HoM *is enriched, tensored, and cotensored over the homotopy category of spaces.*

## 2.4.5     Quillen equivalences between homotopy theories

Two model categories present equivalent homotopy theories if there exists a finite sequence of model categories and a zigzag of Quillen equivalences between them, in a sense we now define. A Quillen adjunction defines a Quillen equivalence just when the derived adjunction of Theorem 2.4.13 defines an **adjoint equivalence**: an adjunction with invertible unit and counit. There are several equivalent characterizations of this situation.

Definition 2.4.31 ([229, §I.4]). A Quillen adjunction between a pair of model categories

$$\mathsf{M} \underset{G}{\overset{F}{\underset{\perp}{\rightleftarrows}}} \mathsf{N}$$

defines a **Quillen equivalence** if any, and hence all, of the following equivalent conditions are satisfied:

(i)  The total left derived functor $\mathbf{L}F\colon \mathsf{Ho}\mathsf{M} \to \mathsf{Ho}\mathsf{N}$ defines an equivalence of categories.

(ii)  The total right derived functor $\mathbf{R}G\colon \mathsf{Ho}\mathsf{N} \to \mathsf{Ho}\mathsf{M}$ defines an equivalence of categories.

(iii)  For every cofibrant object $A \in \mathsf{M}$ and every fibrant object $X \in \mathsf{N}$, a map $f^{\sharp}\colon FA \to X$ is a weak equivalence in $\mathsf{N}$ if and only if its transpose $f^{\flat}\colon A \to GX$ is a weak equivalence in $\mathsf{M}$.

(iv)  For every cofibrant object $A \in \mathsf{M}$, the composite $A \to GFA \to GRFA$ of the unit with fibrant replacement is a weak equivalence in $\mathsf{M}$, and for every fibrant object $X \in \mathsf{N}$, the composite $FQGX \to FGX \to X$ of the counit with cofibrant replacement is a weak equivalence in $\mathsf{N}$.

Famously, the formalism of Quillen equivalences enables a proof that the homotopy theory of spaces is equivalent to the homotopy theory of simplicial sets.

**Theorem 2.4.32** (Quillen [229, §II.3]). *The homotopy theory of simplicial sets is equivalent to the homotopy theory of topological spaces via the geometric realization ⊣ total singular complex adjunction*

$$\mathsf{sSet} \underset{\mathrm{Sing}}{\overset{|-|}{\rightleftarrows}} \perp \mathsf{Top}$$

## 2.4.6    Extending homological algebra to homotopical algebra

Derived functors are endemic to homological algebra. Quillen's homotopical algebra can be understood to subsume classical homological algebra in the following sense. The category of chain complexes of modules over a fixed ring (or valued in an arbitrary abelian category) admits a homotopical structure where the weak equivalences are quasi-isomorphisms. Relative to an appropriately defined model structure, the left and right derived functors of homological algebra can be viewed as special cases of the construction of derived functors of left or right Quillen functors in Corollary 2.4.10 or in the more general context of Theorem 2.4.7.

The following theorem describes an equivalent presentation of the homotopy theory just discussed.

**Theorem 2.4.33** (Schwede–Shipley after Dold–Kan). *The homotopy theory of simplicial modules over a commutative ring, with fibrations and weak equivalences as on underlying simplicial sets, is equivalent to the homotopy theory of non-negatively graded chain complexes of modules, as presented by the projective model structure whose weak equivalences are the quasi-isomorphisms, fibrations are the chain maps which are surjective in positive dimensions, and cofibrations are the monomorphisms with dimensionwise projective cokernel.*

*Proof.* For details of the model structure on simplicial objects see [229, II.4, II.6] and on chain complexes see [130, 2.3.11, 4.2.13]. The proof that the functors $\Gamma, N$ in the Dold–Kan equivalence are each both left and right Quillen equivalences can be found in [265, §4.1] or is safely left as an exercise to the reader. $\qquad\square$

The Dold–Kan Quillen equivalence of Theorem 2.4.33 suggests that simplicial methods might replace homological ones in non-abelian contexts. Let M be any category of "algebras" such as monoids, groups, rings (or their commutative variants), or modules or algebras over a fixed ring; technically M may be any category of models for a *Lawvere theory* [154], which specifies finite operations of any arity and relations between the composites of these operations.

Theorem 2.4.34 (Quillen [229, §II.4]). *For any category* M *of "algebras" — a category of models for a Lawvere theory — the category* $M^{\Delta^{op}}$ *of simplicial algebras admits a simplicial model structure whose*

- *weak equivalences are those maps that are weak homotopy equivalences on underlying simplicial sets,*
- *fibrations are those maps that are Kan fibrations on underlying simplicial sets, and*
- *cofibrations are retracts of free maps.*

## 2.5    Homotopy limits and colimits

Limits and colimits provide fundamental tools for constructing new mathematical objects from existing ones, so it is important to understand these constructions in the homotopical context. There are a variety of possible meanings of a homotopical notion of limit or colimit including:

(i) limits or colimits in the homotopy category of a model category;
(ii) limit or colimit constructions that are "homotopy invariant," with weakly equivalent inputs giving rise to weakly equivalent outputs;
(iii) derived functors of the limit or colimit functors; and finally
(iv) limits or colimits whose universal properties are (perhaps weakly) enriched over simplicial sets or topological spaces.

We will explore these possibilities in turn. We begin in §2.5.1 by observing that the homotopy category has few genuine limits and colimits but does have "weak" ones in the case where the category is enriched, tensored, and cotensored over spaces. For the reason explained in Remark 2.5.5, homotopy limits or colimits rarely satisfy condition (i).

In §2.5.2, we define homotopy limits and colimits as derived functors, which in particular give "homotopy invariant" constructions, and introduce hypotheses on the ambient model category that ensure that these homotopy limit and colimit functors always exist. In §2.5.3 we consider particular diagram shapes, the so-called Reedy categories, for which homotopy limits and colimits exist in any model category. Finally, in §2.5.4 we permit ourselves a tour through the general theory of weighted limits and colimits as a means of elucidating these results and introducing families of Quillen bifunctors that deserve to be better known. This allows us to finally explain the sense in which homotopy limits or colimits in a simplicial model category satisfy properties (ii)-(iv) and in particular have an enriched universal property which may be understood as saying they "represent homotopy coherent cones" over or under the diagram.

## 2.5.1 Weak limits and colimits in the homotopy category

Consider a category M that is enriched over spaces — either topological spaces or simplicial sets will do — meaning that for each pair of objects $x, y$, there is a mapping space $\mathsf{Map}(x, y)$ whose points are the usual set $\mathsf{M}(x, y)$ of arrows from $x$ to $y$. We may define a homotopy category of M using the construction of Definition 2.3.24.

Definition 2.5.1. If M is a simplicially enriched category its **homotopy category** hM has

– objects the same objects as M and
– hom-sets $\mathsf{hM}(x, y) := \pi_0 \mathsf{Map}(x, y)$ taken to be the path components of the mapping spaces.

Thus, a morphism from $x$ to $y$ in hM is a homotopy class of vertices in the simplicial set $\mathsf{Map}(x, y)$, where two vertices are homotopic if and only if they can be connected by a finite zigzag of 1-simplices.

A product of a family of objects $m_\alpha$ in a category M is given by a representation $m$ for the functor displayed on the right:

$$\mathsf{M}(-, m) \xrightarrow{\cong} \prod_\alpha \mathsf{M}(-, m_\alpha).$$

By the Yoneda lemma, a representation consists of an object $m \in \mathsf{M}$ together with maps $m \to m_\alpha$ for each $\alpha$ that are universal in the sense that for any collection $x \to m_\alpha \in \mathsf{M}$, each of these arrows factors uniquely along a common map $x \to m$. But if M is enriched over spaces, we might instead require only that the triangles

$$
\begin{array}{ccc}
 & x & \\
\exists \Big\downarrow & \overset{\simeq}{\searrow} & \\
m & \longrightarrow & m_\alpha
\end{array}
\tag{2.5}
$$

commute "up to homotopy" in the sense of a path in the space $\mathsf{Map}(x, m_\alpha)$ whose underlying set of points is $\mathsf{M}(x, m_\alpha)$. Now we can define the **homotopy product** to be an object $m$ equipped with a natural weak homotopy equivalence

$$\mathsf{Map}(x, m) \to \prod_\alpha \mathsf{Map}(x, m_\alpha)$$

for each $x \in \mathsf{M}$. Surjectivity on path components implies the existence and homotopy commutativity of the triangles (2.5).

*Exercise 2.5.2.* Use the fact that $\pi_0$ commutes with products and is homotopical to show, unusually for homotopy limits, that the homotopy product is a product in the homotopy category hM. Similarly, a homotopy coproduct is a coproduct in the homotopy category.

For non-discrete diagram shapes, the homotopy category of a category enriched in spaces[12] will no longer have genuine limits or colimits but in the presence of tensors in the colimit case and cotensors in the limit case it will have weak ones.

Theorem 2.5.3 ([295, 11.1]). *If* M *is cocomplete and also enriched and tensored over spaces, its homotopy category* hM *has all weak colimits: given any small diagram* $F: D \to hM$, *there is a cone under* F *through which every other cone factors, although not necessarily uniquely.*

In general, the colimit of a diagram $F$ of shape D may be constructed as the reflexive coequalizer of the diagram

$$\coprod_{a,b \in D} D(a,b) \times Fa \; \underset{\substack{\xrightarrow{\text{ev}} \\ \xleftarrow{\;\;\text{id}\;\;} \\ \xrightarrow{\text{proj}}}}{} \; \coprod_{a \in D} Fa$$

Note that this construction does not actually require the diagram $F$ to be a functor; it suffices for the diagram to define a reflexive directed graph in the target category. In the case of a diagram valued in hM, the weak colimit will be constructed as a "homotopy reflexive coequalizer"[13] of a lifted reflexive directed graph in M.

*Proof.* Any diagram $F: D \to hM$ may be lifted to a reflexive directed graph $F: D \to M$, choosing representatives for each homotopy class of morphisms in such a way that identities are chosen to represent identities. Using these lifted maps and writing $I$ for the interval, define the weak colimit of $F: D \to hM$ to be a quotient of the coproduct

$$\left( \coprod_{a,b \in D} D(a,b) \times I \times Fa \right) \sqcup \left( \coprod_{a \in D} Fa \right)$$

modulo three identifications:

$$\coprod_{a,b \in D} (D(a,b) \times \{0\} \times Fa \sqcup D(a,b) \times \{1\} \times Fa) \sqcup \coprod_{a \in D} I \times Fa \xrightarrow{(\text{ev} \sqcup \text{proj}) \sqcup \text{proj}} \coprod_{a \in D} Fa$$

$$\Big\downarrow {\scriptstyle (\text{incl} \sqcup \text{incl}) \sqcup \text{id}} \qquad\qquad\qquad\qquad\qquad \ulcorner \quad \Big\downarrow$$

$$\coprod_{a,b \in D} D(a,b) \times I \times Fa \dashrightarrow \text{wcolim}\, F$$

The right-hand vertical map defines the legs of the colimit cone, which commute in hM via the witnessing homotopies given by the bottom horizontal map.

Now consider a cone in hM under $F$ with nadir $X$. We may regard the data of this

---

[12] Here we can take our enrichment over topological spaces or over simplicial sets, the latter being more general [237, 3.7.15-16].

[13] Succinctly, it may be defined as the weighted colimit of this reflexive coequalizer diagram weighted by the truncated cosimplicial object $* \rightrightarrows I$ whose leftwards maps are the endpoint inclusions into the closed interval $I$; see §2.5.4.

cone as a diagram $D \times 2 \to hM$ that restricts along $\{0\} \hookrightarrow 2$ to $F$ and along $\{1\} \hookrightarrow 2$ to the constant diagram at $X$. This data may be lifted to a reflexive directed graph $D \times 2 \to M$ whose lift over 0 agrees with the previously specified lift $F$ and whose lift over 1 is constant at $X$. This defines a cone under the pushout diagram, inducing the required map $\mathrm{wcolim}F \to X$. □

## 2.5.2 Homotopy limits and colimits of general shapes

In general, limit and colimit constructions in a homotopical category fail to be weak equivalence invariant. Famously the $n$-sphere can be formed by gluing together two disks along their boundary spheres $S^n \cong D^n \cup_{S^{n-1}} D^n$. The diagram

$$
\begin{array}{ccccc}
D^n & \longleftarrow & S^{n-1} & \longrightarrow & D^n \\
\downarrow{\scriptstyle \wr} & & \| & & \downarrow{\scriptstyle \wr} \\
* & \longleftarrow & S^{n-1} & \longrightarrow & *
\end{array}
\tag{2.6}
$$

reveals that the pushout functor fails to preserve componentwise homotopy equivalences.

When a functor fails to be homotopical, the next best option is to replace it by a derived functor. Because colimits are left adjoints, one might hope that $\mathrm{colim} \colon M^D \to M$ has a left derived functor and dually that $\lim \colon M^D \to M$ has a right derived functor, leading us to the following definition:

**Definition 2.5.4.** Let $M$ be a homotopical category and let $D$ be a small category. The **homotopy colimit functor**, when it is exists, is a left derived functor $\mathbb{L}\mathrm{colim} \colon M^D \to M$, while the **homotopy limit functor**, when it exists, is a right derived functor $\mathbb{R}\lim \colon M^D \to M$.

We always take the weak equivalences in the category $M^D$ of diagrams of shape $D$ in a homotopical category $M$ to be defined pointwise. By the universal property of localization, there is a canonical map

$$
\begin{array}{ccc}
M^D & \xrightarrow{\;\gamma^D\;} & (HoM)^D \\
{\scriptstyle \gamma}\downarrow & \nearrow & \\
Ho(M^D) & &
\end{array}
\tag{2.7}
$$

but it is not typically an equivalence of categories. Indeed, some of the pioneering forays into abstract homotopy theory [295, 72, 90] were motivated by attempts to understand the essential image of the functor $Ho(M^D) \to (HoM)^D$, the objects in $(HoM)^D$ being homotopy commutative diagrams while the isomorphism classes of objects in $Ho(M^D)$ being somewhat more mysterious; see §2.6.2.

*Remark 2.5.5.* The diagonal functor $\Delta \colon M \to M^D$ is homotopical and hence acts as its own left and right derived functors. By Theorem 2.4.13 applied to a Quillen adjunction to be constructed in the proof of Theorem 2.5.7, the total derived functor

Lcolim: $\mathrm{Ho}(\mathsf{M}^{\mathsf{D}}) \to \mathrm{HoM}$ is left adjoint to $\Delta \colon \mathrm{HoM} \to \mathrm{Ho}(\mathsf{M}^{\mathsf{D}})$, but unless the comparison of (2.7) is an equivalence, this is not the same as the diagonal functor $\Delta \colon \mathrm{HoM} \to \mathrm{Ho}(\mathsf{M})^{\mathsf{D}}$. Hence, homotopy colimits are not typically colimits in the homotopy category.[14]

In the presence of suitable model structures, Corollary 2.4.10 can be used to prove that the homotopy limit and colimit functors exist.

**Definition 2.5.6.** Let M be a model category and let D be a small category.

(i) The **projective model structure** on $\mathsf{M}^{\mathsf{D}}$ has weak equivalences and fibrations defined pointwise in M.
(ii) The **injective model structure** on $\mathsf{M}^{\mathsf{D}}$ has weak equivalences and cofibrations defined pointwise in M.

When M is a *combinatorial model category*, both model structures are guaranteed to exist. More generally, when M is an *accessible model category* these model structures exist [118, 3.4.1]. Of course, the projective and injective model structures might happen to exist on $\mathsf{M}^{\mathsf{D}}$, perhaps for particular diagram shapes D, in the absence of these hypotheses.

**Theorem 2.5.7.** *Let* M *be a model category and let* D *be a small category.*

(i) *Whenever the projective model structure on* $\mathsf{M}^{\mathsf{D}}$ *exists, the homotopy colimit functor* $\mathbb{L}\mathrm{colim} \colon \mathsf{M}^{\mathsf{D}} \to \mathsf{M}$ *exists and may be computed as the colimit of a projective cofibrant replacement of the original diagram.*
(ii) *Whenever the injective model structure on* $\mathsf{M}^{\mathsf{D}}$ *exists, the homotopy limit functor* $\mathbb{R}\mathrm{lim} \colon \mathsf{M}^{\mathsf{D}} \to \mathsf{M}$ *exists and may be computed as the limit of an injective fibrant replacement of the original diagram.*

*Proof.* This follows from Corollary 2.4.10 once we verify that the colimit and limit functors are respectively left and right Quillen with respect to the projective and injective model structures. These functors are, respectively, left and right adjoint to the constant diagram functor $\Delta \colon \mathsf{M} \to \mathsf{M}^{\mathsf{D}}$, so by Definition 2.4.11 it suffices to verify that this functor is right Quillen with respect to the projective model structure and also left Quillen with respect to the injective model structure. But these model structures are designed so that this is the case. □

*Exercise 2.5.8.*

(i) Show that any pushout diagram $B \longleftarrow\!\!\!\prec A \succ\!\!\!\longrightarrow C$ comprised of a pair of cofibrations between cofibrant objects is projectively cofibrant. Conclude that the pushout of cofibrations between cofibrant objects is a homotopy pushout and use this to compute the homotopy pushout of (2.6).

----

[14] The comparison (2.7) is an equivalence when D is discrete; this is why homotopy products and homotopy coproducts *are* products and coproducts in the homotopy category.

(ii) Argue that for a generic pushout diagram $Y \longleftarrow X \longrightarrow Z$ , its homotopy pushout may be constructed by taking a cofibrant replacement $q\colon X' \to X$ of $X$ and then factoring the composites $hq$ and $kq$ as a cofibration followed by a trivial fibration:

$$
\begin{array}{ccc}
Y' \longleftarrow\!\!\!\!\prec X' \succ\!\!\!\!\longrightarrow Z' \\
\downarrow{\scriptstyle\wr} \qquad {\scriptstyle\wr}\downarrow{\scriptstyle q} \qquad \downarrow{\scriptstyle\wr} \\
Y \longleftarrow X \longrightarrow Z
\end{array}
$$

and then taking the ordinary pushout of this projective cofibrant replacement formed by the top row.

*Exercise 2.5.9.*

(i) Verify that any $\omega$-indexed diagram

$$A_0 \xrightarrow{\ f_{01}\ } A_1 \xrightarrow{\ f_{12}\ } A_2 \xrightarrow{\ f_{23}\ } \cdots$$

of cofibrations between cofibrant objects is projectively cofibrant. Conclude that the sequential colimit of a diagram of cofibrations between cofibrant objects is a homotopy colimit.

(ii) Argue that for a generic sequential diagram

$$X_0 \xrightarrow{\ f_{01}\ } X_1 \xrightarrow{\ f_{12}\ } X_2 \xrightarrow{\ f_{23}\ } \cdots$$

its projective cofibrant replacement may be formed by first replacing $X_0$ by a cofibrant object $Q_0$, then inductively factoring the resulting composite map $Q_n \to X_{n+1}$ into a cofibration followed by a trivial fibration:

$$
\begin{array}{ccc}
G & Q_0 \xrightarrow{\ g_{01}\ } Q_1 \xrightarrow{\ g_{12}\ } Q_2 \xrightarrow{\ g_{23}\ } \cdots \\
\scriptstyle q\downarrow{\scriptstyle\wr} & \scriptstyle q_0\downarrow{\scriptstyle\wr} \qquad \scriptstyle q_1\downarrow{\scriptstyle\wr} \qquad \scriptstyle q_2\downarrow{\scriptstyle\wr} \\
F & X_0 \xrightarrow{\ f_{01}\ } X_1 \xrightarrow{\ f_{12}\ } X_2 \xrightarrow{\ f_{23}\ } \cdots
\end{array}
$$

Conclude that the homotopy sequential colimit is formed as the sequential colimit of this top row.

## 2.5.3    Homotopy limits and colimits of Reedy diagrams

In fact, even if the projective model structures do not exist, certain diagram shapes allow us to construct functorial "projective cofibrant replacements" in any model category nonetheless, for example by following the prescriptions of Exercise 2.5.9. Dual "injective fibrant replacements" for pullback or inverse limit diagrams exist similarly. This happens when the categories indexing these diagrams are *Reedy categories*.

If M is any model category and D is any Reedy category, then the category $\mathsf{M}^{\mathsf{D}}$ of Reedy diagrams admits a model structure. If the indexing category D satisfies the appropriate half of a dual pair of conditions listed in Proposition 2.5.22, then the colimit or limit functors $\operatorname{colim}, \lim\colon \mathsf{M}^{\mathsf{D}} \to \mathsf{M}$ are left or right Quillen. In such contexts, homotopy colimits and homotopy limits can be computed by applying Corollary 2.4.10.

The history of the abstract notion of Reedy categories is entertaining. The category $\Delta^{\mathrm{op}}$ is an example of what is now called a *Reedy category*. The eponymous model structure on simplicial objects taking values in any model category was introduced in an unpublished but widely disseminated manuscript by Reedy [232]. Reedy notes that a dual model structure exists for cosimplicial objects, which, in the case of cosimplicial simplicial sets, coincides with a model structure introduced by Bousfield and Kan to define homotopy limits [58, §X]. The general definition, unifying these examples and many others, is due to Kan and appeared in the early drafts of the book that eventually became [92]. Various draft versions circulated in the mid 1990s and contributed to the published accounts [124, chapter 15] and [130, chapter 5]. The final [92] in turn references these sources in order to "review the notion of a Reedy category" that originated in an early draft of that same manuscript.

Definition 2.5.10.   A **Reedy structure** on a small category A consists of a **degree function** $\deg\colon \mathrm{ob}A \to \omega$ together with a pair of wide subcategories $\overrightarrow{A}$ and $\overleftarrow{A}$ of **degree-increasing** and **degree-decreasing** arrows respectively, so that:

(i) The degree of the domain of every non-identity morphism in $\overrightarrow{A}$ is strictly less than the degree of the codomain, and the degree of the domain of every non-identity morphism in $\overleftarrow{A}$ is strictly greater than the degree of the codomain.

(ii) Every $f \in \mathrm{mor}A$ may be factored uniquely as

$$
\begin{array}{ccc}
\bullet & \xrightarrow{\quad f \quad} & \bullet \\
& {\scriptstyle \overleftarrow{A} \ni f} \searrow \quad \nearrow {\scriptstyle f \in \overrightarrow{A}} & \\
& \bullet &
\end{array}
\tag{2.8}
$$

Example 2.5.11.

(i) Discrete categories are Reedy categories, with all objects having degree zero.

(ii) If A is a Reedy category, then so is $A^{\mathrm{op}}$: its Reedy structure has the same degree function but has the degree-increasing and degree-decreasing arrows interchanged.

(iii) Finite posets are Reedy categories with all morphisms degree-increasing. Declare any minimal element to have degree zero and define the degree of a generic object $d \in D$ to be the length of the maximal-length path of non-identity arrows from an element of degree zero to $d$. This example can be extended without change to include infinite posets such as $\omega$ provided that each object has finite degree.

(iv) The previous example gives the category $b \leftarrow a \to c$ a Reedy structure in which $\deg(a) = 0$ and $\deg(b) = \deg(c) = 1$. There is another Reedy category structure in which $\deg(b) = 0$, $\deg(a) = 1$, and $\deg(c) = 2$.

(iv) The category $a \rightrightarrows b$ is a Reedy category with $\deg(a) = 0$, $\deg(b) = 1$, and both non-identity arrows said to strictly raise degrees.

(vi) The category $\triangle$ of finite non-empty ordinals and the category $\triangle_+$ of finite ordinals and order-preserving maps both support canonical Reedy category structures, for which we take the degree-increasing maps to be the subcategories of face operators (monomorphisms) and the degree-decreasing maps to be the subcategories of degeneracy operators (epimorphisms).

*Exercise 2.5.12.*

(i) Show that every morphism $f$ factors uniquely through an object of minimum degree and this factorization is the **Reedy factorization** of (2.8).

(ii) Show that the Reedy category axioms prohibit any non-identity isomorphisms.

*Remark 2.5.13.* The notion of a Reedy category has been usefully extended by Berger and Moerdijk to include examples such as finite sets or finite pointed sets that do have non-identity automorphisms. All of the results to be described here have analogues in this more general context, but for ease of exposition we leave these details to [38].

To focus attention on our goal, we now introduce the *Reedy model structure*, which serves as motivation for some auxiliary constructions we have yet to introduce.

**Theorem 2.5.14** (Reedy, Kan [245, §7]). *Let* M *be a model category and let* D *be a Reedy category. Then the category* $M^D$ *admits a model structure whose*

– *weak equivalences are the pointwise weak equivalences, and*

– *weak factorization systems* $(\mathcal{C} \cap \mathcal{W}[D], \mathcal{F}[D])$ *and* $(\mathcal{C}[D], \mathcal{F} \cap \mathcal{W}[D])$ *are the Reedy weak factorization systems.*

In the **Reedy weak factorization system** $(\mathcal{L}[D], \mathcal{R}[D])$ defined relative to a weak factorization system $(\mathcal{L}, \mathcal{R})$ on M, a natural transformation $f \colon X \to Y \in M^D$ is in $\mathcal{L}[D]$ or $\mathcal{R}[D]$, respectively, if and only if, for each $d \in D$, the *relative latching map* $X^d \cup_{L^d X} L^d Y \to Y^d$ is in $\mathcal{L}$ or the *relative matching map* $X^d \to Y^d \times_{M^d Y} M^d X$ is in $\mathcal{R}$. The most efficient definition of these latching and matching objects $L^d X$ and $M^d X$ appearing in Example 2.5.17 makes use of the theory of weighted colimits and limits, a subject to which we now turn.

## 2.5.4    Quillen adjunctions for weighted limits and colimits

Ordinary limits and colimits are objects representing the functor of cones with a given summit over or under a fixed diagram. Weighted limits and colimits are defined analogously, except that the cones over or under a diagram might have exotic "shapes." These shapes are allowed to vary with the objects indexing the diagram. More formally, the **weight** — a functor which specifies the "shape" of a cone over a diagram indexed by D or a cone under a diagram indexed by $D^{op}$ — takes the form of a functor in $Set^D$ in the unenriched context or $V^D$ in the V-enriched context.

**Definition 2.5.15** (weighted limits and colimits, axiomatically). For a general small category D and bicomplete category M, the weighted limit and weighted colimit define bifunctors

$$\{-,-\}^D \colon (Set^D)^{op} \times M^D \to M \qquad \text{and} \qquad - *_D - \colon Set^D \times M^{D^{op}} \to M$$

which are characterized by the following pair of axioms.

(i) Weighted (co)limits with representable weights evaluate at the representing object:

$$\{D(d, -), X\}^D \cong X(d) \qquad \text{and} \qquad D(-, d) *_D Y \cong Y(d).$$

(ii) The weighted (co)limit bifunctors are cocontinuous in the weight: for any diagram $X \in M^D$, the functor $- *_D X$ preserves colimits, while the functor $\{-, X\}^D$ carries colimits to limits.

We interpret axiom (ii) to mean that weights can be "made to order": a weight constructed as a colimit of representables — as all Set-valued functors are — will stipulate the expected universal property.

Let M be any locally small category with products and coproducts. For any set $S$, the $S$-fold product and coproduct define cotensor and tensor bifunctors

$$\{-, -\}: \text{Set}^{op} \times M \to M \qquad \text{and} \qquad - * -: \text{Set} \times M \to M,$$

which form a two-variable adjunction with Hom: $M^{op} \times M \to$ Set; cf. Exercise 2.4.26.

Definition 2.5.16 (weighted limits and colimits, constructively). The weighted colimit is a functor tensor product and the weighted limit is a functor cotensor product:

$$\{W, X\}^D \cong \int_{d \in D} \{W(d), X(d)\}, \qquad W *_D Y \cong \int^{d \in D} W(d) * Y(d).$$

The limit $\{W, X\}^D$ of the diagram $X$ weighted by $W$ and the colimit $W *_D Y$ of $Y$ weighted by $W$ are characterized by the universal properties:

$$M(M, \{W, X\}^D) \cong \text{Set}^D(W, M(M, X)), \qquad M(W *_D Y, M) \cong \text{Set}^{D^{op}}(W, M(Y, M)).$$

Example 2.5.17. Let A be a Reedy category and write $A_{\leq n}$ for the full subcategory of objects of degree at most $n$. Restriction along the inclusion $A_{\leq n} \hookrightarrow A$ followed by left Kan extension defines an comonad $\text{sk}_n: \text{Set}^A \to \text{Set}^A$.

Let $a \in A$ be an object of degree $n$ and define

$$\partial A(a, -) := \text{sk}_{n-1} A(a, -) \in \text{Set}^A \qquad \text{and} \qquad \partial A(-, a) := \text{sk}_{n-1} A(-, a) \in \text{Set}^{A^{op}},$$

where $A(a, -)$ and $A(-, a)$ denote the co- and contravariant functors represented by $a$, respectively. For any $X \in M^A$, the **latching** and **matching** objects are defined by

$$L^a X := \partial A(-, a) *_A X \qquad \text{and} \qquad M^a X := \{\partial A(a, -), X\}.$$

*Exercise 2.5.18* (enriched weighted limits and colimits). For the reader who knows some enriched category theory, generalize Definitions 2.5.15 and 2.5.16 to the V-enriched context to define weighted limit and weighted colimit bifunctors

$$\{-, -\}^A: (V^A)^{op} \times M^A \to M \qquad \text{and} \qquad - \otimes_A -: V^A \times M^{A^{op}} \to M$$

in any V-enriched, tensored, and cotensored category M whose underlying unenriched category is complete and cocomplete.

Recall the notion of Quillen two-variable adjunction, the prototypical example being the tensor-cotensor-hom of a V-model category M.

**Theorem 2.5.19** ([239, 7.1]). *Let* A *be a Reedy category and let* $\otimes: K \times L \to M$ *be a left Quillen bifunctor between model categories. Then the functor tensor product*

$$\otimes_A : K^{A^{op}} \times L^A \to M$$

*is left Quillen with respect to the Reedy model structures.*

A dual result holds for functor cotensor products formed relative to a right Quillen bifunctor. In particular, if M is a V-model category, then its tensor, cotensor, and hom define a Quillen two-variable adjunction, and so in particular:

**Corollary 2.5.20.** *Let* M *be a* V*-model category and let* A *be a Reedy category. Then for any Reedy cofibrant weight* $W \in V^A$, *the weighted colimit and weighted limit functors*

$$W *_A - : M^A \to M \qquad and \qquad \{W, -\}^A : M^{A^{op}} \to M$$

*are respectively left and right Quillen with respect to the Reedy model structures on* $M^A$ *and* $M^{A^{op}}$.

**Example 2.5.21** (geometric realization and totalization). The Yoneda embedding defines a Reedy cofibrant weight $\Delta^\bullet \in sSet^\Delta$. The weighted colimit and weighted limit functors

$$\Delta^\bullet *_{\Delta^{op}} - : M^{\Delta^{op}} \to M \qquad and \qquad \{\Delta^\bullet, -\}^\Delta : M^\Delta \to M$$

typically go by the names **geometric realization** and **totalization**. Corollary 2.5.20 proves that if M is a simplicial model category, then these functors are left and right Quillen.

By Exercise 2.4.26, Corollary 2.5.20 also has implications in the case of an unenriched model category M, in which case "Reedy cofibrant" should be read as "Reedy monomorphic." Ordinary limits and colimits are weighted limits and colimits where the weight is the terminal functor, constant at the singleton set.

**Proposition 2.5.22** (homotopy limits and colimits of Reedy shape).

(i) *If* A *is a Reedy category with the property that the constant* A*-indexed diagram at any cofibrant object in any model category is Reedy cofibrant, then the limit functor* $\lim: M^A \to M$ *is right Quillen.*

(ii) *If* A *is a Reedy category with the property that the constant* A*-indexed diagram at any fibrant object in any model category is Reedy fibrant, then the colimit functor* $\text{colim}: M^A \to M$ *is left Quillen.*

*Proof.* Taking the terminal weight 1 in $Set^A$, the weighted limit reduces to the ordinary limit functor. The functor $1 \in Set^A$ is Reedy monomorphic just when, for each $a \in A$, the category of elements for the weight $\partial A(-, a)$ is either empty or connected. This is the case if and only if A has "cofibrant constants," meaning that the constant A-indexed diagram at any cofibrant object in any model category is Reedy cofibrant. Thus, we conclude that if A has cofibrant constants, then the limit functor $\lim \colon \mathsf{M}^{\mathsf{A}} \to \mathsf{M}$ is right Quillen. See [245, §9] for more discussion.          □

There is an analogous result for projective and injective model structures which the author first saw formulated in this way by Gambino in the context of a simplicial model category.

**Theorem 2.5.23** ([101]). *If* M *is a* V*-model category and* D *is a small category, then the weighted colimit functor*

$$- \otimes_{\mathsf{D}} - \colon \mathsf{V}^{\mathsf{D}} \times \mathsf{M}^{\mathsf{D}^{\mathrm{op}}} \to \mathsf{M}$$

*is left Quillen if the domain has the (injective, projective) or (projective, injective) model structure. Similarly, the weighted limit functor*

$$\{-,-\}^{\mathsf{D}} \colon (\mathsf{V}^{\mathsf{D}})^{\mathrm{op}} \times \mathsf{M}^{\mathsf{D}} \to \mathsf{M}$$

*is right Quillen if the domain has the (projective, projective) or (injective, injective) model structure.*

*Proof.* By Definition 2.4.22 we can prove both statements in adjoint form. The weighted colimit bifunctor of Exercise 2.5.18 has a right adjoint (used to express the defining universal property of the weighted colimit)

$$\mathrm{Map}(-,-) \colon (\mathsf{M}^{\mathsf{D}^{\mathrm{op}}})^{\mathrm{op}} \times \mathsf{M} \to \mathsf{V}^{\mathsf{D}}$$

which sends $F \in \mathsf{M}^{\mathsf{D}^{\mathrm{op}}}$ and $m \in \mathsf{M}$ to $\mathrm{Map}(F-, m) \in \mathsf{V}^{\mathsf{D}}$.

To prove the statement when $\mathsf{V}^{\mathsf{D}}$ has the projective and $\mathsf{M}^{\mathsf{D}^{\mathrm{op}}}$ has the injective model structure, we must show that this is a right Quillen bifunctor with respect to the pointwise (trivial) cofibrations in $\mathsf{M}^{\mathsf{D}^{\mathrm{op}}}$, (trivial) fibrations in M, and pointwise (trivial) fibrations in $\mathsf{V}^{\mathsf{D}}$. Because the limits involved in the definition of right Quillen bifunctors are also formed pointwise, this follows immediately from the corresponding property of the simplicial hom bifunctor, which was part of the definition of a simplicial model category. The other cases are similar.          □

The upshot of Theorem 2.5.23 is that there are two approaches to constructing a homotopy colimit: fattening up the diagram, as is achieved by the derived functors of §2.5.2, or fattening up the weight. The famous Bousfield–Kan formulae for homotopy limits and colimits in the context of a simplicial model category define them to be weighted limits and colimits for a particular weight constructed as a projective cofibrant replacement of the terminal weight; see [58] or [237, §11.5]. The Quillen two-variable adjunction of Theorem 2.5.23 can be derived as in Theorem 2.4.29 to express a homotopically enriched universal property of the weighted limit or colimit, as representing "homotopy coherent" cones over or under a diagram, an intuition to be explored in the next section.

## 2.6        Simplicial localizations

Quillen's model categories provide a robust axiomatic framework within which to "do homotopy theory." But the constructions of §2.5 imply that the homotopy theories presented by model categories have all homotopy limits and homotopy colimits, which need not be the case in general. In this section we introduce a framework, originally developed by Dwyer and Kan and re-conceptualized by Bergner, which allows us to extend our notion of equivalence between homotopy theories introduced in §2.4.5 to a more flexible notion of *DK-equivalence* (after Dwyer and Kan) that identifies when any two homotopical categories are equivalent.

A mere equivalence of categories of fractions is insufficient to detect an equivalence of homotopy theories; instead a construction that takes into account the "higher-dimensional" homotopical structure is required. To that end, Dwyer and Kan build, from any homotopical category $(K, \mathcal{W})$, a simplicial category $\mathcal{L}^H(K, \mathcal{W})$ called the *hammock localization* [88] and demonstrate that their construction has a number of good properties that we tour in §2.6.1:

- The homotopy category $h\mathcal{L}^H(K, \mathcal{W})$ is equivalent to the category of fractions $K[\mathcal{W}]^{-1}$ (Proposition 2.6.5).
- If $(K, \mathcal{W})$ underlies a simplicial model category then the Kan complex enriched category $K_{cf} \subset K$ is DK-equivalent to $\mathcal{L}^H(K, \mathcal{W})$ (Proposition 2.6.7).
- More generally, $\mathcal{L}^H(K, \mathcal{W})$ provides a not-necessarily simplicial model category $(K, \mathcal{W})$ with function complexes that have the correct mapping type even if the model structure is not simplicial (Proposition 2.6.6).
- If two model categories are Quillen equivalent, then their hammock localizations are DK-equivalent (Proposition 2.6.8).

The DK-equivalences are those simplicial functors that are bijective on homotopy equivalence classes of objects and define local equivalences of the mapping spaces constructed by the hammock localization. Zooming out a categorical level, the Bergner model structure on simplicially enriched categories gives a presentation of the homotopical category of homotopy theories, with the DK-equivalences as its weak equivalences. This is the subject of §2.6.2.

### 2.6.1    The hammock localization

There are two equivalent ways to present the data of a simplicially enriched category, either as a category equipped with a simplicial set of morphisms between each pair of objects, or a simplicial diagram of categories $K_n$ of $n$-**arrows**, each of which is equipped with a constant set of objects.

*Exercise 2.6.1.* Prove that the following are equivalent:

  (i) A simplicially enriched category with objects obK.
  (ii) A simplicial object $K_\bullet : \triangle^{op} \to \mathsf{Cat}$ in which each of the categories $K_n$ has objects obK and each functor $K_n \to K_m$ is the identity on objects.

We being by introducing the notion of a DK-equivalence between simplicially enriched categories.

**Definition 2.6.2.** A simplicial functor $F\colon \mathsf{K} \to \mathsf{M}$ is a **DK-equivalence** if

(i) It defines an equivalence of homotopy categories $\mathsf{h}F\colon \mathsf{h}\mathsf{K} \to \mathsf{h}\mathsf{M}$.
(ii) It defines a local weak equivalence of mapping complexes: for all $X, Y \in \mathsf{K}$, $\mathrm{Map}_{\mathsf{K}}(X, Y) \xrightarrow{\sim} \mathrm{Map}_{\mathsf{M}}(FX, FY)$.

In the case where $F$ is identity on objects, condition (2.6.2) subsumes condition (2.6.2).

**Definition 2.6.3** ([87, 2.1]). Let $\mathsf{K}$ be a category with a wide subcategory $\mathcal{W}$, containing all the identity arrows. The **hammock localization** $\mathcal{L}^H(\mathsf{K}, \mathcal{W})$ is a simplicial category with the same objects as $\mathsf{K}$ and with the mapping complex $\mathrm{Map}(X, Y)$ defined to be the simplicial set whose $k$-simplices are "reduced hammocks of width $k$" from $X$ to $Y$, these being commutative diagrams

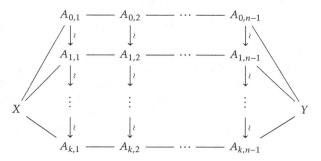

where the length of the hammock is any integer $n \geq 1$, such that

(i) all vertical maps are in $\mathcal{W}$,
(ii) in each column of horizontal morphisms all maps go in the same direction and if they go left then they are in $\mathcal{W}$, and
(iii) the maps in adjacent columns go in different directions.

The graded set of reduced hammocks of width $k$ from $X$ to $Y$ becomes a simplicial set $\mathrm{Map}(X, Y)$ in which

(iv) faces are defined by omitting rows and
 1. degeneracies are defined by duplicating rows.

Composition is defined by horizontally pasting hammocks and then reducing by

(v) composing adjacent columns whose maps point in the same direction and
(vi) omitting any column which contains only identity maps.

There is a canonical functor $\mathsf{K} \to \mathcal{L}^H(\mathsf{K}, \mathcal{W})$ whose image is comprised of dimension zero length 1 hammocks pointing forwards.

*Exercise 2.6.4.* Verify that the composite of the functor $K \to \mathcal{L}^H(K, \mathcal{W})$ just described with the quotient functor $\mathcal{L}^H(K, \mathcal{W}) \to h\mathcal{L}^H(K, \mathcal{W})$ that collapses each mapping space onto its set of path components inverts the weak equivalences in $K$, sending each to an isomorphism in the homotopy category $h\mathcal{L}^H(K, \mathcal{W})$.

In the hammock localization

> cancelation in any dimension is achieved not by "imposing relations" in the same dimension, but by "imposing homotopy relations", i.e. adding maps, in the next dimension, [87, §3]

in contrast with the category of fractions constructed in §2.2. By considering the effect of these "homotopy relations," it is straightforward to see that the induced functor from the category of fractions to the homotopy category of the hammock localization is an isomorphism of categories.

**Proposition 2.6.5** (Dwyer–Kan [88, 3.2]). *The canonical functor* $K \to \mathcal{L}^H(K, \mathcal{W})$ *induces an isomorphism of categories* $K[\mathcal{W}^{-1}] \cong h\mathcal{L}^H(K, \mathcal{W})$.

*Proof.* The comparison functor $K[\mathcal{W}^{-1}] \to h\mathcal{L}^H(K, \mathcal{W})$ induced by Exercise 2.6.4 and by the universal property of Proposition 2.2.2 is clearly bijective on objects and full, homotopy classes in $h\mathcal{L}^H(K, \mathcal{W})$ being represented by zigzags whose backwards maps lie in $\mathcal{W}$. To see that this functor is faithful it suffices to consider a 1-simplex in $\mathsf{Map}(X, Y)$:

and argue that the top and bottom zigzags define the same morphism in $K[\mathcal{W}^{-1}]$. This is an easy exercise in diagram chasing, applying the rules of Definition 2.2.1. □

The previous result applies to a model category $(M, \mathcal{W})$, in which case we see that $\mathcal{L}^H(M, \mathcal{W})$ is a higher-dimensional incarnation of the homotopy category, equipping $M[\mathcal{W}^{-1}]$ with mapping spaces whose path components correspond to arrows in the category of fractions. A further justification that the mapping spaces of the hammock localization have the correct homotopy type, not just the correct sets of path components, proceeds as follows. A **simplicial resolution** of $Y \in M$ is a Reedy fibrant simplicial object $Y_\bullet$ together with a weak equivalence $Y \xrightarrow{\sim} Y_0$. **Cosimplicial resolutions** $X^\bullet \to X$ are defined dually. Every object has a simplicial and cosimplicial resolution, defined as the Reedy fibrant replacement of the constant simplicial object in $M^{\Delta^{\mathrm{op}}}$ and the Reedy cofibrant replacement of the constant cosimplicial object in $M^\Delta$, respectively.

**Proposition 2.6.6** (Dwyer–Kan [88, 4.4]). *For any cosimplicial resolution* $X^\bullet \to X$ *and simplicial resolution* $Y \to Y_\bullet$, *the diagonal of the bisimplicial set* $M(X^\bullet, Y_\bullet)$ *has the same homotopy type of* $\mathsf{Map}_{\mathcal{L}^H(M, \mathcal{W})}(X, Y)$, *and if* $X$ *or* $Y$ *are respectively cofibrant or fibrant the simplicial sets* $M(X, Y_\bullet)$ *and* $M(X^\bullet, Y)$ *do as well.*

As a corollary of this result one can show:

**Proposition 2.6.7** (Dwyer–Kan [88, 4.7, 4.8]). *Let* $(\mathsf{M}, \mathcal{W})$ *be the homotopical category underlying a simplicial model category* $\mathsf{M}$. *Then for cofibrant* $X$ *and fibrant* $Y$, $\mathrm{Map}_{\mathsf{M}}(X, Y)$ *and* $\mathrm{Map}_{\mathcal{L}^H(\mathsf{M}, \mathcal{W})}(X, Y)$ *have the same homotopy type and hence the simplicial categories* $\mathsf{M}_{cf}$ *and* $\mathcal{L}^H(\mathsf{M}, \mathcal{W})$ *are DK-equivalent.*

The statement of this result requires some explanation. If $\mathsf{K}$ is a simplicial category whose underlying category of 0-arrows $\mathsf{K}_0$ has a subcategory of weak equivalences $\mathcal{W}$, then these weak equivalences degenerate to define homotopical categories $(\mathsf{K}_n, \mathcal{W})$ for each category of $n$-arrows in $\mathsf{K}$. For each $n$ we may form the hammock localization $\mathcal{L}^H(\mathsf{K}_n, \mathcal{W})$. As $n$ varies, this gives a bisimplicial sets of mapping complexes for each fixed pair of objects of $\mathsf{K}$. The mapping complexes in the hammock localization $\mathcal{L}^H(\mathsf{K}, \mathcal{W})$ are defined to be the diagonals of these bisimplicial sets. In the case of a simplicial model category $\mathsf{M}$, the hammock localization $\mathcal{L}^H(\mathsf{M}, \mathcal{W})$ is DK-equivalent to the hammock localization $\mathcal{L}^H(\mathsf{M}_0, \mathcal{W})$ of the underlying unenriched homotopical category.

**Proposition 2.6.8** ([88, 5.4]). *A Quillen equivalence*

$$\mathsf{M} \underset{G}{\overset{F}{\underset{\perp}{\rightleftarrows}}} \mathsf{N}$$

*induces DK-equivalences*

$$\mathcal{L}^H(\mathsf{M}_c, \mathcal{W}) \overset{\sim}{\to} \mathcal{L}^H(\mathsf{N}_c, \mathcal{W}) \qquad and \qquad \mathcal{L}^H(\mathsf{N}_f, \mathcal{W}) \overset{\sim}{\to} \mathcal{L}^H(\mathsf{M}_f, \mathcal{W}).$$

*Moreover, for any model category the inclusions*

$$\mathcal{L}^H(\mathsf{M}_c, \mathcal{C} \cap \mathcal{W}) \overset{\sim}{\to} \mathcal{L}^H(\mathsf{M}_c, \mathcal{W}_c) \overset{\sim}{\to} \mathcal{L}^H(\mathsf{M}, \mathcal{W})$$

*are DK-equivalences and hence* $\mathcal{L}^H(\mathsf{M}, \mathcal{W})$ *and* $\mathcal{L}^H(\mathsf{N}, \mathcal{W})$ *are DK-equivalent.*

## 2.6.2    A model structure for homotopy coherent diagrams

Several of Dwyer and Kan's proofs of the results in the previous subsection make use of a model structure on the category of simplicial categories with a fixed set of objects and with identity-on-objects functors. But this restriction to categories with the same objects is somewhat unnatural. The Bergner model structure is the extension of Dwyer and Kan's model structure that drops that restriction, unifying the notions of DK-equivalence, free simplicial category (also known as "simplicial computad"), and Kan complex enriched simplicial category, the importance of which will be made clear in §2.7.

**Theorem 2.6.9** (Bergner [41]). *There exists a model structure on the category of simplicially enriched categories given as follows:*

– *Its equivalences are the DK equivalences.*

- *Its cofibrant objects are the **simplicial computads**: those simplicial categories that, when considered as a simplicial object* $C_\bullet \colon \Delta^{\mathrm{op}} \to \mathsf{Cat}$ *have the property that*
  - *each category* $C_n$ *is freely generated by the reflexive directed graph of its **atomic arrows** (those admitting no non-trivial factorizations) and*
  - *the degeneracy operators* $[m] \twoheadrightarrow [n]$ *in* $\Delta$ *preserve atomic arrows.*
- *Fibrant objects are the **Kan complex enriched categories**: those simplicial categories whose mapping spaces are all Kan complexes.*

More generally, the cofibrations in the Bergner model structure are retracts of relative simplicial computads and the fibrations are those functors that are local Kan fibrations and define isofibrations at the level of homotopy categories; see [41] for more details.

Definition 2.3.24 tells us that maps in the homotopy category of the Bergner model structure from a simplicial category A to a simplicial category K are represented by simplicial functors from a cofibrant replacement of A to a fibrant replacement of K. These are classically studied objects. Cordier and Porter after Vogt define such functors to be **homotopy coherent diagrams** of shape A in K [72].

A particular model for the cofibrant replacement of a strict 1-category A regarded as a discrete simplicial category gives some intuition for the data involved in defining a homotopy coherent diagram. This construction, introduced by Dwyer and Kan under the name "standard resolutions" [89, 2.5], can be extended to the case where A is non-discrete by applying it levelwise and taking diagonals.

**Definition 2.6.10** (free resolutions). There is a comonad $(F, \epsilon, \delta)$ on the category of categories that sends a small category to the free category on its underlying reflexive directed graph. Explicitly $FA$ has the same objects as A and its non-identity arrows are strings of composable non-identity arrows of A.

Adopting the point of view of Exercise 2.6.1, we define a simplicial category $\mathfrak{C}A_\bullet$ with $\mathrm{ob}\,\mathfrak{C}A = \mathrm{ob}\,A$ and with the category of $n$-arrows $\mathfrak{C}A_n := F^{n+1}A$. A non-identity $n$-arrow is a string of composable arrows in A with each arrow in the string enclosed in exactly $n$ pairs of well-formed parentheses. In the case $n = 0$, this recovers the previous description of the non-identity 0-arrows in $FA$, strings of composable non-identity arrows of A.

The required identity-on-objects functors in the simplicial object $\mathfrak{C}A_\bullet$ are defined by evaluating the comonad resolution for $(F, \epsilon, \delta)$ on a small category A:

$$\mathfrak{C}A_\bullet := \qquad FA \underset{\longleftarrow}{\overset{\longleftarrow}{\longrightarrow}} F^2A \; {\overset{\overset{\longleftarrow}{\longrightarrow}}{\underset{\underset{\longleftarrow}{\longrightarrow}}{\longleftarrow}}} \; F^3A \; {\overset{\overset{\overset{\longleftarrow}{\longrightarrow}}{\longleftarrow}}{\underset{\underset{\longleftarrow}{\longrightarrow}}{\longleftarrow}}} \; F^4A \qquad \cdots$$

Explicitly, for $j \geq 1$, the face maps

$$F^k \epsilon F^j \colon F^{k+j+1}A \to F^{k+j}A$$

remove the parentheses that are contained in exactly $k$ others, while $F^{k+j}\epsilon$ composes the morphisms inside the innermost parentheses. For $j \geq 1$, the degeneracy maps

$$F^k \delta F^j : F^{k+j+1}A \to F^{k+j+2}A$$

double up the parentheses that are contained in exactly $k$ others, while $F^{k+j}\delta$ inserts parentheses around each individual morphism.

*Exercise 2.6.11.* Explain the sense in which free resolutions define Bergner cofibrant replacements of strict 1-categories by:

(i) verifying that for any A, the free resolution $\mathfrak{C}A_\bullet$ is a simplicial computad, and
(ii) defining a canonical identity-on-objects augmentation functor $\epsilon : \mathfrak{C}A \to A$ and verifying that it defines a local homotopy equivalence.

The notation $\mathfrak{C}A_\bullet$ for the free resolution is non-standard and will be explained in §2.7.2, where we will gain a deeper understanding of the importance of the Bergner model structure from the vantage point of $(\infty, 1)$-categories.

## 2.7     Quasi-categories as $(\infty, 1)$-categories

Any topological space $Y$ has an associated simplicial set $\mathrm{Sing}(Y)$ called its **total singular complex**. The vertices in $\mathrm{Sing}(Y)$ are the points in $Y$ and the 1-simplices are the paths; in general, an $n$-simplex in $\mathrm{Sing}(Y)$ corresponds to an $n$-simplex in $Y$, that is, to a continuous map $|\Delta^n| \to Y$. In particular, a 2-simplex $|\Delta^2| \to Y$ defines a triangular shaped homotopy from the composite paths along the spine $\Lambda_1^2 \subset \Delta^2$ of the 2-simplex to the direct path from the 0th to the 2nd vertex that is contained in its 1st face.[15] Since the inclusion $|\Lambda_k^n| \to |\Delta^n|$ admits a retraction, $\mathrm{Sing}(Y)$ is a Kan complex.

The total singular complex is a higher-dimensional incarnation of some of the basic invariants of $Y$, which can be recovered by truncating the total singular complex at some level and replacing the top-dimensional simplices with suitably defined "homotopy classes" of such. Its set of path components is the set $\pi_0 Y$ of path components in $Y$. Its homotopy category, in a sense to be defined below, comprised of the vertices and homotopy classes of paths between them, is a groupoid $\pi_1 Y$ called the *fundamental groupoid* of $Y$. By extension, it is reasonable to think of the higher-dimensional simplices of $\mathrm{Sing}(Y)$ as being invertible in a similar sense, with composition relations witnessed by higher cells. In this way, $\mathrm{Sing}(Y)$ models the $\infty$-groupoid associated to the topological space $Y$ and the Quillen equivalence 2.4.32 is one incarnation of Grothendieck's famous "homotopy hypothesis" (the moniker due to Baez), that $\infty$-groupoids up to equivalence should model homotopy types [113].

In the catalog of weak higher-dimensional categories, the $\infty$-groupoids define $(\infty, 0)$-**categories**, weak categories with morphisms in each dimension, all of which are weakly invertible. In §2.7.1, we introduce **quasi-categories**, which provide a particular model for $(\infty, 1)$-categories — infinite-dimensional categories in which every morphism above dimension 1 is invertible — in parallel with the Kan complex model

---

[15] The simplicial $n$-simplex $\Delta^n$, its boundary sphere $\partial\Delta^n$, and its horns $\Lambda_k^n$ are defined in §2.3.5.

for $(\infty, 0)$-categories. We explain the sense in which quasi-categories, which are defined to be simplicial sets with an inner horn lifting property, model $(\infty, 1)$-categories by introducing the homotopy category of a quasi-category and constructing the hom-space between objects in a quasi-category. In §2.7.2, we explain how simplicially enriched categories like those considered in §2.6 can be converted into quasi-categories. Then in §2.7.3, we introduce a model structure whose fibrant objects are the quasi-categories due to Joyal and in this way obtain a suitable notion of (weak) equivalence between quasi-categories.

### 2.7.1   Quasi-categories and their homotopy categories

The **nerve** of a small category D is the simplicial set $D_\bullet$ whose vertices $D_0$ are the objects of D, whose 1-simplices $D_1$ are the morphisms, and whose set of $n$-simplices $D_n$ is the set of $n$ composable pairs of morphisms in D. The simplicial structure defines a diagram in Set

$$\cdots \quad D_3 \rightleftharpoons D_2 \rightleftharpoons D_1 \rightleftharpoons D_0$$

Truncating at level 2 we are left with precisely the data that defines a small category D as a category internal to the category of sets and in fact this higher-dimensional data is redundant in a sense: the simplicial set $D_\bullet$ is 2-coskeletal, meaning any sphere bounding a hypothetical simplex of dimension at least 3 admits a unique filler.

The description of the nerve as an internal category relies on an isomorphism $D_2 \cong D_1 \times_{D_0} D_1$ identifying the set of 2-simplices with the pullback of the domain and codomain maps $D_1 \rightrightarrows D_0$: a composable pair of arrows is given by a pair of arrows such that the domain of the second equals the codomain of the first. Equivalently, this condition asserts that the map

admits a unique filler. In higher dimensions, we can consider the inclusion of the spine $\Delta^1 \cup_{\Delta^0} \cdots \cup_{\Delta^0} \Delta^1 \hookrightarrow \Delta^n$ of an $n$-simplex, and similarly the nerve $D_\bullet$ will admit unique extensions along these maps. From the perspective of an infinite-dimensional category, in which the higher-dimensional simplices represent data and not just conditions on the 1-simplices, it is better to consider extensions along inner horn inclusions $\Lambda^n_k \hookrightarrow \Delta^n$ for the reasons explained by the following exercise.

*Exercise 2.7.1.* Prove that the spine inclusions can be presented as cell complexes (see Definition 2.3.11) built from the inner horn inclusions $\{\Lambda^k_n \hookrightarrow \Delta^n\}_{n \geq 2, 0 < k < n}$ but demonstrate by example that the inner horn inclusions cannot be presented as cell complexes built from the spine inclusions.

The original definition of a simplicial set satisfying the "restricted Kan condition," now called a *quasi-category* (following Joyal [140]) or an $\infty$-*category* (following Lurie [169]), is due to Boardman and Vogt [48]. Their motivating example appears as Corollary 2.7.9.

**Definition 2.7.2.** A **quasi-category** is a simplicial set $X$ such that $X \to *$ has the right lifting property with respect to the inner horn inclusions for each $n \geq 2$, $0 < k < n$.

$$
\begin{array}{ccc}
\Lambda^n_k & \longrightarrow & X \\
\downarrow & \nearrow & \\
\Delta^n & &
\end{array}
\tag{2.9}
$$

Nerves of categories are quasi-categories; in fact in this case each lift (2.9) is unique. Tautologically, Kan complexes are quasi-categories. In particular, the total singular complex of a topological space is a Kan complex and hence a quasi-category. More sophisticated examples of (frequently large) quasi-categories are produced by Theorem 2.7.8 below.

**Definition 2.7.3** (the homotopy category of a quasi-category [48, 4.12]). Any quasi-category $X$ has an associated **homotopy category** $hX$ whose objects are the vertices of $X$ and whose morphisms are represented by 1-simplices, which we consequently depict as arrows $f \colon x \to y$ from their 0th vertex to their 1st vertex. The degenerate 1-simplices serve as identities in the homotopy category, and may be depicted using an equals sign in place of the arrow.

As the name suggests, the morphisms in $hX$ are homotopy classes of 1-simplices, where a pair of 1-simplices $f$ and $g$ with common boundary are **homotopic** if there exists a 2-simplex whose boundary has any of the following forms:

$$\tag{2.10}$$

Indeed, in a quasi-category, if any of the 2-simplices (2.10) exists then there exists a 2-simplex of each type.

Generic 2-simplices in $X$

$$
\begin{array}{ccc}
& \bullet & \\
f \nearrow & \sim & \searrow g \\
\bullet & \xrightarrow{\ h\ } & \bullet
\end{array}
\tag{2.11}
$$

witness that $gf = h$ in the homotopy category. Conversely, if $h = gf$ in $hX$ and $f, g, h$ are any 1-simplices representing these homotopy classes, there exists a 2-simplex (2.11) witnessing the composition relation.

*Exercise 2.7.4.*

(i) Verify the assertions made in Definition 2.7.3 or see [169, §1.2.3].

(ii) Show that $h$ is the left adjoint to the nerve functor:[16]

$$\mathsf{qCat} \underset{N}{\overset{h}{\rightleftarrows}} \bot \; \mathsf{Cat}$$

The mapping space between two objects of a quasi-category $A$ is modeled by the Kan complex defined via the pullback

$$
\begin{array}{ccc}
\mathrm{Map}_A(x,y) & \longrightarrow & A^{\Delta^1} \\
\downarrow & \llcorner & \downarrow \\
\Delta^0 & \underset{(x,y)}{\longrightarrow} & A \times A
\end{array}
$$

The following proposition of Joyal is useful in proving that $\mathrm{Map}_A(x,y)$ is a Kan complex and also characterizes the $\infty$-groupoids in the quasi-categorical model of $(\infty,1)$-categories.

**Proposition 2.7.5** (Joyal [140, 1.4]). *A quasi-category is a Kan complex if and only if its homotopy category is a groupoid.*

**Definition 2.7.6** ([140, 1.6]). A 1-simplex $f$ in a quasi-category $X$ is an **isomorphism** if and only if it represents an isomorphism in the homotopy category, or equivalently if and only if it admits a coherent homotopy inverse:

$$
\begin{array}{ccc}
2 & \xrightarrow{\;f\;} & X \\
\big\downarrow & \nearrow & \\
\mathbb{I} & &
\end{array}
$$

extending along the map $2 \hookrightarrow \mathbb{I}$ including the nerve of the free-living arrow into the nerve of the free-living isomorphism.

### 2.7.2 Quasi-categories found in nature

Borrowing notation from the simplex category $\Delta$, we write $[n] \subset \omega$ for the ordinal category $\mathfrak{m}+1$, the full subcategory spanned by $0,\dots,n$ in the category that indexes a countable sequence:

$$[n] := \quad 0 \longrightarrow 1 \longrightarrow 2 \longrightarrow 3 \longrightarrow \cdots \longrightarrow n$$

These categories define the objects of a diagram $\Delta \hookrightarrow \mathsf{Cat}$ that is a full embedding: the only functors $[m] \to [n]$ are order-preserving maps from $[m] = \{0,\dots,m\}$ to $[n] = \{0,\dots,n\}$. Applying the free resolution construction of Definition 2.6.10 to these categories we get a functor $\mathfrak{C}\colon \Delta \to \mathsf{sCat}$, where $\mathfrak{C}[n]$ is the full simplicial subcategory of $\mathfrak{C}\omega$ spanned by those objects $0,\dots,n$.

---

[16] In fact, this pair defines a Quillen adjunction between the model structure to be introduced in Theorem 2.7.12 and the "folk" model structure on categories [237, 15.3.8].

Definition 2.7.7 (homotopy coherent realization and nerve). The homotopy coherent nerve $\mathfrak{N}$ and homotopy coherent realization $\mathfrak{C}$ are the adjoint pair of functors obtained by applying Kan's construction [237, 1.5.1] to the functor $\mathfrak{C} \colon \Delta \to$ sCat to construct an adjunction

$$\text{sSet} \underset{\mathfrak{N}}{\overset{\mathfrak{C}}{\underset{\perp}{\rightleftarrows}}} \text{sCat}$$

The right adjoint, called the **homotopy coherent nerve**, converts a simplicial category S into a simplicial set $\mathfrak{N}$S whose $n$-simplices are homotopy coherent diagrams of shape $[n]$ in S. That is,

$$\mathfrak{N}S_n := \{\mathfrak{C}[n] \to S\}.$$

The left adjoint is defined by pointwise left Kan extension along the Yoneda embedding:

$$\begin{array}{ccc} \Delta & \xrightarrow{\quad \natural \quad} & \text{sSet} \\ {\scriptstyle \mathfrak{C}} \searrow & {\scriptstyle \cong} & \nearrow {\scriptstyle \mathfrak{C}} \\ & \text{sCat} & \end{array}$$

That is, $\mathfrak{C}\Delta^n$ is defined to be $\mathfrak{C}[n]$ — a simplicial category that we call the **homotopy coherent $n$-simplex** — and for a generic simplicial set $X$, $\mathfrak{C}X$ is defined to be a colimit of the homotopy coherent simplices indexed by the category of simplices of $X$.[17] Because of the formal similarity with the geometric realization functor, another left adjoint defined by Kan's construction, we refer to $\mathfrak{C}$ as **homotopy coherent realization**.

Many examples of quasi-categories fit into the following paradigm.

Theorem 2.7.8 ([72, 2.1]). *If* S *is Kan complex enriched, then* $\mathfrak{N}$S *is a quasi-category.*

In particular, in light of Exercise 2.4.28, the quasi-category associated to a simplicial model category M is defined to be $\mathfrak{N}M_{cf}$.

Recall from §2.6.2 that a **homotopy coherent diagram** of shape A in a Kan complex enriched category S is a functor $\mathfrak{C}A \to$ S. Similarly, a **homotopy coherent natural transformation** $\alpha \colon F \to G$ between homotopy coherent diagrams $F$ and $G$ of shape A is a homotopy coherent diagram of shape $A \times [1]$ that restricts on the endpoints of $[1]$ to $F$ and $G$ as follows:

$$\begin{array}{ccccc} \mathfrak{C}A & \xrightarrow{\ 0\ } & \mathfrak{C}(A \times [1]) & \xleftarrow{\ 1\ } & \mathfrak{C}A \\ & {\scriptstyle F} \searrow & {\scriptstyle \downarrow \alpha} & \swarrow {\scriptstyle G} & \\ & & \text{S} & & \end{array}$$

Note that the data of a pair of homotopy coherent natural transformations $\alpha \colon F \to G$ and $\beta \colon G \to H$ between homotopy coherent diagrams of shape A does not uniquely

---

[17] The simplicial set $X$ is obtained by gluing in a $\Delta^n$ for each $n$-simplex $\Delta^n \to X$ of $X$. The functor $\mathfrak{C}$ preserves these colimits, so $\mathfrak{C}X$ is obtained by gluing in a $\mathfrak{C}[n]$ for each $n$-simplex of $X$.

determine a (vertical) "composite" homotopy coherent natural transformation $F \to H$ because this data does not define a homotopy coherent diagram of shape $A \times [2]$, where $[2] = 0 \to 1 \to 2$. Here $\alpha$ and $\beta$ define a diagram of shape $\mathfrak{C}(A \times \Lambda_1^2)$ rather than a diagram of shape $\mathfrak{C}(A \times [2])$, where $\Lambda_1^2$ is the shape of the generating reflexive directed graph of the category $[2]$. This observation motivated Boardman and Vogt to define, in place of a *category* of homotopy coherent diagrams and natural transformations of shape A, a *quasi-category* of homotopy coherent diagrams and natural transformations of shape A.

For any category A, let $\mathrm{Coh}(A, S)$ denote the simplicial set whose $n$-simplices are homotopy coherent diagrams of shape $A \times [n]$, i.e., are simplicial functors

$$\mathfrak{C}(A \times [n]) \to S.$$

Corollary 2.7.9.  $\mathrm{Coh}(A, S) \cong \Omega S^A$ *is a quasi-category.*

*Proof.* By the adjunction of Definition 2.7.7, a simplicial functor $\mathfrak{C}A \to S$ is the same as a simplicial map $A \to \Omega S$. So $\mathrm{Coh}(A, S) \cong \Omega S^A$ and since the quasi-categories define an exponential ideal in simplicial sets as a consequence of the cartesian closure of the Joyal model structure of Theorem 2.7.12, the fact that $\Omega S$ is a quasi-category implies that $\Omega S^A$ is too. □

*Remark 2.7.10* (all diagrams in homotopy coherent nerves are homotopy coherent). This corollary explains that any map of simplicial sets $X \to \Omega S$ transposes to define a simplicial functor $\mathfrak{C}X \to S$, a homotopy coherent diagram of shape $X$ in S. While not every quasi-category is isomorphic to a homotopy coherent nerve of a Kan complex enriched category, every quasi-category is equivalent to a homotopy coherent nerve; one proof appears as [244, 7.2.2]. This explains the slogan that "all diagrams in quasi-categories are homotopy coherent."

## 2.7.3   The Joyal model structure

In analogy with Quillen's model structure of Theorem 2.3.30, in which the fibrant objects are the Kan complexes and the cofibrations are the monomorphisms, we might hope that there is another model structure on sSet whose fibrant objects are the quasi-categories and with the monomorphisms as cofibrations, and indeed there is one (and by Exercise 2.3.8(2.3.8) it is unique).

The weak equivalences in this hoped-for model structure for quasi-categories can be described using a particularly nice cylinder object. Let $\mathbb{I}$ be the nerve of the free-standing isomorphism $\mathbb{I}$; the name is selected because $\mathbb{I}$ is something like an interval.

Proposition 2.7.11.   *For any simplicial set A, the evident inclusion and projection maps define a cylinder object*

$$A \sqcup A \xrightarrow{(1_A, 1_A)} A$$

with $(i_0, i_1)$ to $A \times \mathbb{I}$ and $\pi$ to $A$.

*Proof.* The map $(i_0, i_1) \colon A \sqcup A \to A \times \mathbb{I}$ is a monomorphism and hence a cofibration. To see that the projection is a trivial fibration, observe that it is a pullback of $\mathbb{I} \to *$ as displayed below left, and hence by Lemma 2.3.9 it suffices to prove that this latter map is a trivial fibration. To that end, we must show that there exist solutions to lifting problems displayed on the right:

$$\begin{array}{ccc} A \times \mathbb{I} & \xrightarrow{\pi} & \mathbb{I} \\ {\scriptstyle \pi}\downarrow & \lrcorner & \downarrow \\ A & \longrightarrow & * \end{array} \qquad \begin{array}{ccc} \partial\Delta^n & \longrightarrow & \mathbb{I} \\ \downarrow & \nearrow & \downarrow \\ \Delta^n & \longrightarrow & * \end{array}$$

When $n = 0$ this is true because $\mathbb{I}$ is non-empty. For larger $n$, we use the fact that $\mathbb{I} \cong \mathrm{cosk}_0 \mathbb{I}$. By adjunction, it suffices to show that $\mathbb{I}$ lifts against $\mathrm{sk}_0 \partial\Delta^n \to \mathrm{sk}_0 \Delta^n$, but for $n > 0$, the 0-skeleton of $\Delta^n$ is isomorphic to that of its boundary. $\square$

The proof of Joyal's model structure has been widely circulated in unpublished notes, and can also be found in [169, 2.2.5.1] or [81, 2.13].

**Theorem 2.7.12 (Joyal).** *There is a cartesian closed model structure on* sSet *whose*

- *cofibrations are monomorphisms,*
- *weak equivalences are those maps $f \colon A \to B$ that induce bijections on the sets*

$$\mathrm{Hom}(B, X)_{/\sim_\ell} \to \mathrm{Hom}(A, X)_{/\sim_\ell}$$

*of maps into any quasi-category $X$ modulo the left homotopy relation relative to the cylinder just defined,*
- *fibrant objects are precisely the quasi-categories, and*
- *fibrations between fibrant objects are the* **isofibrations***, those maps that lift against the inner horn inclusions and also the map $* \to \mathbb{I}$.*

By Proposition 2.3.23, a map between quasi-categories is a weak equivalence, or we say simply **equivalence** of quasi-categories, if and only if it admits an inverse equivalence $Y \to X$ together with an "invertible homotopy equivalence" using the notion of homotopy defined with the interval $\mathbb{I}$. A map between nerves of strict 1-categories is an equivalence of quasi-categories if and only if it is an equivalence of categories, as usually defined. In general, every categorical notion for quasi-categories restricts along the full inclusion $\mathsf{Cat} \subset \mathsf{qCat}$ to the classical notion. This gives another sense in which quasi-categories model the $(\infty, 1)$-categories introduced at the start of this section. However, quasi-categories are not the only model of $(\infty, 1)$-categories, as we shall now discover.

## 2.8    Models of $(\infty, 1)$-categories

An $(\infty, 1)$-category should have a set of objects $X_0$, a space of morphisms $X_1$, together with composition and identities that are at least weakly associative and unital. One idea of how this might be presented, due to Segal [269], is to ask that $X \in \mathrm{sSet}^{\Delta^{\mathrm{op}}}$ is a simplicial space

$$\cdots \quad X_3 \underset{\underset{\longleftarrow}{\overset{\longrightarrow}{\rightleftarrows}}}{\overset{\longrightarrow}{\rightleftarrows}} X_2 \overset{\overset{\longrightarrow}{\rightleftarrows}}{\underset{\longrightarrow}{}} X_1 \rightleftarrows X_0$$

with $X_0$ still a set, so that for all $n$ the map

$$X_n \to X_1 \times_{X_0} \cdots \times_{X_0} X_1 \qquad (2.12)$$

induced on weighted limits from the spine inclusion $\Delta^1 \vee \cdots \vee \Delta^1 \to \Delta^n$ is a weak equivalence in a suitable sense. Segal points out that Grothendieck has observed that, in the case where the spaces $X_n$ are discrete, these so-called Segal maps are isomorphisms if and only if $X$ is isomorphic to the nerve of a category.

In this section we introduce various models of $(\infty, 1)$-categories, many of which are inspired by this paradigm. Before these models make their appearance in §2.8.2, we begin in §2.8.1 with an abbreviated tour of an axiomatization due to Toën that characterizes a homotopy theory of $(\infty, 1)$-categories. In §2.8.3, we then restrict our attention to four of the six models that are better behaved in the sense of providing easy access to the $(\infty, 1)$-category $\mathrm{Fun}(A, B)$ of functors between $(\infty, 1)$-categories $A$ and $B$. These models each satisfy a short list of axioms that we exploit in §2.9 to sketch a natively "model-independent" development of the category theory of $(\infty, 1)$-categories.

## 2.8.1   An axiomatization of the homotopy theory of $(\infty, 1)$-categories

The homotopy theory of $\infty$-groupoids is freely generated under homotopy colimits by the point. We might try to adopt a similar "generators and relations" approach to build the homotopy theory of $(\infty, 1)$-categories, taking the generators to be the category $\Delta$, which freely generates simplicial spaces. The relations assert that the natural maps

$$\Delta^1 \vee \cdots \vee \Delta^1 \to \Delta^n \qquad \text{and} \qquad \mathbb{I} \to \Delta^0 \qquad (2.13)$$

induce equivalences upon mapping into an $(\infty, 1)$-category. This idea motivates Rezk's complete Segal space model, which is the conceptual center of the Toën axiomatization of a model category M whose fibrant objects model $(\infty, 1)$-categories.

For simplicity we assume that M is a combinatorial simplicial model category. In practice, these assumptions are relatively mild: in particular, if M fails to be simplicial it is possible to define a Quillen equivalent model structure on $\mathrm{M}^{\Delta^{\mathrm{op}}}$ that is simplicial [78]. The model category M should be equipped with a functor $C \colon \Delta \to \mathrm{M}$ such that $C(0)$ represents a free point in M while $C(1)$ represents a free arrow. This cosimplicial object is required to be a **weak cocategory**, meaning that the duals of the Segal maps are equivalences

$$C(1) \cup_{C(0)} \cdots \cup_{C(0)} C(1) \overset{\sim}{\to} C(n).$$

We state Toën's seven axioms without defining all the terms because to do so would demand too long of an excursion, and refer the reader to [292] for more details.

Theorem 2.8.1 (Toën [292]). *Let* M *be a combinatorial simplicial model category equipped with a functor* $C \colon \Delta \to$ M *satisfying the following properties.*

(i) *Homotopy colimits are universal over* **0-local objects,** *those X so that*

$$\mathsf{Map}(*, X) \xrightarrow{\sim} \mathsf{Map}(C(1), X).$$

(ii) *Homotopy coproducts are disjoint and universal.*

(iii) *C is an **interval:** meaning* $C(0)$ *and the C-geometric realization of* $\mathbb{I}$ *are contractible.*

(iv) *For any weak category* $X \in$ M$^{\Delta^{\mathrm{op}}}$ *such that* $X_0$ *and* $X_1$ *are 0-local, X is equivalent to the Čech nerve of the map* $X_0 \to |X|_c$.

(v) *For any weak category* $X \in$ M$^{\Delta^{\mathrm{op}}}$ *such that* $X_0$ *and* $X_1$ *are 0-local, the homotopy fiber of* $X \to \mathbb{R}\mathsf{Hom}(C, |X|_c)$ *is contractible.*

(vi) *The point and interval define a **generator:*** $f \colon X \to Y$ *is a weak equivalence in* M *if and only if* $\mathsf{Map}(C(0), X) \xrightarrow{\sim} \mathsf{Map}(C(0), Y)$ *and* $\mathsf{Map}(C(1), X) \xrightarrow{\sim} \mathsf{Map}(C(1), Y)$.

(vii) *C is **homotopically fully faithful:*** $\Delta([n], [m]) \xrightarrow{\sim} \mathsf{Map}(C(n), C(m))$

*Then the functor* $X \mapsto \mathsf{Map}(C(-), X)$ *defines a right Quillen equivalence from* M *to the model structure for complete Segal spaces on the category of bisimplicial sets.*

A similar axiomatization is given by Barwick and Schommer-Pries as a specialization of an axiomatization for $(\infty, n)$-categories [26].

## 2.8.2    Models of $(\infty, 1)$-categories

We now introduce six models of $(\infty, 1)$-categories, each arising as the fibrant objects in a model category that is Quillen equivalent to all of the others. Two of these models — the quasi-categories and the Kan complex enriched categories — have been presented already in Theorems 2.7.12 and 2.6.9.

A **Segal category** is a Reedy fibrant bisimplicial set $X \in$ sSet$^{\Delta^{\mathrm{op}}}$ such that the Segal maps (2.12) are trivial fibrations and $X_0$ is a set.[18]

Theorem 2.8.2 (Hirschowitz–Simpson [125, 278], Pellissier [220], Bergner [40]). *There is a cartesian closed model structure on the category of bisimplicial sets with discrete set of objects whose*

- *cofibrations are the monomorphisms,*
- *fibrant objects are the Segal categories that are Reedy fibrant as simplicial spaces, and*
- *weak equivalences are the DK-equivalences (in a suitable sense).*

A **complete Segal space** is similarly a Reedy fibrant bisimplicial set $X \in$ sSet$^{\Delta^{\mathrm{op}}}$ such that the Segal maps (2.12) are trivial fibrations. In this model, the discreteness condition on $X_0$ is replaced with the so-called **completeness** condition, which is again most elegantly phrased using weighted limits: it asks either that the map $\{\mathbb{I}, X\} \to \{\Delta^0, X\} \cong X_0$ is a trivial fibration or that the map $X_0 \to \{\mathbb{I}, X\}$ is an equivalence. Intuitively this says that the spatial structure of $X_0$ is recovered by the $\infty$-groupoid of $\{\mathbb{I}, X\}$ of isomorphisms in $X$.

---

[18]  In [90, §7] the Reedy fibrancy condition, which implies that the Segal maps are Kan fibrations, is dropped and the Segal maps are only required to be weak equivalences.

**Theorem 2.8.3** (Rezk [233]).   *There is a cartesian closed model structure on the category of bisimplicial sets whose*

- *cofibrations are the monomorphisms,*
- *fibrant objects are the complete Segal spaces, and*
- *weak equivalences are those maps $u : A \to B$ such that for every complete Segal space $X$, the maps $X^B \to X^A$ are weak homotopy equivalences of simplicial sets upon evaluating at 0.*

A **marked simplicial set** is a simplicial set with a collection of marked edges containing the degeneracies; maps must then preserve the markings. A quasi-category is naturally a marked simplicial set whose marked edges are precisely the isomorphisms, described in Definition 2.7.6.

**Theorem 2.8.4** (Verity [293], Lurie [169]).   *There is a cartesian closed model structure on the category of marked simplicial sets whose*

- *cofibrations are the monomorphisms,*
- *fibrant objects are the naturally marked quasi-categories, and*
- *weak equivalences are those maps $A \to B$ so that for all naturally marked quasi-categories $X$ the map $X^B \to X^A$ is a homotopy equivalence of maximal sub Kan complexes.*

A **relative category** is a category equipped with a wide subcategory of weak equivalences. A morphism of relative categories is a homotopical functor. A weak equivalence of relative categories is a homotopical functor $F : (C, \mathcal{W}) \to (D, \mathcal{W})$ that induces a DK-equivalence on hammock localizations $\mathcal{L}^H(C, \mathcal{W}) \to L^H(D, \mathcal{W})$.

**Theorem 2.8.5** (Barwick–Kan [23]).   *There is a model structure for relative categories whose*

- *weak equivalences are the relative DK-equivalences just defined*

*and whose cofibrations and fibrant objects are somewhat complicated to describe.*

Each of these model categories, represented in the diagram below by their sub-categories of fibrant objects, are Quillen equivalent, connected via right Quillen equivalences as follows:[19]

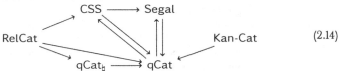

$$(2.14)$$

A nice feature of the simplicial category and relative category models is that their

---

[19] The right Quillen equivalences from relative categories are in [23]. The Quillen equivalences involving complete Segal spaces, Segal categories, and quasi-categories can all be found in [142]. Proofs that the homotopy coherent nerve defines a Quillen equivalence from simplicial categories to quasi-categories can be found in [169] and [81]. A zigzag of Quillen equivalences between simplicial categories and Segal categories is constructed in [42]. The right Quillen equivalence from naturally marked quasi-categories to the Joyal model structure can be found in [169] and [293].

objects and morphisms are strictly-defined, as honest-to-goodness enriched categories in the former case and honest-to-goodness homotopical categories in the latter. From this vantage point it is quite surprising that they are Quillen equivalent to the weaker models. But there are some costs paid to obtain this extra strictness: neither model category is cartesian closed, so both contexts lack a suitable internal hom, whereas the other four models — the quasi-categories, Segal categories, complete Segal spaces, and naturally marked quasi-categories — all form cartesian closed model categories. Consequently, in each of these models the $(\infty, 1)$-categories define an exponential ideal: if $A$ is fibrant and $X$ is cofibrant, then $A^X$ is again fibrant and moreover the maps induced on exponentials by the maps (2.13) are weak equivalences.

## 2.8.3    $\infty$-cosmoi of $(\infty, 1)$-categories

From the cartesian closure of the model categories for quasi-categories, Segal categories, complete Segal spaces, and naturally marked quasi-categories, it is possible to induce a secondary enrichment, in the sense of Definition 2.4.25, on these model categories:

**Theorem 2.8.6** ([240, 2.2.3]). *The model structures for quasi-categories, complete Segal spaces, Segal categories, and naturally marked quasi-categories are all enriched over the model structure for quasi-categories.*

The following definition of an $\infty$-cosmos collects together the properties of the fibrant objects and fibrations and weak equivalences between them in any model category that is enriched over the Joyal model structure and in which the fibrant objects are also cofibrant:

**Definition 2.8.7** ($\infty$-cosmos). An $\infty$-**cosmos** is a simplicially enriched category K whose

- objects we refer to as the $\infty$-**categories** in the $\infty$-cosmos, whose
- hom simplicial sets $\mathsf{Fun}(A, B)$ are all quasi-categories,

and that is equipped with a specified subcategory of **isofibrations**, denoted by "$\twoheadrightarrow$", satisfying the following axioms:

(i) (completeness) As a simplicially enriched category, K possesses a terminal object 1, cotensors $A^U$ of objects $A$ by all[20] simplicial sets $U$, and pullbacks of isofibrations along any functor.[21]

(ii) (isofibrations) The class of isofibrations contains the isomorphisms and all of the functors $!: A \twoheadrightarrow 1$ with codomain 1; is stable under pullback along all

---

[20] For most purposes, it suffices to require only cotensors with finitely presented simplicial sets (those with only finitely many non-degenerate simplices).

[21] For the theory of homotopy coherent adjunctions and monads developed in [241], limits of towers of isofibrations are also required, with the accompanying stability properties of (ii). These limits are present in all of the $\infty$-cosmoi we are aware of, but will not be required for any results discussed here.

functors; and if $p\colon E \twoheadrightarrow B$ is an isofibration in $\mathsf{K}$ and $i\colon U \hookrightarrow V$ is an inclusion of simplicial sets then the Leibniz cotensor $\widehat{\{i,p\}}\colon E^V \twoheadrightarrow E^U \times_{B^U} B^V$ is an isofibration. Moreover, for any object $X$ and isofibration $p\colon E \twoheadrightarrow B$, $\mathsf{Fun}(X,p)\colon \mathsf{Fun}(X,E) \twoheadrightarrow \mathsf{Fun}(X,B)$ is an isofibration of quasi-categories.

The underlying category of an $\infty$-cosmos $\mathsf{K}$ has a canonical subcategory of equivalences, denoted by "$\xrightarrow{\sim}$", satisfying the two-of-six property. A functor $f\colon A \to B$ is an **equivalence** just when the induced functor $\mathsf{Fun}(X,f)\colon \mathsf{Fun}(X,A) \to \mathsf{Fun}(X,B)$ is an equivalence of quasi-categories for all objects $X \in \mathsf{K}$. The **trivial fibrations**, denoted by "$\xrightarrow{\sim}\!\!\!\twoheadrightarrow$", are those functors that are both equivalences and isofibrations. It follows from 2.8.7(i)-(ii) that:

(iii) (cofibrancy) All objects are **cofibrant**, in the sense that they enjoy the left lifting property with respect to all trivial fibrations in $\mathsf{K}$:

$$
\begin{array}{ccc}
 & & E \\
 & \nearrow & \downarrow{\scriptstyle\wr} \\
A & \longrightarrow & B
\end{array}
$$

(iv) (trivial fibrations) The trivial fibrations define a subcategory containing the isomorphisms; are stable under pullback along all functors; and the Leibniz cotensor $\widehat{\{i,p\}}\colon E^V \xrightarrow{\sim}\!\!\!\twoheadrightarrow E^U \times_{B^U} B^V$ of an isofibration $p\colon E \twoheadrightarrow B$ in $\mathsf{K}$ and a monomorphism $i\colon U \hookrightarrow V$ between presented simplicial sets is a trivial fibration when $p$ is a trivial fibration in $\mathsf{K}$ or $i$ is a trivial cofibration in the Joyal model structure on $\mathsf{sSet}$. Moreover, for any object $X$ and trivial fibration $p\colon E \xrightarrow{\sim}\!\!\!\twoheadrightarrow B$, $\mathsf{Fun}(X,p)\colon \mathsf{Fun}(X,E) \xrightarrow{\sim}\!\!\!\twoheadrightarrow \mathsf{Fun}(X,B)$ is a trivial fibration of quasi-categories.

(v) (factorization) Any functor $f\colon A \to B$ may be factored as $f = pj$:

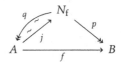

where $p\colon N_f \twoheadrightarrow B$ is an isofibration and $j\colon A \xrightarrow{\sim} N_f$ is right inverse to a trivial fibration $q\colon N_f \xrightarrow{\sim}\!\!\!\twoheadrightarrow A$.

It is a straightforward exercise in enriched model category theory to verify that these axioms are satisfied by the fibrant objects in any model category that is enriched over the Joyal model structure on simplicial sets, at least when all of these objects are cofibrant. Consequently:

**Theorem 2.8.8** (Joyal–Tierney, Verity, Lurie, Riehl–Verity [240]). *The full subcategories* $\mathsf{qCat}$, $\mathsf{CSS}$, $\mathsf{Segal}$, *and* $\mathsf{qCat}_{\natural}$ *all define $\infty$-cosmoi.*

Moreover, each of the model categories referenced in Theorem 2.8.8 is a closed monoidal model category with respect to the cartesian product. It follows that each of these four $\infty$-cosmoi is **cartesian closed** in the sense that it satisfies the extra axiom:

(vi) (cartesian closure) The product bifunctor $- \times -\colon \mathsf{K} \times \mathsf{K} \to \mathsf{K}$ extends to a simplicially enriched two-variable adjunction

$$\mathsf{Fun}(A \times B, C) \cong \mathsf{Fun}(A, C^B) \cong \mathsf{Fun}(B, C^A).$$

A **cosmological functor** is a simplicial functor $F\colon \mathsf{K} \to \mathsf{L}$ preserving the class of isofibrations and all of the limits enumerated in Definition 2.8.7(i). A cosmological functor is a **biequivalence** when it is:

(i) surjective on objects up to equivalence: i.e., if for every $C \in \mathsf{L}$, there is some $A \in \mathsf{K}$ so that $FA \simeq C \in \mathsf{L}$;

(ii) a local equivalence of quasi-categories: i.e., if for every pair $A, B \in \mathsf{K}$, the map $\mathsf{Fun}(A, B) \xrightarrow{\sim} \mathsf{Fun}(FA, FB)$ is an equivalence of quasi-categories.

The inclusion $\mathsf{Cat} \hookrightarrow \mathsf{qCat}$ defines a cosmological functor but not a biequivalence, since it fails to be essentially surjective. Each right Quillen equivalence of

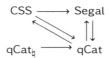

defines a cosmological biequivalence.

As discussed in the next section, Theorem 2.8.8 together with an additional observation — that the $\infty$-cosmoi of quasi-categories, Segal categories, complete Segal spaces, and naturally marked simplicial sets are biequivalent — forms the lynchpin of an approach to develop the basic theory of $(\infty, 1)$-categories in a model-independent fashion. In fact, most of that development takes places in a strict 2-category that we now introduce.

**Definition 2.8.9** (the homotopy 2-category of $\infty$-cosmos). The **homotopy 2-category** of an $\infty$-cosmos $\mathsf{K}$ is a strict 2-category $\mathfrak{h}\mathsf{K}$ such that

– the objects of $\mathfrak{h}\mathsf{K}$ are the objects of $\mathsf{K}$, i.e., the $\infty$-**categories**,
– the 1-cells $f\colon A \to B$ of $\mathfrak{h}\mathsf{K}$ are the vertices $f \in \mathsf{Fun}(A, B)$ in the mapping quasi-categories of $\mathsf{K}$, i.e., the $\infty$-**functors**, and

– a 2-cell $A \underset{g}{\overset{f}{\Rightarrow\alpha}} B$ in $\mathfrak{h}\mathsf{K}$, which we call an $\infty$-**natural transformation**, is represented by a 1-simplex $\alpha\colon f \to g \in \mathsf{Fun}(A, B)$, where a parallel pair of 1-simplices in $\mathsf{Fun}(A, B)$ represent the same 2-cell if and only if they bound a 2-simplex whose remaining outer face is degenerate.

Put concisely, the homotopy 2-category is the 2-category $\mathfrak{h}\mathsf{K}$ defined by applying the homotopy category functor $h\colon \mathsf{qCat} \to \mathsf{Cat}$ to the mapping quasi-categories of the $\infty$-cosmos; the hom-categories in $\mathfrak{h}\mathsf{K}$ are defined by the formula

$$\mathrm{Hom}(A, B) := h\mathsf{Fun}(A, B)$$

to be the homotopy categories of the mapping quasi-categories in $\mathsf{K}$.

As we shall see in the next section, much of the theory of $(\infty, 1)$-categories can be developed simply by considering them as objects in the homotopy 2-category using an appropriate weakening of standard 2-categorical techniques. A key to the feasibility of this approach is that the standard 2-categorical notion of equivalence, reviewed in Definition 2.9.2 below, coincides with the representably-defined notion of equivalence present in any $\infty$-cosmos. The proof of this result should be compared with Quillen's Proposition 2.3.23.

**Proposition 2.8.10.** *An $\infty$-functor $f : A \to B$ is an equivalence in the $\infty$-cosmos $\mathsf{K}$ if and only if it is an equivalence in the homotopy 2-category $\mathfrak{h}\mathsf{K}$.*

*Proof.* By definition, any equivalence $f : A \xrightarrow{\sim} B$ in the $\infty$-cosmos induces an equivalence $\mathrm{Fun}(X, A) \xrightarrow{\sim} \mathrm{Fun}(X, B)$ of quasi-categories for any $X$, which becomes an equivalence of categories $\mathrm{Hom}(X, A) \xrightarrow{\sim} \mathrm{Hom}(X, B)$ upon applying the homotopy category functor $\mathfrak{h} : \mathsf{qCat} \to \mathsf{Cat}$. Applying the Yoneda lemma in the homotopy 2-category $\mathfrak{h}\mathsf{K}$, it follows easily that $f$ is an equivalence in the standard 2-categorical sense.

Conversely, as the map $\mathbb{I} \to \Delta^0$ of simplicial sets is a weak equivalence in the Joyal model structure, an argument similar to that used to prove Proposition 2.7.11 demonstrates that the cotensor $B^{\mathbb{I}}$ defines a path object for the $\infty$-category $B$:

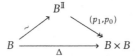

It follows from the two-of-three property that any $\infty$-functor that is isomorphic in the homotopy 2-category to an equivalence in the $\infty$-cosmos is again an equivalence in the $\infty$-cosmos. Now it follows immediately from the two-of-six property for equivalences in the $\infty$-cosmos, plus the fact that the class of equivalences includes the identities, that any 2-categorical equivalence is an equivalence in the $\infty$-cosmos.  $\square$

A consequence of Proposition 2.8.10 is that any cosmological biequivalence in particular defines an biequivalence of homotopy 2-categories, which explains the choice of terminology.

## 2.9    Model-independent $(\infty, 1)$-category theory

We now develop a small portion of the theory of $\infty$-categories in any $\infty$-cosmos, thereby developing a theory of $(\infty, 1)$-categories that applies equally to quasi-categories, Segal categories, complete Segal spaces, and naturally marked quasi-categories. The definitions of the basic $(\infty, 1)$-categorical notions presented here might be viewed as "synthetic," in the sense that they are blind to which model is being considered, in contrast with the "analytic" theory of quasi-categories first outlined in Joyal's [140] and later greatly expanded in his unpublished works and Lurie's [169, 168]. In §2.9.1, we introduce adjunctions and equivalences between $\infty$-categories, which generalize the notions of Quillen adjunction and Quillen equivalence between model categories from §2.4.3 and §2.4.5. Then in §2.9.2, we develop the theory of limits and colimits in an $\infty$-category, which correspond to the homotopy limits and colimits of §2.5.

Our synthetic definitions specialize in the $\infty$-cosmos of quasi-categories to notions that precisely recapture the Joyal–Lurie analytic theory; the proofs that this is the case are not discussed here, but can be found in [243, 240]. Considerably more development along these lines can be found in [242].

## 2.9.1    Adjunctions and equivalences

In any 2-category, in particular in the homotopy 2-category $\mathfrak{h}K$ of an $\infty$-cosmos, there are standard definitions of adjunction or equivalence, which allow us to define adjunctions and equivalences between $\infty$-categories.

**Definition 2.9.1.** An **adjunction** between $\infty$-categories consists of:

- a pair of $\infty$-categories $A$ and $B$;
- a pair of $\infty$-functors $f : B \to A$ and $u : A \to B$; and
- a pair of $\infty$-natural transformations $\eta : \mathrm{id}_B \Rightarrow uf$ and $\epsilon : fu \Rightarrow \mathrm{id}_A$

so that the triangle equalities hold:

$$
\begin{array}{ccc}
B \xequal B & B & B \xequal B \qquad B \\
u \Vert\epsilon\ f\ \Downarrow\eta\ u & = u\left(=\right)u & f\ \Downarrow\eta\ u\ \Vert\epsilon\ f \qquad = \left(=\right)f \\
A \xequal A & A & A \xequal A \qquad A
\end{array}
$$

We write $f \dashv u$ to assert that the $\infty$-functor $f : B \to A$ is **left adjoint** to the $\infty$-functor $u : A \to B$, its **right adjoint**.

**Definition 2.9.2.** An **equivalence** between $\infty$-categories consists of:

- a pair of $\infty$-categories $A$ and $B$;
- a pair of $\infty$-functors $f : B \to A$ and $g : A \to B$; and
- a pair of natural isomorphisms $\eta : \mathrm{id}_B \cong gf$ and $\epsilon : fg \cong \mathrm{id}_A$.

An $\infty$-**natural isomorphism** is a 2-cell in the homotopy 2-category that admits a vertical inverse 2-cell.

We write $A \simeq B$ and say that $A$ and $B$ are **equivalent** if there exists an equivalence between $A$ and $B$. The direction for the $\infty$-natural isomorphisms comprising an equivalence is immaterial. Our notation is chosen to suggest the connection with adjunctions conveyed by the following exercise.

*Exercise 2.9.3.* In any 2-category, prove that:

(i) Adjunctions compose: given adjoint $\infty$-functors

$$
C \underset{u'}{\overset{f'}{\rightleftarrows}} \perp B \underset{u}{\overset{f}{\rightleftarrows}} \perp A \qquad \rightsquigarrow \qquad C \underset{u'u}{\overset{ff'}{\rightleftarrows}} \perp A
$$

the composite $\infty$-functors are adjoint.

(ii) Any equivalence can always be promoted to an **adjoint equivalence** by modifying one of the $\infty$-natural isomorphisms. That is, show that the $\infty$-natural isomorphisms in an equivalence can be chosen so as to satisfy the triangle equalities. Conclude that if $f$ and $g$ are inverse equivalences then $f \dashv g$ and $g \dashv f$.

The point of Exercise 2.9.3 is that there are various diagrammatic 2-categorical proofs that can be taken off the shelf and applied to the homotopy 2-category of an $\infty$-cosmos to prove theorems about adjunctions and equivalence between $(\infty, 1)$-categories.

## 2.9.2   Limits and colimits

We now introduce definitions of limits and colimits for diagrams valued inside an $\infty$-category. We begin by defining terminal objects, or as we shall call them "terminal elements," to avoid an overproliferation of the generic name "objects."

Definition 2.9.4. A **terminal element** in an $\infty$-category $A$ is a right adjoint $t: 1 \to A$ to the unique $\infty$-functor $!: A \to 1$. Explicitly, the data consists of

- an element $t: 1 \to A$ and
- a $\infty$-natural transformation $\eta: \mathrm{id}_A \Rightarrow t!$ whose component $\eta t$ at the element $t$ is an isomorphism.[22]

Several basic facts about terminal elements can be deduced immediately from the general theory of adjunctions.

*Exercise 2.9.5.*

(i) Terminal elements are preserved by right adjoints and by equivalences.
(ii) If $A' \simeq A$ then $A$ has a terminal element if and only if $A'$ does.

Terminal elements are limits of empty diagrams. We now turn to limits of generic diagrams whose indexing shapes are given by 1-categories. For any $\infty$-category $A$ in an $\infty$-cosmos K, there is a 2-functor $A^{(-)}: \mathrm{Cat}^{\mathrm{op}} \to \mathfrak{h}K$ defined by forming simplicial cotensors with nerves of categories. Using these simplicial cotensors, if $J$ is a 1-category and $A$ is an $\infty$-category, the $\infty$-**category of $J$-indexed diagrams in** $A$ is simply the cotensor $A^J$.[23]

*Remark 2.9.6.* In the cartesian closed $\infty$-cosmoi of Definition 2.8.7(vi), we also permit the indexing shape $J$ to be another $\infty$-category, in which case the internal hom $A^J$ defines the $\infty$-**category of $J$-indexed diagrams in** $A$. The development of the theory of limits indexed by an $\infty$-category in a cartesian closed $\infty$-cosmos entirely parallels the development for limits indexed by 1-categories, a parallelism we highlight by conflating the notation of 2.8.7(i) and 2.8.7(vi).

---

[22] If $\eta$ is the unit of the adjunction $! \dashv t$, then the triangle equalities demand that $\eta t = \mathrm{id}_t$. However, by a 2-categorical trick, to show that such an adjunction exists, it suffices to find a 2-cell $\eta$ such that $\eta t$ is an isomorphism.

[23] More generally, this construction permits arbitrary simplicial sets as indexing shapes for diagrams in an $\infty$-category $A$. In either case, the elements of $A^J$ are to be regarded as homotopy coherent diagrams along the lines of Remark 2.7.10.

In analogy with Definition 2.9.4, we have:

**Definition 2.9.7.** An $\infty$-category $A$ **admits all limits of shape** $J$ if the constant diagram $\infty$-functor $\Delta \colon A \to A^J$, induced by the unique $\infty$-functor $! \colon J \to 1$, has a right adjoint:

$$A \underset{\lim}{\overset{\Delta}{\rightleftarrows}} A^J$$

From the vantage point of Definition 2.9.7, the following result is easy:

*Exercise 2.9.8.* Using the general theory of adjunctions, show that a right adjoint $\infty$-functor $u \colon A \to B$ between $\infty$-categories that admit all limits of shape $J$ necessarily preserves them, in the sense that the $\infty$-functors

$$\begin{array}{ccc}
A^J & \xrightarrow{\ u^J\ } & B^J \\
{\scriptstyle \lim}\downarrow & \cong & \downarrow{\scriptstyle \lim} \\
A & \xrightarrow[\ u\ ]{} & B
\end{array}$$

commute up to isomorphism.

The problem with Definition 2.9.7 is that it is insufficiently general: many $\infty$-categories will have certain, but not all, limits of diagrams of a particular indexing shape. With this in mind, we will now re-express Definition 2.9.7 in a form that permits its extension to cover this sort of situation. For this, we make use of the 2-categorical notion of an **absolute right lifting**, which is the "op"-dual (reversing the 1-cells but not the 2-cells) of the notion of absolute right Kan extension introduced in Definition 2.4.1.

*Exercise 2.9.9.* Show that in any 2-category, a 2-cell $\epsilon \colon fu \Rightarrow \mathrm{id}_A$ defines the counit of an adjunction $f \dashv u$ if and only if

$$\begin{array}{c}
 & B & \\
{\scriptstyle u}\nearrow & {\scriptstyle \Downarrow \epsilon} & \downarrow{\scriptstyle f} \\
A & =\!=\!= & A
\end{array}$$

defines an absolute right lifting diagram.

Applying Exercise 2.9.9, Definition 2.9.7 is equivalent to the assertion that the **limit cone**, our term for the counit of $\Delta \dashv \lim$, defines an absolute right lifting diagram:

$$\begin{array}{c}
 & A & \\
{\scriptstyle \lim}\nearrow & {\scriptstyle \Downarrow \epsilon} & \downarrow{\scriptstyle \Delta} \\
A^J & =\!=\!= & A^J
\end{array} \tag{2.15}$$

Recall that the appellation "absolute" means "preserved by all functors," in this case by restriction along any $\infty$-functor $X \to A^J$. In particular, an absolute right lifting diagram (2.15) restricts to define an absolute right lifting diagram on any subobject of the $\infty$-category of diagrams. This motivates the following definition.

Definition 2.9.10 (limit). A **limit** of a $J$-indexed diagram in $A$ is an absolute right lifting of the diagram $d$ through the constant diagram $\infty$-functor $\Delta\colon A \to A^J$

$$
\begin{array}{ccc}
 & & A \\
 & \overset{\lim d}{\nearrow} & \downarrow{\scriptstyle \Delta} \\
 & {\scriptstyle \Downarrow \lambda} & \\
1 & \underset{d}{\longrightarrow} & A^J
\end{array}
\tag{2.16}
$$

the 2-cell component of which defines the **limit cone** $\lambda\colon \Delta \lim d \Rightarrow d$.

If $A$ has all $J$-indexed limits, the restriction of the absolute right lifting diagram (2.15) along the element $d\colon 1 \to A^J$ defines a limit for $d$. Interpolating between Definitions 2.9.10 and 2.9.7, we can define a **limit of a family of diagrams** to be an absolute right lifting of the family $d\colon K \to A^J$ through $\Delta\colon A \to A^J$. For instance:

Theorem 2.9.11 ([243, 5.3.1]). *For every cosimplicial object in an $\infty$-category that admits a coaugmentation and a splitting, the coaugmentation defines its limit. That is, for every $\infty$-category $A$, the $\infty$-functors*

$$
\begin{array}{ccc}
 & & A \\
 & \overset{\mathrm{ev}_{[-1]}}{\nearrow} & \downarrow{\scriptstyle \Delta} \\
 & {\scriptstyle \Downarrow \lambda} & \\
A^{\Delta_\perp} & \underset{\mathrm{res}}{\longrightarrow} & A^{\Delta}
\end{array}
$$

*define an absolute right lifting diagram.*

Here $\Delta$ is the usual simplex category of finite non-empty ordinals and order-preserving maps. It defines a full subcategory of $\Delta_+$, which freely appends an initial object $[-1]$, and this in turn defines a subcategory of $\Delta_\perp$, which adds an "extra degeneracy" map between each pair of consecutive ordinals. Diagrams indexed by $\Delta \subset \Delta_+ \subset \Delta_\perp$ are, respectively, called **cosimplicial objects, coaugmented cosimplicial objects**, and **split cosimplicial objects**. The limit of a cosimplicial object is often called its **totalization**.

*Proof sketch.* In Cat, there is a canonical 2-cell

$$
\begin{array}{ccc}
\Delta & \lhook\joinrel\longrightarrow & \Delta_\perp \\
{\scriptstyle !}\downarrow & {\scriptstyle \Uparrow \lambda}\nearrow & \\
 & \nearrow{\scriptstyle [-1]} & \\
\mathbb{1} & &
\end{array}
$$

because $[-1] \in \Delta_\perp$ is initial. This data defines an absolute right extension diagram that is moreover preserved by any 2-functor, because the universal property of the functor $[-1]\colon \mathbb{1} \to \Delta_\perp$ and the 2-cell $\lambda$ is witnessed by a pair of adjunctions. The 2-functor $A^{(-)}\colon \mathrm{Cat}^{\mathrm{op}} \to \mathfrak{h}K$ converts this into the absolute right lifting diagram of the statement. $\qquad\square$

The most important result relating adjunctions and limits is of course this:

Theorem 2.9.12 ([243, 5.2.13]). *Right adjoints preserve limits.*

Our proof will closely follow the classical one. Given a diagram $d\colon 1 \to A^J$ and a right adjoint $u\colon A \to B$ to some $\infty$-functor $f$, a cone with summit $b\colon 1 \to B$ over $u^J d$ transposes to define a cone with summit $fb$ over $d$, which factors uniquely through the limit cone. This factorization transposes back across the adjunction to show that $u$ carries the limit cone over $d$ to a limit cone over $u^J d$.

*Proof.* Suppose that $A$ admits limits of a diagram $d\colon 1 \to A^J$ as witnessed by an absolute right lifting diagram (2.16). Since adjunctions are preserved by all 2-functors, an adjunction $f \dashv u$ induces an adjunction $f^J \dashv u^J$. We must show that

$$
\begin{array}{ccc}
 & A & \xrightarrow{u} B \\
\lim d \nearrow & \downarrow\!\Delta & \downarrow\!\Delta \\
 \Downarrow\lambda & & \\
1 \xrightarrow{d} & A^J \xrightarrow{u^J} & B^J
\end{array}
$$

is again an absolute right lifting diagram. Given a square

$$
\begin{array}{ccc}
X & \xrightarrow{b} & B \\
!\downarrow & \Downarrow\chi & \downarrow\Delta \\
1 & \xrightarrow[d]{} A^J \xrightarrow[u^J]{} & B^J
\end{array}
$$

we first "transpose across the adjunction," by composing with $f$ and the counit.

$$
\left(
\begin{array}{c}
X \xrightarrow{b} B \xrightarrow{f} A \\
!\downarrow \quad \Downarrow\chi \quad \downarrow\Delta \quad \downarrow\Delta \\
1 \xrightarrow{d} A^J \xrightarrow{u^J} B^J \xrightarrow{f^J} A^J \\
\Downarrow\epsilon^J
\end{array}
\right)
=
\left(
\begin{array}{c}
X \xrightarrow{b} B \xrightarrow{f} A \\
!\downarrow \;\exists!\Downarrow\zeta\; \lim d \quad \Downarrow\lambda \quad \downarrow\Delta \\
1 \xrightarrow{d} A^J
\end{array}
\right)
$$

The universal property of the absolute right lifting diagram $\lambda\colon \Delta\lim \Rightarrow d$ induces a unique factorization $\zeta$, which may then be "transposed back across the adjunction" by composing with $u$ and the unit.

$$
\begin{array}{c}
X \xrightarrow{b} B \xrightarrow{f} A \xrightarrow{u} B \\
\exists!\Downarrow\zeta \quad \lim d \;\Downarrow\lambda\; \downarrow\Delta \quad \downarrow\Delta \\
1 \xrightarrow{d} A^J \xrightarrow{u^J} B^J
\end{array}
=
\begin{array}{c}
X \xrightarrow{b} B \xrightarrow{f} A \xrightarrow{u} B \\
!\downarrow \;\Downarrow\chi\; \downarrow\Delta \quad \downarrow\Delta \quad \downarrow\Delta \\
1 \xrightarrow{d} A^J \xrightarrow{u^J} B^J \xrightarrow{f^J} A^J \xrightarrow{u^J} B^J
\end{array}
$$

$$
=
\begin{array}{c}
X \xrightarrow{b} B \quad B \\
!\downarrow \;\Downarrow\chi\; \downarrow\Delta \quad \downarrow\Delta \\
1 \xrightarrow{d} A^J \xrightarrow{u^J} B^J \xrightarrow{f^J} A^J \xrightarrow{u^J} B^J \\
\Downarrow\epsilon^J
\end{array}
=
\begin{array}{c}
X \xrightarrow{b} B \\
!\downarrow \;\Downarrow\chi\; \downarrow\Delta \\
1 \xrightarrow{d} A^J \xrightarrow{u^J} B^J
\end{array}
$$

Here the second equality is a consequence of the 2-functoriality of the simplicial cotensor, while the third is an application of a triangle equality for the adjunction $f^J \dashv u^J$. The pasted composite of $\zeta$ and $\eta$ is the desired factorization of $\chi$ through $\lambda$.

The proof that this factorization is unique, which again parallels the classical argument, is left to the reader: the essential point is that the transposes defined via these pasting diagrams are unique.                                                        □

Colimits are defined "co"-dually, by reversing the direction of the 2-cells but not the 1-cells. There is no need to repeat the proofs however: any $\infty$-cosmos K has a co-dual $\infty$-cosmos $K^{co}$ with the same objects but in which the mapping quasi-categories are defined to be the opposites of the mapping quasi-categories in K.

## 2.10   Epilogue

A category K equipped with a class of "weak equivalences" $\mathcal{W}$ — perhaps saturated in the sense of containing all of the maps inverted by the Gabriel–Zisman localization functor or perhaps merely generating the class of maps to be inverted in the category of fractions — defines a "homotopy theory," a phrase generally used to refer to the associated homotopy category together with the homotopy types of the mapping spaces, as captured for instance by the Dwyer–Kan hammock localization. We have studied two common axiomatizations of this abstract notion: Quillen's model categories, which present homotopy theories with all homotopy limits and homotopy colimits, and $(\infty, 1)$-categories, which might be encoded using one of the models introduced in §2.8 or worked with model-independently in the sense outlined in §2.9.

From the point of view of comparing homotopy categories, the model-independent theory of $(\infty, 1)$-categories has some clear advantages: equivalences between homotopy theories are directly definable (see Definition 2.9.2) instead of being presented as zigzags of DK- or Quillen equivalences. The formation of diagram categories (see Remark 2.9.6) is straightforward and homotopy limit and colimit functors become genuine adjoints (see Definition 2.9.7) and homotopy limits and colimits become genuine limits and colimits — at least in the sense appropriate to the theory of $(\infty, 1)$-categories. So from this vantage point it is natural to ask: Do we still need model categories?[24] While some might find this sort of dialog depressing, in our view it does not hurt to ask.

Chris Schommer-Pries has suggested a useful analogy to contextualize the role played by model categories in the study of homotopy theories that are complete and cocomplete:

$$\begin{cases} \text{model category} :: (\infty, 1)\text{-category} \\ \text{basis} :: \text{vector space} \\ \text{local coordinates} :: \text{manifold} \end{cases}$$

A precise statement is that combinatorial model categories present those $(\infty, 1)$-categories that are complete and cocomplete and more generally (locally) presentable; this

---

[24]  See https://mathoverflow.net/questions/78400/do-we-still-need-model-categories.

result is proven in [169, A.3.7.6] by applying a theorem of Dugger [77].[25] In general having coordinates is helpful for calculations. In particular, when working inside a particular homotopy theory as presented by a model category, you also have access to the non-bifibrant objects. For instance, the Bergner model structure of §2.6.2 is a useful context to collect results about homotopy coherent diagrams, which are defined to be maps from the cofibrant (and not typically fibrant) objects to the fibrant ones (which are not typically cofibrant).

But Quillen himself was somewhat unsatisfied with the paradigm-shifting abstract framework that he introduced, writing:

> This definition of the homotopy theory associated to a model category is obviously unsatisfactory. In effect, the loop and suspension functors are a kind of primary structure on HoM and the families of fibration and cofibration sequences are a kind of secondary structure since they determine the Toda bracket .... Presumably there is higher order structure ... on the homotopy category which forms part of the homotopy theory of a model category, but we have not been able to find an inclusive general definition of this structure with the property that this structure is preserved when there are adjoint functors which establish an equivalence of homotopy theories. [229, pp. 3–4]

Quillen was referring to a model category that is pointed, in the sense of having a zero object, like the role played by the singleton space in Top$_*$. A more modern context for the sort of stable homotopy theory that Quillen is implicitly describing is the category of spectra, the $(\infty, 1)$-category of which has many pleasant properties collected together in the notion of a *stable $\infty$-category*. We posit that these notions, which are the subject of Chapter 4 of this volume, might fulfill Quillen's dream.

---

[25] Morally, in the sense discussed in §2.3.2, all model categories are Quillen equivalent to locally presentable ones. More precisely, the result that every cofibrantly generated (in a suitable sense of this term) model category is Quillen equivalent to a combinatorial one has been proven by Raptis and Rosicky to be equivalent to a large cardinal axiom called Vopěnka's principle [254].

# 3    Stable categories and spectra via model categories

by Daniel Dugger

## 3.1    Introduction

The first popular model category of spectra was due to Bousfield–Friedlander [56], and for many years it was the only one in common use (a previous model due to K. Brown [61] never really caught on). But this category does not admit a suitable smash product on the model category level. Following an early but limited attempt by Robinson [248], in the late 1990s several new model categories of spectra appeared that fixed this problem. These days a working topologist should know a little about each of these models, and about their various advantages and disadvantages.

Here is a list of the main players:

         (1) Bousfield–Friedlander spectra
         (2) Symmetric spectra
         (3) Orthogonal spectra
         (4) EKMM spectra
         (5) $\Gamma$-spaces (which only model connective spectra)
         (6) $\mathcal{W}$-spaces (generalizing "functors with smash product")

While it would be nice to pick out one model and say *this* is the one everyone should learn, life is not that simple. An algebraic topologist is likely to encounter each of the above models at some point, and some will have advantages over others depending on the context. For example, at this point there is a developing consensus that orthogonal spectra work best for equivariant homotopy theory; but some constructions — like Waldhausen $K$-theory — naturally produce a symmetric spectrum, not an orthogonal one. Functors with smash product (FSPs) have largely disappeared from the stage, being eclipsed by (2) and (3), but they are still worth a passing familiarity. In this survey we concentrate on (1)–(4), with (5) and (6) only making a quick appearance at the end.

To describe the organization of this survey, it is helpful to use an analogy from daily life: the automobile. For most of us, an automobile is a box with wheels that has certain behaviors when we turn the steering wheel or step on the pedals. That very primitive level of understanding is sufficient for most day-to-day functioning, and it is rare that any of us have to actually look under the hood. To some extent, the same holds true of spectra. Much of daily life can be covered just by knowing that there exists a model category of spectra with a smash product satisfying a small list of basic properties. This kind of superficial knowledge is fine for driving around town, but unlike the automobile analogy my experience has been that nearly every trip on the homotopy-theory highway requires one or two stops to mess around with the engine. It bothers me that this is so, and I usually find myself cursing at the injustice when I have to do it, but this seems to be the nature of the subject.

To continue beating our analogy to death, when one *is* messing around under the hood there is simply no substitute for the technical manuals. For spectra these are [94], [133], [178], [267], and [132]. The present survey cannot replace them. Instead, we concentrate on two aims. The first is to give a kind of "driver's manual" to the world of stable model categories, monoidal model categories, and general properties that are satisfied by all the commonly used model categories of spectra. This takes roughly the first half of the chapter. The second goal is to give enough of a technical introduction to the different categories that readers can confidently go open up the manuals and feel that they have a fighting chance.

Before moving on let's state the definitions of the basic objects:

1.  A **classical spectrum** is a collection of pointed spaces $X_n$ for $n \geq 0$ together with structure maps $\sigma_n \colon S^1 \wedge X_n \to X_{n+1}$. The notion of a spectrum originated with Lima [158], but the first model structure was developed by Bousfield–Friedlander. The phrase "Bousfield–Friedlander spectra" sometimes gets used for these objects, even though the definition of the objects themselves came much earlier. They are also sometimes called "prespectra", mainly in the work of Peter May and his collaborators. A **suspension spectrum** is a spectrum where the structure maps are all identity maps, and an $\Omega$-spectrum (read "omega spectrum") is one where the adjoints $X_n \to \Omega X_{n+1}$ of the structure maps $\sigma_n$ are weak equivalences.

2.  A **symmetric spectrum** is a classical spectrum where each $X_n$ comes equipped with an action of the symmetric group $\Sigma_n$, and where each of the iterated structure maps

$$\sigma^p \colon (S^1)^{\wedge(p)} \wedge X_q \to X_{p+q}$$

    is $\Sigma_p \times \Sigma_q$-equivariant. Here $\sigma^p$ is actually a composite of associativity maps with $p$ different applications of $\sigma$, the $\Sigma_p \times \Sigma_q$-action on the domain is the evident one, and the action on the target comes from the embedding of groups $\Sigma_p \times \Sigma_q \hookrightarrow \Sigma_{p+q}$ where the image consists of permutations that permute the first $p$ elements and last $q$ elements without mixing the two blocks.

3.  An **orthogonal spectrum** is an assignment that sends each finite-dimensional real inner product space $V$ to a pointed space $X_V$ equipped with an action of the orthogonal group $O(V)$, together with structure maps $\sigma_{V,W} \colon S^V \wedge X_W \to X_{V \oplus W}$ that are $O(V) \times O(W)$-equivariant (with $S^V$ the one-point compactification of $V$). In addition, to any isometry $V \to W$ is assigned (continuously) a homeomorphism $X_V \to X_W$, and these must be compatible with all the previous structure. Finally,

the structure maps must satisfy some evident unital and associativity conditions. (If we drop the orthogonal group actions then the assignment $V \mapsto X_V$ together with the structure maps is often called a **coordinate-free spectrum**).

4. The definition of an **EKMM spectrum** cannot be given in a few lines, but the following words at least give a rough idea. An EKMM spectrum is a coordinate-free $\Omega$-spectrum where the adjoints of the structure maps are all homeomorphisms, together with an action of a certain linear isometries monad on this spectrum, and satisfying an extra "$S$-unital" condition.

5. For each $n \geq 0$ write $\mathbf{n}^+ = \{0, 1, \dots, n\}$ for the pointed set with 0 as basepoint. Let $\mathcal{F}$ be the category whose objects are all the $\mathbf{n}^+$ and whose morphisms are the based maps. A **$\Gamma$-space** is simply a functor $\mathcal{F} \to \mathcal{T}op_*$.

6. Let $\mathcal{W}$ be the category of pointed spaces homeomorphic to finite $CW$-complexes. Regard this as a category enriched over topological spaces. A **$\mathcal{W}$-space** is just an enriched functor $\Phi \colon \mathcal{W} \to \mathcal{T}op_*$. Note that for every $X$ and $Y$ there is a natural map $X \to \mathcal{T}op_*(Y, X \wedge Y)$ (adjoint to the identity); composing with the map $\mathcal{T}op_*(Y, X \wedge Y) \to \mathcal{T}op_*(\Phi(Y), \Phi(X \wedge Y))$ and taking the adjoint therefore gives a family of natural structure maps

$$X \wedge \Phi(Y) \to \Phi(X \wedge Y).$$

These maps are broad generalizations of the structure maps for classical spectra — for example, we could get a classical spectrum by setting $\Phi_n = \Phi(S^n)$ and letting $X = S^1$, or more generally by fixing $Y$ and setting $\Phi_n^Y = \Phi(S^n \wedge Y)$. The notion of $\mathcal{W}$-space is roughly equivalent to that of "simplicial functor", and the objects classically called "functors with smash product" are the monoids in this category.

*Remark 3.1.1.* What we here call "EKMM spectra" were called "$S$-modules" when first introduced, and are often still called that. Unfortunately, both symmetric spectra and orthogonal spectra are also $S$-modules, just in different contexts. So the phrase "$S$-module" is now very ambiguous, whereas "EKMM spectrum" cannot be confused with anything else.

From a historical perspective, the objects in (1) and (5) date to the 1960s and 1970s and vastly predate all of the others in the list. The objects in (2), (3), (4), and (6) all appeared in the 1990s, and their importance is that they admit a symmetric monoidal smash product on the model category level (sometimes colloquially referred to as the "point-set level"), rather than just on the associated homotopy category — see Section 3.1.3 below for more discussion of this. (The objects in (6) actually first appeared in the 1970s, but didn't enter the limelight until the 1990s with the other models).

Having such a point-set level smash product quickly led to a flurry of advances, and nowadays this is a standard part of any algebraic topologist's toolkit. But because there are four models and not just one, learning to use the toolkit also means learning what the different models do best, and how to navigate between them. The different models come with their own advantages and disadvantages, or pros and cons. These terms don't feel quite right, though, because the pros and cons are so closely linked.

If something good only happens because of something bad, is the "bad" thing really all that bad? Rather than delve into this philosophical quagmire, we take the elementary-school approach in the table below (focusing only on the three most commonly used models):

|  | Things that make us happy | Things that make us sad |
|---|---|---|
| EKMM spectra | All objects are fibrant. Weak equivalences are easy. Plays well with the linear isometries operad. | The unit is not cofibrant. Definition of the category is quite hard, with several layers of machinery. |
| Symmetric spectra | Easy definition of the objects. The unit is cofibrant. | Weak equivalences are hard to understand. Need fibrant replacement, and this can destroy other structure. One can make a theory of genuine $G$-spectra, but it feels a bit unnatural. |
| Orthogonal spectra | Works well for $G$-spectra. Unit is cofibrant. Weak equivalences are easy. Objects are not as easy as symmetric spectra, but not hard. | Need fibrant replacement. |

By "weak equivalences are easy" we mean that they coincide with the $\pi_*$-isomorphisms on the underlying classical spectrum. The issue of whether every object is fibrant has a surprisingly large simplifying effect on how one ends up handling certain monoidal phenomena — we discuss this more in Section 3.3.2.

For the rest of this introduction I am going to do something a bit unusual. Mathematical narratives tend to have two sides: one consists of the definitions and theorems, and the other is the *story* behind those definitions and theorems (sometimes called *motivation*). The latter might try to answer why a certain definition is the right one, or why a certain theorem should be expected. It is an odd phenomenon that these two sides of mathematical narration sometimes end up getting in the way of each other.

To help try to combat this, for the rest of this introduction I am going to give a series of mathematical vignettes that attempt to highlight various important issues or ideas behind the "story" of spectra. These come in no particular order, and are also by no means exhaustive. The hope is that a reader can get some basic picture from the vignettes right away, even if they don't make complete sense on first reading. Be assured that we will return to each of these ideas in more formal ways later in the text.

### 3.1.1    Why use model categories?

Let me begin by painting a picture. Somewhere up in the heavens is a wondrous paradise where lives the homotopy theory of spectra. You are welcome to think of this realm as an infinity-category if you like, but I will intentionally keep things more vague. Regardless, it is a magical shangri-la where the theories of associative and commutative ring spectra, their modules, equivariant analogs, and so forth all work out easily and naturally. The gods who walk that land are happy and content, and can do many fine things.

Most of us mortals cannot inhabit this kingdom directly, and so instead we gain limited access by choosing a *model*. As with all attempts at creating paradise down on earth, this doesn't entirely succeed. These models are not canonical, different models come with different pros and cons, and no one model seems to be completely satisfactory for everything. But such is the price we pay for our mortality. Dan Kan used to compare choosing a model to choosing coordinates on a manifold, and Jeff Smith once remarked that model categories give a way of bringing infinity-categorical phenomena down into the realm of 1-categories. These are good ways of thinking about the situation.

As one reaches for more and more sophisticated structures, any fixed model seems to inevitably run its course. Early models of spectra adequately capture the homotopy category but fail to admit a point-set-level smash product. Other models capture the smash product but fail to give an adequate theory of commutative ring spectra, or of equivariant spectra. Recent work [221] suggests that none of the existing models can handle coalgebra spectra correctly. The homotopy theorists' version of Murphy's Law is that after choosing any particular model for spectra, a topologist will eventually want to do something where the model seems to get in the way and make things harder than they should be.

This picture so far gives a somewhat skewed view, because the heavenly paradise is not always one's main goal. Down here on earth we have concrete objects like manifolds, chain complexes, and differential graded algebras, and often at the end of the day we are trying to prove theorems about these concrete things. The more one ascends into the heavens, the more blurred these objects become in their very existence. It is not always clear what infinity-categorical theorems are actually saying about our concrete objects, and this is another place where model categories turn out to be helpful. In addition to giving us a view into heavenly realms, model categories are also a tool for taking theorems from those realms and applying them down here on earth.

### 3.1.2    Where do models come from?

There is no one answer to this question, but the following schema covers very many cases. Recall that for any two objects $X$ and $Y$ in a "homotopy theory" there is a homotopy mapping space $\mathrm{hMap}(X, Y)$, well-defined up to weak homotopy equivalence. If $X$ and $X'$ are related in some homotopy-theoretic sense, then there will be some corresponding relation between $\mathrm{hMap}(X, Y)$ and $\mathrm{hMap}(X', Y)$. The simplest example is that if there is a map $X \to X'$ then there should be an associated $\mathrm{hMap}(X', Y) \to \mathrm{hMap}(X, Y)$.

If $\mathcal{C}$ is a collection of "test objects" in our homotopy theory, we can attempt to understand an object $Y$ by remembering the collection of all function spaces $\mathrm{hMap}(U, Y)$ for $U \in \mathcal{C}$. That is, we understand $Y$ by remembering how all of our test objects map into it. That's the basic idea. If there are some relations between our test objects, we should remember the corresponding relations between our mapping spaces. In this way we are attempting to model our homotopy theory as certain functions $\mathcal{C} \to \mathcal{T}op$. Often $\mathcal{C}$ will be a category, and so we actually look at functors $\mathcal{C}^{op} \to \mathcal{T}op$.

For example, the homotopy theory of spectra should have objects $S^{-n}$ for $n \geq 0$, together with equivalences $\Sigma(S^{-n}) \simeq S^{-(n-1)}$. If we take these as our test objects, then a spectrum $Y$ will be modeled by the collection of spaces $Y_n = \mathrm{hMap}(S^{-n}, Y)$ together with the relations $\Omega Y_n \simeq Y_{n-1}$. In this way we arrive at the classical definition of an $\Omega$-spectrum.

Instead of starting with the objects $S^{-n}$ we could just start with $S^{-1}$ together with the spectra $I_n = (S^{-1})^{\wedge(n)}$. The symmetric group $\Sigma_n$ acts on $I_n$, and so there will be an induced action on the function complexes $\mathrm{Map}(I_n, Y)$. This perspective leads directly to the notion of a symmetric spectrum.

Likewise, the fact that the orthogonal group $O(n)$ acts on $S^n$ might lead one to believe that it should also act on $S^{-n}$, in which case there would be an induced action of $O(n)$ on $Y_n = \mathrm{Map}(S^{-n}, Y)$. Thus one is led to the notion of an orthogonal spectrum.

### 3.1.3    The smash product

Let's go back to the most basic model of a spectrum: a collection of pointed spaces $X_n$ for $n \geq 0$ with structure maps $\sigma_n \colon S^1 \wedge X_n \to X_{n+1}$. Given spectra $X$ and $Y$, how could we make a spectrum that deserves to be called $X \wedge Y$? In level 0 there is only one thing that makes sense, which is $X_0 \wedge Y_0$. We will need a structure map $\Sigma(X_0 \wedge Y_0) \to (X \wedge Y)_1$, and there are two obvious choices: we could use $\sigma_X$ to get into $X_1 \wedge Y_0$, or we could use $\sigma_Y$ to get into $X_0 \wedge Y_1$. There is no reason for choosing one over the other, so let's randomly choose $(X \wedge Y)_1 = X_0 \wedge Y_1$. Similar reasoning leads to choices for $(X \wedge Y)_n$ for each $n$, and it's not hard to believe that we will be fine as long as we don't keep making the same choice over and over again: that is, we should make sure to use each of $\sigma_X$ and $\sigma_Y$ infinitely many times. These considerations do indeed produce a spectrum $X \wedge Y$, but because of all the choices it is far from canonical. In fact we have an uncountable collection of models for $X \wedge Y$. In the old days these were called handcrafted smash products. One can prove that they all are homotopy equivalent, thereby giving a well-defined smash product on the homotopy category, but clearly this is not a very good state of affairs. Still, this at least shows immediately that there is some kind of smash product around.

Rather than constructing $X \wedge Y$ by making these arbitrary choices, another approach is to build all the choices into the spectrum from the beginning. All the modern incarnations of the smash product involve some form of this, but let us start by exploring the most naive. We still take $(X \wedge Y)_0 = X_0 \wedge Y_0$, but now for $(X \wedge Y)_1$ we might first make the guess $(X_0 \wedge Y_1) \vee (X_1 \wedge Y_0)$. The suspension operators $\sigma_X$ and $\sigma_Y$ then take us into opposite wedge summands, which is no good, so we fix this by identifying them in an appropriate way:

$$(X \wedge Y)_1 = \text{pushout of } [X_0 \wedge Y_1 \longleftarrow S^1 \wedge (X_0 \wedge Y_0) \longrightarrow X_1 \wedge Y_0],$$

where the maps are the evident ones coming from $\sigma_Y$ and $\sigma_X$. Note that the left-pointing map must involve the twist map, used to commute the $S^1$ and the $X_0$. We leave the reader to derive the definition for $(X \wedge Y)_n$ for $n \geq 2$, along the same lines.

This definition does not give us what we want, but it is informative to see why. The first problem one encounters is that the sphere spectrum $S$ is not a unit (recall that $S$ is the suspension spectrum of $S^0$). To see this, let us compute $S \wedge S$. One readily checks that $(S \wedge S)_0 = S^0$ and $(S \wedge S)_1 = S^1$, but $(S \wedge S)_2$ is the colimit of the diagram

Replacing each parenthesized $(S^i \wedge S^j)$ in the diagram with $(X_i \wedge Y_j)$ gives the diagram for $(X \wedge Y)_2$ and helps one understand the various maps. Each map in the diagram uses associativity, twist, and the structure maps from $S$ in the evident way — for example, the left map in the bottom row commutes the second $S^1$ past the $S^0$ and then uses the structure map on the rightmost two terms. Upon analyzing these maps, one finds that they are all canonical identifications (labeled $\gamma$ in the diagram), except for one: this last map involves the twist map on $S^1$ and so ends up being $-\gamma$. Consequently, the colimit of this diagram is the coequalizer of $(id, -id): S^2 \rightrightarrows S^2$, which is $\mathbb{R}P^2$. So we see that $S \wedge S \neq S$.

*Exercise 3.1.2.* For an arbitrary spectrum $Y$, convince yourself that under the above definition $(S \wedge Y)_2$ is the colimit of the following diagram:

Working through the simple example preceding Exercise 3.1.2 already suggests the key for fixing the situation. The problem is that we are not keeping track of the "twists" that occur when we apply our structure maps, so we need to build in some machinery for doing so. This is what symmetric spectra do, by building in symmetric groups. In symmetric spectra, $(X \wedge Y)_2$ is made from $X_0 \wedge Y_2$, $X_2 \wedge Y_0$, and *two* copies of $X_1 \wedge Y_1$ (indexed by the elements of the symmetric group $\Sigma_2$), and then one quotients by the same kind of relations we saw above. This fixes the problem. See Section 3.7.2 to find this worked out in detail.

Orthogonal spectra solve the problem in an even more elegant way (though secretly it is really the same way). Here spectra are indexed on the category of finite-dimensional inner product spaces, and the direct sum operation on this category already has twist maps built into it. If $X$ is an orthogonal spectrum then $X_{V \oplus W}$ and $X_{W \oplus V}$ are different objects, though the twist $t \colon V \oplus W \to W \oplus V$ gives a homeomorphism between them. The moral here is that indexing things on inner product spaces forces one to keep track of the relevant twists in the very notation.

There is another way to see that symmetric groups should come into the picture. Let us imagine that we have a homotopy theory of spectra (off in some shangri-la) and we are attempting to model spectra $X$ by the collection of mapping spaces $X_n = \mathrm{Map}(I^{\wedge(n)}, X)$ where $I$ is a model for $S^{-1}$. We need to ask ourselves: if we have all the $\{X_n\}$ and all the $\{Y_n\}$, what is the best we can do in terms of approximating the spaces $\{(X \wedge Y)_n\}$? Clearly if $p + q = n$ we will have maps

$$\mathrm{Map}(I^{\wedge(p)}, X) \wedge \mathrm{Map}(I^{\wedge(q)}, Y) \to \mathrm{Map}(I^{\wedge(p+q)}, X \wedge Y) = \mathrm{Map}(I^{\wedge(n)}, X \wedge Y) \quad (3.1.4)$$

induced by the shangri-la smash product. However, this kind of process only gives maps $I^{\wedge(n)} \to X \wedge Y$ which send the first set of "coordinates" into $X$ and the second set into $Y$. Not all maps will look this way! Indeed, the action of $\Sigma_n$ on $I^{\wedge(n)}$ induces an action on $\mathrm{Map}(I^{\wedge(n)}, X \wedge Y)$ and lets us scramble the "coordinates" any way we want. This suggests, though, that if we use the maps in (3.1.4) *together with* a superimposed symmetric group action, then we might get a sensible approximation to $\mathrm{Map}(I^{\wedge(n)}, X \wedge Y)$. This leads us to write down the space

$$\left[ \bigvee_{p+q=n} (\Sigma_n)_+ \wedge_{\Sigma_p \times \Sigma_q} (X_p \wedge Y_q) \right] \Big/ \sim$$

as a model for $\mathrm{Map}(I^{\wedge(n)}, X \wedge Y)$, where the equivalence relation just comes from thinking about the evident ways that the maps (3.1.4) interact with symmetric group actions and the structure maps. We have just invented the smash product for symmetric spectra!

### 3.1.5     Coordinate-free spectra

The world of classical spectra provides inverses (under the smash product) for the standard spheres $S^n$. If $V$ is a finite-dimensional real vector space then its one-point compactification $S^V$ is isomorphic to $S^{\dim V}$, and so of course $S^V$ has an inverse in this world as well. But this inverse is not canonical, because the isomorphism $V \cong \mathbb{R}^{\dim V}$ is not canonical. This might seem like a small point, but in some constructions (like Pontryagin–Thom) it is very convenient to have a canonical inverse for $S^V$.

A larger issue arises in the setting of $G$-equivariant homotopy theory. Here one has different spheres $S^V$ for each finite-dimensional $G$-representation $V$, so to introduce inverses for these it is not enough to just work with the standard spheres $S^n$. Thus, for various reasons we are led to the need for a notion of "coordinate-free" spectra.

The first idea of what a coordinate-free spectrum should be is an assignment $V \mapsto X_V$ that sends every finite-dimensional vector space to a pointed space. For $V \subseteq W$ there should be structure maps $S^{??} \wedge X_V \to X_W$, but already one runs into trouble as far as what sphere to put in the domain. This sphere should be related to the complement of $V$ in $W$, but there is no canonical such complement. To get around this, we assume that the vector spaces have inner products on them so that we can take orthogonal complements. If $W - V$ denotes the orthogonal complement of $V$ in $W$, then our structure map should have the form $S^{W-V} \wedge X_V \to X_W$.

Finally, since the collection of *all* finite-dimensional inner product spaces is not a set, we prefer to set things up so that there is an intrinsic bound to where these live — an underlying "universe". To be precise, define a **May universe** to be a real inner product space of countably infinite dimension. Any universe $\mathcal{U}$ is isometric to $\mathbb{R}^\infty$ with the dot product, but not canonically. Then a **coordinate-free spectrum on** $\mathcal{U}$ is defined to be an assignment $V \mapsto X_V$ for finite-dimensional $V \subseteq \mathcal{U}$, together with maps $S^{W-V} \wedge X_V \to X_W$ for every pair $V \subseteq W \subseteq \mathcal{U}$. These must satisfy some evident unital and associativity conditions.

Example 3.1.3. The definitions of some familiar classical spectra immediately generalize to give coordinate-free spectra:

(a) The sphere spectrum is $V \mapsto S^V$.
(b) If $A$ is an abelian group, the Eilenberg–MacLane spectrum $HA$ is the spectrum $V \mapsto C(S^V; A)$ where for any pointed space $X$ the space $C(X; A)$ is the Dold–Thom space of finite configurations of points on $X$ labeled by elements of $A$.
(c) The real cobordism spectrum $MO$ is $V \mapsto \mathrm{Th}(EO(V) \times_{O(V)} V \to BO(V))$, where $O(V)$ is the group of isometries of $V$ (with its natural topology) and $\mathrm{Th}(E \to B)$ is the Thom space. This is also commonly written in the form $V \mapsto EO(V)_+ \wedge_{O(V)} S^V$.

For orthogonal spectra, it is important that we are able to form the direct sum of our inner product spaces. That is to say, if $X$ is an orthogonal spectrum we need $X_{V \oplus W}$ to make sense when $X_V$ and $X_W$ do. For this reason we cannot restrict ourselves to subspaces of a universe $\mathcal{U}$ anymore. To avoid set-theoretical issues we must either fix a small skeletal subcategory of the category of finite-dimensional inner product spaces, or else fix some Grothendieck universe at the very beginning. See Remark 3.5.4 for more details.

## 3.1.6    Rings, modules, and algebras

Let $(\mathcal{C}, \otimes, S)$ be a symmetric monoidal category. A monoid in $\mathcal{C}$ is an object $R$ together with a unit map $S \to R$ and a product $R \otimes R \to R$ satisfying the evident associativity and unital actions. A monoid in $(\mathcal{A}b, \otimes, \mathbb{Z})$ is just a ring, and for this reason we will sometimes call monoids in other symmetric monoidal categories "rings" as well.

If $R$ is a ring in $\mathcal{C}$ then one likewise has notions of left and right $R$-modules, and if $R$ is a commutative ring then one can talk about $R$-algebras. The definitions are all the obvious ones.

In the 1970s, after Boardman had constructed the symmetric monoidal structure on Ho($Spectra$), one could apply the above ideas and talk about ring- and module-spectra. Nowadays these would probably be called "homotopy ring spectra", or "naive ring spectra", to differentiate them from more rigid notions. Suppose that $R$ is one of these homotopy ring spectra and that $f : M \to N$ is a map of left $R$-modules. One would like for the homotopy cofiber $Cf$ to be again a left $R$-module in a canonical way, but this doesn't work out. Try it: there is a diagram in the homotopy category that looks like

$$
\begin{array}{ccc}
R \wedge M \longrightarrow R \wedge N \longrightarrow R \wedge Cf \\
\downarrow \qquad\qquad \downarrow \\
M \longrightarrow N \longrightarrow Cf
\end{array}
$$

and both rows are homotopy cofiber sequences, so there does indeed exist an extension $\mu : R \wedge Cf \to Cf$ (apply $[-, Cf]$ to the top cofiber sequence and use the resulting long exact sequence). However, the homotopy class of $\mu$ is not unique and moreover one cannot prove that $\mu$ satisfies the necessary associativity condition.

So this is a deficiency. Using the naive definitions of rings and modules in Ho($Spectra$) does not lead to a situation where we can do homotopy theory for $R$-modules. The problem is the usual one: the homotopy category itself is not robust enough for most purposes. The above problem with cofibers is coming from the fact that the homotopy category doesn't have colimits.

This was one of the motivations for desiring a symmetric monoidal smash product on the model category level. Assuming that one has a model category $Spectra$ with a smash product that commutes with colimits in either variable, it follows at once that colimits of left $R$-modules are again left $R$-modules in a canonical way. One would hope that the adjoint functors

$$R \wedge (-) \colon Spectra \rightleftarrows R\text{–Mod} \colon U$$

would lift the model category structure on $Spectra$ to a corresponding model structure on the category of left $R$-modules. Similarly, if $R$ is a commutative ring spectrum then one might hope for a model category structure on $R$-algebras, and also one on commutative $R$-algebras.

In short, the hope would be that the model structure on $Spectra$ could be passed to various categories of algebraic structures on spectra. This basically works out, but it doesn't work out for free. One approach was developed in [94] for topological model categories where all objects are fibrant, which reduced things down to their so-called "Cofibration Hypothesis". For more general model categories another approach was developed by Schwede–Shipley [267], who identified the need for a separate axiom they called the "Monoid Axiom". The Monoid Axiom is one of those things that is safely left under the hood on regular days, but that one needs to be prepared to play with when the car breaks down.

We discuss the Monoid Axiom and its applications to model categories of modules and algebras in Section 3.3.2.

## 3.1.7    The Lewis enigma

In 1991, before the advent of the modern categories of spectra, Lewis discovered an argument showing that some of the expected properties of such categories were mutually inconsistent [156]. It is worth understanding this argument not only to see how the modern categories of spectra interface with it, but also because this same argument explains some of the complications in various theories of commutative ring spectra.

Let $\mathcal{S}$ be a category with the following properties:

(A1) There exists a symmetric monoidal functor $\wedge \colon \mathcal{S} \times \mathcal{S} \to \mathcal{S}$.
(A2) There exists an adjoint pair $\Sigma^{\infty} \colon \mathcal{T}op_* \rightleftarrows \mathcal{S} \colon \Omega^{\infty}$.
(A3) There is a natural transformation

$$\eta_{X,Y} \colon \Sigma^{\infty}(X \wedge Y) \to \Sigma^{\infty} X \wedge \Sigma^{\infty} Y$$

that is compatible with the associativity and commutativity isomorphisms for $(\mathcal{T}op_*, \wedge)$ and $(\mathcal{S}, \wedge)$.
(A4) $\Sigma^{\infty} S^0$ is the unit for $\wedge$, and $\eta$ is compatible with the unital isomorphism.
(A5) There is a natural weak equivalence $\Omega^{\infty} \Sigma^{\infty} X \simeq QX$, where as usual one defines $QX = \text{hocolim}_n \Omega^n \Sigma^n X$.

Putting $X = \Omega^{\infty} E$ and $Y = \Omega^{\infty} F$ into (A3) and using the counit of the adjunction gives a natural transformation $\epsilon_{E,F} \colon \Omega^{\infty} E \wedge \Omega^{\infty} F \to \Omega^{\infty}(E \wedge F)$, and this will also be compatible with the associativity and commutativity isomorphisms.

Given such a category, set $S = \Sigma^{\infty} S^0$. The unit isomorphism $S \wedge S \to S$ makes $S$ into a commutative ring spectrum. Then $\epsilon \colon \Omega^{\infty} S \wedge \Omega^{\infty} S \to \Omega^{\infty} S$ makes $\Omega^{\infty} S$ into a commutative monoid. So its identity component is a generalized Eilenberg-MacLane space. But this contradicts (A5), which says $\Omega^{\infty} S = \Omega^{\infty} \Sigma^{\infty} S^0 \simeq QS^0$. So the conclusion is that (A1)–(A5) are mutually incompatible.

Symmetric and orthogonal spectra satisfy (A1)–(A4), but get around the problem via the failure of (A5). Here $\Sigma^{\infty} S^0$ is not fibrant and so $\Omega^{\infty} \Sigma^{\infty} S^0$ has the "wrong" homotopy type; said differently, (A5) must be modified to say that $\Omega^{\infty} \mathcal{F} \Sigma^{\infty} X \simeq QX$, where $\mathcal{F}$ is a fibrant replacement functor.

The EKMM setup gets around this problem by having two sets of adjoint functors, called here $(\Sigma_S^{\infty}, \Omega_S^{\infty})$ and $(\Sigma^{\infty}, \Omega^{\infty})$ (see Section 3.9 for more details). There is a natural transformation $\Sigma_S^{\infty} \to \Sigma^{\infty}$ that is a weak equivalence on cofibrant pointed spaces, and there is its adjoint $\Omega^{\infty} \to \Omega_S^{\infty}$. The pair $(\Sigma_S^{\infty}, \Omega_S^{\infty})$ is the one with homotopical meaning (it turns out to be a Quillen pair, with the right model category structures), whereas $(\Sigma^{\infty}, \Omega^{\infty})$ is the one with the good monoidal properties. So $\Sigma^{\infty}$ satisfies (A3) and (A4), but $\Omega^{\infty} \Sigma^{\infty}$ does not satisfy (A5); whereas $\Omega_S^{\infty} \Sigma_S^{\infty}$ satisfies (A5), but $\Sigma_S^{\infty}$ does not satisfy (A3) and (A4).

Returning to the simpler setting of symmetric spectra, replacing (A5) with its derived version is not the end of the story. Even with this modified (A5), Lewis's argument shows that if $R$ is a fibrant spectrum with a commutative and associative product then $\Omega^{\infty} R$ (which is already appropriately derived) must be a generalized

Eilenberg–MacLane space. This is obviously a matter of concern, since we would like spectra such as $S$, $K$, $MO$, and $MU$ to have models which are commutative ring spectra on the nose. That is not prohibited, but such models cannot *also* be fibrant in the usual model structure for symmetric (or orthogonal) spectra. The standard way for dealing with this is to use a different model structure called the **positive model category structure**. We will discuss this briefly in Section 3.10.5.

### 3.1.8    Organization of the chapter

We assume a basic familiarity with model categories, as provided by sources like [91], [124], [130], and [229]. See also Chapter 2 of this volume. Specifically, we assume the reader is familiar with the model category axioms, cylinder and path objects, the homotopy category, Quillen functors, derived functors, the small object argument, simplicial model categories, and the notion of cofibrant-generation.

We occasionally assume the reader has a passing acquaintance with the classical aspects of spectra and their connection to (co)homology theories, as represented for example in any of [1], [2, Part III], and [288].

We also assume the reader has a basic knowledge of closed symmetric monoidal categories; MacLane's book [174] is a good source. Finally, we use enriched categories to a certain extent. Not much more is needed than the basic definition and the notion of enriched functor, which are essentially obvious; but consult [148] for any needed background here.

With homotopy-theoretic machinery, there is the usual issue of whether to take as foundation simplicial sets or topological spaces. For the most part we have tried to present results in a way that applies to either situation, but this is not always convenient. To avoid having to constantly work in two situations at once, we choose topological spaces as our main framework. The reader who prefers to work simplicially should be able to make the necessary modifications to the exposition with little trouble.

### 3.1.9    Notation and terminology

When $C$ is a category we write $C(X, Y)$ for $\mathrm{Hom}_C(X, Y)$. If $C$ is a category enriched over some symmetric monoidal category $\mathcal{V}$, we write $\underline{C}(X, Y)$ for the corresponding $\mathcal{V}$-mapping object. We write $\mathcal{T}op_*$ for the category of pointed topological spaces. We fix $S^1 = I/\partial I$ and define $S^n = S^1 \wedge (S^1 \wedge (S^1 \wedge (\cdots \wedge S^1)))$.

### 3.1.10    Acknowledgments

The author would like to thank the editors for numerous useful comments, and Andrew Blumberg in particular for assistance with some questions about EKMM spectra. The author is especially grateful to Mike Mandell, who gave a "cursory" reading that somehow produced several pages of valuable suggestions.

## 3.2     Stable model categories

A model category is called stable when the suspension functor is a self-equivalence on the homotopy category. The homotopy categories of stable model categories enjoy several nice properties: they are additive, triangulated, and the notions of homotopy cofiber and fiber sequences are the same. These simply stated facts take a nontrivial amount of effort to set up and prove carefully. Most of Chapters 6 and 7 of [130] are devoted to this. We aim to give a quick tour for those who are new to this machinery, partly because the depth of the results in [130] make them a bit of a maze. We hope the treatment here can serve as a guide through that material.

A category $\mathcal{M}$ is called **pointed** if it has an initial object, a terminal object, and the two are isomorphic. Quillen [229, Chapter I.2] showed that if $\mathcal{M}$ is a pointed model category then the homotopy category $\mathrm{Ho}(\mathcal{M})$ comes equipped with a special pair of adjoint functors

$$\Sigma \colon \mathrm{Ho}(\mathcal{M}) \rightleftarrows \mathrm{Ho}(\mathcal{M}) \colon \Omega,$$

called **suspension** and **loop** functors. If $X$ is a cofibrant object, factor $X \to *$ as $X \rightarrowtail CX \xrightarrow{\sim} *$. Then $\Sigma X$ can be defined to be the pushout of $* \leftarrow X \to CX$. Likewise, if $Z$ is a fibrant object then factor $* \to Z$ as $* \xrightarrow{\sim} PZ \twoheadrightarrow Z$ and define $\Omega Z$ as the pullback of $* \to Z \leftarrow PZ$. It is easy to show that these homotopy types do not depend on the choice of $CX$ or $PZ$, and moreover that these definitions extend to give the desired functors. (Note that "$C$" and "$P$" stand for "cone" and "path object").

Let $X$ be cofibrant and consider the diagram

$$
\begin{array}{ccccc}
CX & \longleftarrow\!\!\!\prec & X & \succ\!\!\!\longrightarrow & CX \\
{\scriptstyle \simeq}\downarrow & & \| & & \| \\
* & \longleftarrow & X & \succ\!\!\!\longrightarrow & CX
\end{array}
$$

Taking pushouts gives a map $CX \amalg_X CX \to \Sigma X$, and the model category axioms force this to be a weak equivalence (see [232, Corollary to Theorem B]). But collapsing $X$ gives $CX \amalg_X CX \to \Sigma X \vee \Sigma X$, and so we have constructed a map $\Sigma X \to \Sigma X \vee \Sigma X$ in $\mathrm{Ho}(\mathcal{M})$. A little work shows that this makes $\Sigma X$ into a cogroup object in $\mathrm{Ho}(\mathcal{M})$, and that $\Sigma^2 X$ is a cocommutative cogroup object. Similarly, when $Y$ is fibrant, $\Omega Y$ is a group object in $\mathrm{Ho}(\mathcal{M})$ and $\Omega^2 Y$ is a commutative group object. It follows that $[\Sigma^2 X, Z]$ and $[A, \Omega^2 Y]$ have natural structures of abelian groups, where from now on we will write $[-,-]$ for maps in $\mathrm{Ho}(\mathcal{M})$.

**Definition 3.2.1.** A pointed model category $\mathcal{M}$ is called **stable** if the suspension functor $\Sigma \colon \mathrm{Ho}(\mathcal{M}) \to \mathrm{Ho}(\mathcal{M})$ is an equivalence of categories.

The $(\Sigma, \Omega)$ adjunction shows that it is equivalent to require that $\Omega$ be an equivalence. Moreover, when $\mathcal{M}$ is stable the functors $\Sigma$ and $\Omega$ will be inverses. The following is an easy exercise:

**Proposition 3.2.2.** *Let $\mathcal{M}$ be a pointed model category. The following conditions are equivalent:*

(a) $\mathcal{M}$ *is stable.*

(b) *For all objects* $X$ *and* $Y$ *the maps* $\Sigma\Omega X \to X$ *and* $Y \to \Omega\Sigma Y$ *are isomorphisms in* $\mathrm{Ho}(\mathcal{M})$.

If $\mathcal{M}$ is a stable model category then every object in $\mathrm{Ho}(\mathcal{M})$ is a double suspension (and a double loop space), and so the hom sets are all abelian groups and composition is additive in both variables. The homotopy category inherits coproducts and products from $\mathcal{M}$, so $\mathrm{Ho}(\mathcal{M})$ is additive. In particular, it follows formally that the canonical map $i\colon A \vee B \to A \times B$ is an isomorphism in $\mathrm{Ho}(\mathcal{M})$. We recall the brief proof: If $j_A\colon A \to A \vee B$ and $\pi_A\colon A \times B \to A$ are the canonical inclusions and projections, then $j_A\pi_A + j_B\pi_B$ is a two-sided inverse to $i$.

When $\mathcal{M}$ is a pointed model category Quillen also showed that $\mathrm{Ho}(\mathcal{M})$ comes equipped with special "triangles" called homotopy fiber and cofiber sequences. An **$\Omega$-triangle** is a diagram $\Omega C \to A \to B \to C$ in $\mathrm{Ho}(\mathcal{M})$ such that the composition of any two maps is zero, and a **$\Sigma$-triangle** is a diagram $A \to B \to C \to \Sigma A$ with the same property. A map of $\Omega$-triangles is a commutative diagram

$$\begin{array}{ccccccc} \Omega C & \longrightarrow & A & \longrightarrow & B & \longrightarrow & C \\ \downarrow{\scriptstyle\Omega h} & & \downarrow{\scriptstyle f} & & \downarrow{\scriptstyle g} & & \downarrow{\scriptstyle h} \\ \Omega C' & \longrightarrow & A' & \longrightarrow & B' & \longrightarrow & C' \end{array}$$

and an isomorphism of $\Omega$-triangles is a map where all the vertical maps are isomorphisms. We use similar notions for maps and isomorphisms of $\Sigma$-triangles.

*Exercise 3.2.3.* Check that changing the signs of two maps in an $\Omega$-triangle (or $\Sigma$-triangle) produces an isomorphic triangle.

If $p\colon X \twoheadrightarrow Y$ is a fibration between fibrant objects, there exists a lifting in the square

$$\begin{array}{ccc} * & \longrightarrow & X \\ {\scriptstyle\simeq}\downarrow & \nearrow{\scriptstyle\lambda} & \downarrow \\ PY & \longrightarrow & Y \end{array}$$

and therefore an induced map $\Omega Y \to F$, where $F$ is the fiber of $X \twoheadrightarrow Y$. We leave it as an exercise to check that a different choice for $\lambda$ gives the same map $\Omega Y \to F$ in $\mathrm{Ho}(\mathcal{M})$. The $\Omega$-triangle $\Omega Y \to F \to X \to Y$ is called the **homotopy fiber sequence** corresponding to $p$. More generally:

Definition 3.2.4. An $\Omega$-triangle is called a **homotopy fiber sequence** if it is isomorphic to the homotopy fiber sequence corresponding to some fibration between fibrant objects $p\colon X \to Y$.

*Remark 3.2.5.* It is a common abuse of terminology to say things like "$F \to X \to Y$ is a homotopy fiber sequence", leaving the map $\Omega Y \to F$ implicit.

We leave the reader to write down the dual notion of a homotopy cofiber sequence, which yields a special class of $\Sigma$-triangles.

*Remark 3.2.6.* In addition to the map $\Omega F \to Y$ we constructed above, one can show that there is a map $\gamma \colon \Omega F \times Y \to Y$ giving an action of $\Omega F$ on $Y$ in $\mathrm{Ho}(\mathcal{M})$. Our map $\Omega F \to Y$ is the restriction of $\gamma$ along $\Omega F \times * \to \Omega F \times Y$. The notion of "homotopy fiber sequence" should really include this map $\gamma$ as part of the data. But when $\mathcal{M}$ is stable $\Omega F \vee Y \to \Omega F \times Y$ is an equivalence, and the restriction of $\gamma$ to the $Y$ summand is just the identity. So in this case there is no more information in $\gamma$ than in our map $\Omega F \to Y$. We refer to [130, Chapter 6.3] or [229, Chapter I.3] for careful studies of homotopy fiber and cofiber sequences in the unstable setting.

From now on assume that $\mathcal{M}$ is stable. The first result about homotopy cofiber and fiber sequences is the following:

**Proposition 3.2.7.** *Let $\mathcal{M}$ be a stable model category and let $T$ be any object.*

(a) *For any homotopy fiber sequence $\Omega Y \to F \to X \to Y$, the induced sequence of abelian groups*

$$[T, \Omega Y] \to [T, F] \to [T, X] \to [T, Y]$$

*is exact at the two middle spots.*

(b) *For any homotopy cofiber sequence $A \to B \to C \to \Sigma A$, the induced sequence of abelian groups*

$$[\Sigma A, T] \to [C, T] \to [B, T] \to [A, T]$$

*is exact at the two middle spots.*

If $X \xrightarrow{f} Y \xrightarrow{g} Z \xrightarrow{h} \Sigma X$ is a homotopy cofiber sequence, we get associated maps

$$\Omega Z \xrightarrow{\Omega h} \Omega \Sigma X \cong X \quad \text{and} \quad Y \xrightarrow{g} Z \cong \Sigma \Omega Z,$$

where the two isomorphisms are the unit and counit of the $\Sigma - \Omega$ adjunction. One might expect the evident sequence $\Omega Z \to X \to Y \to \Sigma \Omega Z$ made from these maps to be a homotopy cofiber sequence, but this is not correct — there is a sign issue. To get a homotopy cofiber sequence one must negate one of the maps.

The following proposition gives several results of this form. Rather than give names to all the maps, we adopt the convention that a minus sign by itself means "take the negative of the evident map one would get by using $\Sigma$, $\Omega$, and the adjunctions".

**Proposition 3.2.8.** *Let $\mathcal{M}$ be a stable model category.*

(a) *Given a diagram in $\mathrm{Ho}(\mathcal{M})$ of the form*

$$
\begin{array}{ccccccc}
A & \longrightarrow & B & \longrightarrow & C & \longrightarrow & \Sigma A \\
\downarrow & & \downarrow & & & & \downarrow{\scriptstyle -} \\
\Omega Z & \longrightarrow & X & \longrightarrow & Y & \longrightarrow & Z
\end{array}
$$

*in which the top row is a homotopy cofiber sequence and the bottom row is a homotopy fiber sequence, there is a map $C \to Y$ making the diagram commute.*

(b) *Given a diagram in* $\text{Ho}(\mathcal{M})$ *of the form*

$$
\begin{array}{ccccccc}
A & \longrightarrow & B & \longrightarrow & C & \longrightarrow & \Sigma A \\
\downarrow & & \downarrow & & \downarrow & & \downarrow^{-} \\
\Omega Z & \longrightarrow & X & \longrightarrow & Y & \longrightarrow & Z
\end{array}
$$

*in which the top row is a homotopy cofiber sequence and the bottom row is a homotopy fiber sequence, there is a map* $B \to X$ *making the diagram commute.*

(c) *Given a diagram in* $\text{Ho}(\mathcal{M})$ *of the form*

$$
\begin{array}{ccccccc}
A & \longrightarrow & B & \longrightarrow & C & \longrightarrow & \Sigma A \\
\downarrow & & \downarrow & & \downarrow & & \downarrow \\
A' & \longrightarrow & B' & \longrightarrow & C' & \longrightarrow & \Sigma A'
\end{array}
$$

*in which both rows are homotopy cofiber sequences, there is a map* $C \to C'$ *making the diagram commute. The dual statement for homotopy fiber sequences holds as well.*

(d) *If any of the following* $\Sigma$*-triangles are homotopy cofiber sequences, then so are the others:*

(i) $X \longrightarrow Y \longrightarrow Z \longrightarrow \Sigma X,$      (ii) $Y \longrightarrow Z \longrightarrow \Sigma X \dashrightarrow \Sigma Y$

(iii) $\Sigma X \longrightarrow \Sigma Y \longrightarrow \Sigma Z \dashrightarrow \Sigma^2 X,$      (iv) $\Omega Z \dashrightarrow X \longrightarrow Y \longrightarrow \Sigma \Omega Z.$

(e) *If any of the following* $\Omega$*-triangles are homotopy fiber sequences, then so are the others:*

(i) $\Omega Z \longrightarrow X \to Y \longrightarrow Z,$      (ii) $\Omega Y \dashrightarrow \Omega Z \longrightarrow X \longrightarrow Y,$

(iii) $\Omega^2 Z \dashrightarrow \Omega X \longrightarrow \Omega Y \longrightarrow \Omega Z,$      (iv) $\Omega \Sigma X \dashrightarrow Y \longrightarrow Z \longrightarrow \Sigma X.$

Reading this extensive list of results is a bit tedious, but having it around is very useful. It captures several of the main points from [130, Chapter 6]. A good (but challenging) exercise is to try to prove all of these facts from first principles yourself. If you get stuck, parts (a) and (b) are the content of [130, Proposition 6.3.7], and (c) is [130, Proposition 6.3.5]. The equivalence of (i) and (ii) in parts (d,e) is covered in [130, Proposition 6.3.4], and the equivalence with (iii) comes from repeatedly applying (i) $\Longleftrightarrow$ (ii) and using Exercise 3.2.3. Finally, the equivalence with (iv) is an easy exercise using the other parts.

*Remark 3.2.9.* Although it is necessary to get the signs right in cofiber or fiber sequences, in practice one almost always passes at some point to a long exact sequence of homotopy classes. In these long exact sequences, one can indiscriminately alter the signs on the maps without changing exactness. This is why one can sometimes get away with a cavalier attitude about some of these sign issues.

Part (c) of the following result is a lynchpin of the theory of stable model categories. It is often phrased colloquially as saying that in a stable model category the classes of homotopy fiber sequences and homotopy cofiber sequences are the same. We include the proof here because of the key nature of the result, and because it takes a bit of work to extract it from [130].

*Proposition 3.2.10. Let* $\mathcal{M}$ *be a stable model category.*

(a) *If $X \to Y \to Z \to \Sigma X$ is a homotopy cofiber sequence and $T$ is any object, then*

$$[T,X] \to [T,Y] \to [T,Z] \to [T,\Sigma X]$$

*is exact in the middle two spots.*

(b) *More generally, given a homotopy cofiber sequence $X \xrightarrow{f} Y \xrightarrow{g} Z \xrightarrow{h} \Sigma X$ and an object $T$,*

$$\cdots \to [T,\Omega Y] \to [T,\Omega Z] \to [T,X] \to [T,Y] \to [T,Z] \to [T,\Sigma X] \to \cdots$$

*is a long exact sequence, where each map is the obvious one obtained by applying $\Sigma$ and $\Omega$ to $f$, $g$, or $h$ and (if necessary) using the unit and counit of the adjunction.*

(c) *The triangle $\Omega Z \longrightarrow X \longrightarrow Y \longrightarrow Z$ is a homotopy fiber sequence if and only if $\Omega Z \longrightarrow X \longrightarrow Y \xrightarrow{-} \Sigma\Omega Z$ is a homotopy cofiber sequence, or equivalently if and only if $X \longrightarrow Y \longrightarrow Z \longrightarrow \Sigma X$ is a homotopy cofiber sequence.*

*Proof.* Denote the maps by $X \xrightarrow{f} Y \xrightarrow{g} Z \xrightarrow{h} \Sigma X$. For (a), suppose $u: T \to Y$ is such that $gu = *$ (we work always in the homotopy category). Rotate the cofiber sequence and construct the following diagram:

$$
\begin{array}{ccccccc}
Y & \xrightarrow{g} & Z & \xrightarrow{h} & \Sigma X & \xrightarrow{-\Sigma f} & \Sigma Y \\
\uparrow{\scriptstyle u} & & \uparrow & & & & \uparrow{\scriptstyle \Sigma u} \\
T & \longrightarrow & * & \longrightarrow & \Sigma T & \xrightarrow{id} & \Sigma T
\end{array}
$$

Both rows are homotopy cofiber sequences, so by Proposition 3.2.8(c) there is a fill-in $v: \Sigma T \to \Sigma X$. But $\Sigma: [T,X] \to [\Sigma T, \Sigma X]$ is an isomorphism, so let $\bar{v}$ be a preimage of $v$. Then $f \circ \bar{v} = -u$, so $-\bar{v}$ is the desired lift of $u$ in our sequence. Exactness at $[T,Z]$ can be proven by rotating the homotopy cofiber sequence and then applying what we just proved.

Part (b) is a direct consequence of (a) and stability. We can iteratively rotate the homotopy cofiber sequence to get the Puppe sequence

$$X \longrightarrow Y \longrightarrow Z \longrightarrow \Sigma X \xrightarrow{-} \Sigma Y \xrightarrow{-} \Sigma Z \xrightarrow{-} \Sigma^2 X \longrightarrow \cdots$$

(where each four terms are a homotopy cofiber sequence), and then apply $[T,-]$. But we can also apply $[\Sigma T,-]$ and then use both adjunction and stability to rewrite this as

$$[T,\Omega X] \to [T,\Omega Y] \to [T,\Omega Z] \to [T,X] \to \cdots$$

Similarly, we repeatedly extend the long exact sequence to the left by applying $[\Sigma^N T,-]$ to our Puppe sequence. The signs can be neglected because leaving them off does not change exactness.

For (c) we just prove one direction as the other is similar. Assume given that $\Omega Z \longrightarrow X \longrightarrow Y \xrightarrow{-} \Sigma\Omega Z$ is a homotopy cofiber sequence. Let $\Omega Z \to F \to Y \to Z$ be a homotopy fiber sequence and consider the diagram

$$
\begin{array}{ccccccc}
\Omega Z & \longrightarrow & X & \longrightarrow & Y & \xrightarrow{-} & \Sigma\Omega Z \\
\downarrow{\scriptstyle id} & & & & \downarrow{\scriptstyle id} & & \downarrow{\scriptstyle -} \\
\Omega Z & \longrightarrow & F & \longrightarrow & Y & \longrightarrow & Z
\end{array}
$$

By Proposition 3.2.8(b) there is a fill-in $u: X \to F$. Now let $T$ be any object and consider the diagram below:

$$
\begin{array}{ccccccccc}
[T,\Omega Y] & \longrightarrow & [T,\Omega Z] & \longrightarrow & [T,X] & \longrightarrow & [T,Y] & \overset{-}{\longrightarrow} & [T,\Sigma\Omega Z] \\
\downarrow{\scriptstyle id} & & \downarrow{\scriptstyle id} & & \downarrow{\scriptstyle u_*} & & \downarrow{\scriptstyle id} & & \downarrow{\scriptstyle -}{\scriptstyle\cong} \\
[T,\Omega Y] & \longrightarrow & [T,\Omega Z] & \longrightarrow & [T,F] & \longrightarrow & [T,Y] & \longrightarrow & [T,Z]
\end{array}
$$

Here we have mostly just applied $[T,-]$ to our diagram in $\mathrm{Ho}(\mathcal{M})$, but we have used (b) to extend the top sequence to the left by one term. The top row is exact by (b), and the bottom row is exact by Proposition 3.2.7(a). The Five Lemma then implies that $u_*$ is an isomorphism. Since this holds for all $T$ we conclude that $u$ itself was an isomorphism.

Finally, consider the commutative diagram

$$
\begin{array}{ccccccc}
\Omega Z & \longrightarrow & X & \longrightarrow & Y & \longrightarrow & Z \\
\downarrow{\scriptstyle id} & & \downarrow{\scriptstyle u} & & \downarrow{\scriptstyle id} & & \downarrow{\scriptstyle id} \\
\Omega Z & \longrightarrow & F & \longrightarrow & Y & \longrightarrow & Z
\end{array}
$$

The bottom row was a homotopy fiber sequence by construction, and $u$ is an isomorphism, so the top row is a homotopy fiber sequence as well.

For the last statement in (c), use Proposition 3.2.8(d).    □

We refer the reader to [297, Chapter 10.2] for the axioms of a triangulated category. The culmination of the above line of work is the following:

**Proposition 3.2.11.** *Let $\mathcal{M}$ be a stable model category. Then the suspension functor and the class of homotopy cofiber sequences make $\mathrm{Ho}(\mathcal{M})$ into a triangulated category.*

*Proof.* Axiom TR1 is routine, and TR2 is Proposition 3.2.8(d). Axiom TR3 is Proposition 3.2.8(c). So the only part that requires additional work is TR4, the Octahedral Axiom. The main point of this final axiom is to relate the homotopy cofiber sequence for a composition $fg$ to the homotopy cofiber sequences for $f$ and $g$. The reader can find a proof of this axiom (in the unstable version) in [130, Proposition 6.3.6].    □

## 3.3    Monoidal machinery

This section concerns categorical (and model categorical) material that is not specific to the theory of spectra, mostly centering around monoidal structures. We survey some basic facts about monoidal categories and monoidal model categories, as well as invertible objects.

### 3.3.1    Sufficiently combinatorial model categories

A common issue in model categories is that one wants to take a model structure on a given category $\mathcal{M}$ and produce an associated model structure on a related category

$\mathcal{M}'$. The first example is where $\mathcal{M}'$ is diagrams (of a fixed shape) inside of $\mathcal{M}$, but we will see others as well. There are almost no general theorems along these lines; in most cases some extra structure is required on $\mathcal{M}$ or $\mathcal{M}'$ or both. These structures typically take the form of sets of generating maps where the domains and codomains satisfy certain smallness properties — whatever one needs to run the small object argument.

The first notion of this type is that of a *cofibrantly generated* model category; see [124]. This notion works well for some purposes, but is too weak for others. Later notions are that of a *cellular model category* (also in [124]), and Jeff Smith's notion of a *combinatorial model category*. A combinatorial model category is one that is cofibrantly generated and where the underlying category is locally presentable; see [35] and [77] for written accounts. The combinatorial setting is especially appealing, because here all objects are small (with respect to large enough cardinals) and this property passes to most associated categories.

Most model categories built in some way starting from $s\mathcal{S}et$ or $\mathcal{T}op$ are cofibrantly generated, and the ones built from $s\mathcal{S}et$ are almost all combinatorial. Jeff Smith observed that one can make combinatorial forms of $\mathcal{T}op$-based model categories by replacing $\mathcal{T}op$ with the category of $\Delta$-generated spaces.

In this chapter we will sometimes want to phrase results in a way that applies both to categories of spectra based on simplicial sets and those made from topological spaces. The safe thing is to always assume the categories in question are combinatorial, but this does not apply to the category of compactly generated spaces used in [94]. To cut the Gordian knot, we will use the phrase **sufficiently combinatorial** as an intentionally imprecise stand-in for "assume enough hypotheses so that the smallness conditions necessary for the arguments actually work".

## 3.3.2    Monoids and models

Let $(\mathcal{M}, \otimes, I)$ be a monoidal category ($I$ is the unit). Recall that a monoid in this category is an object $R$ together with unit map $I \to R$ and multiplication $R \otimes R \to R$ satisfying the evident axioms. The monoids in $(\mathcal{A}b, \otimes, \mathbb{Z})$ are usually called rings, and in stable homotopy contexts the monoids are often called rings as well. For this reason we will use the word "ring" as a synonym for "monoid", although the latter is really the correct term.

If $R$ is a ring in $\mathcal{M}$, a left $R$-module is an object $X$ together with a map $R \otimes X \to X$ satisfying the evident axioms. One similarly defines right-modules and bimodules. By convention, we mean "left $R$-module" whenever we say "$R$-module" without qualification. Recall that if $M$ is a right $R$-module and $N$ is a left $R$-module then one defines $M \otimes_R N$ to be the coequalizer (if it exists) of the two action maps $M \otimes R \otimes N \rightrightarrows M \otimes N$.

When $\mathcal{M}$ is a symmetric monoidal category we can talk about commutative rings in $\mathcal{M}$, and for such rings there is an evident way of turning any left module into a right module, and vice versa. If $R$ is a commutative ring then we define an **$R$-algebra** to be a ring map $f\colon R \to W$ such that $R$ is central in $W$, meaning that the diagram

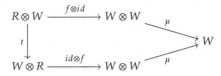

is commutative. Observe that if $\mathcal{M}$ has coproducts and the tensor distributes over them, then we have the expected "tensor algebra" functor $T\colon R\text{–Mod} \to R\text{–Alg}$ given by $T(V) = R \amalg V \amalg (V \otimes_R V) \amalg \cdots$ with the evident multiplication. This gives an adjoint pair $T\colon R\text{–Mod} \rightleftarrows R\text{–Alg}\colon U$, where $U$ is the forgetful functor.

We will be interested in the question of when certain structures on $\mathcal{M}$ pass to the category of $R$-modules. For example, if $\mathcal{M}$ is complete then so is $R$–Mod. To see this, let $\{M_\alpha\}$ be a diagram of $R$-modules and write $\lim_\alpha M_\alpha$ for the limit in $\mathcal{M}$. The canonical map $R \otimes (\lim_\alpha M_\alpha) \to \lim_\alpha(R \otimes M_\alpha)$ makes $\lim_\alpha M_\alpha$ into an $R$-module, and one readily checks that this has the properties of the limit in the category $R$–Mod. To say the same thing in fancier language, the forgetful functor $U\colon R\text{–Mod} \to \mathcal{M}$ is right adjoint to the free $R$-module functor $X \mapsto R \otimes X$ and therefore preserves all limits.

The situation for colimits is a little more challenging. Here the canonical map $\operatorname{colim}_\alpha(R \otimes M_\alpha) \to R \otimes \operatorname{colim}_\alpha M_\alpha$ goes in the wrong direction, and so does not give an $R$-module structure on $\operatorname{colim}_\alpha M_\alpha$. However, in many cases the functor $R \otimes (-)$ is a left adjoint and hence preserves colimits; so in these cases the above map *is* an isomorphism and everything works as before.

A symmetric monoidal category $(\mathcal{M}, \otimes, I)$ is called **closed** if there exists a cotensor (or "internal hom") functor $\mathcal{F}\colon \mathcal{M}^{op} \times \mathcal{M} \to \mathcal{M}$ together with natural adjunctions

$$\mathcal{M}(A \otimes B, C) \cong \mathcal{M}(A, \mathcal{F}(B, C)).$$

Note that this implies that $(-) \otimes (-)$ commutes with colimits in both variables.

**Proposition 3.3.1.** *Suppose $(\mathcal{M}, \otimes, I, \mathcal{F})$ is a closed symmetric monoidal category. Then both $R$-Mod and $R$-Alg are complete and cocomplete.*

*Proof.* We have already discussed the situation for $R$–Mod. For $R$–Alg, limits are created by the forgetful functor $U$ in the adjoint pair $T\colon R\text{–Mod} \rightleftarrows R\text{–Alg}\colon U$. Colimits in $R$–Alg are more complicated, but by [53, Proposition 4.3.6] the category is cocomplete provided that the tensor functor $T(-)$ preserves filtered colimits. The latter condition is immediate from the fact that $\otimes$ preserves colimits in each variable. $\quad\square$

See Section 5.6 in Chapter 5 of this volume for a more detailed discussion of limits and colimits in categories of operadic algebras.

We will next discuss the issue of compatibility between a monoidal structure and a model structure.

**Definition 3.3.2.** A **monoidal model category** is a model category $\mathcal{M}$ equipped with a monoidal structure $(\otimes, I)$ satisfying the following two axioms:

(1) [Pushout-Product Axiom] For any two cofibrations $f: A \rightarrowtail B$ and $j: K \rightarrowtail L$ in $\mathcal{M}$, the induced map

$$f \,\square\, j: (A \otimes L) \amalg_{A \otimes K} (B \otimes K) \longrightarrow B \otimes L$$

is a cofibration. Moreover, $f \,\square\, j$ is a weak equivalence if either $f$ or $j$ is a trivial cofibration.

(2) [Unit Axiom] There exists a cofibrant replacement $QI \xrightarrow{\sim} I$ having the property that for all cofibrant $X$ the map $QI \otimes X \to I \otimes X$ is a weak equivalence.

The notion of monoidal model category was introduced in [130]. The Pushout-Product Axiom is analogous to one common form of Quillen's SM7 axiom for simplicial model categories; it is the standard axiom for compatibility of a tensor with the model structure. In the presence of the Pushout-Product Axiom, the Unit Axiom is equivalent to requiring that *every* cofibrant replacement $QI \xrightarrow{\sim} I$ has the stated property. This axiom is automatically satisfied if the unit $I$ is itself cofibrant.

It is an easy exercise to verify that in a monoidal model category the derived functor of $\otimes$ descends to give a monoidal structure on the homotopy category.

By a **closed symmetric monoidal model category** we simply mean a monoidal model category where the underlying monoidal category is symmetric and closed. It is an easy exercise in adjoint functors to check the following:

**Proposition 3.3.3.** *Let $\mathcal{M}$ be a closed symmetric monoidal model category. If $f: A \rightarrowtail B$ and $g: X \twoheadrightarrow Y$ are maps in $\mathcal{M}$ then the induced map*

$$\mathcal{F}(B, X) \to \mathcal{F}(A, X) \times_{\mathcal{F}(A,Y)} \mathcal{F}(B, Y)$$

*is a fibration, and moreover it is a weak equivalence if either $f$ or $g$ is so.*

We next consider when a model category structure on $\mathcal{M}$ induces an associated model structure for $R$–Mod and for $R$–Alg. Suppose given a model category $\mathcal{M}$ together with an adjoint pair $L: \mathcal{M} \rightleftarrows \mathcal{N}: U$. In good cases one can put a model category structure on $\mathcal{N}$ where a map $f$ is a weak equivalence (respectively, fibration) if and only if $Uf$ is a weak equivalence (respectively, fibration). The cofibrations are forced to be the maps with the left lifting property with respect to the trivial fibrations, but often this is about all one can say about them. When such a model structure on $\mathcal{N}$ exists, one refers to it as the model structure **created by** the right adjoint $U$.

The main result on such structures is Kan's Recognition Theorem [124, Theorem 11.3.2], which says that $U$ creates a model structure on $\mathcal{N}$ if the following conditions are satisfied:

(1) $\mathcal{M}$ is cofibrantly generated.
(2) The images under $L$ of the generating cofibrations and trivial cofibrations permit the small object argument.
(3) If $J$ denotes the set of generating trivial cofibrations for $\mathcal{M}$, then $U$ takes all maps in $\widehat{LJ}$ to weak equivalences, where $\widehat{LJ}$ is the class of maps obtained from $L(J)$ by taking cobase changes and transfinite compositions.

Conditions (1) and (2) are technical conditions that are always satisfied in the cases of interest; we will bundle them into the "sufficiently combinatorial" adjective. Condition (3) is where the real content is.

Let $\mathcal{M}$ be a monoidal model category and let $R$ be a monoid in $\mathcal{M}$. Then we have adjoint functors

$$\mathcal{M} \underset{U}{\overset{F_R}{\rightleftarrows}} R\text{-Mod}$$

where $U$ is the forgetful functor and $F_R(X) = R \otimes X$. If we are lucky, then $U$ will create a model category structure on $R$-Mod. Here are some general conditions where this happens:

Proposition 3.3.4. *Let $\mathcal{M}$ be a sufficiently combinatorial monoidal model category.*

(a) *If $R$ is cofibrant in $\mathcal{M}$, then $R$-Mod has the model structure created by $U$.*

(b) *Start with the collection of maps $f \otimes id_R \colon R \otimes A \to R \otimes B$, where $f \colon A \overset{\sim}{\rightarrowtail} B$ is a trivial cofibration. Let $\mathcal{S}$ be the collection of maps obtained from the original collection using cobase change and transfinite composition. If every element of $\mathcal{S}$ is a weak equivalence, then $R$-Mod has the model structure created by $U$.*

*Proof.* In (b), the stated hypothesis exactly verifies condition (3) from Kan's Recognition Theorem. For (a), the point is that when $R$ is cofibrant the functor $R \otimes (-)$ preserves trivial cofibrations by the Pushout-Product Axiom. Since trivial cofibrations are closed under cobase change and transfinite composition, the condition from (b) is automatically satisfied. □

Now assume that $\mathcal{M}$ is a closed symmetric monoidal model category. This allows us to talk about *commutative* monoids in $\mathcal{M}$. Let $R$ be a commutative monoid and let $M$ and $N$ be $R$-modules (we will identify left and right $R$-modules, as usual). Define

$$M \otimes_R N = \operatorname{coeq}(M \otimes R \otimes N \rightrightarrows M \otimes N),$$

where the two maps in the coequalizer come from the $R$-module structures on $M$ and $N$. Then $\otimes_R$ is a symmetric monoidal product on $R$-Mod with unit $R$. Likewise, define

$$\mathcal{F}_R(M,N) = \operatorname{eq}(\mathcal{F}(M,N) \rightrightarrows \mathcal{F}(R \otimes M, N)),$$

where the two maps in the equalizer are the adjoints to the two evident maps $F(M,N) \otimes R \otimes M \to N$ (twist-evaluate-multiply and multiply-evaluate). It follows by quite general considerations that these definitions give a closed symmetric monoidal structure on $R$-Mod with unit $R$. We can hope that this makes $R$-Mod into a closed symmetric monoidal model category.

Finally, let us turn to algebras. If $R$ is a commutative monoid in $\mathcal{M}$ then we have the adjoint functors $T_R \colon R$-Mod $\rightleftarrows R$-Alg$\colon U$. We can again hope that $U$ creates a model structure on $R$-Alg.

We now bundle all of these "hopes" into the following definition:

Definition 3.3.5. Let $\mathcal{M}$ be a closed symmetric monoidal model category. We say that $\mathcal{M}$ satisfies the **Algebraic Creation Property** if

(1) for every monoid $R$ in $\mathcal{M}$, the forgetful functor $R$–Mod $\to \mathcal{M}$ creates a model structure on $\mathcal{M}$;

(2) when $R$ is a commutative monoid, $\otimes_R$ and $\mathcal{F}_R(-,-)$ make $R$–Mod into a closed symmetric monoidal model category; and

(3) when $R$ is a commutative monoid, the forgetful functor $R$–Alg $\to R$–Mod creates a model structure on $R$–Alg.

There are essentially two separate circumstances where the Algebraic Creation Property is known to hold. The first is when all objects of $\mathcal{M}$ are fibrant, and a few other conditions are satisfied — this kind of case was treated in [94, Chapter VII], though some of the ideas go back as far as [229]. When it is not true that all objects of $\mathcal{M}$ are fibrant, the situation is more delicate; it was first analyzed in [267]. The following proposition, though somewhat awkward, brings together these different threads.

Proposition 3.3.6. *Let $(\mathcal{M}, \otimes, I)$ be a symmetric monoidal model category that is sufficiently combinatorial and consider the following hypotheses:*

(1) *For some cofibrant replacement $QI \xrightarrow{\sim} I$ and any object $X$, the map $QI \otimes X \to I \otimes X$ is a weak equivalence.*

(2) *All objects of $\mathcal{M}$ are fibrant, and $\mathcal{M}$ is a simplicial or topological model category.*

(3) *[The Monoid Axiom] For any trivial cofibration $A \rightarrowtail B$ and any object $X$, the map $A \otimes X \to B \otimes X$ is a weak equivalence. Additionally, all maps obtained from the class*

$$\{A \otimes X \to B \otimes X \mid A \to B \text{ is a trivial cofibration and } X \text{ is any object}\}$$

*by cobase change and transfinite composition are also weak equivalences.*

*Assume that* (1) *holds and that* **either** (2) *or* (3) *holds. Then $\mathcal{M}$ satisfies the Algebraic Creation Property.*

*Remark 3.3.7.* Condition (1) is automatic if the unit is cofibrant. In general condition (1) seems much too strong, but it is not clear how to weaken it. Condition (3) was isolated by Schwede–Shipley [267] and christened by them.

*Proof of Proposition 3.3.6.* Condition (2) implies that the appropriate model structures are created on $R$–Mod and $R$–Alg; this is by [267, Lemma 2.3(2)] and the fact that the simplicial (or topological) structure on $\mathcal{M}$ gives canonical path objects on both $R$–Mod and $R$–Alg. See also [267, Remark 4.5].

Condition (3) also implies that the appropriate model structures are created on $R$–Mod and $R$–Alg. For $R$–Mod this is automatic, because the condition of Proposition 3.3.4(b) is a special case of (3). For $R$–Alg this is a little more difficult, but was worked out in [267, Theorem 4.1(3)].

It remains to prove that $R$–Mod is a monoidal model category. For the Pushout-Product Axiom, as in [267, Theorem 4.1(2)] it suffices to check this on the generating cofibrations and trivial cofibrations of $R$-Mod. But these are of the form $id_R \otimes f$, where $f$ is a generating cofibration or trivial cofibration of $\mathcal{M}$, and the pushout-product is readily analyzed. The necessary condition follows at once from the Pushout–Product Axiom on $\mathcal{M}$.

The trouble arises with the Unit Axiom for $R$-Mod. This was not dealt with in [267]. Let $QI \to I$ be a cofibrant replacement in $\mathcal{M}$. Hypothesis (1) implies that $R \otimes QI \to R \otimes I = R$ is a weak equivalence, and of course $R \otimes QI$ is cofibrant in $R$-Mod. So we must check that for every cofibrant $R$-module $M$, the map $(R \otimes QI) \otimes_R M \to R \otimes_R M$ is a weak equivalence. This is just the map $QI \otimes M \to M$, and so hypothesis (1) completes the verification.          $\square$

If we have model categories on $R$-Mod and $R$-Alg, we should of course be concerned with the extent to which they depend on the homotopy type of $R$. If $R \to T$ is a map of monoids then there is an adjoint pair

$$T \otimes_R (-): R\text{-Mod} \rightleftarrows T\text{-Mod}: V, \qquad (3.3.3)$$

where here the right adjoint $V$ is restriction of scalars, and this will be a Quillen pair if the categories have the model structures created by $U$ (because $V$ will preserve both fibrations and trivial fibrations).

Similarly, if $R \to T$ is a map of commutative monoids then $T \otimes_R (-)$ takes $R$-algebras to $T$-algebras and we have a similar Quillen pair

$$T \otimes_R (-): R\text{-Alg} \rightleftarrows T\text{-Alg}: V. \qquad (3.3.4)$$

In both cases, if $R \to T$ is a weak equivalence one would hope that the above adjoint pairs are Quillen equivalences. Unfortunately, this does not work out for free and is not known without various unsatisfying extra hypotheses. To sweep some of these under the rug, we make the following definition:

Definition 3.3.8. Let $\mathcal{M}$ be a symmetric monoidal model category that satisfies the Algebraic Creation Property. Then $\mathcal{M}$ satisfies the **Algebraic Invariance Property** if for every weak equivalence of monoids $R \to T$ the Quillen pair of (3.3.3) is a Quillen equivalence, and if for every weak equivalence of commutative monoids $R \to T$ the pair (3.3.4) is a Quillen equivalence.

The following result is basically Theorems 4.3 and 4.4 of [267]. It follows readily from Quillen's criterion for checking that an adjoint pair is a Quillen equivalence. The proof is an easy exercise.

Proposition 3.3.9. *Let $\mathcal{M}$ be a symmetric monoidal model category satisfying the Algebraic Creation Property. Suppose further that*

(1) *for every monoid $R$ and every cofibrant $R$-module $M$, the functor $(-) \otimes_R M$ preserves all weak equivalences, and*

(2) *every cofibration $R \to T$ in $R$-Alg is a cofibration in $R$-Mod as well.*

*Then $\mathcal{M}$ satisfies the Algebraic Invariance Property.*

The conditions in this proposition seem like a lot to check, and in some sense they are. But they have been verified for all the modern model categories of spectra. Condition (1) turns out to be surprisingly important, and deserves its own name:

**Definition 3.3.10.** Let $\mathcal{M}$ be a symmetric monoidal model category satisfying the Algebraic Creation Property. Say that $\mathcal{M}$ satisfies the **Strong Flatness Property** if for every monoid $R$ in $\mathcal{M}$ and every cofibrant $R$-module $M$, the functor $(-) \otimes_R M$ preserves all weak equivalences of right $R$-modules.

While this property seems somewhat unnatural from the perspective of model category theory, it nevertheless is a crucial element of all the modern model categories of spectra. It automatically implies condition (1) of Proposition 3.3.6, using the Unit Axiom. One of the lessons of this whole section is that when it comes to model structures on categories of modules and algebras in a monoidal model category, none of the existing theory works out quite as naturally as one would like.

*Remark 3.3.11.* Lewis and Mandell in [157] have some interesting things to say about the Algebraic Invariance Property. Define an object $C$ of $\mathcal{M}$ to be **semi-cofibrant** if $\mathcal{F}(C,-)$ preserves fibrations and trivial fibrations; by adjointness this is equivalent to saying that $C \otimes (-)$ preserves cofibrations and trivial cofibrations. Every cofibrant object is semi-cofibrant, but the converse does not necessarily hold. Lewis–Mandell prove that if one has a weak equivalence of monoids $R \to T$, where $R$ and $T$ are semi-cofibrant, then the Quillen pair of (3.3.3) is a Quillen equivalence. The same paper has many other interesting results about the homotopy theory of module categories.

*Remark 3.3.12.* If $T$ is a monad on $\mathcal{M}$, one can consider the category of $T$-algebras $\mathcal{M}[T]$ and again ask whether the forgetful functor $U \colon \mathcal{M}[T] \to \mathcal{M}$ creates a model structure on $\mathcal{M}[T]$. This question generalizes the specific cases of $R$–Mod and $R$–Alg we have considered in this section. While we will not address the general version here, we refer the reader to [94, Chapter VII.4] for techniques that apply to the case where $\mathcal{M}$ is a topological model category where all objects are fibrant. The task of creating the model structures is essentially reduced to verifying two criteria, embodied in the so-called "Cofibration Hypothesis" [94, Remark IV.4.12].

See also Section 5.8 in Chapter 5 of this volume for a detailed discussion of model structures on operadic algebras more generally.

## 3.3.5     Invertible objects

If one had to describe the idea of spectra in a single sentence, one approach would be to say that it is a modification of $\mathcal{T}op_*$ that makes the spheres invertible in the homotopy category. So it is good to know a little about the general theory of invertible objects.

Let $(\mathcal{C}, \otimes, I)$ be a symmetric monoidal category. An object $X$ in $\mathcal{C}$ is **invertible** if the functor $X \otimes (-) \colon \mathcal{C} \to \mathcal{C}$ is an equivalence of categories. This is equivalent to saying that there exists an object $Y$ and an isomorphism $\alpha \colon I \xrightarrow{\cong} Y \otimes X$, and here we say that the pair $(Y, \alpha)$ is an inverse for $X$. Note that $\alpha$ is not unique, since given one choice one can make others by precomposing with automorphisms of $I$. Likewise, $Y$ is unique up to isomorphism but not up to *unique* isomorphism. However, given an inverse $(Y, \alpha_Y)$ and another inverse $(Z, \alpha_Z)$ it is easy to check that there is a unique map $f : Y \to Z$ making the diagram

commute, and moreover $f$ is an isomorphism.

Note that the tensor product of invertible objects is again invertible.

In a symmetric monoidal category, the endomorphisms of the unit always form a *commutative* monoid: this is an easy exercise using that if $f$ and $g$ are any two maps then $f \otimes g = (f \otimes id)(id \otimes g) = (id \otimes g)(f \otimes id)$. Given any object $X$ in $\mathcal{C}$, there is a map of monoids $\Gamma_X : \mathrm{End}(I) \to \mathrm{End}(X)$ that sends $f : I \to I$ to the composite

$$ X \xrightarrow{\cong} I \otimes X \xrightarrow{f \otimes id} I \otimes X \xrightarrow{\cong} X. $$

When $X$ is invertible, the map $\Gamma_X$ is an isomorphism. So in particular, the endomorphisms of an invertible object are always commutative. One checks that if $(Y, \alpha)$ is an inverse to $X$ and $f : X \to X$ then $\Gamma_X^{-1}(f)$ is the composite

$$ I \xrightarrow{\alpha} Y \otimes X \xrightarrow{id \otimes f} Y \otimes X \xrightarrow{\alpha^{-1}} I. $$

Now let $X$ be any object in $\mathcal{C}$. For $n \geq 0$ set $X^{\otimes n} = X \otimes (X \otimes (X \otimes \cdots \otimes X))$. Let $\sigma \in \Sigma_n$ and consider natural transformations

$$ X_1 \otimes (X_2 \otimes (X_3 \otimes \cdots \otimes X_n)) \longrightarrow X_{\sigma^{-1}(1)} \otimes (X_{\sigma^{-1}(2)} \otimes (X_{\sigma^{-1}(3)} \otimes \cdots \otimes X_{\sigma^{-1}(n)})), $$

where the domain and codomain are considered as functors $\mathcal{C}^{\times n} \to \mathcal{C}$. MacLane's Coherence Theorem for symmetric monoidal categories says that all natural transformations of this form, made from composites of associativity and commutativity isomorphisms, are identical; see [174, Theorem XI.1.1]. So we have a canonical such transformation. Evaluating at the case where all $X_i$ equal $X$ gives a map $\sigma_* : X^{\otimes n} \to X^{\otimes n}$, and one readily checks that this gives a group homomorphism $\Sigma_n \to \mathrm{Aut}(X^{\otimes n})$. If $X$ is invertible then so is $X^{\otimes n}$, which means $\mathrm{Aut}(X^{\otimes n})$ is abelian and therefore this map factors through the abelianization of $\Sigma_n$ (which is $\mathbb{Z}/2$). In particular, every commutator in $\Sigma_n$ acts as the identity on $X^{\otimes n}$. The first interesting case is $n = 3$, where the commutator subgroup is generated by the cyclic permutation $(123)$. Moreover, via block sum of permutations and conjugation this case generates the relations for all higher $n$ as well.

**Proposition 3.3.13** (The cyclic permutation condition). *If $X$ is an invertible object in a symmetric monoidal category then the composite*

$$ X \otimes (X \otimes X) \xrightarrow{id \otimes t} X \otimes (X \otimes X) \xrightarrow{a} (X \otimes X) \otimes X \xrightarrow{t \otimes id} (X \otimes X) \otimes X \xrightarrow{a} X \otimes (X \otimes X) $$

*must equal the identity, where all maps labeled $a$ and $t$ are associativity and commutativity isomorphisms, respectively.*

The cyclic permutation condition seems to have first been identified by Voevodsky, when attempting to construct symmetric spectra in motivic homotopy theory. See the discussion preceding Theorem 4.3 in [294].

Invertible objects are, in particular, examples of *dualizable* objects. Self-maps of dualizable objects have a *trace*. We will not recount the general theory here, but just give a very streamlined version suitable for our present context. For the general theory, see [155, Section III.1] or the survey in [76].

Assume $X$ is invertible and $(Y, \alpha)$ is a chosen inverse. Then there is a unique map $\hat{\alpha} \colon X \otimes Y \to I$ with the property that the composite

$$X \xrightarrow{\cong} X \otimes I \xrightarrow{id \otimes \alpha} X \otimes (Y \otimes X) \xrightarrow{a} (X \otimes Y) \otimes X \xrightarrow{\hat{\alpha} \otimes id} I \otimes X \xrightarrow{\cong} X$$

equals the identity. If $f \colon X \to X$ then the **trace** of $f$ is the element $\operatorname{tr}(f) \in \operatorname{End}(I)$ defined by the composite

$$I \xrightarrow{\alpha} Y \otimes X \xrightarrow{id \otimes f} Y \otimes X \xrightarrow{t} X \otimes Y \xrightarrow{\hat{\alpha}} I.$$

Given $f \colon X \to X$ we now have two ways to extract an element of $\operatorname{End}(I)$: via $\Gamma_X^{-1}(f)$ and via $\operatorname{tr}(f)$. These don't always give the same element! The following results explain the relation between them. They certainly must be classical, but see [76] for a written account:

**Proposition 3.3.14.** *Let $X$ be an invertible object in a symmetric monoidal category, and let $\tau_X = \operatorname{tr}(id_X) \in \operatorname{End}(I)$.*

(a) $\tau_X = \Gamma_{X \otimes X}^{-1}(t_X) = \operatorname{tr}(t_X)$ *where* $t_X \colon X \otimes X \to X \otimes X$ *is the twist.*

(b) $\tau_X^2 = id$.

(c) *For any* $f \colon X \to X$, $\Gamma_X^{-1}(f) = \tau_X \cdot \operatorname{tr}(f)$.

(d) *If* $Y$ *is another invertible object then* $\tau_{X \otimes Y} = \tau_X \tau_Y$.

The elements $\tau_X$ should be thought of as "generalized signs". They appear as control factors in commutation formulas involving $X$, in the same way that $\pm 1$ terms appear in the standard formulas of topology.

**Example 3.3.15.** Fix a field $k$ and consider the category of $\mathbb{Z}$-graded vector spaces, equipped with the graded tensor product, standard associativity isomorphism, and the twist isomorphism that incorporates the Koszul sign rule. Write $k[n]$ for the graded vector space consisting of a single $k$ in degree $n$ and zero in all other degrees. We identify $k$ with $\operatorname{End}(k[0])$ by letting $x \in k$ correspond to multiplication by $x$.

The object $k[1]$ is invertible. For an inverse we may choose $k[-1]$ and the map $\alpha \colon k[0] \to k[-1] \otimes k[1]$ sending 1 to $1 \otimes 1$. The map $\hat{\alpha} \colon k[1] \otimes k[-1] \to k[0]$ then sends $1 \otimes 1$ to 1. If $x \in k$ and $\rho_x \colon k[1] \to k[1]$ is multiplication by $x$, we leave it as an exercise to check that $\Gamma_X^{-1}(\rho_x) = x$ and $\operatorname{tr}(\rho_x) = -x$. In particular, $\tau_{k[1]} = -1$ here.

## 3.4    Spectra for Sulu and Chekov

For many applications one needs a model category of spectra but doesn't care much about the inner workings, other than a few basic properties. In the words of one eloquent topologist, "Sometimes one just needs to drive the Enterprise, not necessarily be Mr. Scott." The goal of this section is to supply a list of properties that are shared by most of the existing models, and to give some standard examples of how they can be used. These examples were all originally worked out in [94].

In this section we assume the existence of a pointed category $\mathcal{S}pectra$ equipped with a closed symmetric monoidal smash product $\wedge$ with unit $S$ and cotensor $\mathcal{F}(-,-)$. Additionally, we suppose given adjoint functors $\Sigma^\infty \colon \mathcal{T}op_* \rightleftarrows \mathcal{S}pectra \colon \Omega^\infty$ and a stable model category structure on $\mathcal{S}pectra$. We assume the following properties:

1. $\Sigma^\infty \colon \mathcal{T}op_* \rightleftarrows \mathcal{S}pectra \colon \Omega^\infty$ is a Quillen pair.
2. The smash product makes $\mathcal{S}pectra$ into a monoidal model category. So we have
   (a) the pushout-product axiom: given cofibrations $f \colon A \rightarrowtail B$ and $g \colon C \rightarrowtail D$, the induced map
   $$f \,\square\, g \colon (A \wedge D) \amalg_{A \wedge C} (B \wedge C) \to B \wedge D$$
   is a cofibration, and additionally it is a weak equivalence if either $f$ or $g$ is so. And
   (b): for every cofibrant object $X$ and every cofibrant replacement $QS \xrightarrow{\sim} S$, the induced map $QS \wedge X \to S \wedge X$ is a weak equivalence.
3. There exists a weak equivalence $\epsilon \colon \Sigma^\infty S^0 \to S$ and a natural transformation
   $$\eta \colon \Sigma^\infty(X \wedge Y) \to \Sigma^\infty X \wedge \Sigma^\infty Y$$
   that is oplax monoidal: this says that the evident associativity and unital squares commute. Additionally, $\eta$ is a weak equivalence when $X$ and $Y$ are cofibrant.
4. $(\mathcal{S}pectra, \wedge)$ satisfies the Algebraic Creation and Invariance Properties (see Definitions 3.3.5 and 3.3.8).
5. $(\mathcal{S}pectra, \wedge)$ satisfies the Strong Flatness Condition of Definition 3.3.10. In particular, for any cofibrant spectrum $A$ and any weak equivalence of spectra $X \to Y$, the induced map $A \wedge X \to A \wedge Y$ is a weak equivalence.
6. There is an equivalence of triangulated categories between $\mathrm{Ho}(\mathcal{S}pectra)$ and the homotopy category of Bousfield–Friedlander spectra that carries the spectra $\Sigma^\infty(S^n)$ to the standard $n$-sphere.
7. For any directed system $X_0 \to X_1 \to X_2 \to \cdots$ in $\mathcal{S}pectra$ and any $n \geq 0$, the canonical map
   $$\mathrm{colim}_k[\Sigma^\infty(S^n), X_k] \to [\Sigma^\infty(S^n), \mathrm{hocolim}_k X_k]$$
   is an isomorphism, and similarly sequences indexed by other transfinite ordinals.

All these properties are satisfied by the categories of symmetric spectra, orthogonal spectra, and $\mathcal{W}$-spaces (all to be defined in subsequent sections). Actually (7) is a consequence of (6) (using the smallness of spheres in $\mathcal{T}op$), but is included separately here for emphasis. Note also that $\Gamma$-spaces are eliminated from the discussion because they are not a stable model category, but except for this and the related property (6) all the other properties are satisfied.

*Remark 3.4.1.* EKMM spectra are a special case as they do NOT satisfy property (3), although they satisfy all of the others. Instead, in EKMM spectra there are two pairs of adjoint functors called $(\Sigma_S^\infty, \Omega_S^\infty)$ and $(\Sigma^\infty, \Omega^\infty)$ together with natural maps $\Sigma_S^\infty X \to \Sigma^\infty X$ which are weak equivalences whenever $X$ is cofibrant as a pointed space. The pair $(\Sigma_S^\infty, \Omega_S^\infty)$ satisfies (1), and the pair $(\Sigma^\infty, \Omega^\infty)$ satisfies (3). But if we use the pair $(\Sigma_S^\infty, \Omega_S^\infty)$ for (1)-(7) then we can replace (3) above with (3′) stating that there is a contractible space of choices for an $\eta$, giving an oplax symmetric monoidal map in the homotopy category. Keeping this small variation in mind, all of the arguments in the remainder of this section apply to EKMM spectra as well. (It is unfortunate that the EKMM $(\Sigma^\infty, \Omega^\infty)$ notation conflicts with what we use above, but we will just live with this).

## 3.4.1   Homotopy groups of spectra

Write $S^0 = \Sigma^\infty(S^0)$ and $S^1 = \Sigma^\infty(S^1)$. For $p > 1$ define the stable sphere $S^p$ recursively by $S^p = S^1 \wedge S^{p-1}$, so that

$$S^p = S^1 \wedge (S^1 \wedge (S^1 \wedge \cdots))).$$

Note that $S^1$ is cofibrant by property (1), and then $S^p$ is cofibrant by the Pushout–Product Axiom. Also we see using property (3) that there is a canonical weak equivalence $\eta \colon \Sigma^\infty(S^p) \to S^p$. Some authors prefer to adopt $\Sigma^\infty(S^p)$ as the *definition* of the stable sphere, but $\eta$ shows that for homotopical purposes this is equivalent to our approach.

Since $\Sigma$ is an autoequivalence of the homotopy category, there exists a desuspension of $S^0$. Let $S^{-1}$ be any chosen cofibrant spectrum for which there exists an isomorphism $\alpha \colon S \to S^{-1} \wedge S^1$ in Ho($\mathcal{Spectra}$). For $p \geq 1$ inductively define $S^{-p} = S^{-1} \wedge S^{1-p}$. Let $\hat{\alpha} \colon S^1 \wedge S^{-1} \to S$ be the dual map to $\alpha$ in Ho($\mathcal{Spectra}$) as defined after Proposition 3.3.13.

Under these definitions, there are canonical isomorphisms in Ho($\mathcal{Spectra}$) of the form

$$\gamma \colon S^k \wedge S^l \to S^{k+l}$$

for any $k, l \in \mathbb{Z}$. If $k, l > 0$ then we define $\gamma$ as a composite of associativity isomorphisms, and MacLane's Coherence Theorem for monoidal categories says that all choices for such associativity isomorphisms lead to the same map $\gamma$. Similar remarks apply when $k, l < 0$. When $k = 0$ we use

$$S^0 \wedge S^l \xrightarrow{\epsilon \wedge id} S \wedge S^l \cong S^l,$$

which uses property (3) and also property (2) to know that the first map is an isomorphism. Likewise for $l = 0$. When $k < 0$ and $l > 0$ we use associativity isomorphisms together with repeated applications of the map $\alpha^{-1}$ and the unit map. Again, one can prove that the exact choice of maps here does not affect the final composite. Finally, when $k > 0$ and $l < 0$ we do the same thing but using $\hat{\alpha}$ instead of $\alpha$.

It is a theorem that these specified isomorphisms are compatible, in the sense that the evident pentagon containing $S^k \wedge (S^l \wedge S^n)$ and $S^{k+l+n}$ is commutative in the homotopy category. More generally, any two composites derived from these canonical maps (but having the same domain and range) are identical (again, in the homotopy category). See [76] for a complete discussion.

Here is why this tedious discussion is actually important. For any spectrum $X$ we write $\pi_p(X)$ for $\mathrm{Ho}(\mathcal{S}pectra)(S^p, X)$. If $X$ is a ring spectrum and $f: S^p \to X$ and $g: S^q \to X$ we may form the composite

$$S^{p+q} \xrightarrow{\gamma} S^p \wedge S^q \xrightarrow{f \wedge g} X \wedge X \xrightarrow{\mu} X,$$

and this determines a pairing $\pi_p(X) \otimes \pi_q(X) \to \pi_{p+q}(X)$. Also, the composite map $S^0 \xrightarrow{\epsilon} S \to X$ determines a special element $1 \in \pi_0(X)$.

**Lemma 3.4.2.** *When $X$ is a ring spectrum, $\pi_*(X)$ is a ring. If $M$ is a left $X$-module then $\pi_*(M)$ is a left $\pi_*(X)$-module.*

*Proof.* Left to the reader as an exercise, but note that the properties of the canonical maps $\gamma$ are important here. See [76] for details and generalizations.  $\square$

### 3.4.2   Homotopy groups of tensors and cotensors

Let $R$ be a commutative ring spectrum and let $M$ and $N$ be $R$-modules. We will construct a spectral sequence of the form

$$\mathrm{Tor}_{p,q}^{\pi_* R}(\pi_* M, \pi_* N) \Rightarrow \pi_{p+q}(M \wedge_R^{\mathbb{L}} N),$$

where $\wedge_R^{\mathbb{L}}$ denotes the derived version of $\wedge_R$. When $M = R \wedge X$ and $N = R \wedge Y$ this gives the Künneth spectral sequence $\mathrm{Tor}^{\pi_* R}(R_*(X), R_*(Y)) \Rightarrow R_*(X \wedge Y)$.

The following argument can be made almost entirely in the homotopy category $\mathrm{Ho}(R\text{--Mod})$, using only the triangulated structure. However, the model structure on $R$--Mod is key to setting up this homotopy category to begin with. The model structure also plays a small role in the following lemma:

**Lemma 3.4.3.** *Let $R$ be a commutative ring spectrum and let $M$ be an $R$-module. Then there exists an $R$-module $F$ of the form $F = \bigvee_i R \wedge S^{n_i}$ together with a map $F \to M$ in $\mathrm{Ho}(R\text{--Mod})$ that is surjective on homotopy groups.*

*Proof.* Let $M \to M^{fib}$ be a fibrant replacement in $R$--Mod. Choose a set of $\pi_* R$-module generators $\alpha_i \in \pi_*(M)$, together with representative maps $\alpha_i: S^{n_i} \to M^{fib}$ in $\mathcal{S}pectra$. We then get $R$-module maps $R \wedge S^{n_i} \to M^{fib}$ using the adjoint pair $\mathcal{S}pectra \rightleftarrows R\text{--Mod}$. Let $F = \bigvee_i R \wedge S^{n_i}$ and let $\alpha: F \to M^{fib}$ be the evident map.

Since $\alpha$ is a map of $R$-modules, $\pi_* \alpha$ is a map of $\pi_* R$-modules. So to see that $\pi_* \alpha$ is surjective we only need argue that each $\alpha_i$ is in the image. This follows from the

commutative diagram

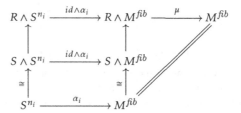

Let $R$ be a commutative ring spectrum and let $M$ be an $R$-module. The following argument takes place entirely in the category $\mathrm{Ho}(R\text{–Mod})$. Set $X_0 = M$. Using Lemma 3.4.3 choose an $R$-module $F_0 = \bigvee_i R \wedge S^{n_i}$ and a map $F_0 \to X_0$ that is a surjection on $\pi_*(-)$. Let $X_1 \to F_0 \to X_0$ be a homotopy fiber sequence in $\mathrm{Ho}(R\text{–Mod})$ (see the discussion of fiber and cofiber sequences in Section 3.2, and in particular Remark 3.2.5).

Repeat this process inductively to likewise construct homotopy fiber sequences $X_n \to F_{n-1} \to X_{n-1}$ where $F_{n-1}$ is a wedge of suspensions of $R$ and $F_{n-1} \to X_{n-1}$ is surjective on homotopy groups. One way to present all this information is through the diagram

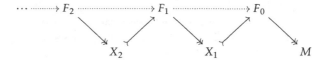

where double-headed arrows represent maps that induce surjections on homotopy groups and tailed arrows represent maps that induce injections on homotopy groups. Observe that the induced sequence $\pi_*(F_\bullet)$ is a free $\pi_* R$-resolution of $\pi_* M$. (There are some subtleties in justifying this last claim, which for the moment we leave for the reader to try to uncover. But see Section 3.4.4 below.)

Our diagram can also be restructured as a diagram of homotopy fiber sequences. We rotate the fiber sequence $X_n \to F_{n-1} \to X_{n-1}$ to become $X_{n-1} \to \Sigma X_n \to \Sigma F_{n-1}$ and suspend $n-1$ times to get

$$
\begin{array}{ccc}
\Sigma F_0 & & \Sigma^2 F_1 \\
\uparrow & & \uparrow \\
M = X_0 \longrightarrow & \Sigma X_1 \longrightarrow & \Sigma^2 X_2 \longrightarrow \cdots
\end{array}
$$

where every "layer" is a homotopy fiber sequence (note that we are being cavalier about signs, but that will be okay for our application). Now apply the derived functor $(-)\wedge_R^{\mathbb{L}} N$. This is still taking place entirely within $\mathrm{Ho}(R\text{–Mod})$, but observe that we know this derived functor exists because of model category machinery. For convenience we will drop the derived "$\mathbb{L}$" in all smash products and write our new tower of homotopy fiber sequences as

$$\begin{array}{ccc} & \Sigma F_0 \wedge_R N & & \Sigma^2 F_1 \wedge_R N \\ & \uparrow & & \uparrow \\ M \wedge_R N == X_0 \wedge_R N \longrightarrow & \Sigma X_1 \wedge_R N & \longrightarrow & \Sigma^2 X_2 \wedge_R N \longrightarrow \cdots \end{array}$$

Every layer of this tower induces a long exact sequence in homotopy groups, because homotopy fiber sequences of $R$-modules are also homotopy fiber sequences of spectra (the forgetful functor from $R$-modules to spectra is a right adjoint and preserves all weak equivalences, so is its own right derived functor). These long exact sequences braid together to give a spectral sequence in the usual way, taking the form

$$E^1_{a,b} = \pi_a(\Sigma^{b+1} F_b \wedge_R N) \Rightarrow \pi_{a-1}(M \wedge_R N), \qquad d^r : E^r_{a,b} \to E^r_{a-1,b-r}$$

(and recall once more that all smash products are derived).

Finally, observe that $F_b \wedge_R N = \bigvee_i (R \wedge S^{n_i}) \wedge_R N = \bigvee_i S^{n_i} \wedge N$, and so $\pi_*(F_b \wedge_R N)$ is a direct sum of shifted copies of $\pi_*(N)$. Said in the most canonical way possible, for any $R$-module $W$ we have a natural map

$$\pi_*(W) \otimes_{\pi_*(R)} \pi_*(N) \to \pi_*(W \wedge_R N)$$

and when $W$ is $R \wedge S^n$ or a wedge of such things this map is an isomorphism. This identifies the $E_1$-term of our spectral sequence as $\pi_*(F_\bullet) \otimes_{\pi_*(R)} \pi_*(N)$, and a little thought shows the $d^1$ maps are the boundary maps in this complex. So the $E_2$-term is $\mathrm{Tor}^{\pi_*R}(\pi_*M, \pi_*N)$, as desired. Specifically, $E^2_{a,b} = \mathrm{Tor}^{\pi_*R}_{b,a-b-1}(\pi_*M, \pi_*N)$ and this converges to $\pi_{a-1}(M \wedge_R N)$. Recoordinatizing the spectral sequence by setting $b = p$ and $a - b - 1 = q$ yields the following:

**Theorem 3.4.4.** *Let $R$ be a commutative ring spectrum and let $M$ and $N$ be $R$-modules. Then there is a spectral sequence*

$$E^2_{p,q} = \mathrm{Tor}^{\pi_*R}_{p,q}(\pi_*M, \pi_*N) \Rightarrow \pi_{p+q}(M \wedge^{\mathbb{L}}_R N)$$

*with differentials of the form $d^r : E^r_{p,q} \to E^r_{p-r,q+r-1}$.*

The construction of a spectral sequence for $\pi_* \mathcal{F}_R(M, N)$ is entirely similar. Start with the same tower of homotopy fiber sequences and apply $\mathcal{F}_R(-, N)$. The key part of the calculation is that

$$\mathcal{F}_R(R \wedge S^n, N) \simeq \mathcal{F}(S^n, N) \simeq \Sigma^{-n} N,$$

and so $\pi_*(\mathcal{F}_R(F_q, N)) \cong \mathrm{Hom}_{\pi_*R}(\pi_*F_q, \pi_*N)$. We leave the reader to work out the details for the following:

**Theorem 3.4.5.** *Let $R$ be a commutative ring spectrum and let $M$ and $N$ be $R$-modules. Write $\mathbb{R}\mathcal{F}(M, N)$ for the derived cotensor. Then there is a spectral sequence*

$$E^{p,q}_2 = \mathrm{Ext}^{p,q}_{\pi_*R}(\pi_*M, \pi_*N) \Rightarrow \pi_{-(p+q)}\mathbb{R}\mathcal{F}_R(M, N)$$

*with differentials of the form $d_r : E^{p,q}_r \to E^{p+r,q-r+1}_r$.*

For more about the above two spectral sequences, see [94, Chapter IV.4].

### 3.4.3    Constructing Morava $K$-theory

For each prime $p$ the $n$-th Morava $K$-theory spectrum is a certain ring spectrum $K(n)$ having the property that $\pi_* K(n) = \mathbb{Z}/p[v_n^{\pm 1}]$, where $|v_n| = 2(p^n - 1)$. In addition to those properties it can be characterized by the existence of a map $MU \to K(n)$ having a prescribed behavior on homotopy groups (where $MU$ is the usual complex cobordism spectrum). As a demonstration of the model-category-theoretic tools we have been describing, we show how they lead to a construction of the spectrum $K(n)$ starting with $MU$.

We start with the assumption that there is a commutative ring spectrum $MU$ in our category $Spectra$ and a ring isomorphism $\pi_*(MU) \cong \mathbb{Z}[x_1, x_2, \ldots]$ with $|x_i| = 2i$ for all $i$. Let $MU \to X$ be a fibrant replacement in the category of $MU$-modules, and recall that this implies $X$ is fibrant in $Spectra$.

Fix a prime $p$. Since $\pi_0(MU) = \mathbb{Z}$ and $X$ is fibrant, there exists a map $S^0 \to X$ that represents the element $p \in \pi_0(MU)$. Consider the composite $MU \wedge S^0 \to MU \wedge X \xrightarrow{\mu} X$, and let $MU_1$ be the homotopy cofiber in the category $MU$–Mod. This is also a homotopy cofiber in $Spectra$: the forgetful functor from $MU$-modules to spectra is its own right derived functor and therefore preserves homotopy fiber sequences, and homotopy cofiber and fiber sequences are the same by Proposition 3.2.10(c). The long exact sequence on homotopy groups immediately shows that $\pi_*(MU_1) = \mathbb{Z}/p[x_1, x_2, \ldots]$. (Note: There is a subtlety here! The reader may try to uncover it, or see the end of Section 3.4.4.)

Now let $MU_1 \to X_1$ be a fibrant replacement of $MU$-modules, and choose a map $S^2 \to X_1$ that represents $x_1$. Let $MU_2$ be the homotopy cofiber in $MU$–Mod of the composite $MU \wedge S^2 \to MU \wedge X_1 \to X_1$, and verify that $\pi_*(MU_2) = \mathbb{Z}/p[x_2, x_3, \ldots]$.

The only thing we are ever using is that we are quotienting by an element $x_i$ which is a nonzerodivisor on homotopy groups, so we can continue to do this for whichever $x_i$ we choose. Fix an $n$ and successively kill off all the $x_i$ except for $x_{p^n - 1}$. For convenience set $r = p^n - 1$. This produces a sequence in $\mathrm{Ho}(MU$–Mod$)$ of the form

$$MU = MU_0 \to MU_1 \to MU_2 \to \cdots \to MU_{r-1} \to MU_{r+1} \to \cdots$$

Lift this to a directed system in $MU$–Mod, and let $Z$ be the homotopy colimit in $MU$–Mod. Then $Z$ sits in a homotopy cofiber sequence $\bigvee_n MU_n \to \bigvee_n MU_n \to Z$, where the first map is the difference between the identity and the shift map. This is also a homotopy fiber sequence, by Proposition 3.2.10(c), and that property is preserved after applying the forgetful functor to $Spectra$. So $Z$ is the homotopy colimit of the $MU_n$ in $Spectra$, not just in $MU$–Mod. We then know by property (7) that $\pi_*(Z) = \mathrm{colim}_n \pi_*(MU_n)$, and so $\pi_*(Z) \cong \mathbb{Z}/p[x_r]$.

Now consider the composite map $Z \wedge S^{2r} \longrightarrow Z \wedge MU \xrightarrow{t} MU \wedge Z \xrightarrow{\mu} Z$. This is a map of left $MU$-modules, using that $MU$ is commutative. Applying $(-) \wedge S^{-2r}$ gives a map of $MU$-modules $Z \to Z \wedge S^{-2r}$. On homotopy groups this is multiplication by $x_r$. Consider the sequence in $\mathrm{Ho}(MU$–Mod$)$

$$Z \to Z \wedge S^{-2r} \to Z \wedge S^{-2r} \wedge S^{-2r} \to \cdots$$

then lift it to $MU$–Mod, and let $W$ be the homotopy colimit. It follows again from property (7) that $\pi_*(W) = \mathbb{Z}/p[x_r^{\pm 1}]$.

In this way we have constructed an $MU$-module spectrum $W$ whose homotopy groups make it look like $W$ is the $n$-th Morava $K$-theory spectrum. The construction has also produced a map $MU \to W$ which does the right thing on homotopy groups, so $W$ really *is* Morava $K$-theory.

Note that we have not constructed $W$ as a ring spectrum, only as an $MU$-module spectrum. In Chapters V.3 and V.4 of [94] (see especially Theorem V.4.1) it is explained how to construct a product $W \wedge W \to W$ making $W$ into a homotopy ring spectrum, but this is much weaker than what is desired. To construct $W$ as a ring spectrum one seems to need the full force of $A_\infty$-obstruction theory, which we will not recount here.

*Remark 3.4.6* (historical note). All of the arguments in this section first appeared in [94]. They needed very little of the inner workings of EKMM-spectra, however, and as we have seen here they work in any of the modern model categories of spectra.

### 3.4.4    Loose ends

In the course of the argument from Section 3.4.3 we had a homotopy cofiber sequence $MU \wedge S^0 \to MU \to MU_1$ and wanted to compute the homotopy groups of $MU_1$ using the long exact sequence. This required us to know $\pi_*(MU \wedge S^0)$ — but how exactly do we know these groups? Recall that $S^0 \to S$ is a cofibrant replacement, and so it is tempting to use property (2) to say that $MU \wedge S^0 \to MU \wedge S = MU$ is a weak equivalence. But that works only if $MU$ is cofibrant as a spectrum, which we have not assumed!

To try to get around this issue, let $\widetilde{MU} \xrightarrow{\sim} MU$ be a cofibrant replacement in *Spectra*. We certainly know $\widetilde{MU} \wedge S^0 \simeq \widetilde{MU} \simeq MU$ by property (2), so we know the homotopy groups of $\widetilde{MU} \wedge S^0$. We could go back to the beginning and try to do the entire construction with $\widetilde{MU}$ replacing $MU$, except we do not know that $\widetilde{MU}$ is a ring spectrum. The lifting diagram

produces a multiplication, but in general it will only be associative up to homotopy. If $\widetilde{MU}$ is only a homotopy ring spectrum we do not have a good homotopy theory of $\widetilde{MU}$-modules, so we are again defeated.

What saves us here is the amazing property (5). Since $S^0$ is cofibrant this property guarantees that $\widetilde{MU} \wedge S^0 \to MU \wedge S^0$ is still an equivalence, and so we have $MU \wedge S^0 \simeq \widetilde{MU} \wedge S^0 \simeq \widetilde{MU} \simeq MU$. This analysis is actually needed at each stage of the construction, since at the $n$-th stage we need to know the homotopy groups of $MU \wedge S^{2n}$ and it is only property (5) that allows these to be identified with the homotopy groups of $MU \wedge^{\mathbb{L}} S^{2n}$ (which we know are just a shifted version of the homotopy groups of $MU$).

A similar subtle issue came up in §3.4.2. There we had a spectrum $X = \bigvee_\alpha (R \wedge S^{n_\alpha})$ and wanted to conclude that $\pi_*(X) \cong \bigoplus_\alpha \pi_{*-n_\alpha}(R)$. Given cofibrant spectra $E_\alpha$, general model category considerations show that $\bigvee_\alpha E_\alpha$ is the homotopy colimit of a directed system $E_{\alpha_1} \to E_{\alpha_1} \vee E_{\alpha_2} \to \cdots$ (possibly indexed by an ordinal larger than $\omega$). So property (7) implies that $\pi_*(\bigvee_\alpha E_\alpha) \cong \bigoplus_\alpha \pi_*(E_\alpha)$. The spectra $R \wedge S^{n_\alpha}$ need not be cofibrant, but if $\tilde{R} \to R$ is a cofibrant replacement in $Spectra$ then we can write

$$X = \bigvee_\alpha (R \wedge S^{n_\alpha}) \cong R \wedge \left( \bigvee_\alpha S^{n_\alpha} \right) \simeq \tilde{R} \wedge \left( \bigvee_\alpha S^{n_\alpha} \right) \cong \bigvee_\alpha (\tilde{R} \wedge S^{n_\alpha})$$

where we have used property (5) for the weak equivalence in the middle. Since the spectra $\tilde{R} \wedge S^{n_\alpha}$ are cofibrant, we can use the previously mentioned result to see that $\pi_*(X)$ is as desired.

Though not necessarily the most important applications of property (5), these are good examples of how that property can come to the rescue at key moments.

## 3.5    Diagram categories and spectra

With the exception of the EKMM model, all of the common model categories of spectra are built on the foundation of diagram categories. It is perhaps not immediately apparent from the classical definition, but a spectrum is a kind of diagram. The goal of this section is to survey the general theory of model structures on diagram categories, and then to explain how spectra can be regarded as diagrams. This whole "diagrammatic" perspective is one of the main points of [178].

### 3.5.1    Model category structures on diagram categories

Let $\mathcal{M}$ be a category and let $I$ be a small category. We write $\mathcal{M}^I$ for the category whose objects are the functors $X \colon I \to \mathcal{M}$ and whose morphisms are natural transformations. Such functors are also called $I$-**diagrams** in $\mathcal{M}$. When $\mathcal{M}$ has a notion of weak equivalence, $\mathcal{M}^I$ can be equipped with the **objectwise weak equivalences**, namely the maps $X \to Y$ such that $X_i \to Y_i$ is a weak equivalence for every object $i$ in $I$. These are sometimes called **levelwise weak equivalences** as well.

If $\mathcal{M}$ has a model structure then one might expect there to be an associated model structure on $\mathcal{M}^I$ built around the objectwise weak equivalences, but unfortunately this doesn't seem to work out unless one assumes some extra hypotheses on $\mathcal{M}$.

**Theorem 3.5.1.** *Let $\mathcal{M}$ be a model category and let $I$ be a small category.*

(a) *If $\mathcal{M}$ is cofibrantly generated then there is a model category structure on $\mathcal{M}^I$ in which a map $f \colon X \to Y$ is a weak equivalence (resp., fibration) if and only if $f_i \colon X_i \to Y_i$ is a weak equivalence (resp., fibration) for all objects $i$ in $I$. This is called the **projective model structure** on $\mathcal{M}^I$. The cofibrations are forced to be those maps satisfying the left lifting property with respect to the trivial fibrations.*

(b) *If $\mathcal{M}$ is combinatorial (cofibrantly generated and locally presentable) then there is a model category structure on $\mathcal{M}^I$ in which a map $f \colon X \to Y$ is a weak equivalence (resp.,* *cofibration) if and only if $f_i\colon X_i \to Y_i$ is a weak equivalence (resp., cofibration) for all objects $i$ in $I$. This is called the* **injective** *model structure on $\mathcal{M}^I$. The fibrations are forced to be those maps satisfying the right lifting property with respect to the trivial cofibrations.*

Both parts (a) and (b) were proven by Heller [117, Theorem II.4.5] in the case $\mathcal{M} = s\mathcal{S}et$, with (b) also following from work of Jardine in this case [135]. For part (a) in the above generality, see [124, Theorem 11.6.1]. Part (b) in the above generality is due to Jeff Smith; it follows from [35, Theorem 1.7 and Propositions 1.15, 1.18], using the forgetful functor $\mathcal{M}^I \to \prod_{i \in I} \mathcal{M}$ as the "detection functor" for Beke's Proposition 1.18.

Let us say a little about how Theorem 3.5.1 is proven, since the main idea is easy and also useful in a variety of situations. For each $i$ in $I$ there are adjoint functors

$$F_i\colon \mathcal{M} \rightleftarrows \mathcal{M}^I : \mathrm{Ev}_i,$$

where the right adjoint $\mathrm{Ev}_i$ is the "evaluation at $i$" functor. The diagram $F_i X$ is the free diagram generated by starting with an $X$ at spot $i$. One readily checks that for each $X$ in $\mathcal{M}$ and $j$ in $I$,

$$(F_i X)(j) = \coprod_{I(i,j)} X.$$

That is, $(F_i X)(j)$ is a coproduct of copies of $X$ indexed by $I(i,j)$. For $T$ a set it is convenient to write $T \odot X$ for the coproduct of copies of $X$ generated by $T$, so that $(F_i X)(j) = I(i,j) \odot X$.

Start with sets $\{f_\alpha\colon A_\alpha \rightarrowtail B_\alpha\}$ and $\{\tilde{f}_\alpha\colon \tilde{A}_\alpha \xrightarrow{\sim} \tilde{B}_\alpha\}$ of generating cofibrations and trivial cofibrations for $\mathcal{M}$. The collections $\mathcal{I} = \{F_i(f_\alpha)\}_{i,\alpha}$ and $\mathcal{J} = \{F_i(\tilde{f}_\alpha)\}_{i,\alpha}$ are potential sets of generating cofibrations and trivial cofibrations for $\mathcal{M}^I$: the maps with the right lifting property with respect to $\mathcal{I}$ and $\mathcal{J}$ are clearly the objectwise trivial fibrations and the objectwise fibrations, respectively. The only thing nontrivial in setting up the projective model category structure is the factorization axiom, and this can be proven by the small object argument — it works in $\mathcal{M}^I$ as long as it worked in $\mathcal{M}$, which is the cofibrant-generation assumption. This proves (a).

Another way of describing the proof of (a) is to package all the pairs $(F_i, \mathrm{Ev}_i)$ into a single adjoint pair

$$F\colon \prod_{i \in I} \mathcal{M} \rightleftarrows \mathcal{M}^I : \mathrm{Ev}.$$

Kan's Recognition Theorem [124, Theorem 11.3.2] immediately yields that the right adjoint Ev creates the projective model structure on $\mathcal{M}^I$.

The proof of (b) works a little differently; it is a direct descendant of the classical proof that categories of sheaves have enough injectives. Here one fixes a large cardinal $\lambda$ (depending on $I$ and $\mathcal{M}$) and looks at a skeletal set of all objectwise cofibrations (or objectwise trivial cofibrations) where the domain and codomain are both $\lambda$-small. The $\lambda$-small conditions guarantee that the isomorphism classes of such things actually form a set and not a proper class. By choosing $\lambda$ large enough, one can show that these give generating cofibrations and trivial cofibrations for the desired injective model structure.

*Remark 3.5.2.* The cofibrations in the projective model structure on $\mathcal{M}^I$ are often called "projective cofibrations". For general $I$ they are hard to identify explicitly, but for some special classes of indexing categories $I$ this can be done. One such class consists of the "upwards-directed Reedy categories", that is, categories whose objects can be assigned a degree in $\mathbb{N}$ in such a way that all non-identity maps raise degree. Maps of diagrams over such categories can be built inductively, degree by degree, and this is what makes it easy to identify the projective cofibrations. See Corollary 3.5.8 below for an example, or [75, Section 14] for a detailed discussion.

*Remark 3.5.3* (Comparing diagram categories). Suppose $f : I \rightarrow J$ is a functor between small categories. Then there is an induced "restriction" map $f_* : \mathcal{M}^J \rightarrow \mathcal{M}^I$, obtained by precomposition with $f$. The functor $f_*$ has a left adjoint $f^*$ given by left Kan extension, and the pair $(f^*, f_*)$ is a Quillen pair between the projective model structures (since $f_*$ clearly preserves objectwise fibrations and trivial fibrations). We will often make use of this Quillen pair.

We will not have need of the following, but note that $f_*$ also has a right adjoint $f_!$ given by right Kan extension, and the pair $(f_*, f_!)$ is a Quillen pair when $\mathcal{M}^I$ and $\mathcal{M}^J$ are given the injective model structures.

*Remark 3.5.4.* We have assumed $I$ is a small category, otherwise we run into set-theoretic difficulties in constructing $\mathcal{M}^I$. However, in applications one often wants to apply these ideas to non-small categories as well. One typical approach is to fix a Grothendieck universe and to redefine "small" to mean "small with respect to the universe". Then one can still construct $\mathcal{M}^I$ for non-small $I$, but at the expense of passing to a larger universe.

If $I_0 \hookrightarrow I$ is a small skeletal subcategory, the adjoint functors from Remark 3.5.3 give an equivalence between $\mathcal{M}^I$ and $\mathcal{M}^{I_0}$. So one could instead just use $\mathcal{M}^{I_0}$ as a substitute for $\mathcal{M}^I$ and thereby *avoid* passing to the larger universe.

In practice a combination of these two ideas is often used, mostly without explanation. When $I$ has a small skeletal subcategory one *can* stay on firm ground by using $\mathcal{M}^{I_0}$, and common practice is to regard this as allowing one to use $\mathcal{M}^I$ with impunity.

## 3.5.2    Enriched diagrams

If $I$ is a category enriched over $s\mathcal{S}et$ and $\mathcal{M}$ is a simplicial model category, then one can look at enriched diagrams $X : I \rightarrow \mathcal{M}$. These are collections of objects $X_i$ for $i \in I$ together with maps of simplicial sets $I(i,j) \rightarrow \underline{\mathcal{M}}(X_i, X_j)$ that satisfy the evident unital and associativity axioms. Here we will write $\mathcal{M}^I$ for the category of enriched diagrams, with the comment that in practice this abuse of notation never leads to any confusion. The analog of Theorem 3.5.1 still holds for enriched diagrams, and the proof is the same. The only modification is to realize that here one has $(F_i X)(j) = I(i,j) \otimes X$, where the simplicial tensor now replaces the previous $\odot$ symbol.

Similar results hold when $\mathcal{M}$ is a model category enriched over $\mathcal{T}op$ (satisfying the analog of SM7) and $I$ is a $\mathcal{T}op$-enriched category, or the same with $\mathcal{T}op$ replaced by $\mathcal{T}op_*$. This will be the case most relevant to spectra.

### 3.5.3        Spectra and diagram categories

Classically, a spectrum is a sequence of pointed spaces $X_n$ together with maps $\Sigma X_n \to X_{n+1}$. Such an object does not manifestly suggest a diagram, but it turns out that spectra are *precisely* certain enriched diagrams. The key here is to realize that a map $\Sigma X_n \to X_{n+1}$ corresponds under the usual adjunctions to a pointed map $S^1 \to \underline{Top}_*(X_n, X_{n+1})$ (where $\underline{Top}_*(A, B)$ denotes the *space* of maps from $A$ to $B$).

Define a $Top_*$-enriched category $\Theta$ where the objects are non-negative integers $n$, and where

$$\Theta(k, n) = \begin{cases} * & \text{if } k > n, \\ (S^1)^{\wedge(n-k)} & \text{if } k \leq n. \end{cases}$$

The pairings $\Theta(l, n) \wedge \Theta(k, l) \to \Theta(k, n)$ are the canonical maps obtained from the associativity isomorphisms for the smash product in $Top_*$, and the identity maps in $\Theta(n, n)$ are given by the non-basepoint in $S^0$. It is routine to check that this really is a $Top_*$-enriched category. Here is a depiction of the first few levels of $\Theta$:

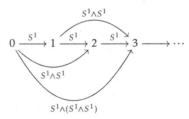

At this point it is an exercise to check that a classical spectrum is the same as an enriched diagram $\Theta \to Top_*$.

### 3.5.4        The level model structure on classical spectra

We are going to construct this model category in two ways: by brute force (as is normally done) and then by the diagrammatic perspective. The two ways are really the same, but it is informative to see that firsthand.

So for the moment let us pause and start from scratch. A spectrum $X$ is a sequence of pointed spaces $X_n$ for $n \geq 0$ together with structure maps $\sigma_n \colon \Sigma X_n \to X_{n+1}$. A map of spectra $f \colon X \to Y$ is a collection of pointed maps $f_n \colon X_n \to Y_n$ such that the diagrams

$$
\begin{array}{ccc}
\Sigma X_n & \xrightarrow{\sigma_X} & X_{n+1} \\
{\scriptstyle f_n}\downarrow & & \downarrow{\scriptstyle f_{n+1}} \\
\Sigma Y_n & \xrightarrow{\sigma_Y} & Y_{n+1}
\end{array}
$$

all commute. Let $Sp^{\mathbb{N}}$ denote the resulting category.

Let $Ev_n \colon Sp^{\mathbb{N}} \to Top_*$ be the functor $X \mapsto X_n$. This has a left adjoint which takes a pointed space $W$, puts it in level $n$, and generates a spectrum from that information in the freest way possible. Specifically, one readily checks that

$$(F_n W)_k = \begin{cases} * & \text{if } k < n, \\ \Sigma^{k-n} W & \text{if } k \geq n, \end{cases}$$

with the evident structure maps.

*Exercise 3.5.5.* Check that $\text{Ev}_n$ also has a right adjoint $I_n \colon \mathcal{T}op_* \to \text{Sp}^{\mathbb{N}}$, that sends a pointed space $W$ to the spectrum with

$$(I_n W)_k = \begin{cases} \Omega^{n-k} W & \text{if } k \leq n, \\ * & \text{if } k > n, \end{cases}$$

again with the evident structure maps.

As another exercise with these adjoints, observe that there are canonical maps $F_{n+1}(\Sigma W) \to F_n W$ and $I_n W \to I_{n-1}(\Omega W)$. The first is an isomorphism in degrees larger than $n$, and the second is an isomorphism in degrees lower than $n$.

**Theorem 3.5.6.** *There exists a model category structure on $\text{Sp}^{\mathbb{N}}$ in which a map $f \colon X \to Y$ is a weak equivalence (resp., fibration) if and only if $f_n \colon X_n \to Y_n$ is a weak equivalence (resp., fibration) for all $n$. This is called the **projective, level model structure** on $\text{Sp}^{\mathbb{N}}$.*

*Additionally, the adjoint pairs*

$$\mathcal{T}op_* \xrightarrow[\text{Ev}_n]{F_n} \text{Sp}^{\mathbb{N}} \qquad \text{and} \qquad \text{Sp}^{\mathbb{N}} \xrightarrow[I_n]{\text{Ev}_n} \mathcal{T}op_*$$

*are Quillen pairs (with the left adjoint always drawn on top, going left to right).*

*Proof.* We explain the proof in two ways. The first is to take the generating cofibrations and trivial cofibrations in $\mathcal{T}op_*$ and apply all the functors $F_n$ to them, thereby getting generating sets for $\text{Sp}^{\mathbb{N}}$. The model structure then basically constructs itself, using the small object argument. The second way, which says the same thing, is to use the observation that $\text{Sp}^{\mathbb{N}}$ is secretly the category $\mathcal{T}op_*^{\Theta}$ and then use Theorem 3.5.1(a).

For the statements about Quillen pairs, the right adjoints $\text{Ev}_n$ and $I_n$ clearly preserve fibrations and trivial fibrations. $\square$

*Remark 3.5.7.* Using Theorem 3.5.1(b) there is also a "level, injective" model category structure on $\text{Sp}^{\mathbb{N}}$, which is sometimes useful. However, the model structures derived from the projective one end up having better properties when we get to symmetric and orthogonal spectra. See Remark 3.7.8(2).

The category $\Theta$ acts like an upwards-directed Reedy category, in the sense that all the interesting maps raise degree. As in Remark 3.5.2, this is a case where we can explicitly identify the projective cofibrations:

**Corollary 3.5.8.** *A map of spectra $f \colon X \to Y$ is a cofibration in the projective, level model structure if and only if the evident maps*

$$X_n \sqcup_{\Sigma X_{n-1}} \Sigma Y_{n-1} \longrightarrow Y_n$$

*are cofibrations for all $n$, where by convention we set $X_{-1} = Y_{-1} = *$.*

*Sketch of proof.* Let $W \xrightarrow{\sim} Z$ be a levelwise trivial fibration of spectra, and suppose given a square

$$
\begin{array}{ccc}
X & \longrightarrow & W \\
\downarrow & & \downarrow \\
Y & \longrightarrow & Z
\end{array}
$$

We will attempt to produce a lifting $Y \to W$ by constructing it inductively on the levels. At level 0 we have the diagram

$$
\begin{array}{ccc}
X_0 & \longrightarrow & W_0 \\
\downarrow & & \downarrow{\scriptstyle\simeq} \\
Y_0 & \longrightarrow & Z_0
\end{array}
$$

and so get a lifting if $X_0 \to Y_0$ is a cofibration. At level 1 we have a similar diagram, but we can't just take any lifting — because we need the map $Y_1 \to W_1$ to be compatible with the already chosen $Y_0 \to W_0$. This compatibility is encoded by the diagram

$$
\begin{array}{ccc}
X_1 \amalg_{\Sigma X_0} (\Sigma Y_0) & \longrightarrow & W_1 \\
\downarrow & & \downarrow{\scriptstyle\simeq} \\
Y_1 & \longrightarrow & Z_1
\end{array}
$$

and we will get a lift provided the vertical map on the left is a cofibration. Continuing inductively in the evident manner, one sees that a map satisfying the conditions started in the corollary is a cofibration in the projective level model structure.

For the converse, assume $X \to Y$ is a projective cofibration and suppose given a lifting diagram

$$
\begin{array}{ccc}
X_n \amalg_{\Sigma X_{n-1}} \Sigma Y_{n-1} & \longrightarrow & W \\
\downarrow & & \downarrow{\scriptstyle\simeq} \\
Y_n & \longrightarrow & Z
\end{array}
$$

This adjoints over to a diagram

$$
\begin{array}{ccc}
X & \longrightarrow & I_n W \\
\downarrow & & \downarrow \\
Y & \longrightarrow & I_n Z \times_{I_{n-1}(\Omega Z)} I_{n-1}(\Omega W)
\end{array}
$$

and the right vertical map is a levelwise trivial fibration by inspection, so there is a lifting. Now adjoint back. $\square$

*Remark 3.5.9.* The level model structure is a rather formal thing, not capturing any kind of stabilization phenomenon. It treats spectra as mere diagrams, and not as true stable objects. For example, a spectrum $X$ and its truncation $\{*, X_1, X_2, \ldots\}$ should represent the same "stable object", but the level model structure sees them as different. The suspension functor on $\mathrm{Sp}^{\mathbb{N}}$ just applies suspension objectwise, and clearly this is not an equivalence on the homotopy category level — so we do not have a stable model category. In Section 3.6 we will see how to impose relations into the level model structure that incorporate stability.

### 3.5.5 The level model structure on coordinate-free spectra

This is an easy modification of what we have already done. Fix a May universe $\mathcal{U}$, as in Section 3.1.5. For $V \subseteq W \subseteq \mathcal{U}$, write $W - V$ for the orthogonal complement of $V$ in $W$. Define a **coordinate-free spectrum** to be an assignment $V \mapsto X_V$ for $V \subseteq \mathcal{U}$ a finite-dimensional subspace, together with maps $S^{W-V} \wedge X_V \to X_W$ for every pair $V \subseteq W$, subject to the evident unital and associativity conditions. Write $\mathrm{Sp}^{\mathcal{U}}$ for the evident category of coordinate-free spectra on $\mathcal{U}$.

Define a $\mathcal{T}op_*$-enriched category $\Theta_{\mathcal{U}}$ whose objects are the finite-dimensional subspaces of $\mathcal{U}$. Let the morphisms be given by

$$\Theta_{\mathcal{U}}(V, W) = \begin{cases} S^{W-V} & \text{if } V \subseteq W, \\ * & \text{otherwise.} \end{cases}$$

For $V \subseteq W \subseteq Z$, the evident isomorphism $S^{Z-W} \wedge S^{W-V} \to S^{Z-V}$ gives a composition map for $\Theta$ that is readily checked to be unital and associative. Observe that an enriched diagram $\Theta_{\mathcal{U}} \to \mathcal{T}op_*$ is the same as a coordinate-free spectrum defined on $\mathcal{U}$.

The projective model structure on the diagram category $\mathcal{T}op_*^{\Theta_{\mathcal{U}}}$ is called the projective, level model structure on $\mathrm{Sp}^{\mathcal{U}}$.

To compare this construction to classical spectra, pick an orthonormal basis $e_1, e_2, \ldots$ for $\mathcal{U}$ and let $\mathbb{R}^n$ be the span of the first $n$ basis elements. Consider the particular linear map $\mathbb{R} \to \mathbb{R}^{n+1} - \mathbb{R}^n$ sending $1$ to $e_{n+1}$, so that compactifying gives us a preferred homeomorphism $S^1 \cong S^{(\mathbb{R}^{n+1} - \mathbb{R}^n)}$. If $X$ is a coordinate-free spectrum then the assignment $[n] \mapsto X_{\mathbb{R}^n}$ gives a classical spectrum. Let $U \colon \mathrm{Sp}^{\mathcal{U}} \to \mathrm{Sp}^{\mathbb{N}}$ denote this forgetful functor. From the diagrammatic viewpoint we have described an embedding $j \colon \Theta \hookrightarrow \Theta_{\mathcal{U}}$ and $U$ is just restriction along this embedding. Category theory automatically tells us that $U$ has a left adjoint $G$: it sends a spectrum $X \colon \Theta \to \mathcal{T}op_*$ to its left Kan extension along $j$. Note that $(GX)_V$ is an appropriate (enriched) colimit over the category of all $\mathbb{R}^n$ contained in $V$. One easy but important fact is that the map $X_n \to (UGX)_n$ is an isomorphism, for all $n$.

It is formal that the pair $G \colon \mathrm{Sp}^{\mathbb{N}} \rightleftarrows \mathrm{Sp}^{\mathcal{U}} \colon U$ is a Quillen pair, since $U$ preserves fibrations and trivial fibrations. It is of course not a Quillen equivalence, because we are using the levelwise model structures. This will change when we pass to the stable model structures in the next section.

*Remark 3.5.10* (Change of universe). Suppose that $\mathcal{U}$ and $\mathcal{U}'$ are two May universes, and $f \colon \mathcal{U} \to \mathcal{U}'$ is an isometry (which will necessarily be injective, but possibly not surjective). Then there is an enriched functor $\Theta_{\mathcal{U}} \to \Theta_{\mathcal{U}'}$ that on objects behaves as $V \mapsto f(V)$ and on maps as $S^{W-V} \mapsto S^{f(W)-f(V)}$ (induced by $f$). We therefore get a restriction functor $f_* \colon \mathcal{T}op_*^{\Theta_{\mathcal{U}'}} \to \mathcal{T}op_*^{\Theta_{\mathcal{U}}}$ and its left adjoint $f^*$ as in Remark 3.5.3.

Again, these are not Quillen equivalences — but their analogs will become Quillen equivalences after stabilization.

## 3.6     Localization and the stable model structures on spectra

In this section we will see how to modify the level model structure on spectra in a way that captures true stable phenomena. This uses a technique that is now called Bousfield localization, although it of course did not have this name when it first appeared back in [56]. Here we review the relevant model category theoretic techniques and then we repeat the work of [56] to obtain the stable structure on spectra. This works in both the classical and coordinate-free contexts. See also Chapter 7 of this volume for more on Bousfield localization.

### 3.6.1     Homotopy mapping spaces

Let $\mathcal{M}$ be a model category. To any two objects $X$ and $Y$ in $\mathcal{M}$ one can associate a homotopy mapping space $\underline{\mathcal{M}}(X, Y)$, also sometimes called a homotopy function complex. This is a simplicial set, well defined up to weak homotopy equivalence, which only depends on the weak homotopy types of $X$ and $Y$. Given maps $X \to X'$ and $Y \to Y'$ one can construct models for these function complexes that come with maps $\underline{\mathcal{M}}(X', Y) \to \underline{\mathcal{M}}(X, Y)$ and $\underline{\mathcal{M}}(X, Y) \to \underline{\mathcal{M}}(X, Y')$.

Here are four standard ways to construct models of $\underline{\mathcal{M}}(X, Y)$:

(1) Replace $X$ by a cosimplicial resolution $QX^*$, choose a fibrant replacement $Y \to RY$, and use the simplicial set $\mathcal{M}(QX^*, RY)$ obtained by applying $\mathcal{M}(-, RY)$ to $QX^*$.

(2) Choose a cofibrant replacement $QX \to X$, a simplicial resolution $Y \to RY_*$, and use the simplicial set $\mathcal{M}(QX, RY_*)$.

(3) Use nerves of categories of zig-zags from $X$ to $Y$ to form the so-called *hammock localization space* $L_H \mathcal{M}(X, Y)$.

(4) When $\mathcal{M}$ is a simplicial model category, choose a cofibrant replacement $QX \to X$ and a fibrant replacement $Y \to RY$ and use the simplicial mapping space from $QX$ to $RY$.

See [124] and [130] for more on (1) and (2), and [88] or Chapter 2 of this volume for (3). But all the model categories considered in this chapter are simplicial, so feel free to focus on (4). Depending on the context one might also write $\mathrm{Map}(X, Y)$ or $\mathrm{hMap}(X, Y)$ as a synonym for $\underline{\mathcal{M}}(X, Y)$.

### 3.6.2     Localization of model categories

Given a model category $\mathcal{M}$ and a collection of maps $T$ in $\mathcal{M}$, one sometimes wants to construct a new model category structure that is obtained from $\mathcal{M}$ by adjoining the maps in $T$ to the already existing weak equivalences. This will likely force even more maps to be weak equivalences (at the very least one has to close up the set under two-out-of-three), and at least one of the notions of cofibration/fibration will have to change as well. The main technique for accomplishing this is called *Bousfield localization*.

**Definition 3.6.1.** Let $\mathcal{M}$ be a model category and let $T$ be a set of maps in $\mathcal{M}$.

(a) An object $X$ in $\mathcal{M}$ is $T$-**local** if, for all $f : A \to B$ in $T$, the induced map $\underline{\mathcal{M}}(B, X) \to \underline{\mathcal{M}}(A, X)$ is a weak equivalence.
(b) A map $f : A \to B$ is a $T$-**local equivalence** if, for all $T$-local objects $X$, the induced map $\underline{\mathcal{M}}(B, X) \to \underline{\mathcal{M}}(A, X)$ is a weak equivalence.

Briefly, an object $X$ is $T$-local if it sees all the maps in $T$ as weak equivalences, where "see" amounts to looking at things from the perspective of $\underline{\mathcal{M}}(-, X)$. Likewise, the $T$-local equivalences are the maps that are seen as weak equivalences by all the $T$-local objects. So the $T$-local equivalences include all of $T$, but will usually include other maps as well.

The following result is due to Hirschhorn [124] in the cellular case, and to Jeff Smith in the combinatorial case (see [35] for a written account):

**Theorem 3.6.2.** *Let $\mathcal{M}$ be a cellular or combinatorial model category, and let $T$ be a set of maps in $\mathcal{M}$. Then there exists a new model structure $T^{-1}\mathcal{M}$ on the same underlying category as $\mathcal{M}$ such that*

(i) *the cofibrations in $T^{-1}\mathcal{M}$ are the same as the cofibrations in $\mathcal{M}$,*
(ii) *the weak equivalences in $T^{-1}\mathcal{M}$ are the $T$-local equivalences,*
(iii) *the fibrations are the maps with the right lifting property with respect to cofibrations that are $T$-local equivalences.*

*Moreover, an object $X$ is fibrant in $T^{-1}\mathcal{M}$ if and only if $X$ is fibrant in $\mathcal{M}$ and $X$ is $T$-local. Finally, if $X$ and $Y$ are $T$-local then a map $f : X \to Y$ is a weak equivalence in $T^{-1}\mathcal{M}$ if and only if it is a weak equivalence in $\mathcal{M}$.*

When it exists, the model category $T^{-1}\mathcal{M}$ is called the **left Bousfield localization** of $\mathcal{M}$ at the set $T$. A fibrant replacement functor in $T^{-1}\mathcal{M}$ is called a $T$-**localization** functor.

*Remark 3.6.3.* It is useful to know a bit about how Theorem 3.6.2 is proven and about the construction of the localization functor. For each map in $T$ choose a model $f : A \rightarrowtail B$ that is a cofibration. For each simplicial horn $j : \Lambda^{n,k} \hookrightarrow \Delta^n$ consider the box product $j \square f$, which is the map

$$ j \square f : (\Lambda^{n,k} \otimes B) \amalg_{(\Lambda^{n,k} \otimes A)} (\Delta^n \otimes A) \longrightarrow \Delta^n \otimes B. $$

Here the tensor refers to the simplicial tensor if $\mathcal{M}$ is a simplicial model category, or more generally it refers to a version built using cosimplicial frames (see [124] for details). Formally add these maps $j \square f$ (for every $j$ and $f$) to the set of generating trivial cofibrations of $\mathcal{M}$, and then repeat the small object argument using this new set to get the required factorization. In particular, the localization functor is obtained as a transfinite composition of cobase changes of the generating trivial cofibrations in $\mathcal{M}$ together with the maps $j \square f$.

### 3.6.3          Bousfield–Friedlander spectra

If $X$ is a spectrum and $n \geq 0$, define the $n$-truncation $\tau_{\geq n} X$ to be the spectrum $\{*, *, \ldots, *, X_n, X_{n+1}, \ldots\}$. There is a natural map $\tau_{\geq n} X \to X$. Our basic goal will be to localize the level, projective model structure on spectra at the class $\{\tau_{\geq n} X \to X \mid n, X\}$. However, this class is not a set and so the first task is to pare it down somewhat. To this end, define

$$\mathcal{T} = \{\tau_{\geq (n+1)} F_n(S^k) \to F_n(S^k) \mid n, k \geq 0\}.$$

Observe that $\tau_{\geq (n+1)} F_n(X)$ is canonically isomorphic to $F_{n+1}(\Sigma X)$, so we can also describe the set as

$$\mathcal{T} = \{F_{n+1}(S^{k+1}) \to F_n(S^k) \mid n, k \geq 0\},$$

where the map in question is adjoint to the identity $S^{k+1} \to \mathrm{Ev}_{n+1}(F_n S^k)$.

**Definition 3.6.4.** The **stable projective model structure** on $\mathrm{Sp}^{\mathbb{N}}$ is the localization of the level projective model structure at the set $\mathcal{T}$.

Let us analyze the $\mathcal{T}$-local objects. Here the relevant observation is that

$$\underline{\mathrm{Sp}^{\mathbb{N}}}\big(F_n(S^k), X\big) \simeq \underline{\mathcal{T}op}_*(S^k, X_n)$$

by adjoint functors. If $f$ denotes our map $F_{n+1}(S^{k+1}) \to F_n(S^k)$ then on mapping spaces this is

$$
\begin{array}{ccc}
\underline{\mathrm{Sp}^{\mathbb{N}}}\big(F_n S^k, X\big) & \longrightarrow & \underline{\mathrm{Sp}^{\mathbb{N}}}\big(F_{n+1} S^{k+1}, X\big) \\
\simeq \downarrow & & \simeq \downarrow \\
\underline{\mathcal{T}op}_*(S^k, X_n) & \longrightarrow & \underline{\mathcal{T}op}_*(S^{k+1}, X_{n+1}) = \underline{\mathcal{T}op}_*(S^k, \Omega X_{n+1})
\end{array}
$$

and one checks that the lower horizontal composite is induced by the structure map $X_n \to \Omega X_{n+1}$. So a spectrum $X$ is $\mathcal{T}$-local precisely when it is an $\Omega$-spectrum.

*Remark 3.6.5.* We only needed $k = 0$ to make this argument. So the maps in $\mathcal{T}$ where $k > 0$ represent redundant information, and we could throw them out of $\mathcal{T}$ and still get the same localization.

For the following result, recall that if $X$ is a spectrum and $n \in \mathbb{Z}$ then

$$\pi_n(X) = \mathrm{colim}_k \, \pi_{n+k}(X_k)$$

where the maps in the colimit system are induced by the structure maps of $X$.

**Proposition 3.6.6.** *In the stable projective model structure on* $\mathrm{Sp}^{\mathbb{N}}$,

(a) *the fibrant objects are the levelwise fibrant $\Omega$-spectra, and*

(b) *a map* $f\colon X \to Y$ *is a weak equivalence if and only if it induces isomorphisms* $\pi_n(X) \to \pi_n(Y)$ *for all* $n \in \mathbb{Z}$.

Note that the levelwise fibrant condition is vacuous if we are defining spectra in terms of topological spaces, but not if we are doing so in terms of simplicial sets.

*Proof.* We have already proven (a). For (b), first note that for a map of $\Omega$-spectra the notions of level weak equivalence, $\pi_*$-isomorphism, and stable equivalence all coincide: level equivalence = stable equivalence by the last line of Theorem 3.6.2, and level equivalence = $\pi_*$-isomorphism by inspection.

Next consider the map $f_{n,k}\colon F_{n+1}(S^{k+1}) \to F_n(S^k)$. This is an isomorphism in levels $n+1$ and higher, so this same property passes to any cobase change. Hence any cobase change of an $f_{n,k}$ is a $\pi_*$-isomorphism. Similarly, for any set of horns $j\colon \Lambda^{p,r} \hookrightarrow \Delta^p$ the box product $j \,\square\, f_{n,k}$ is also an isomorphism in levels $n+1$ and higher. It follows that any map obtained from these box products by cobase changes and transfinite compositions is a $\pi_*$-isomorphism. In particular, the fibrant replacement functors $X \to RX$ in the stable projective structure are made this way (see Remark 3.6.3) and are therefore $\pi_*$-isomorphisms.

Finally, suppose given a map of spectra $g\colon X \to Y$ and consider the square

$$
\begin{array}{ccc}
X & \longrightarrow & RX \\
{\scriptstyle g}\downarrow & & \downarrow{\scriptstyle Rg} \\
Y & \longrightarrow & RY
\end{array}
$$

The horizontal maps are both stable equivalences and $\pi_*$-isomorphisms. So $g$ is a stable equivalence (resp., $\pi_*$-isomorphism) if and only if $Rg$ is so. But $RX$ and $RY$ are $\Omega$-spectra, so the conditions of $Rg$ being a stable equivalence or $\pi_*$-isomorphism are equivalent; hence, the same must hold for $g$. □

In general, it can be very hard to give a nice description for the fibrations in a Bousfield localization. In the present case one can actually do it, though. Note that since there are more trivial cofibrations in $\mathcal{T}^{-1}\mathcal{M}$ than in $\mathcal{M}$, there will be fewer fibrations.

**Proposition 3.6.7.** *For a spectrum $X$, let $QX = \mathrm{hocolim}_n \Omega^n X_n$. Then a map of spectra $X \to Y$ is a fibration in the projective stable structure on $\mathrm{Sp}^{\mathbb{N}}$ if and only if it is a levelwise fibration and for every $n \geq 0$ the square*

$$
\begin{array}{ccc}
X_n & \longrightarrow & QX \\
\downarrow & & \downarrow \\
Y_n & \longrightarrow & QY
\end{array}
$$

*is homotopy Cartesian.*

*Proof.* See [56]. In that paper the projective stable category is not constructed by Bousfield localization, but directly by brute force. The cofibrations and weak equivalences, however, match the ones in our structure, and fibrations are always determined by the trivial cofibrations, so the two structures are in fact the same. □

### 3.6.4    The coordinate-free setting

Recall the coordinate-free setting of Section 3.5.5. Here we localize at the maps $F_W(S^{W-V} \wedge S^k) \to F_V(S^k)$ for all $k$ and all pairs $V \subseteq W \subseteq \mathcal{U}$. The functor $G$ from Section 3.5.5 sends the maps in $\mathcal{J}$ to these kinds of maps, so by general localization theory the adjoint pair $(G, U)$ descends to give Quillen functors between the resulting stable model categories:

$$G \colon \mathrm{Sp}^{\mathbb{N}}_{stable} \rightleftarrows \mathrm{Sp}^{\mathcal{U}}_{stable} \colon U.$$

By the same arguments that we have seen for $\mathrm{Sp}^{\mathbb{N}}$, the stable equivalences in $\mathrm{Sp}^{\mathcal{U}}$ are all $\pi_*$-isomorphisms. Since $X \to UGX$ is a levelwise isomorphism (see Section 3.5.5), it follows at once that the above pair is a Quillen equivalence.

We leave the reader to think about change of universe in this setting, building off of Remark 3.5.10.

## 3.7    Symmetric spectra

The definitions and basic results about symmetric spectra are very elegant and beautiful. Understanding the details of what is going on beneath the surface is a different matter. Our approach here will be to quickly survey the basic theory from [133] and then go back and work on some motivation afterwards.

**Definition 3.7.1.** A **symmetric sequence** in a category $\mathcal{C}$ is a collection of objects $X_n$ together with group homomorphisms $\Sigma_n \to \mathrm{Aut}(X_n)$, for each $n \geq 0$.

It will be convenient to have a more diagrammatic way of phrasing this definition. Let $\Sigma I$ be the subcategory of $\mathcal{S}et$ consisting of the objects $\underline{n} = \{1, 2, \ldots, n\}$ for $n \geq 0$ (with $\underline{0} = \emptyset$) together with all automorphisms. A symmetric sequence in $\mathcal{C}$ is simply a functor $X \colon \Sigma I \to \mathcal{C}$. As usual, we write $\mathcal{C}^{\Sigma I}$ for the category of all such functors.

Now assume that $(\mathcal{C}, \otimes, I, \mathcal{F})$ is closed symmetric monoidal and also cocomplete. Given symmetric sequences $X$ and $Y$, define a new symmetric sequence $X \otimes Y$ by

$$(X \otimes Y)_n = \coprod_{p+q=n} (\Sigma_n)_+ \odot_{\Sigma_p \times \Sigma_q} (X_p \otimes X_q).$$

To explain the $\odot$ notation, regard any group $G$ as a groupoid with one object and $G$ as its endomorphism group. If $H \leq G$ and $W$ is an object with a left $H$-action, then $G \odot_H W$ is the left Kan extension in the diagram

Equivalently, one can write

$$G \odot_H W = \mathrm{coeq}\Big( \coprod_{G \times H} W \rightrightarrows \coprod_G W \Big),$$

where the top map sends the copy of $W$ indexed by $(g, h)$ to the copy of $W$ indexed by $g$ via left multiplication by $h$, and the bottom map sends the copy of $W$ indexed by $(g, h)$ to the copy of $W$ indexed by $gh$ via the identity. The action of $G$ on $\coprod_G W$ by permutation of the factors descends to give a left action of $G$ on $G \odot_H W$.

There is a self evident, though tedious to write down, associativity isomorphism $X \otimes (Y \otimes Z) \cong (X \otimes Y) \otimes Z$. Define the twist isomorphism $\tau_{X,Y} : X \otimes Y \to Y \otimes X$ on level $n$ to be the coproduct of maps $\Sigma_n \odot_{\Sigma_a \times \Sigma_b} (X_a \otimes Y_b) \to \Sigma_n \odot_{\Sigma_b \times \Sigma_a} (Y_b \otimes X_a)$ (where $a + b = n$) sending $[\alpha, X_a \otimes Y_b]$ to $[\alpha \rho_{b,a}, Y_b \otimes X_a]$ via the twist map from $\mathcal{C}$, where $\rho_{b,a} \in \Sigma_n$ is the evident $(b, a)$-shuffle. It is a good exercise to check that without $\rho_{b,a}$ in the formula this is not a well-defined map, as it does not exhibit the required $\Sigma_a \times \Sigma_b$-equivariance; indeed, check that one needs to include a permutation $\rho$ having the property that $(\beta_a | \gamma_b) \circ \rho = \rho \circ (\gamma_b | \beta_a)$ for every $\beta_a \in \Sigma_a$, $\gamma_b \in \Sigma_b$. The only permutation that does the job is $\rho = \rho_{b,a}$. (For a general schema that helps one quickly determine the correct permutation to use in situations like this, see Remark 3.7.9).

When $\mathcal{C}$ is complete one can also define a cotensor $X, Y \mapsto \mathcal{F}(X, Y)$ for symmetric sequences. Before giving the definition, let us record the basic property it should have:

**Lemma 3.7.2.** *Let $X$, $Y$, and $Z$ be symmetric sequences in $\mathcal{C}$. There are natural bijections between the following three sets:*

(1) $\mathcal{C}^{\Sigma I}(X \otimes Y, Z)$;
(2) *collections of $\Sigma_p \times \Sigma_q$-equivariant maps $X_p \otimes Y_q \to Z_{p+q}$ for all $p, q \geq 0$;*
(3) $\mathcal{C}^{\Sigma I}(X, \mathcal{F}(Y, Z))$.

Parts (2) and (3) of the lemma lead one directly to the definition of the cotensor. For $X$ and $Y$ in $\mathcal{C}^{\Sigma I}$ define $\mathcal{F}(X, Y)$ by

$$\mathcal{F}(X, Y)_n = \prod_q \mathcal{F}(X_q, Y_{n+q})^{\Sigma_q},$$

where the $\Sigma_q$ action is as follows. If $\alpha \in \Sigma_q$ then we have maps $\alpha_X : X_q \to X_q$ and $(id_n | \alpha)_Y : Y_{n+q} \to Y_{n+q}$, where $(id_n | \alpha) \in \Sigma_{n+q}$ is the map that permutes the last $q$ elements according to $\alpha$. Then $\alpha$ acts on $\mathcal{F}(X_q, Y_{n+q})$ via the composite

$$\mathcal{F}(X_q, Y_{n+q}) \xrightarrow{((id_n|\alpha)_Y)_*} \mathcal{F}(X_q, Y_{n+q}) \xrightarrow{(\alpha_X^{-1})^*} \mathcal{F}(X_q, Y_{n+q}).$$

This gives an action of $\Sigma_q$, and $\mathcal{F}(X_q, Y_{n+q})^{\Sigma_q}$ is the fixed object (the limit of the corresponding functor $\Sigma_q \to \mathcal{C}$). The action of $\Sigma_n$ on $Y_{n+q}$ coming from permutation of the first block of $n$ elements descends to an action of $\Sigma_n$ on $\mathcal{F}(X_q, Y_{n+q})^{\Sigma_q}$.

The following is a routine exercise:

**Proposition 3.7.3.** *With the above associativity and twist isomorphisms, the tensor product on $\mathcal{C}^{\Sigma I}$ is closed symmetric monoidal with unit $\mathbb{I} = \{I, \emptyset, \emptyset, \ldots\}$ and cotensor $\mathcal{F}(-, -)$.*

Now fix any object $X$ in $\mathcal{C}$. Recall from Section 3.3.5 that $X^{\otimes n}$ is defined inductively by $X^{\otimes n} = X \otimes X^{\otimes(n-1)}$, and that there is a natural left action of $\Sigma_n$ on $X^{\otimes n}$. Define $\mathbb{X}$ to be the symmetric sequence $\mathbb{X}_n = X^{\otimes n}$, and let $\mathbb{I} \to \mathbb{X}$ be the unique map that is the identity in level 0.

The associativity maps give natural isomorphisms $\mu_{a,b} \colon \mathbb{X}_a \otimes \mathbb{X}_b \to \mathbb{X}_{a+b}$. We use these to define a pairing $\mathbb{X} \otimes \mathbb{X} \to \mathbb{X}$ that on level $n$ is the coproduct of maps

$$\Sigma_n \otimes_{\Sigma_a \times \Sigma_b} (X^{\otimes a} \otimes X^{\otimes b}) \to X^{\otimes(a+b)},$$

which on the summand $[\alpha, X^{\otimes a} \otimes X^{\otimes b}]$ equals $\alpha \circ \mu_{a,b}$. One readily checks that this is well-defined and makes $\mathbb{X}$ into a commutative monoid. The category of left $\mathbb{X}$-modules then inherits a closed symmetric monoidal structure as in Section 3.3.2, where for example the tensor is $(-) \otimes_{\mathbb{X}} (-)$.

**Definition 3.7.4.** A **symmetric $X$-spectrum** is a left $\mathbb{X}$-module.

Unwinding the definitions, a left $\mathbb{X}$-module $M$ is a sequence of objects $M_n$ in $\mathcal{C}$ together with an action of $\Sigma_n$ on $M_n$ and structure maps

$$\alpha_{p,q} \colon X^{\otimes p} \otimes M_q \to M_{p+q}$$

that are $\Sigma_p \times \Sigma_q$-equivariant. The unital condition says that $\alpha_{0,q}$ is the identity, and associativity says that for $p = a + b$ one has $\alpha_{p,q} = \alpha_{a,b+q} \circ (id \otimes \alpha_{b,q})$; that is, the diagram

$$
\begin{array}{ccc}
X^{\otimes a} \otimes (X^{\otimes b} \otimes M_q) & \xrightarrow{id \otimes \alpha_{b,q}} X^{\otimes a} \otimes M_{b+q} \xrightarrow{\alpha_{a,b+q}} & M_{a+b+q} \\
{\scriptstyle\cong}\downarrow & & \\
(X^{\otimes a} \otimes X^{\otimes b}) \otimes M_q & \xrightarrow{\ \cong\ } X^{\otimes(a+b)} \otimes M_q & \\
\end{array}
$$

is commutative. This implies that the maps $\alpha_{p,q}$ with $p > 1$ can be built up from the $\alpha_{1,*}$ maps.

So at the end of the day, a symmetric $\mathbb{X}$-spectrum is a collection of objects $M_n$ in $\mathcal{C}$ equipped with a left $\Sigma_n$-action and structure maps $\alpha \colon X \otimes M_n \to M_{n+1}$ having the property that the iterated structure maps

$$X^{\otimes p} \otimes M_n \to M_{n+p}$$

are $\Sigma_p \times \Sigma_n$-equivariant, for all $n, p \geq 0$. Here "iterated structure map" means an evident composition of associativity maps with the structure maps $\alpha$.

### 3.7.1 The model category of symmetric spectra

We now specialize to the case where $\mathcal{C}$ is $\mathcal{T}op_*$ and $X = S^1$. The spectrum $\mathbb{X} = \{S^0, S^1, S^2, \ldots\}$ is called the sphere spectrum and denoted simply by $S$. So symmetric spectra are just left $S$-modules. Write $Sp^{\Sigma}$ for the category of symmetric spectra. The evaluation map $Ev_n \colon Sp^{\Sigma} \to \mathcal{T}op_*$ has a left adjoint $F_n$ given by

$$(F_n X)_k = \begin{cases} * & \text{if } k < n, \\ \Sigma_k \otimes_{\Sigma_{k-n}} (S^{k-n} \wedge X) & \text{if } k \geq n, \end{cases}$$

where in the second line $\Sigma_{k-n}$ sits in $\Sigma_k$ as permutations of the front $(k-n)$-block. Note that there are canonical maps

$$F_{n+1}(S^1 \wedge X) \to F_n(X)$$

that are equal to the identity in level $n + 1$. (Warning: Unlike the case of Bousfield–Friedlander spectra, these maps are not isomorphisms in degrees larger than $n + 1$. See the discussion below for an example.)

**Proposition 3.7.5.** *There is a model category structure on* $\mathrm{Sp}^\Sigma$ *where the weak equivalences and fibrations are objectwise. This is called the **level, projective model structure**.*

*Proof.* One can do this directly using the functors $F_n$ and Kan's Recognition Theorem, just as we did for Bousfield–Friedlander spectra. Alternatively, one can realize that symmetric spectra are just certain enriched functors and use Theorem 3.5.1(a). See Section 3.7.4 below for more on this perspective. □

**Definition 3.7.6.** The **projective stable model structure** on $\mathrm{Sp}^\Sigma$ is the left Bousfield localization of the projective level model category structure at the set of maps $\{F_{n+1}(S^1 \wedge S^k) \to F_n(S^k) \mid n, k \geq 0\}$.

Say that a symmetric spectrum is an $\Omega$-spectrum if its underlying classical spectrum is an $\Omega$-spectrum. Here is the main foundational result about symmetric spectra, pulling together various statements from [133]:

**Theorem 3.7.7.**

(a) *The projective stable model structure on* $\mathrm{Sp}^\Sigma$ *is a stable, closed symmetric monoidal model category satisfying the Monoid Axiom as well as the Algebraic Creation and Invariance Properties.*

(b) *The fibrant objects are the objectwise fibrant $\Omega$-spectra.*

(c) *The forgetful functor* $U \colon \mathrm{Sp}^\Sigma \to \mathrm{Sp}^\mathbb{N}$ *has a left adjoint $G$, and the adjoint pair $G \colon \mathrm{Sp}^\mathbb{N} \rightleftarrows \mathrm{Sp}^\Sigma \colon U$ is a Quillen equivalence between the projective stable model structures.*

*Remarks 3.7.8.*

(1) Part (b) is automatic from the way we choose the maps to localize, just as for Bousfield–Friedlander spectra.

(2) In (a) it suffices to verify the Pushout-Product Axiom for box products of generating cofibrations and trivial cofibrations. This is where it is finally important that we started with the projective level structure and not the injective level structure. In the former, the generating maps are well understood and it is easy to analyze their box products. In the latter, there are far too many cofibrations and in fact the Pushout-Product Axiom does not hold.

(3) The Quillen equivalence in part (c) is not unexpected, but it is not as easy as one might think. The left adjoint just comes as in Remark 3.5.3, and the fact that it is a Quillen pair is easy. But the equivalence part takes a bit of work. See Section 3.10.3 for further discussion.

(4) The precise references for the different parts of Theorem 3.7.7 are these: monoidal model category, [133, 5.3.8]; monoid axiom, [133, 5.4.1]; Algebraic Creation Property, [133, 5.4.2 and 5.4.3]; Algebraic Invariance Property, [133, 5.4.5]; Strong Flatness Property, [133, 5.4.4]; Quillen equivalence with $\mathrm{Sp}^{\mathbb{N}}$, [133, 4.2.5].

The derived functors of the Quillen equivalence from Theorem 3.7.7(c) give an equivalence of categories

$$\mathrm{Ho}(\mathrm{Sp}^{\mathbb{N}}) \xrightleftharpoons[\mathbb{R}U]{\mathbb{L}G} \mathrm{Ho}(\mathrm{Sp}^{\Sigma}).$$

A common misconception is to confuse $\mathbb{R}U$ and $U$. That is, if $E$ is a symmetric spectrum, it is tempting to believe that the image of $E$ in $\mathrm{Ho}(\mathrm{Sp}^{\mathbb{N}})$ is represented by the underlying classical spectrum $UE$. This is false in general — an example is $E = F_1(S^1)$, discussed below. Two other related issues are these:

(1) The functor $U$ does *not* preserve all stable weak equivalences.

(2) If $X$ is a symmetric spectrum then define

$$\pi_n^{\mathrm{naive}}(X) = \pi_n(UX) = \operatorname{colim}_k \pi_{n+k}(X_k).$$

It is *not* true that all stable weak equivalences induce isomorphisms on $\pi_*^{\mathrm{naive}}(-)$. In particular, the groups $\pi_*^{\mathrm{naive}}(X)$ are not guaranteed to be the "correct" homotopy groups unless $X$ is fibrant.

One source of confusion here is that $\pi_*^{\mathrm{naive}}(X)$ sometimes *are* the correct homotopy groups even when $X$ is not fibrant. The paper [261] gives a detailed discussion of which spectra $X$ are well-behaved in this regard.

The following example from [133, Example 3.1.10] demonstrates (1) and (2) above. It is worth examining in some detail. Consider the canonical map $f : F_1(S^1) \to F_0(S^0)$ that is the identity in level 1. This is one of the maps we localized to form the stable model structure, so it is a stable weak equivalence by definition. Note that $F_0(S^0)$ is just the sphere spectrum $S$. For $X$ any pointed space, $(F_1 X)_n = \Sigma_n \odot_{\Sigma_{n-1}} ((S^1)^{\wedge(n-1)} \wedge X)$ for $n \geq 1$; in particular, $(F_1 S^1)_n = \Sigma_n \odot_{\Sigma_{n-1}} S^n$. As a space, this is a wedge of $n$ copies of $S^n$, and the copies may be regarded as indexed by the set of permutations $T_n = \{Id, (1n), (2n), \ldots, (n-1, n)\}$ (these are coset representatives for $\Sigma_n/\Sigma_{n-1}$). Our map $f$ takes the form

$$
\begin{array}{ccccc}
S^0 & S^1 & S^2 & S^3 & \cdots \\
\uparrow & \uparrow{\scriptstyle =} & \uparrow & \uparrow & \\
* & S^1 & \bigvee_{T_2}(S^1 \wedge S^1) & \bigvee_{T_3}(S^1 \wedge S^1 \wedge S^1) & \cdots
\end{array}
$$

where in each level the component indexed by $\alpha \in T_n$ is mapped into $S^n$ via the canonical identification followed by $\alpha$.

Of course we know $\pi_0^{\text{naive}}(S) = \mathbb{Z}$. The colimit system for $\pi_0^{\text{naive}}(F_1 S^1)$ is

$$0 \to \mathbb{Z} \hookrightarrow \mathbb{Z}^2 \hookrightarrow \mathbb{Z}^3 \hookrightarrow \mathbb{Z}^4 \hookrightarrow \cdots,$$

where in each case the group includes into the next as a direct summand. So $\pi_0^{\text{naive}}(F_1 S^1)$ is an infinite direct sum of copies of $\mathbb{Z}$. In particular, we see that $Uf$ is not a stable equivalence and (equivalently) that $f$ does not induce isomorphisms on $\pi_*^{\text{naive}}(-)$. Note that $\pi_*^{\text{naive}}(-)$ gives the "correct" answer for $S$, but not for $F_1 S^1$.

## 3.7.2    Understanding the smash product

Let us open up the definition of the smash product and look inside. If $X$ and $Y$ are symmetric spectra (left $S$-modules) recall that $X \wedge Y$ (also known as $X \wedge_S Y$) is the coequalizer of $X \otimes S \otimes Y \rightrightarrows X \otimes Y$. Note that here $X$ is being implicitly converted from a left $S$-module into a right $S$-module via the twist map. Looking level by level, we find that $(X \wedge Y)_n$ is the coequalizer of

$$\bigvee_{a+b+c=n} \Sigma_n \otimes_{\Sigma_a \times \Sigma_b \times \Sigma_c} (X_a \wedge (S^1)^{\otimes b} \wedge Y_c) \rightrightarrows \bigvee_{p+q=n} \Sigma_n \otimes_{\Sigma_p \times \Sigma_q} (X_p \wedge Y_q).$$

This looks scary, but we can tame things a bit by adopting a more algebraic notation, which we now pause to explain.

If $a + b + c + d + e = n$ write $\rho_{a[b]c[d]e}$ for the permutation in $\Sigma_n$ that interchanges the $b$-block and the $d$-block and otherwise maintains the internal order within all 5 blocks. When $a$ or $c$ or $e$ is zero we will drop them from the notation. Also, if $\alpha \in \Sigma_p$ and $\beta \in \Sigma_q$ write $\alpha | \beta \in \Sigma_{p+q}$ for the permutation that is $\alpha$ on the front $p$-block and $\beta$ on the back $q$-block.

Let us denote elements of symmetric groups by Greek letters, elements of $(S^1)^{\wedge n}$ by capital Roman letters, and elements of $X_*$ and $Y_*$ by lowercase Roman letters. In addition, we write subscripts $x_n$ to denote elements of degree $n$, e.g., $x_n \in X_n$. Denote the iterated structure map $(S^1)^{\wedge n} \wedge X_p \to X_{p+n}$ by $(A_n, x_p) \mapsto A_n x_p$, and the $\Sigma_n$ action on $X_n$ by $(\alpha_n, x_n) \mapsto \alpha_n x_n$. Observe that the equivariance of the structure map is the relation

$$(\alpha_n A_n)(\beta_p x_p) = (\alpha_n | \beta_p)(A_n x_p). \tag{3.7.3}$$

We claim that the spaces $(X \wedge Y)_n$ consist of all elements $\alpha_n[x_p \wedge y_q]$ for $p + q = n$ subject to the following relations:

(1) $(\alpha_n(\beta_p | \gamma_q))[x_p \wedge y_q] = \alpha_n[\beta_p x_p \wedge \gamma_q y_q]$ for $p + q = n$.

(2) $A_k[x_r \wedge y_s] = A_k x_r \wedge y_s = \rho_{[r][k]s}[x_r \wedge A_k y_s]$.

(3) $(\alpha_k A_k)(\gamma_{r+s}[x_r \wedge y_s]) = (\alpha_k | \gamma_{r+s})(A_k x_r \wedge y_s) = (\alpha_k | \gamma_{r+s})\rho_{[r][k]s}[x_r \wedge A_k y_s]$

Relation (2) is a special case of (3); we have listed it separately because it is easier to absorb in this simpler form. Also, relation (3) is really just relation (2) plus equivariance.

*Remark 3.7.9.* There is a procedure for determining the permutations $\rho$ appearing in formulas like the ones above. For an equation of the form "$\rho$(formula $P$) = formula $Q$", regard each subscript $u$ in $P$ as a block of $u$ symbols. Then $\rho$ is the permutation that rearranges the blocks as listed in $P$ into the order listed in $Q$. For example, in equation (2) consecutive blocks of length $r$, $k$, and $s$ must be reordered by bringing the $k$-block in front of the $r$-block.

As an example of how to use the above notation, we work out $X \wedge Y$ in the first three levels. Level 0 is easy, as there are no relations: $(X \wedge Y)_0 = X_0 \wedge Y_0$. Level 1 has $(X \wedge Y)_1 = [(X_0 \wedge Y_1) \wedge (X_1 \wedge Y_0)]/\sim$, where the relation is $A_1(x_0 \wedge y_0) = (A_1 x_0) \wedge y_0 = x_0 \wedge (A_1 y_0)$. If desired we can translate this back into categorical language and say that $(X \wedge Y)_1$ is the pushout of the diagram

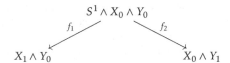

with $f_1(A_1, x_0, y_0) = A_1 x_0 \wedge y_0$ and $f_2(A_1, x_0, y_0) = x_0 \wedge A_1 y_0$.

In general, for $(X \wedge Y)_n$ one writes down a big wedge of $X_p \wedge Y_q$ (with extra symmetric groups out front) and then quotients by relations coming from structure maps out of lower levels. So for $n = 2$ we start with

$$(X_2 \wedge Y_0) \vee (X_1 \wedge Y_1) \vee (12)(X_1 \wedge Y_1) \vee (X_0 \wedge Y_2),$$

where $(12)$ is the generator of $\Sigma_2$ and appears here as a bookkeeping factor. The relations are

$$(A_2 x_0) \wedge y_0 = x_0 \wedge A_2 y_0, \quad A_1 x_0 \wedge y_1 = x_0 \wedge A_1 y_1, \quad A_1 x_1 \wedge y_0 = \rho_{[1],[1]} x_1 \wedge A_1 y_0.$$

Translating again to categorical language, $(X \wedge Y)_2$ is the colimit of a diagram

$$S^1 \wedge X_1 \wedge Y_0 \qquad\qquad S^2 \wedge X_0 \wedge Y_0 \qquad\qquad S^1 \wedge X_0 \wedge Y_1$$
$$\downarrow \qquad\qquad\qquad\qquad\qquad\qquad\qquad\qquad\qquad\qquad \downarrow$$
$$X_2 \wedge Y_0 \qquad (12)(X_1 \wedge Y_1) \qquad\qquad\qquad X_1 \wedge Y_1 \qquad X_0 \wedge Y_2$$

where the maps are easily written down from the algebraic relations. As an exercise, check that when $Y = S$ this colimit gives exactly $X_2$. Note that this fixes the problem we saw in our naive attempt back in Section 3.1.3, where the factors $X_1 \wedge Y_1$ and $(12)(X_1 \wedge Y_1)$ were compressed into a single term.

This discussion also leads to the following useful fact:

**Proposition 3.7.10.** *Let $X$, $Y$, and $Z$ be symmetric spectra. To give a map of symmetric spectra $X \wedge Y \to Z$ is equivalent to giving maps $X_p \wedge Y_q \to Z_{p+q}$ for all $p, q \geq 0$ that are $\Sigma_p \times \Sigma_q$-equivariant and satisfy the identities*

$$A_k(x_p \cdot y_q) = A_k x_p \cdot y_q = \rho_{[p],[k],q}(x_p \cdot A_k y_q).$$

*A pairing $X \wedge X \to Z$ is commutative if it also satisfies the identity*

$$x_p \cdot x'_q = \rho_{[q],[p]}(x'_q \cdot x_p).$$

*Proof.* For the first claim, note that relation (3) is a consequence of the listed relations and the equivariance of the structure maps in $Z$. The second claim is routine.  □

This would be a good moment to see some examples of symmetric ring spectra, but most of the standard examples are also examples of orthogonal ring spectra and it is clearer to discuss them in that context. The curious reader might wish to look ahead at Section 3.8.8.

### 3.7.4 Symmetric spectra and diagram categories

Let $\mathcal{C}$ be a closed symmetric monoidal category and let $X$ be an invertible object in $\mathcal{C}$. Let $X^*$ and $\alpha: I \xrightarrow{\cong} X^* \otimes X$ be a choice for inverse, and recall the dual map $\hat{\alpha}: X \otimes X^* \to I$ from Section 3.3.5. The adjoint of $\hat{\alpha}$ is a map $X \to \mathcal{F}(X^*, I)$, and more generally we get canonical maps

$$X^{\otimes(k)} \to \mathcal{F}\left((X^*)^{\otimes(n+k)}, (X^*)^{\otimes(n)}\right) \tag{3.7.5}$$

adjoint to the map $X^{\otimes(k)} \otimes (X^*)^{\otimes(n+k)} \to (X^*)^{\otimes(n)}$ that reverses the order of the tensor factors in $X^{\otimes(k)}$ and then uses $\hat{\alpha}$ repeatedly to eliminate adjacent factors of $X$ and $X^*$ (note that there are various associativity isomorphisms as well, but we are ignoring them). This leads to the following picture of elements in $\mathcal{C}$ and canonical "maps" between them, where an arrow from $A$ to $B$ labeled $Z$ means a map $Z \to \mathcal{F}(A, B)$ and $\Sigma_n$ acts on the *right* of $(X^*)^{\otimes(n)}$ by permutation of the factors:

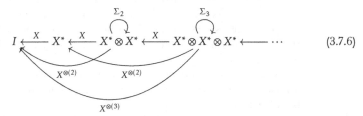

$$\tag{3.7.6}$$

*Remark 3.7.11.* There are canonical isomorphisms $(X^*)^{\otimes(k)} \to \mathcal{F}(X^{\otimes(k)}, I)$ induced by the tensoring operation $\mathcal{F}(A, B) \otimes \mathcal{F}(C, D) \to \mathcal{F}(A \otimes C, B \otimes D)$, and the above descriptions might make more sense if one uses these isomorphisms to replace every appearance of the domain by the codomain. The usual left action of $\Sigma_n$ on $X^{\otimes(n)}$ (see Section 3.3.5) gives a right action on $\mathcal{F}(X^{\otimes(n)}, I)$, and the maps in (3.7.5) were set up so that the adjoints generalize the evaluations $X^{\otimes(k)} \otimes \mathcal{F}(X^{\otimes(k)}, I) \to I$.

To capture the picture in (3.7.6) more formally, define a category $\Sigma_X^{op}$ enriched over $\mathcal{C}$ as follows (apologies for the mysterious "op" but it will become clear in a moment). $\Sigma_X^{op}$ has one object $[n]$ for every $n \geq 0$, and

$$\Sigma_X^{op}([n], [k]) = \begin{cases} \emptyset & \text{if } k > n, \\ X^{\otimes(n-k)} \odot_{\Sigma_{n-k}} \Sigma_n & \text{if } k \leq n. \end{cases}$$

In the last line, $\Sigma_{n-k}$ sits in $\Sigma_n$ as permutations of the first $n-k$ elements, and the notation means the evident analog of $\Sigma_n \odot_{\Sigma_{n-k}} X^{\otimes(n-k)}$ obtained by reversing left and right. To define this as a category we need to explain how to compose maps, and we will do this using algebraic notation as in the last section. If maps from $[n]$ to $[k]$ are written $B_1 \ldots B_{n-k}\beta_n$ then the rule is

$$(B_1 \ldots B_{k-l}\beta_k)(C_1 \ldots C_{n-k}\gamma_n) = C_1 \ldots C_{n-k}B_1 \ldots B_{k-l}(id_{n-k}|\beta_k)\gamma_n \qquad (3.7.7)$$

(the switching of the $B$'s and $C$'s seems annoying but works itself out when we move from $\Sigma_X^{op}$ to $\Sigma_X$). This rule comes from reading off how compositions work in (3.7.6). For example, pretend $X$ is a one-dimensional vector space and $\hat\alpha$ is evaluation. The left-hand side of (3.7.7) takes a tensor product of functionals $\phi_1 \otimes \cdots \otimes \phi_n$ on $X$, permutes them into the new tensor $\phi_{\gamma(1)} \otimes \cdots \otimes \phi_{\gamma(n)}$, evaluates the first $n-k$ of these on the $C$'s to get $[\phi_{\gamma(1)}(C_1)\phi_{\gamma(2)}(C_2)\cdots] \cdot \phi_{\gamma(n-k+1)} \otimes \cdots \otimes \phi_{\gamma(n)}$, permutes the remaining functionals according to $\beta$, and then evaluates the first $k-l$ of these at the $B$'s. One readily verifies that the right-hand side of (3.7.7) does the same thing.

So we have a category $\Sigma_X^{op}$ and (3.7.6) amounts to the observation that our choice of $(X^*, \hat\alpha)$ determines a canonical (enriched) functor $\Sigma_X^{op} \to \mathcal{C}$ sending $[n]$ to $(X^*)^{\otimes(n)}$. This in turn means that if $Z$ is any object in $\mathcal{C}$ then we get an (enriched) functor $\Sigma_X \to \mathcal{C}$ by $[n] \mapsto \mathcal{F}((X^*)^{\otimes(n)}, Z)$.

A brief amount of thought reveals that enriched functors $\Sigma_X \to \mathcal{C}$ are precisely symmetric $X$-spectra. Note that in $\Sigma_X$ rule (3.7.7) becomes instead

$$\gamma_n^{-1}C_1 \cdots C_{n-k} \circ \beta_k^{-1}B_1 \cdots B_{k-l} = \gamma_n^{-1}(id_{n-k}|\beta_k^{-1})C_1 \cdots C_{n-1}B_1 \cdots B_{k-l}$$

which could be made prettier by removing all of the inverses.

To paraphrase this discussion, the category $\Sigma_X^{op}$ in some sense encodes the universal structure an inverse of $X$ would have in $\mathcal{C}$. Symmetric $X$-spectra arise by "remembering" how all the inverses of $X$ map into some given object. This is how one could re-invent the notion of symmetric spectra, if one were trapped on a desert island and forgot how it all worked.

Let us push these ideas a little further. The subcategory of $\mathcal{C}$ pictured in (3.7.6) is symmetric monoidal, and this structure can be lifted back to $\Sigma_X^{op}$. Define the tensor by $[k] \otimes [l] = [k+l]$, let the associativity isomorphism be the identity, and let the symmetry isomorphism $t: [k] \otimes [l] \to [l] \otimes [k]$ be the permutation $\rho_{[k],[l]}$. We also have to define the tensor product of maps, and this is done using the formula

$$A_1 \ldots A_k\alpha_s \otimes B_1 \ldots B_l\beta_t = A_1 \ldots A_kB_1 \ldots B_l\rho_{k,[s-k],[l],t-l}(\alpha_s|\beta_t). \qquad (3.7.8)$$

This formula is again easily derived by thinking about vector spaces and functionals. The left-hand side is the operation that takes functions $\phi_1, \ldots, \phi_s, \mu_1, \ldots, \mu_t$, permutes the first set according to $\alpha$ and the second set according to $\beta$, then successively evaluates the first part of each set at the $A$'s and $B$'s in order (with the first $A$ getting plugged into the first $\phi$, and so forth). The right-hand side also does the $\alpha$ and $\beta$ scrambling but then moves the first group of $\mu$'s in front of the last group of $\phi$'s, before plugging in the $A$'s and $B$'s. These are clearly the same operation.

It is a good exercise to check that $\Sigma_X^{op}$, thus defined, is indeed symmetric monoidal.

The symmetric monoidal structure on $\Sigma_X^{op}$ yields a corresponding structure on $\Sigma_X$, and then this passes to a symmetric monoidal structure on the functor category $F(\Sigma_X, \mathcal{C})$ through a process called *Day convolution*. Briefly, given two functors $Y, Z : \Sigma_X \to \mathcal{C}$ one forms the diagram

$$
\begin{array}{ccc}
\Sigma_X \times \Sigma_X & \xrightarrow{Y \times Z} & \mathcal{C} \times \mathcal{C} \xrightarrow{\otimes} \mathcal{C} \\
\otimes \downarrow & \nearrow_{Y \otimes Z} & \\
\Sigma_X & &
\end{array}
\qquad (3.7.9)
$$

and $Y \otimes Z$ is the (enriched) left Kan extension. The fact that the tensors on $\Sigma_X$ and $\mathcal{C}$ are both symmetric monoidal yields that the tensor product of functors is symmetric monoidal as well.

To summarize this discussion, we could have defined symmetric spectra as follows:

Definition 3.7.12 (Symmetric spectra, approach #2). Let $\Sigma$ denote the category $\Sigma_{S^1}$, as defined above. This is a category enriched over $\mathcal{T}op_*$. A symmetric spectrum is simply an enriched functor $\Sigma \to \mathcal{T}op$.

This approach provides a useful perspective on the difference between classical spectra and symmetric spectra. Classical spectra are diagrams indexed by the evident subcategory $\mathbb{N}S^1$ of $\Sigma_{S^1}$. The monoidal structure on $\Sigma_{S^1}$ does not descend to this subcategory: to define the tensor product of two maps one needs the $\rho$-permutations as in (3.7.8), and these are not available in $\mathbb{N}S^1$. This seems to be the core reason that classical spectra do not have a smash product at the model category level.

## 3.8    Orthogonal spectra

The development of orthogonal spectra proceeds along lines very similar to what we did for symmetric spectra, and so we will be able to cover it fairly quickly. We describe the two (equivalent) approaches, one going through $S$-modules and the other via enriched diagrams. In each case there are some annoying technicalities to be dealt with at the beginning, but after that everything works much as for symmetric spectra. Certain formulas that were a little complicated in symmetric spectra — because they required an introduction of a permutation — have an easier counterpart in the orthogonal case, because the machinery in some sense keeps track of the permutation for us. The theory of orthogonal spectra was developed in [178].

Very briefly, an orthogonal spectrum assigns to each finite-dimensional inner product space $V$ a pointed space $X_V$, and to every linear isometric inclusion $f : V \hookrightarrow W$ a natural structure map $\sigma_f : S^{W - f(V)} \wedge X_V \to X_W$, where $W - f(V)$ is the orthogonal complement of $f(V)$ in $W$. The extra complication is that these structure maps must be continuous in $f$ in an appropriate sense. Some other things are as expected: if $f$ is an isomorphism then by naturality the structure map will be an isomorphism $X_V \xrightarrow{\cong} X_W$, in particular showing that the orthogonal group $O(V)$ of self-isometries will act on each $X_V$.

Why bother with orthogonal spectra? There are at least three reasons. As mentioned, the theory works out a bit more naturally, with simpler formulas. Secondly, orthogonal spectra adapt easily to the setting of equivariant spectra (see [177] or Appendix A of [120]). Finally, unlike symmetric spectra, orthogonal spectra have the nice property that the weak equivalences are just the maps inducing isomorphisms on stable homotopy groups.

In this section we will in fact discuss four types of spectra, interrelated thus:

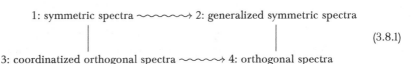

$$(3.8.1)$$

(Types 2 and 3 on the anti-diagonal seem to lack standard names; these are our own.) Our development will proceed in the order $1 \longrightarrow 2 \longrightarrow 4 \longrightarrow 3$, although other orders of navigation are also possible.

### 3.8.2     Prelude: generalized symmetric spectra

The generalized symmetric spectra we are about to introduce do not typically get much airtime, as there is little payoff for the extra work and they are not truly "coordinate-free". But they are a useful prelude to orthogonal spectra, and only a slight modification of the symmetric spectra story we saw in Section 3.7. They come up, for example, in Remark 2.1.5 of [133].

For any finite set $T$ consider the real vector space $\mathbb{R}\langle T \rangle$ with basis $T$, together with its one-point compactification $S^T = S^{\mathbb{R}\langle T \rangle}$. Let $\Sigma(T)$ denote the group of permutations of $T$; it acts naturally on $S^T$. Write $\mathbf{n}$ for the set $\{1, 2, \ldots, n\}$, so that $\Sigma_n = \Sigma(\mathbf{n})$.

A generalized symmetric spectrum should be, in part, a functor $T \mapsto X_T$ defined on the category of finite sets with isomorphisms, taking values in the category of pointed spaces. Functoriality will give each $X_T$ a $\Sigma(T)$-action. In addition, the spectrum should assign to every subset inclusion $T \subseteq U$ a structure map

$$\sigma_{T,U} \colon S^{U-T} \wedge X_T \to X_U$$

that is $\Sigma(U-T) \times \Sigma(T)$-equivariant, with the assignment being compatible with the isomorphisms $X_J \cong X_{J'}$ for $J \cong J'$. By restricting to the special sets $\mathbf{n}$ and inclusions $\mathbf{n} \hookrightarrow \mathbf{k}$ for $n \leq k$, we get a (classical) symmetric spectrum $\tilde{X}$. If $|T| = n$ then every bijection $T \to \mathbf{n}$ induces a homeomorphism $X_T \to X_{\mathbf{n}}$, and one can check that there is really no more information in $X$ than in $\tilde{X}$. But what we have accomplished here is to produce a notion of symmetric spectrum that avoids any dependence on the particular choice of finite sets $\mathbf{n}$, which after all are a bit unnatural.

*Remark 3.8.1.* In fact the above yields structure maps for *any* inclusion $f \colon T \hookrightarrow U$, of the form

$$\sigma_f \colon S^{U-f(T)} \wedge X_T \to X_U,$$

via composition:

$$S^{U-f(T)} \wedge X_T \xrightarrow{id \wedge X_f} S^{U-f(T)} \wedge X_{f(T)} \xrightarrow{\sigma_{f(T),U}} X_U.$$

Just as for symmetric spectra, we can follow two approaches for setting up the generalized version. Let $\mathbf{\Sigma I}$ denote the category of finite sets and isomorphisms.

**Approach #1**: Define a $\mathbf{\Sigma I}$-sequence to be a functor $\mathbf{\Sigma I} \to \mathcal{T}op_*$. Define the tensor product of $\mathbf{\Sigma I}$-sequences $X$ and $Y$ by

$$(X \otimes Y)_U = \bigvee_{T \subseteq U} X_T \wedge Y_{U-T}. \tag{3.8.3}$$

For the $\Sigma(U)$-action, an $\alpha \in \Sigma(U)$ maps the summand $X_T \wedge Y_{U-T}$ to $X_{\alpha(T)} \wedge Y_{\alpha(U-T)}$ via $X_{\alpha|_T} \wedge X_{\alpha|_{U-T}}$. The twist map in the symmetric monoidal structure carries the summand $X_T \wedge Y_{U-T}$ (indexed by $T \subseteq U$) to $Y_{U-T} \wedge X_T$ (indexed by $U - T \subseteq U$) via the usual twist map from $\mathcal{T}op_*$.

The sphere spectrum $S$ is the $\mathbf{\Sigma I}$-sequence $T \mapsto S^T$, which can be checked to be a commutative monoid. We define a generalized symmetric spectrum to be an $S$-module.

Unfortunately, because $\mathbf{\Sigma I}$ is not a small category we cannot form the category of $\mathbf{\Sigma I}$-sequences without running into set-theoretic issues. See Remark 3.5.4 for the common ways to get around this: for example, we can choose a skeletal subcategory $\mathbf{\Sigma I}_{skel} \hookrightarrow \mathbf{\Sigma I}$ together with a retraction $r$, and then transplant all the definitions for $\mathbf{\Sigma I}$-sequences to $\mathbf{\Sigma I}_{skel}$-sequences. One choice for skeletal subcategory is precisely the category $\Sigma I$ from Section 3.7, leading to the previous (ungeneralized) notion of symmetric spectra.

The monoidal product on $\mathbf{\Sigma I}$-sequences is another example of Day convolution (see (3.7.9)): the category $\mathbf{\Sigma I}$ has the symmetric monoidal structure $\sqcup$ given by disjoint union, and $X \otimes Y$ is the left Kan extension in the diagram

$$\begin{array}{ccc} \mathbf{\Sigma I} \times \mathbf{\Sigma I} & \xrightarrow{X \wedge Y} & \mathcal{T}op_* \\ {\scriptstyle \sqcup} \downarrow & \nearrow & \\ \mathbf{\Sigma I} & \raisebox{1ex}{$\scriptstyle X \otimes Y$} & \end{array}$$

The most natural formula for this left Kan extension is

$$(X \otimes Y)(U) = \mathrm{colim}_{[A \sqcup B \to U]} X_A \wedge Y_B,$$

where the indexing category consists of triples $(A, B, f : A \sqcup B \to U)$ for $f$ a map in $\mathbf{\Sigma I}$ and therefore an isomorphism. The maps between triples are the evident ones. This indexing category is not small, but again it has a small skeleton and so the colimit still exists. By associating the triple $(A, B, f)$ with the image $f(A) \subseteq U$, one readily identifies the above colimit with the expression in (3.8.3).

**Approach #2**: For finite sets $A$ and $B$ define a category $[A, B]$ whose objects are sets $C$ such that $A \subseteq C$ and $|C| = |B|$; morphisms $C \to C'$ are bijections $g : C \to C'$ which are the identity on $A$. Next define a category $\Sigma$ enriched over $\mathcal{T}op_*$ whose objects are the finite sets and where the morphisms are given by

$$\Sigma(A,B) = \mathrm{colim}_{[A,B]}\left[\mathrm{Isom}(C,B)_+ \wedge S^{C-A}\right]$$

(and $\mathrm{Isom}(C,B)$ is the set of bijections from $C$ to $B$). The category $[A,B]$ indexing the colimit consists only of isomorphisms, and so the colimit can be identified with the co-invariants of the group of automorphisms acting on any spot of the diagram. In particular, for any subset $A \subseteq C$ such that $|C| = |B|$ one has

$$\Sigma(A,B) \cong \mathrm{Isom}(C,B)_+ \wedge_{\Sigma(C-A)} S^{C-A}.$$

We can also regard $\Sigma(A,B)$ as the subset of $\mathrm{Hom}(A,B)_+ \wedge S^B$ consisting of all pairs $(f,x)$ where $f$ is an injection and $x \in S^{B-f(A)}$; it is easy to check that the above colimit maps to this space in the evident way. If we do this, the composition is easy to describe: $\Sigma(B,C) \times \Sigma(A,B) \longrightarrow \Sigma(A,C)$ is the map

$$((g,y),\,(f,x)) \mapsto (gf, y \wedge g(x)).$$

In this approach, a generalized symmetric spectrum is simply an enriched functor $\Sigma \to \mathcal{T}op_*$. Just as in Approach #1, one runs into the difficulty that $\Sigma$ is not a small category — and one way of dealing with this is to replace $\Sigma$ with a skeletal subcategory, such as the category $\Sigma$ from Definition 3.7.12.

### 3.8.4    Orthogonal spectra

Generalized symmetric spectra were built around the vector spaces $\mathbb{R}\langle A \rangle$, where $A$ ranged over all finite sets. So these are vector spaces with a choice of basis, and one is naturally led to wonder about a basis-free approach. That is essentially what orthogonal spectra are. The role of the symmetric groups $\Sigma(A)$ is instead played by orthogonal groups $O(V)$.

Let $\mathcal{O}I$ be the category of finite-dimensional real inner product spaces, with linear isometric isomorphisms for the maps. This category only has maps from $V$ to $W$ when $\dim V = \dim W$, and all such maps are isomorphisms. We regard $\mathcal{O}I$ as being enriched over $\mathcal{T}op$, with $\mathcal{O}I(V,W)$ having the usual subspace topology induced by the compact-open topology on the space of all continuous maps $W^V$. For $W \in \mathrm{ob}\,\mathcal{O}I$ define $O(W) = \mathcal{O}I(W,W)$ to be the space of isometries from $W$ to itself. If $V \subseteq W$ write $W - V$ for the orthogonal complement of $V$ in $W$. Then we have a canonical inclusion $O(V) \hookrightarrow O(W)$: isometries of $V$ extend to $W$ by having them act as the identity on $W - V$. We will write $\mathrm{Isom}(U,V)$ for space of linear isometric inclusions from $U$ into $V$, so when $\dim U = \dim V$ we have $\mathrm{Isom}(U,V) = \mathcal{O}I(U,V)$.

**Approach #1:** An $\mathcal{O}I$-sequence is simply an enriched functor $\mathcal{O}I \to \mathcal{T}op_*$. The symmetric monoidal structure $\oplus$ on $\mathcal{O}I$ induces a symmetric monoidal structure on $\mathcal{O}I$-sequences by Day convolution. Specifically, if $X$ and $Y$ are $\mathcal{O}I$-sequences then

$X \otimes Y$ is the (enriched) left Kan extension

$$\mathbb{O}I \times \mathbb{O}I \xrightarrow{X \times Y} \mathcal{T}op_* \times \mathcal{T}op_* \xrightarrow{\wedge} \mathcal{T}op_*$$

with vertical map $\otimes$ from $\mathbb{O}I \times \mathbb{O}I$ to $\mathbb{O}I$ and diagonal $X \otimes Y$.

and can be given by the (enriched) colimit formula

$$(X \otimes Y)_W = \operatorname{colim}_{A \oplus B \to W} (X_A \wedge Y_B). \tag{3.8.5}$$

Here the indexing category has objects consisting of tuples $(A, B, f : A \oplus B \to W)$, where $f$ is a map in $\mathbb{O}I$, and the evident morphisms (once again this is not a small category, but has a small skeleton). The enriched colimit is the coequalizer in $\mathcal{T}op$ of the two evident arrows

$$\coprod_{A,B,A',B'} \operatorname{Isom}(A, A') \times \operatorname{Isom}(B, B') \times \operatorname{Isom}(A \oplus B, W) \times (X_A \wedge Y_B)$$

$$\downarrow\downarrow$$

$$\coprod_{A,B} \operatorname{Isom}(A \oplus B, W) \times (X_A \wedge Y_B)$$

and so in particular the topology on $(X \otimes Y)_W$ comes from the topology on both $\operatorname{Isom}(A \oplus B, W)$ and on $X_A \wedge Y_B$. As a set (ignoring the topology) we can write

$$(X \otimes Y)_W = \bigvee_{V \subseteq W} X_Y \wedge Y_{W-V}. \tag{3.8.6}$$

by associating to every isometric isomorphism $f : A \oplus B \to W$ the subspace $f(A) \subseteq W$ (but this precisely ignores the topology on $\operatorname{Isom}(A \oplus B, W)$). Note that in this picture an isometry $h : W \to W'$ acts on this wedge by sending the summand $X_V \wedge Y_{W-V}$ to $X_{h(V)} \wedge Y_{h(W-V)}$ using the maps $X(h|_V)$ and $Y(h|_{W-V})$. The description in (3.8.5) readily gives the continuity of the maps

$$\mathbb{O}I(W, W') \times (X \otimes Y)_W \to (X \otimes Y)_{W'}.$$

The indexing category for the colimit in (3.8.5) has the property that all maps are isomorphisms; it follows formally that the colimit can be identified with the wedge of the co-invariants of the groups of automorphisms corresponding to every connected component of the category. So if we choose one $V_p \subseteq W$ of dimension $p$ for every $0 \le p \le \dim W$ then we can write

$$(X \otimes Y)_W \cong \bigvee_p O(W)_+ \wedge_{O(V_p) \times O(W - V_p)} [X_{V_p} \wedge Y_{W-V_p}]. \tag{3.8.7}$$

This is correct as topological spaces but is non-canonical because of the choices of $V_p$. The bijection from (3.8.7) to (3.8.6) sends a tuple $(\alpha, x \wedge y \in X_{V_p} \wedge Y_{W-V_p})$ to $\alpha_*(x) \wedge \alpha_*(y) \in X_{\alpha(V_p)} \wedge Y_{\alpha(W-V_p)}$.

This tensor gives a closed symmetric monoidal product on the category of $\mathbb{O}I$-sequences, where the symmetry isomorphism $t : X \otimes Y \to Y \otimes X$ sends $x \wedge y \in X_A \wedge Y_B$ to $y \wedge x \in Y_B \wedge X_A$, using the description of (3.8.5).

Let $S$ denote the $\mathcal{O}I$-sequence defined by $V \mapsto S^V$. It is easy to check that the maps $S^V \wedge S^W \to S^{V \oplus W}$ make $S$ into a commutative monoid in the category of $\mathcal{O}I$-sequences. Define an **orthogonal spectrum** to be a left $S$-module. If $X$ and $Y$ are orthogonal spectra then their smash product is $X \wedge Y = X \otimes_S Y$.

We will write $\mathrm{Sp}^{\mathcal{O}}$ for the category of orthogonal spectra.

*Remark 3.8.2.* In wanting to consider all enriched functors $\mathcal{O}I \to \mathcal{T}op_*$ as a category, we run into the usual problem that $\mathcal{O}I$ is not small. To circumvent this using a small skeletal subcategory, as in Remark 3.5.4, we can take for such a subcategory the Euclidean spaces $(\mathbb{R}^n, \cdot)$ with standard dot product, for each $n \geq 0$. This leads to a spectrum being an assignment $n \mapsto X_n$, where $X_n$ is a pointed space with an $O(n)$-action, together with structure maps $S^1 \wedge X_n \to X_{n+1}$ such that the iterated maps $S^p \wedge X_n \to X_{n+p}$ are $O(p) \times O(n)$-equivariant. Such an object could be called a "coordinatized orthogonal spectrum", and completes our tour of the square (3.8.1).

**Approach #2:** Here we define a $\mathcal{T}op_*$-enriched category $\mathcal{O}$ having the same objects as $\mathcal{O}I$ and where $\mathcal{O}(V, W)$ is supposed to parameterize the various suspension maps from $X_V$ to $X_W$ in a spectrum $X$. Recall that for every isometry $f: V \to W$ (which will necessarily be injective) we are supposed to have a suspension map $\sigma_f: S^{W-f(V)} \wedge X_V \to X_W$. The tricky part here is that there is not one single sphere involved in these maps: the sphere varies continuously with $f$. So to this end, let $\mathrm{Isom}(V, W)$ be the space of isometries from $V$ into $W$ and let $W - V \to \mathrm{Isom}(V, W)$ denote the bundle whose fiber over $f: V \to W$ is $W - f(V)$. Define

$$\mathcal{O}(V, W) = \mathrm{Th}(W - V \to \mathrm{Isom}(V, W)),$$

the Thom space of the bundle $W - V$. Note that if $|V| > |W|$ then $\mathrm{Isom}(V, W)$ is empty and this Thom space is a single point.

A point in $\mathcal{O}(V, W)$ can be represented by a pair $(f, x)$ consisting of an isometry $f: V \to W$ and $x \in S^{W-f(V)}$. Using this notation, if $(g, y) \in \mathcal{O}(W, Z)$ then composition in $\mathcal{O}$ is given by the formula

$$(g, y) \circ (f, x) = (gf, g(x) + y),$$

the sum-of-vectors map $(g(W) - gf(V)) \times (Z - g(W)) \to Z - gf(V)$ being extended to the one-point compactifications in the usual way.

We can make the following identifications:

$$\mathcal{O}(V, W) = \begin{cases} O(W)_+ \wedge_{O(W-V)} S^{W-V} & \text{if } V \subseteq W, \\ \mathrm{Isom}(V, W) & \text{if } \dim V = \dim W, \\ \mathrm{Isom}(U, W)_+ \wedge_{O(U-V)} S^{U-V} & \text{if } \dim V \leq \dim W \text{ and } V \subseteq U \cong W, \\ * & \text{if } \dim W < \dim V. \end{cases}$$

The first two lines are actually special cases of the third, but are included separately for pedagogical purposes. For the third line use the map $\mathrm{Isom}(U, W)_+ \wedge_{O(U-V)} S^{U-V} \to \mathrm{Th}(W - V)$ given by $(h, x) \mapsto (h|_V, h(x))$.

The point to remember in the above descriptions is that when $\dim V = \dim W$ we have exactly $\mathrm{Isom}(V, W)$ as the space of maps from $V$ to $W$. When $V \subseteq W$ we put an $S^{W-V}$ into the space of maps from $V$ to $W$, and then allow post-compositions with our $O(W)$ maps from $W$ to itself — this accounts for the $O(W)_+ \wedge_{O(W-V)} S^{W-V}$ term. When $V$ and $W$ are incomparable we choose $U \supseteq V$ such that $\dim U = \dim W$ and allow compositions between our $S^{U-V}$ maps from $V$ to $U$ and our $\mathrm{Isom}(U, W)$ maps from $U$ to $W$, accounting for the $\mathrm{Isom}(U,W)_+ \wedge_{O(U-V)} S^{U-V}$ term.

In this approach an **orthogonal spectrum** is simply an enriched functor $\mathbb{O} \to \mathcal{T}op_*$. Unraveling this definition, an orthogonal spectrum $X$ consists of

- a functor $X \colon \mathbb{O}I \to \mathcal{T}op_*$, and
- for every pair $V \subseteq W$ a structure map

$$\sigma_{V,W} \colon S^{W-V} \wedge X_V \to X_W$$

that is $O(W - V) \times O(V)$-equivariant.

These structure maps must satisfy unital and associativity conditions that are easy to work out.

We leave the reader to justify the following analog of Proposition 3.7.10. Note that the isometry $\rho$ that appears here is naturally forced upon us, since the second equality does not even make sense without it. In this sense the situation is a bit simpler than for symmetric spectra.

**Proposition 3.8.3.** *Let $X$, $Y$, and $Z$ be orthogonal spectra. Giving a pairing $X \wedge Y \to Z$ is equivalent to giving a collection of maps $X_V \wedge Y_W \to Z_{V \oplus W}$ that are $O(V) \times O(W)$-equivariant and satisfy the identities*

$$A_U(x_V y_W) = (A_U x_V) y_W = \rho(x_V \cdot (A_U y_W)),$$

*where $\rho$ is the evident isometry $V \oplus (U \oplus W) \to (U \oplus V) \oplus W$ that is natural in the three variables. (Here we are using the algebraic notation from (3.7.2), adapted in the obvious way to the present context.) A pairing $X \wedge X \to Z$ is commutative if it also satisfies the identities $x_V \cdot y_W = \rho(y_W \cdot x_V)$, where $\rho$ is the twist isometry $W \oplus V \to V \oplus W$.*

### 3.8.8    Examples

We now give several standard examples of orthogonal and symmetric ring spectra.

(a) Let $R$ be a ring and let $HR$ be the spectrum $V \mapsto R\langle S^V \rangle$, where the latter is the free $R$-module on the set $S^V$ with an appropriate topology (and where the basepoint is equal to zero). It is convenient to think of points in $R\langle S^V \rangle$ as finite configurations on $S^V$ with labels in $R$, written formally as $\sum_i r_i x_i$ with $r_i \in R$, $x_i \in S^V$. The maps $S^W \wedge R\langle S^V \rangle \to R\langle S^{W \oplus V} \rangle$ send $(x, \sum r_i y_i) \mapsto \sum r_i (x \wedge y_i)$. The product maps $R\langle S^V \rangle \wedge R\langle S^W \rangle \to R\langle S^{V \oplus W} \rangle$ send $(\sum r_i x_i, \sum s_j y_j) \mapsto \sum_{i,j} r_i s_j [x_i \wedge y_j]$, and the unit maps $S^W \to R\langle S^W \rangle$ send $x \mapsto 1 \cdot x$.

(b) Let $MO$ be the spectrum $V \mapsto MO_V = EO(V)_+ \wedge_{O(V)} S^V$. Here we take $EG$ to be the geometric realization of the standard simplicial space $[n] \mapsto G^{n+1}$ with projections as face maps. Note that this comes with canonical maps $EH \to EG$ for $H \to G$ and

$EG_1 \times EG_2 \xrightarrow{\cong} E(G_1 \times G_2)$, and that $G$ acts on $EG$ from both the left and the right via its diagonal action on the $G^{n+1}$ terms. The $O(V)$ action on $MO_V$ comes from the left action on $EO(V)$.

The maps $S^W \wedge MO_V \to MO_{W \oplus V}$ are $(x, (\alpha, y)) \mapsto (\alpha, x \wedge y)$, where by abuse we write $\alpha$ for both an element of $EO(V)$ and its image in $EO(W \oplus V)$. It is informative to check the $O(W) \times O(V)$-equivariance. The $O(V)$-equivariance is clear, but the $O(W)$-equivariance looks wrong at first. One must use that $O(W)$ and $O(V)$ commute inside of $O(W \oplus V)$!

The pairings $MO_V \wedge MO_W \to MO_{V \oplus W}$ are the evident ones: $(\alpha, x) \wedge (\beta, y) \mapsto (\alpha\beta, x \wedge y)$, where $\alpha\beta$ refers to the pairing $EO(V) \times EO(W) \to EO(V \oplus W)$. The unit maps $S^V \to MO_V$ send $x$ to $(Id_V, x)$. We leave the reader to check the necessary relations to see that this is indeed a commutative ring spectrum.

(c) Constructing $MU$ as an orthogonal ring spectrum is a little tricky. One can mimic our construction of $MO$ using complexifications and unitary groups and write $MU(V) = EU(V_{\mathbb{C}})_+ \wedge_{U(V_{\mathbb{C}})} S^{V_{\mathbb{C}}}$, where $V_{\mathbb{C}}$ is the complexification of $V$, but then one only gets suspension operators by $S^{W_{\mathbb{C}} - V_{\mathbb{C}}}$ when one wants $S^{W-V}$. So this doesn't quite work. To explain the fix, if $W$ is a Hermitian inner product space define

$$MU_W^{Herm} = EU(W)_+ \wedge_{U(W)} S^W.$$

This has a left $U(W)$-action coming from the left action on $EU(W)$. This construction satisfies all the analogous properties to (b) above, but only for Hermitian spaces. For a real inner product space $V$ define $MU_V = \mathrm{Map}(S^{iV}, MU_{V_{\mathbb{C}}}^{Herm})$, where $iV$ is the imaginary part of $V_{\mathbb{C}}$. Note that $O(V)$ acts on $S^{iV}$ in the evident way, on $MU_{V_{\mathbb{C}}}^{Herm}$ through the map $O(V) \to U(V_{\mathbb{C}})$, and then on the mapping space via conjugation.

It is an easy exercise to check that one gets natural maps $S^V \wedge MU_W \to MU_{V \oplus W}$ making $MU$ into an orthogonal $\Omega$-spectrum. Moreover, smashing of maps gives the pairings

$$
\begin{array}{c}
MU_V \wedge MU_W \;=\!=\; \mathrm{Map}(S^{iV}, MU_{V_{\mathbb{C}}}^{Herm}) \wedge \mathrm{Map}(S^{iW}, MU_{W_{\mathbb{C}}}^{Herm}) \\[2mm]
\Big\downarrow {\scriptstyle (f,g) \mapsto f \wedge g} \\[2mm]
\mathrm{Map}(S^{iV \oplus iW}, MU_{V_{\mathbb{C}}}^{Herm} \wedge MU_{W_{\mathbb{C}}}^{Herm}) \\[2mm]
\Big\downarrow \\[2mm]
\mathrm{Map}(S^{iV \oplus iW}, MU_{(V \oplus W)_{\mathbb{C}}}^{Herm}) \;=\!=\!=\; MU_{V \oplus W}
\end{array}
$$

which make $MU$ into an orthogonal commutative ring spectrum.

(d) Real $K$-theory was written down as a *symmetric* commutative ring spectrum by Joachim [136]. It is not completely obvious how to do this, but Joachim found a way using spaces of Fredholm operators. The $\Sigma_n$-actions come from the action on a tensor product of Hilbert spaces $\mathcal{H}^{\otimes n}$. This construction can be adapted to complex $K$-theory using techniques similar to those in (c), but it does not immediately yield an orthogonal spectrum.

(e) (Waldhausen $K$-theory). Let $\mathcal{C}$ be an exact category in the sense of [224] (or alternatively, a category with cofibrations and weak equivalences in the sense of Waldhausen). Waldhausen's $S_{\bullet}$-construction produces a spectrum $K(\mathcal{C})$ called the **Waldhausen $K$-theory spectrum** of $\mathcal{C}$. Geisser and Hesselholt observed in [102, Section 6] that if one sets things up carefully then this construction actually produces a symmetric spectrum, and that if $\mathcal{C}$ has a well-behaved tensor product then $K(\mathcal{C})$ is in fact a symmetric ring spectrum. While it would take us too far afield to give a rigorous development of these ideas, by doing a bit of handwaving we can nevertheless give the general idea. In this example we work entirely simplicially, mostly just to avoid the excess step of needing to apply geometric realization constantly.

The $S_{\bullet}$-construction applied to $\mathcal{C}$ gives a simplicial set $[n] \mapsto S_n\mathcal{C}$, where an element of $S_n\mathcal{C}$ is, roughly speaking, a filtered object $A_1 \hookrightarrow A_2 \hookrightarrow \cdots \hookrightarrow A_n$ in $\mathcal{C}$ together with a particular choice for every quotient $A_i/A_j$ with $j \leq i$. We will refer to this as a "filtered object with quotient data". For $i \geq 1$ the face map $d_i$ omits $A_i$ from the filtration, whereas $d_0$ sends the filtered object to $A_2/A_1 \hookrightarrow A_3/A_1 \hookrightarrow \cdots \hookrightarrow A_n/A_1$. Note that $S_0\mathcal{C} = *$ by convention, and $S_1\mathcal{C}$ is the set of objects in $\mathcal{C}$.

Define $K(\mathcal{C})_0 = *$ and $K(\mathcal{C})_1 = S_{\bullet}\mathcal{C}$. We will extend this to a generalized symmetric spectrum (as discussed in Section 3.8.2) by defining $K(\mathcal{C})_Q$ for every finite set $Q$. To do this we need the notion of a $Q$-simplicial set. Recall that $\Delta$ denotes the simplicial indexing category, and define $\Delta^Q$ to be the product category $\prod_Q \Delta$ — a product of copies of $\Delta$ indexed by the set $Q$. An object in $\Delta^Q$ is a $Q$-tuple $\underline{n} = (n_q)_{q \in Q}$, or equivalently a function $Q \to \mathbb{N}$. We define a $Q$-simplicial set to be a functor $(\Delta^Q)^{op} \to \mathcal{S}et$. If $|Q| = k$, a $Q$-simplicial set is the same as a $k$-fold multi-simplicial set, but we think of the different simplicial directions as being indexed by $Q$.

If $X$ is a $Q$-simplicial set, define $\text{diag}(X)$ to be the simplicial set $[n] \mapsto X_{(n,n,\dots,n)}$, where the subscript indicates the constant $Q$-tuple whose value is $n$. We will also need the notion of skeleton: if $T \subseteq Q$ and $r \geq 0$, define the $(T, r)$-skeleton of $X$ to be the $Q$-simplicial set given by

$$(\text{sk}_{(T,r)}X)_{(\underline{n})} = X_{(\underline{n}')}, \quad \text{where} \quad n_q' = \begin{cases} n_q & \text{if } q \notin T, \\ \min\{n_q, r\} & \text{if } q \in T. \end{cases}$$

Despite the cumbersome definition, this just says that whenever $q \in T$ we replace the simplicial $q$-direction of $X$ by its usual $r$-skeleton.

Let $S^Q$ be the smash product of copies of $S^1 = \Delta^1/\partial\Delta^1$ indexed by the set $Q$. In simplicial degree $k$ the set $(S^Q)_k$ consists of $k + 1$ elements, which correspond to the basepoint together with the $k$ possible degeneracies of the 1-simplex $[01]$.

The following strange result turns out to be the key to producing our desired symmetric spectrum.

**Proposition 3.8.4.** *Let $Q$ and $Q'$ be finite sets, and let $X$ be a $Q \sqcup Q'$-simplicial set. Assume that $\text{sk}_{(Q',0)}X = *$. Then there is a natural map of simplicial sets*

$$S^{Q'} \wedge \text{diag}(\text{sk}_{(Q',1)}X) \longrightarrow \text{diag}(X).$$

*Proof.* This is a combinatorial exercise left to the reader. The main point is that the non-basepoint elements of $(S^{Q'})_k$ can be thought of as exactly corresponding to the $k$ different ways of applying degeneracies in the $Q'$-directions to move from simplicial degree 1 up to simplicial degree $k$. The desired map is defined to consist exactly of these degeneracy maps.                    □

With these tools in hand, we return to Waldhausen $K$-theory. Recall that every $[n]$ in $\Delta$ may be regarded as a category, in which there is a unique map from $i$ to $j$ whenever $i < j$. Filtered objects of length $n$ in $C$ may be identified with functors $[n] \to C$ that send 0 to the zero object of $C$. Likewise, we associate the tuple $\underline{n} = (n_q)_{q \in Q}$ to the product category $[\underline{n}] = \prod_{q \in Q}[n_q]$, and define an $\underline{n}$-filtered object to be a functor $[\underline{n}] \to C$ which sends every tuple containing 0 to the zero object. For example, a $(1,1)$-filtered object is the same as an object of $C$, and a $(2,3)$-filtered object is a diagram of the form

$$
\begin{array}{ccccc}
X_{11} & \longrightarrow & X_{12} & \longrightarrow & X_{13} \\
\downarrow & & \downarrow & & \downarrow \\
X_{21} & \longrightarrow & X_{22} & \longrightarrow & X_{23}
\end{array}
$$

For each finite set $Q$, define $S_\bullet^Q C$ to be the $Q$-simplicial set which in multidegree $(\underline{n})$ consists of all $\underline{n}$-filtered objects of $C$ satisfying certain cofibration conditions together with particular choices for various quotient objects (again, we are being intentionally vague and only giving the basic idea). Define $K(C)_Q = \mathrm{diag}(S_\bullet^Q C)$. Note that $\Sigma(Q)$ acts naturally on this construction, by permutation of the factors.

Observe that $\mathrm{sk}_{(Q',1)}(S_\bullet^{Q \sqcup Q'} C) = S_\bullet^Q C$. So Proposition 3.8.4 gives maps

$$S^{Q'} \wedge K(C)_Q \to K(C)_{Q \sqcup Q'}$$

which are readily checked to be $\Sigma(Q') \times \Sigma(Q)$-equivariant. Thus, we have a generalized symmetric spectrum. Note that there does not seem to be any obvious approach for producing an orthogonal spectrum here.

If in addition $C$ has a well-behaved tensor product — one that preserves cofibrations and exactness — then we can take an $(n_q)_{q \in Q}$-filtered object $X$ and an $(k_s)_{s \in Q'}$-filtered object $Y$ and tensor them together to get a $(\underline{n} \sqcup \underline{k})_{Q \sqcup Q'}$-filtered object $X \otimes Y$. This yields maps

$$K(C)_Q \wedge K(C)_{Q'} \to K(C)_{Q \sqcup Q'}$$

making $K(C)$ into a symmetric ring spectrum.

We again refer to [102, Section 6.1] for a detailed treatment of this material.

### 3.8.9    Model structures for orthogonal spectra

We now turn to the development of the commonly used model category structures for orthogonal spectra. By now the following series of results will be very familiar.

**Proposition 3.8.5.** *There exists a model category structure on* $\mathrm{Sp}^{\mathbb{O}}$ *where the weak equivalences and fibrations are levelwise. This is called the **level, projective model structure**.*

*Proof.* Direct application of Theorem 3.5.1(a) in the setting of enriched diagrams. □

The evaluation functors $\mathrm{Ev}_V \colon \mathrm{Sp}^{\mathcal{O}} \to \mathcal{T}op_*$ have left adjoints $F_V$ given by

$$(F_V X)_W = \mathrm{Th}\big(W - V \to \mathrm{Isom}(V, W)\big) \wedge X$$

$$\cong \begin{cases} O(W)_+ \wedge_{O(W-V)} (S^{W-V} \wedge X) & \text{if } V \subseteq W, \\ \mathcal{O}I(U, W)_+ \wedge_{O(U-V)} (S^{U-V} \wedge X) & \text{if } V \subseteq U \text{ and } \dim U = \dim W, \\ * & \text{if } \dim W < \dim V. \end{cases}$$

If $V \subseteq W$ there is a canonical map $F_W(S^{W-V} \wedge X) \to F_V(X)$.

Definition 3.8.6. The **stable projective model structure** on $\mathrm{Sp}^{\mathcal{O}}$ is the Bousfield localization of the level projective model category structure at the set of maps

$$\big\{ F_W(S^{W-V} \wedge S^0) \to F_V(S^0) \, \big| \, V \subseteq W \big\}.$$

There is a simple comparison map between orthogonal spectra and symmetric spectra. Let $e_1, \ldots, e_n$ be the standard basis for $\mathbb{R}^n$, so that we have the usual inclusion $\mathbb{R}^n \subseteq \mathbb{R}^{n+1}$. The choice of vector $e_{n+1}$ gives a map $\mathbb{R} \to \mathbb{R}^{n+1} - \mathbb{R}^n$ (sending 1 to $e_{n+1}$) and therefore an induced homeomorphism $S^1 \to S^{(\mathbb{R}^{n+1}-\mathbb{R}^n)}$. Permutation of basis elements gives a group map $\Sigma_n \to O(\mathbb{R}^n)$.

There is a forgetful functor $U \colon \mathrm{Sp}^{\mathcal{O}} \to \mathrm{Sp}^{\Sigma}$ that sends an orthogonal spectrum $X$ to the symmetric spectrum $[n] \mapsto X_{\mathbb{R}^n}$, where the $\Sigma_n$-action on $X_{\mathbb{R}^n}$ comes from restricting the $O(\mathbb{R}^n)$-action and the structure maps come from those in $X$ via the identification $S^1 \cong S^{(\mathbb{R}^{n+1}-\mathbb{R}^n)}$.

The following results are all proven in [178]:

Proposition 3.8.7.

(a) *The stable projective structure on $\mathrm{Sp}^{\mathcal{O}}$ is a stable, closed symmetric monoidal model category satisfying the Monoid Axiom, the Algebraic Creation and Invariance Properties and the Strong Flatness Property.*

(b) *The fibrant objects in $\mathrm{Sp}^{\mathcal{O}}$ are the levelwise fibrant $\Omega$-spectra, meaning orthogonal spectra for which the adjoints to the structure maps $X_V \to \Omega^{W-V} X_W$ are all weak equivalences for $V \subseteq W$.*

(c) *The forgetful functor $U \colon \mathrm{Sp}^{\mathcal{O}} \to \mathrm{Sp}^{\Sigma}$ has a left adjoint $G$ and the pair $(G, U)$ is a Quillen equivalence.*

(d) *A map $f \colon X \to Y$ in $\mathrm{Sp}^{\mathcal{O}}$ is a stable weak equivalence if and only if $Uf$ is a weak equivalence in $\mathrm{Sp}^{\mathbb{N}}$ (slightly abusing our use of $U$ here).*

*Proof.* The precise references for the different parts are: model structure, [178, 9.2]; monoidal properties, [178, 12.1 (take $R = S$)]; Algebraic Creation Property, [178, 12.1(i)]; Algebraic Invariance, [178, 12.1vi,vii]; Strong Flatness, [178, 12.3, 12.7]; Quillen Equivalence, [178, 10.4]; $U$ detects stable weak equivalences, [178, 8.7]. □

Statement (d) is something of a surprise, as this is not true when orthogonal spectra are replaced with symmetric spectra. The topology of the orthogonal groups turns out to be what makes this work, as we now explain. If $X$ is an orthogonal spectrum define $\pi_k(X) = \text{colim}_n \pi_{n+k}(X_{\mathbb{R}^n})$. These are precisely the homotopy groups of the underlying Bousfield–Friedlander spectrum. One might think to include other $X_V$ in the colimit system, but there is no point as $X_{\mathbb{R}^n} \cong X_V$ when dim $V = n$. Part (d) of Proposition 3.8.7 is equivalent to the statement that the stable equivalences of orthogonal spectra are just the $\pi_*$-isomorphisms.

The key to understanding this is to look at the map $F_{n+1}(S^1 \wedge A) \to F_n(A)$, where we now write $F_n$ as short for $F_{\mathbb{R}^n}$. We claim this is a $\pi_*$-isomorphism (the analog was false for symmetric spectra). In level $n + k$ this map is

$$O(n+k)_+ \wedge_{O(k-1)} (S^{k-1} \wedge S^1 \wedge A) \longrightarrow O(n+k)_+ \wedge_{O(k)} (S^k \wedge A).$$

The $A$ comes out on both sides as a smash factor, so we might as well throw it away. Also, we won't change the stable homotopy groups (except for a shift) if we smash both sides with $S^n$, and this gives

$$O(n+k)_+ \wedge_{O(k-1)} S^{n+k} \longrightarrow O(n+k)_+ \wedge_{O(k)} S^{n+k}.$$

Now, if $X$ is a left $G$-space and $H \leq G$ then

$$G_+ \wedge_H X \cong G_+ \wedge_H (G_+ \wedge_G X) \cong (G_+ \wedge_H G_+) \wedge_G X \cong (G/H_+ \wedge G_+) \wedge_G X \cong G/H_+ \wedge X.$$

In our case $O(n+k)$ acts on $S^{n+k}$, so the map simplifies to

$$O(n+k)/O(k-1)_+ \wedge S^{n+k} \to O(n+k)/O(k)_+ \wedge S^{n+k}.$$

Since $O(k)/O(k-1) \cong S^{k-1}$, the map $O(n+k)/O(k-1) \to O(n+k)/O(k)$ is $(k-1)$-connected and so the smash with $S^{n+k}$ is $(n+2k-1)$-connected. As this goes to infinity with $k$, we have our isomorphism on stable homotopy groups.

## 3.9     EKMM spectra

Unpacking the definitions of [94] takes time and energy. There are several layers to unravel, with quite a bit of intricate mathematics. Anything close to a complete account would involve reproducing a big chunk of the book [94]. Since our aim is only to survey this material, we will content ourselves with a very incomplete account, outlining the main steps but omitting the details behind them.

We first explain the basic idea. Start with the notion of a spectrum defined on a May universe $\mathcal{U}$. This is basically the idea of Bousfield–Friedlander spectra, but done in a coordinate-free way. If $M$ and $N$ are two such spectra, then the smash product $M \wedge N$ seems to be most naturally defined as a spectrum on the universe $\mathcal{U} \oplus \mathcal{U}$. To get a spectrum on $\mathcal{U}$ we can choose an isomorphism $\mathcal{U} \cong \mathcal{U} \oplus \mathcal{U}$, but this involves a choice. The space of all choices is contractible, so in some sense the choice doesn't matter. But if we want a smash product that is commutative and associative on the point-set level, we can't afford to make a single choice.

To get around this, one adopts a definition that builds all the choices in from the beginning. An EKMM-spectrum is (approximately) a coordinate-free spectrum that comes bundled together with its images under all possible changes of universe. The smash product of two such things gives a "bundle" (in a very non-technical sense) of spectra on $\mathcal{U} \oplus \mathcal{U}$, and then changing back to $\mathcal{U}$ in all possible ways just creates another bundle. No choices have been made, but at the expense of introducing extra complexity into the objects themselves.

It is informative to contrast symmetric (or orthogonal) spectra with EKMM-spectra. For the former, the category itself is fairly concrete and easy to understand. The complexities appear in the model structure, where the fibrant objects and weak equivalences are complicated. With EKMM-spectra all the complexity is built into the objects themselves. They are "flabby" enough to all be fibrant in the model structure, and the weak equivalences are quite simple to understand.

## 3.9.1    Outline for the EKMM approach

Fix a May universe $\mathcal{U}$, by which we mean a real inner product space isomorphic to $\mathbb{R}^\infty$ with the dot product. For subspaces $V \subseteq W \subseteq \mathcal{U}$ write $W - V$ for the orthogonal complement of $V$ in $W$. Let $S^V$ be the one-point compactification of $V$, and for $X$ a pointed space write $\Omega^V X$ for the pointed function space $\mathcal{F}_*(S^V, X)$.

It is important to understand that the machinery we describe below was developed over a long time in the works of May and his collaborators. We note especially [155], [93], and [94], but there are plenty of precursors in [63] and [199] as well.

(1) A **prespectrum** is an assignment $V \mapsto E_V$ that sends finite-dimensional subspaces of $\mathcal{U}$ to pointed spaces, together with suspension maps $S^{W-V} \wedge E_V \to E_W$ for every pair $V \subseteq W$. These maps must satisfy an associativity condition and be the identity when $V = W$. Write $\mathcal{P}\mathcal{U}$ for the category of prespectra on $\mathcal{U}$, with the evident maps.

(2) A **spectrum** is a prespectrum where the adjoints $E_V \to \Omega^{W-V} E_W$ are homeomorphisms. Write $\mathcal{S}\mathcal{U}$ for the category of spectra on $\mathcal{U}$.

(3) There are adjoint functors $L \colon \mathcal{P}\mathcal{U} \rightleftarrows \mathcal{S}\mathcal{U} \colon i$ where the right adjoint $i$ is the evident inclusion. The functor $L$ is called "spectrification". (This functor is more mysterious than one might first guess, and having control over colimits in $\mathcal{S}\mathcal{U}$ is entirely dependent on having a good working knowledge of $L$, as provided by Lewis in [155, Appendix].)

(4) For universes $\mathcal{U}$, $\mathcal{U}'$ there is an external smash product $\wedge_{pre} \colon \mathcal{S}\mathcal{U} \times \mathcal{S}\mathcal{U}' \to \mathcal{P}(\mathcal{U} \oplus \mathcal{U}')$ defined as follows. For $M$ and $N$ in $\mathcal{S}\mathcal{U}$, define

$$(M \wedge_{pre} N)(V \oplus V') = M_V \wedge N_{V'}.$$

This only defines $M \wedge_{pre} N$ on subspaces of $\mathcal{U} \oplus \mathcal{U}$ of the form $V \oplus V'$, but these are cofinal amongst all subspaces; so extend $M \wedge_{pre} N$ to all subspaces in any reasonable way. For example, this can be done inductively on the dimension: given an arbitrary finite-dimensional subspace $W \subseteq \mathcal{U}$, choose $V$ and $V'$ with $W \subseteq V \oplus V'$ and define

$$(M \wedge_{pre} N)(W) = \Omega^{(V \oplus V') - W}(M_V \wedge N_{V'}).$$

Finally, define the external smash product $\wedge_{ext} \colon \mathcal{S}\mathcal{U} \times \mathcal{S}\mathcal{U}' \to \mathcal{S}(\mathcal{U} \oplus \mathcal{U}')$ by

$$M \wedge_{ext} N = L(M \wedge_{pre} N).$$

The choices involved in the definition of $M \wedge_{pre} N$ get ironed out by the spectrification functor $L$, and one can check that $M \wedge_{ext} N$ is well-defined.

(5) [Change of universe] By an isometry $f : \mathcal{U} \to \mathcal{U}'$ we mean a linear isometric embedding, not necessarily surjective. Given an isometry $f : \mathcal{U} \to \mathcal{U}'$ and a spectrum $M$ on $\mathcal{U}'$, there is an induced spectrum $f^*M$ given by $V \mapsto M_{f(V)}$. The functor $f^* : \mathcal{SU}' \to \mathcal{SU}$ has a left adjoint $f_*$, defined as follows. For $W \subseteq \mathcal{U}'$ write $W_f = W \cap \text{im}(f)$. For a spectrum $E$ defined on $\mathcal{U}$, define a prespectrum $f_*^{pre} E$ by

$$(f_*^{pre} E)(W) = S^{W-W_f} \wedge E_{f^{-1}(W_f)}.$$

We leave the reader the pleasant exercise of working out the structure maps. Then define $f_* M = L(f_*^{pre} M)$. See [155, II.1] for more details.

(6) Let $\mathcal{I}(\mathcal{U}, \mathcal{U}')$ denote the space of linear isometries from $\mathcal{U}$ to $\mathcal{U}'$. This is a contractible space. One would therefore hope that if $f, g \in \mathcal{I}(\mathcal{U}, \mathcal{U}')$ and $E$ is a spectrum on $\mathcal{U}$ then $f_* E$ and $g_* E$ are weakly equivalent spectra on $\mathcal{U}'$. This is not known in general, but there is a special class of spectra for which it does hold. Define a spectrum $E$ to be $\Sigma$-**cofibrant** if the structure maps $S^W \wedge E_V \to E_{V \oplus W}$ are all cofibrations, and define $E$ to be **tame** if it is homotopy equivalent to a $\Sigma$-cofibrant spectrum. It is known that if $E$ is tame then $f_* E$ and $g_* E$ are homotopy equivalent [94, I.2.5]. We will need to study all these different pushforwards at once.

(7) Given a space $A$, a map $\alpha : A \to \mathcal{I}(\mathcal{U}, \mathcal{U}')$, and a spectrum $E$ on $\mathcal{U}$, there is a construction $A \ltimes E$ which is a spectrum on $\mathcal{U}'$. It is called the "twisted half-smash product". It depends on $\alpha$, but this is omitted from the notation. Loosely speaking, $A \ltimes E$ contains all the ways of constructing a pushforward of $E$ from $\mathcal{U}$ to $\mathcal{U}'$, as parameterized by the map $\alpha$, all bundled together. When $A$ is contractible and $E$ is tame, this has the same homotopy type as the simple pushforwards $f_* E$.

(8) Write $\mathcal{L}(j) = \mathcal{I}(\mathcal{U}^j, \mathcal{U})$ where $\mathcal{U}^j$ is the direct sum of $j$ copies of $\mathcal{U}$. The spaces $\mathcal{L}(j)$ together form an operad $\mathcal{L}$, called the **linear isometries operad**.

(9) Let $\mathbb{L} : \mathcal{SU} \to \mathcal{SU}$ denote the monad $\mathbb{L}(E) = \mathcal{L}(1) \ltimes E$. Then the composition map $\mathcal{L}(1) \times \mathcal{L}(1) \to \mathcal{L}(1)$ induces the natural transformation $\mu : \mathbb{LL} E \to \mathbb{L} E$, and the identity element $id \in \mathcal{L}(1)$ induces the unit $\eta : E \to \mathbb{L} E$.

(10) An $\mathbb{L}$-**spectrum** is an $\mathbb{L}$-algebra: that is, an $\mathbb{L}$-spectrum is a spectrum $X$ together with a map $\mathbb{L}X \to X$ making the usual diagrams commute.

(11) Given $\mathbb{L}$-spectra $M$ and $N$, we define the smash product by

$$M \wedge_{\mathcal{L}} N = \mathcal{L}(2) \ltimes_{\mathcal{L}(1) \times \mathcal{L}(1)} (M \wedge_{ext} N).$$

Note that $M \wedge_{ext} N$ is a spectrum on $\mathcal{U}^2$. The object on the right in this definition is a coequalizer of certain evident maps coming from the $\mathbb{L}$-algebra structures on $M$ and $N$ and the operad maps in $\mathcal{L}$. The smash product $\wedge_{\mathcal{L}}$ turns out to be associative and symmetric (see [94, I.5]), but not unital.

(12) The sphere spectrum $S$ is the spectrification of the prespectrum $V \mapsto S^V$. It turns out that $S$ is an $\mathbb{L}$-algebra in a natural way, and that for any $\mathbb{L}$-spectrum $M$ there is a natural map $\lambda_M : S \wedge_{\mathcal{L}} M \to M$. Define an **EKMM-spectrum** to be an $\mathbb{L}$-spectrum

$M$ for which $\lambda_M$ is an isomorphism. Denote the category of EKMM-spectra by EKMM$_S$. The spectrum $S$ is itself an EKMM spectrum.

*Remark.* EKMM-spectra are called "$S$-modules" in [94]. While not a terrible name, it conflicts with the notions of $S$-modules that one has in other categories like symmetric spectra and orthogonal spectra. The name "EKMM-spectra" seems to lead to less confusion.

(13) The smash product of EKMM-spectra $M$ and $N$ is defined as $M \wedge_S N = M \wedge_{\mathcal{L}} N$. This gives a symmetric monoidal smash product on EKMM$_S$ with unit $S$.

(14) Now suppress the universe and abbreviate $\mathcal{S}\mathcal{U}$ to just $\mathcal{S}$. There are adjunctions

$$\mathcal{S} \underset{\mathbb{U}}{\overset{\mathbb{L}(-)}{\rightleftarrows}} (\mathbb{L}-\mathcal{S}pectra) \underset{\mathcal{F}_{\mathcal{L}}(S,-)}{\overset{S \wedge_{\mathcal{L}} (-)}{\rightleftarrows}} \text{EKMM}_S$$

where $\mathbb{U}$ is the forgetful functor and the left adjoints both point left to right.

(15) For each $V \subseteq \mathcal{U}$, the evaluation map $\mathrm{Ev}_V \colon \mathcal{S} \to \mathcal{T}op_*$ has a left adjoint, denoted $F_V$. We also write $\Sigma^\infty$ for the functor $F_0$.

(16) For a map $f$ in $\mathcal{S}$, say that $f$ is a weak equivalence if $f$ is a $\pi_*$-isomorphism on underlying spectra. Since the objects of $\mathcal{S}$ are all $\Omega$-spectra, we can also characterize the weak equivalences as maps inducing objectwise weak equivalences in $\mathcal{T}op_*$ on application of $\mathrm{Ev}_V$ (for all $V$).

If $i\colon \text{EKMM}_S \hookrightarrow \mathbb{L}-\mathcal{S}pectra$ denotes the inclusion then for any $M$ in EKMM$_S$ there is a canonical map $iM \to \mathcal{F}_{\mathcal{L}}(S,M)$ and this map is always a weak equivalence. So up to homotopy the functors $i$ and $\mathcal{F}_{\mathcal{L}}(S,-)$ are really the same; as a consequence, a map in EKMM$_S$ is a weak equivalence if and only if $\mathcal{F}_{\mathcal{L}}(S,-)$ is a weak equivalence.

Say that $f$ is a fibration if it has the right lifting property with respect to all maps $F_n(I^k \times \{0\}) \to F_n(I^k \wedge I_+)$, for all $n$ and $k$.

Then $\mathcal{S}$ has a model category structure with the weak equivalences and fibrations defined above, and the right adjoints $\mathbb{U}$ and $\mathcal{F}_{\mathcal{L}}(S,-)$ create induced model category structures on $\mathbb{L}-\mathcal{S}pectra$ and EKMM$_S$. Note that since all objects are fibrant in $\mathcal{T}op_*$, the same holds in each of the categories $\mathcal{S}$, $\mathbb{L}-\mathcal{S}pectra$, and EKMM$_S$.

Moreover, the two pairs of adjoint functors from (14) are both Quillen equivalences.

(17) For any pointed space $X$ we define

$$\Sigma_S^\infty X = S \wedge_{\mathcal{L}} \mathbb{L}(\Sigma^\infty X).$$

This is just the composite of the left adjoints in the diagram

$$\mathcal{T}op_* \underset{\mathrm{Ev}_0}{\overset{\Sigma^\infty}{\rightleftarrows}} \mathcal{S} \underset{\mathbb{U}}{\overset{\mathbb{L}(-)}{\rightleftarrows}} (\mathbb{L}-\mathcal{S}pectra) \underset{\mathcal{F}_{\mathcal{L}}(S,-)}{\overset{S \wedge_{\mathcal{L}}(-)}{\rightleftarrows}} \text{EKMM}_S$$

and so in particular is a left Quillen functor. Write $\Omega_S^\infty$ for the composition of the right adjoints in the above diagram. For $n \geq 0$ write

$$S_S^n = \Sigma_S^\infty(S^n) = S \wedge_{\mathcal{L}} (\mathbb{L}(\Sigma^\infty S^n)).$$

We regard $S_S^n$ as a "stable $n$-sphere", and from this we can define the notion of $CW$-spectra for EKMM$_S$ in the usual way. Such spectra will all be cofibrant.

(18) Now we come to a major point. We have the object $S = \Sigma^\infty S^0$, which is an EKMM-spectrum (see (12)) and the unit for the smash product. But we also have the stable 0-sphere $S_S^0 = \Sigma_S^\infty S^0 = S \wedge_{\mathcal{L}} \mathbb{L}S$. The $\mathbb{L}$-algebra structure on $S$ is a map $\mathbb{L}S \to S$, which induces the canonical map

$$S_S^0 = S \wedge_{\mathcal{L}} \mathbb{L}S \to S \wedge_{\mathcal{L}} S = S.$$

This map is a weak equivalence, but it is *not* an isomorphism. In fact it turns out that $S$ is not cofibrant in EKMM$_S$, and so $S_S^0$ is a cofibrant replacement for $S$.

The fact that $S$ is not cofibrant, and the distinction between $S_S^0$ and $S$, is one of the major differences between EKMM-spectra and symmetric (or orthogonal) spectra.

(19) For any pointed space $X$, the spectrum $\Sigma^\infty X$ (from (15) above) turns out to be an $\mathbb{L}$-spectrum in a natural way and also an EKMM-spectrum. So we can think of $\Sigma^\infty$ as a functor $\mathcal{T}op_* \to$ EKMM$_S$. It has a right adjoint $\Omega^\infty$. It is dangerous to confuse $\Sigma_S^\infty$ and $\Sigma^\infty$. The first is a left Quillen functor, but the second is not. We have the comparison map

$$\Sigma_S^\infty X = S \wedge_{\mathcal{L}} \mathbb{L}(\Sigma^\infty X) \longrightarrow S \wedge_{\mathcal{L}} \Sigma^\infty X \cong \Sigma^\infty X,$$

with the middle map coming from the $\mathbb{L}$-structure on $\Sigma^\infty X$, and the last isomorphism being because $\Sigma^\infty X$ is an $S$-module. This comparison map is a weak equivalence whenever $X$ is nondegenerately based (i.e., $* \to X$ is a cofibration).

The functor $\Sigma^\infty$ has good monoidal properties, such as a natural isomorphism $\Sigma^\infty(X \wedge Y) \cong (\Sigma^\infty X) \wedge_S (\Sigma^\infty Y)$ compatible with associativity and commutativity isomorphisms.

The work in [94] shows the following:

**Theorem 3.9.1.** *The category* EKMM$_S$ *is a stable, closed symmetric monoidal model category satisfying the Algebraic Creation and Invariance Properties as well as the Strong Flatness Property. As a model category it is Quillen equivalent to the stable projective model structure on* $\mathrm{Sp}^\mathbb{N}$.

*Proof.* We sketch a proof here, since there seems to be no simple reference where this can be just looked up. Let $\mathbb{F}_n : \mathcal{T}op_* \to$ EKMM$_S$ be the functor $\mathbb{F}_n(X) = S \wedge_{\mathcal{L}} \mathbb{L}F_n(X)$.

In [94] the closed symmetric monoidal structure is established, as well as the model structure. The latter comes with the set $\{\mathbb{F}_m(S^n) \to \mathbb{F}_m(D^{n+1}) \mid m, n \geq 0\}$ of generating cofibrations and the set $\{\mathbb{F}_m(D^n) \to \mathbb{F}_m(D^n \wedge I_+) \mid m, n \geq 0\}$ of generating trivial cofibrations (see [94, VII.5.6–5.8]).

To prove the Pushout-Product Axiom, it suffices to check it on generating cofibrations and trivial cofibrations. So we need to analyze the box product of $\mathbb{F}_m(f)$ and $\mathbb{F}_n(g)$ for $f : A \rightarrowtail B$ and $g : C \rightarrowtail D$ cofibrations in $\mathcal{T}op_*$. The key point is then that a choice of homeomorphism $\mathcal{U}^2 \cong \mathcal{U}$ induces a homeomorphism $\mathcal{L}(2) \cong \mathcal{L}(1)$ and thus an identification $\mathbb{F}_m(f) \square \mathbb{F}_n(g) \cong \mathbb{F}_{m+n}(f \square g)$; the Pushout-Product Axiom then follows. (See [45, 4.21] for a version of this argument in the context of spaces.)

There is a canonical map $\mathbb{L}S \to S$, and the induced map $\alpha : S \wedge_{\mathcal{L}} \mathbb{L}S \to S \wedge_{\mathcal{L}} S \cong S$ is a cofibrant-approximation in $\mathrm{EKMM}_S$. Note that the domain is $\Sigma_S^\infty(S^0)$. We must show for any $M$ in $\mathrm{EKMM}_S$ that $(S \wedge_{\mathcal{L}} \mathbb{L}S) \wedge_S M \to S \wedge_S M = M$ is a weak equivalence. Remembering that $\wedge_S = \wedge_{\mathcal{L}}$, consider the diagram

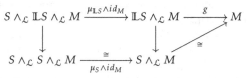

The diagonal map is an isomorphism by the definition of $\mathrm{EKMM}_S$. The map $g$ is a weak equivalence by [94, I.6.2], and $\mu_{\mathbb{L}S} \wedge id_M$ is a weak equivalence by [94, I.8.5(iii)]. It follows that every map in the diagram is a weak equivalence, and this verifies the Unit Axiom in the definition of monoidal model category. It also verifies condition (1) in Proposition 3.3.6.

Condition (2) of Proposition 3.3.6 also holds, since $\mathrm{EKMM}_S$ is a topological model category where all objects are fibrant. So Proposition 3.3.6 yields the Algebraic Creation Property.

The Strong Flatness Property follows from [94, III.3.8] together with the fact that every cofibrant $R$-module is a retract of a cell-module. For the Algebraic Invariance Property we verify the conditions of Proposition 3.3.9: condition (1) is the Strong Flatness Property, and condition (2) is [94, VII.6.2].

For the Quillen equivalence between $\mathrm{EKMM}_S$ and $\mathrm{Sp}^\mathbb{N}$, it is easiest to go through $\mathrm{Sp}^\mathbb{O}$ or $\mathrm{Sp}^\Sigma$. The Quillen equivalence with $\mathrm{Sp}^\mathbb{O}$ is in [177], and the equivalence with $\mathrm{Sp}^\Sigma$ is in [262].                                              $\square$

## 3.10 Afterthoughts

One of the drawbacks of a survey like this is that there is never enough time or space to say everything that one would like. This final section will give a blitz treatment of various topics that are important and should not go unmentioned.

### 3.10.1    Functors with smash product

This was an early attempt at a strict model for ring spectra, due to Bökstedt and used by him in his work on topological Hochschild homology [52]. In modern times these have been eclipsed by ring objects in either symmetric or orthogonal spectra, but it is still good to know the basic idea.

Let $\mathcal{W}$ be the category of pointed spaces that are homeomorphic to a finite $CW$-complex, Regard $\mathcal{W}$ as a $\mathcal{T}op_*$-enriched category. A $\mathcal{W}$-**sequence** is an enriched functor $\mathcal{W} \to \mathcal{T}op_*$ (these are also called $\mathcal{W}$-spaces sometimes). Day convolution, as in (3.7.9), gives a symmetric monoidal product on $\mathcal{W}$-sequences.

There is a "sphere sequence" $S$ given by the inclusion $\mathcal{W} \hookrightarrow \mathcal{T}op_*$, and this is a commutative monoid. We define a **W-spectrum** to be a left $S$-module. Unraveling this, a $W$-spectrum is an enriched functor $\Phi\colon W \to \mathcal{T}op_*$ together with structure maps $X \wedge \Phi(Y) \to \Phi(X \wedge Y)$ satisfying unital and associativity conditions. However, these extra structure maps do not provide new information — they are an automatic consequence of being an enriched functor, as was explained back in Section 3.1. So in this case $W$-sequences and $W$-spectra are the same thing.

There is a functor $\mathcal{O}I \to W$ given by $V \mapsto S^V$, and restriction along this functor takes $W$-spectra to orthogonal spectra. One can restrict further along the composite $\Sigma I \to \mathcal{O}I \to W$ to get a symmetric spectrum.

The model category story works out in the same way as for orthogonal spectra. See [178].

A "functor with smash product" (FSP) is a monoid in the category of $W$-spectra. This amounts to an enriched functor $\Phi\colon W \to \mathcal{T}op_*$ equipped with maps $X \to \Phi(X)$ and $\Phi(X) \wedge \Phi(Y) \to \Phi(X \wedge Y)$ satisfying various properties that are not hard to work out.

*Remark 3.10.1.* We saw in Section 3.1.2 that the notion of a classical spectrum comes from the idea of "remembering" the mapping spaces $E_n = \mathrm{Map}(S^{-n}, E)$ for a fantasy stable object $E$. In a similar vein, a pointed finite CW-complex $X$ should give rise to a stable object $\Sigma^\infty X$, which should have a Spanier–Whitehead dual $(\Sigma^\infty X)^*$. The idea of $W$-spectra is that they "remember" the mapping spaces $E(X) = \mathrm{Map}((\Sigma^\infty X)^*, E)$.

We remark that the notion of $W$-sequence is essentially equivalent (homotopically speaking) to the notion of a simplicial functor from $sSet$ to $sSet$. The connection between these kinds of functors and spectra was initially raised by Anderson [3]. Lydakis [171] first produced (in the simplicial setting) a model category structure as well as the symmetric monoidal product, showed the Quillen equivalence with Bousfield–Friedlander spectra, and identified the ring objects with Bökstedt's FSPs.

## 3.10.2   Γ-spaces

Let $\Gamma^{op}$ be the category of finite based sets $\mathbf{n}_+ = \{0, 1, \ldots, n\}$ (based at 0) and based maps. A functor $\Gamma^{op} \to \mathcal{T}op_*$ is called a **Γ-space**. The smash product of based sets induces a symmetric monoidal product on $\Gamma^{op}$: specifically, we identify $\mathbf{m}_+ \wedge \mathbf{n}_+$ with $(\mathbf{m} \cdot \mathbf{n})_+$ using the lexicographic ordering. Day convolution then gives a monoidal structure on the category of Γ-spaces.

Γ-spaces were introduced by Segal [268], who showed that the homotopy category is equivalent to the full subcategory of the stable homotopy category consisting of the connective spectra. The first model category structure on Γ-spaces goes back to Bousfield–Friedlander [56] (note that no such model category could be stable, given that the suspension functor on the homotopy category is not an equivalence). Lydakis [172] introduced the symmetric monoidal product on Γ-spaces and showed that it models the smash product of spectra, and [264] produced a model category structure on the ring objects. See also the discussion in [178].

The idea behind Γ-spaces comes from considerations similar to those made in

Remark 3.10.1. In any homotopy theory of spectra we would have objects $\Sigma^\infty T$ for every pointed set $T$ (this will just be a wedge of copies of the sphere spectrum $S$, indexed by the non-basepoints in $T$). Therefore we would also have Spanier–Whitehead duals $(\Sigma^\infty T)^*$. The assignment $T \mapsto (\Sigma^\infty T)^*$ would be a contravariant functor defined on $\Gamma^{op}$, and for a stable object $E$ the assignment $T \mapsto \operatorname{Map}((\Sigma^\infty T)^*, E)$ would therefore be a $\Gamma$-space.

If $T = [n]$ then $\Sigma^\infty T = \bigvee_{i=1}^n S$, and so $(\Sigma^\infty T)^*$ can be identified with the product $\prod_{i=1}^n S$ (using that $S^* = S$). So another way to say the above is that a $\Gamma$-space comes from remembering what a spectrum looks like through the eyes of the finite products $*, S, S \times S, S \times S \times S$, and so forth. That is to say, if $E$ is a spectrum we remember $[n] \mapsto E_n = \operatorname{Map}(S^{\times n}, E)$. As finite products are weakly equivalent to finite wedges in spectra, it's clear that this data can only remember the connective part of a spectrum.

In fact, since $\prod_{i=1}^n S \simeq \bigvee_{i=1}^n S$ we would additionally have the relations

$$E_n = \operatorname{Map}\Big( \prod_{i=1}^n S, E \Big) \simeq \operatorname{Map}\Big( \bigvee_{i=1}^n S, E \Big) \simeq \prod_{i=1}^n \operatorname{Map}(S, E) = \prod_{i=1}^n E_1.$$

This suggests that what we really care about are $\Gamma$-spaces $X$ such that a canonical map $X_n \to \prod_{i=1}^n X_1$ is an equivalence (and when $n = 0$ this should be interpreted as $X_0 \simeq *$). These were called "special" $\Gamma$-spaces in [56]. This turns out to equip $\pi_0(X_1)$ with the structure of an abelian monoid via the multiplication

$$\pi_0(X_1) \times \pi_0(X_1) \xleftarrow{\cong} \pi_0(X_2) \xrightarrow{\mu} \pi_0(X_1),$$

where $\mu$ is induced by the map $[2]_+ \to [1]_+$ sending $1, 2 \mapsto 1$. But if $X_1 = \operatorname{Map}(S, E)$ then we should have $X_1 \simeq \Omega^2 \operatorname{Map}(S^{-2}, E)$, which means $\pi_0(X_1)$ would actually be an abelian group. Adding on this condition yields what [56] called "very special" $\Gamma$-spaces. The pleasant surprise is that there are no further "relations" that one has to keep track of here: that is, the model category structure on $\Gamma$-spaces is set up so that the fibrant objects are precisely these very special $\Gamma$-spaces, and this is enough to get the Quillen equivalence with connective spectra. See also [79, Example 5.7] for another perspective on these "relations".

The inclusion of categories $\Gamma^{op} \hookrightarrow \mathcal{W}$, regarding every pointed set as a discrete topological space, yields comparison functors between $\mathcal{W}$-spaces and $\Gamma$-spaces in the usual way. See Remark 3.5.3.

Segal introduced $\Gamma$-spaces in [268] because they were a natural receptor for a certain version of algebraic $K$-theory. We outline this briefly. Let $\mathcal{C}$ be a category with finite coproducts. For a finite set $T$ write $\mathcal{P}(T)$ for the category whose elements are the subsets of $T$ and whose maps are subset inclusions. Let $\mathcal{C}(T)$ be the category whose objects are functors $F : \mathcal{P}(T) \to \mathcal{C}$ having the property that whenever $A_1, \ldots, A_n \subseteq T$ are disjoint the set of maps $\{F(A_i) \to F(\cup_i A_i)\}$ induces an isomorphism

$$\coprod_i F(A_i) \xrightarrow{\cong} F\Big( \bigcup_i A_i \Big).$$

When $n = 0$ this property implies that $F(\emptyset)$ is an initial object in $\mathcal{C}$.

If $T$ is a pointed set, let $(K\mathcal{C})(T) = B\mathcal{C}(T - *)$, where $B(-)$ denotes the usual classifying space of a small category (that is, the geometric realization of the nerve).

If $f\colon T \to U$ is a map of pointed sets, there is an induced map $\mathcal{P}(U-*) \to \mathcal{P}(T-*)$ sending $A \mapsto f^{-1}(A) \cap (T-*)$, and this in turn induces a functor $\mathcal{C}(T-*) \to \mathcal{C}(U-*)$. So $K\mathcal{C}$ is a $\Gamma$-space. (The basepoint is playing the role of a "sink" here, in the sense that pointed maps $f\colon T \to U$ are the same as pairs $(A \subseteq T, A \to U)$, where in the correspondence one has $T - A = f^{-1}(*)$. The reader is advised to work out the maps in $K\mathcal{C}$ where $T$ and $U$ are $\{0,1\}$ and $\{0,1,2\}$ — in either order — to get a feeling for what is happening here.)

Note that an object in $\mathcal{C}(T)$ can be thought of as a $T$-indexed collection of objects in $\mathcal{C}$ together with consistent choices of coproducts for all subsets of $T$. Compare the description of Waldhausen $K$-theory from Section 3.8.8.

### 3.10.3   Spectra in other settings

Let $\mathcal{M}$ be a symmetric monoidal model category and let $K$ be a cofibrant object. Just as spectra stabilize $\mathcal{T}op_*$ under the operation of smashing with $S^1$, one might want to stabilize $\mathcal{M}$ under the operation of tensoring with $K$. Under mild "sufficiently combinatorial" hypotheses on $\mathcal{M}$, this works out just fine. Hovey [132] showed that one can form both Bousfield–Friedlander and symmetric spectra in this generalized setting, and all the basic model structures work out just as expected.

Standard applications include stabilizing the model category of $G$-spaces along a representation sphere $S^V$, or stabilizing a model category of motivic spaces along the motivic sphere $S^{2,1}$.

Hovey in fact showed that the Bousfield–Friedlander construction is really about inverting a *functor* $G\colon \mathcal{M} \to \mathcal{M}$, whereas (as discussed in Section 3.7.4) the symmetric spectrum construction is about making an object invertible in the symmetric monoidal sense. This difference has consequences for the comparison of the two constructions $\mathrm{Sp}^{\mathbb{N},\wedge K}$ and $\mathrm{Sp}^{\Sigma,K}$. In the latter, the suspension spectrum of $K$ is an invertible object and so must satisfy the cyclic permutation condition (3.3.13). In the former, where we are only inverting the functor $(-) \wedge K$ and don't necessarily have a monoidal product around anymore, there is no guarantee that this holds. So there is no reason to suspect a Quillen equivalence here: in general, $\mathrm{Sp}^{\Sigma,K}$ has more "relations" than $\mathrm{Sp}^{\mathbb{N},\wedge K}$. Hovey [132] has some results showing that in the presence of the cyclic permutation condition these two constructions *are* Quillen equivalent, but he also observes that the results are perhaps not as general as one would like.

A version of $\mathcal{W}$-spaces (or simplicial functors) for model categories satisfying certain technical hypotheses has also been developed, by Dundas–Röndigs–Østvaer [84].

### 3.10.4   $G$-spectra

Let $G$ be a compact Lie group, but feel free to think only of a finite group if desired. There should of course be a model category of genuine $G$-spectra, where one stabilizes with respect to all finite-dimensional representation spheres. The associated homotopy category was first developed in [155], and is nicely summarized in [190].

To construct an appropriate model category via symmetric spectra, one could pick representatives $V_1, V_2, \ldots, V_n$ for all finite-dimensional irreducible $G$-representations and set $V = V_1 \oplus \cdots \oplus V_n$. Performing the symmetric spectra construction on $G$-spaces using the object $S^V$ makes a perfectly good model category of genuine $G$-spectra. Although this is fine for some purposes, it is a little unnatural. The fact that all finite-dimensional $G$-representations aren't inherently built into the machinery can make some things more trouble than they should be.

The construction of orthogonal spectra works right out of the box for $G$-spaces, requiring only the obvious modifications. See [177] or [120, Appendix A] for details. Currently this is the preferred setting for $G$-equivariant spectra.

The equivariant version of EKMM spectra is developed in [177]. One starts with a $G$-universe $\mathcal{U}$ that is "complete" in the sense that it contains infinitely many copies of every irreducible representation. One of the surprises is that there are *two* naturally arising model category structures on $G$-equivariant EKMM-spectra, both having the same notion of stable weak equivalence. One has cofibrations built from cellular inclusions based on cells of the form $F_n(G/H_+ \wedge S^k)$ for $n, k \geq 0$, and the other has cofibrations built from cells of the form $F_V(G/H_+ \wedge S^k)$ with $k \geq 0$ and $V$ a $G$-representation. These model structures are Quillen equivalent, but different. We refer to [177, Chapter IV.2] for details.

When $G$ is finite, versions of equivariant symmetric spectra have been produced by Mandell [183] and Hausmann [115]. Ostermayr [218] developed a model structure for equivariant $\Gamma$-spaces. A model category structure for an equivariant version of $\mathcal{W}$-spaces is developed in [84] (see also [43]).

## 3.10.5    Model categories for commutative algebras

Let $(Spectra, \wedge, S)$ be a closed symmetric monoidal model category of spectra that satisfies the Algebraic Creation Property. Let $R$ be a commutative ring spectrum, and write $R$–ComAlg for the category of commutative $R$-algebras. The forgetful functor $U \colon R\text{–ComAlg} \to R\text{–Mod}$ has a left adjoint Sym given by the symmetric algebra functor

$$\text{Sym}(M) = R \vee M \vee (M \wedge_R M)/\Sigma_2 \vee (M \wedge_R M \wedge_R M)/\Sigma_3 \vee \cdots .$$

We can ask if the forgetful functor creates a model structure on $R$–ComAlg.

In $\text{EKMM}_S$, this works with no trouble — in part because all objects are fibrant. See [94, VII.4.7–4.10]. In contrast, for symmetric and orthogonal spectra there is a difficulty and such a model structure *cannot* exist in general. For example, it cannot exist when $R = S$: as we saw in Section 3.1.7, there cannot exist a commutative ring spectrum that is weakly equivalent to $S$ and whose underlying spectrum is fibrant.

One solution to this problem is via the *positive model structure* on symmetric (or orthogonal) spectra, suggested originally by Jeff Smith. Basically, go back and mimic the development of the level and stable structures but remove all references to what happens in level 0. Change the levelwise weak equivalences to maps that are weak equivalences in levels greater than zero, and so forth. The fibrant objects in the positive stable model structure are then spectra $X$ with the property that $X_n \to \Omega X_{n+1}$ is a weak equivalence for all $n \geq 1$ (these are called "positive $\Omega$-spectra"). This model structure is Quillen equivalent to the one we already had, and it is also monoidal and satisfies all the nice properties we are used to.

The adjoint to the $\Sigma^\infty$ functor is $\mathrm{Ev}_0$ as always, but note that $\mathrm{Ev}_0$ no longer has the behavior of $\Omega^\infty$ for fibrant objects. So there is no problem with having a model for $S$ that is a commutative ring spectrum and is fibrant in the positive model structure.

The positive model structures on symmetric and orthogonal spectra are developed in [178], which also shows that if one uses these structures the forgetful functor does create a model structure on $R$–ComAlg for any commutative ring spectrum $R$.

For more work related to these issues, including yet another model structure on symmetric spectra, see [274].

As another application, the positive model structure on $\mathrm{Sp}^\Sigma$ is used in [262] to get a monoidal Quillen equivalence between $\mathrm{Sp}^\Sigma$ and $\mathrm{EKMM}_S$.

Commutative ring spectra are discussed in more detail in Chapter 6 of this volume.

### 3.10.6    Stable categories and categories of modules

This is only a very brief remark, but if you want to better understand stable model categories in general and how they interact with the modern monoidal categories of spectra, go read [266]. That paper provides a basic technique that is pervasive in how we approach these categories.

# 4  Stable homotopy theory via ∞-categories

## by Clark Barwick

The task before us is to investigate stable homotopy theory — and stable homotopy theories more generally — through the lens of ∞-category theory. Of necessity, this chapter is somewhat ahistorical; we refer the reader to the historical discussions outlined in Chapter 3 for background on the development of modern categories of spectra. However, the reader who is familiar with that story will have a keen appreciation for the foundational problems that become much cleaner in this framework.

Let us assume familiarity with elementary ∞-category theory as presented by Jacob Lurie in [169] — most particularly, the theory of limits, colimits, adjunctions, and presentability. In particular, Chapter 5 of [169] will be frequently cited, but this is the upper limit: nothing of the later chapters or of any more advanced text will be needed here. We have tried to be systematic in our citations.

Our exposition is largely a gentle introduction to some of the material in [168] and subsequent papers, and of course much of our understanding of spectra was informed by this remarkable and beautiful text. We hope that this presentation will appeal to mathematicians both within and without homotopy theory.

I offer my sincere thanks to Andrew Blumberg for his enormous assistance in making my writing palatable.

## 4.1  Spectra

Let $X$ be a pointed simplicial set. An old observation of Dan Kan provides a simple way to extract the reduced homology of the geometric realisation $|X|$ from $X$. Namely, we let $\widetilde{\mathbb{Z}}\{X\}$ be the simplicial abelian group in which $\widetilde{\mathbb{Z}}\{X\}_n = \widetilde{\mathbb{Z}}\{X_n\}$ is freely generated by the *pointed* set $X_n$ (so that the point of $X_n$ becomes the zero element of $\widetilde{\mathbb{Z}}\{X\}_n$). One then has

$$\widetilde{H}_n(|X|, \mathbb{Z}) \cong \pi_n \widetilde{\mathbb{Z}}\{X\}.$$

More precisely, the simplicial abelian group $\widetilde{\mathbb{Z}}\{X\}$ corresponds, under Dold–Kan, to the chain complex $\widetilde{C}_*(|X|, \mathbb{Z})$.

Let us disregard the abelian group structure and regard $\widetilde{\mathbb{Z}}\{X\}$ merely as a pointed simplicial set. In fact, the functor $X \mapsto \widetilde{\mathbb{Z}}\{X\}$ preserves weak equivalences of pointed spaces, so we are entitled to think of this assignment as a functor from the ∞-category of pointed spaces to itself. We can also deduce the following properties:

1. The functor $X \mapsto \widetilde{\mathbb{Z}}\{X\}$ is *reduced*. That is, if $X$ is contractible, then so is the simplicial abelian group $\widetilde{\mathbb{Z}}\{X\}$. Thus $\widetilde{H}_i(*) = 0$.

2. The functor $X \mapsto \widetilde{\mathbb{Z}}\{X\}$ is *unital*. In other words, $\widetilde{\mathbb{Z}}\{S^0\}$ is the constant simplicial set with value $\mathbb{Z}$; under the Dold–Kan correspondence, it corresponds to the complex $\mathbb{Z}[0]$ concentrated in degree 0. Thus $\widetilde{H}_0(S^0) = \mathbb{Z}$, and $\widetilde{H}_i(S^0) = 0$ for $i > 0$.

3. The functor $X \mapsto \widetilde{\mathbb{Z}}\{X\}$ is *excisive*: for any homotopy pushout

$$
\begin{array}{ccc}
U & \xrightarrow{\ i\ } & V \\
\downarrow & & \downarrow \\
W & \longrightarrow & X
\end{array}
$$

(e.g., any "honest" pushout in which $i$ is a monomorphism of simplicial sets), the square

$$
\begin{array}{ccc}
\widetilde{\mathbb{Z}}\{U\} & \longrightarrow & \widetilde{\mathbb{Z}}\{V\} \\
\downarrow & & \downarrow \\
\widetilde{\mathbb{Z}}\{W\} & \longrightarrow & \widetilde{\mathbb{Z}}\{X\}
\end{array}
$$

is homotopy cartesian, so that one obtains a long exact sequence

$$
\cdots \to \widetilde{H}_n(|U|, \mathbb{Z}) \to \widetilde{H}_n(|V|, \mathbb{Z}) \oplus \widetilde{H}_n(|W|, \mathbb{Z}) \to \widetilde{H}_n(|X|, \mathbb{Z}) \to \widetilde{H}_{n-1}(|U|, \mathbb{Z}) \to \cdots.
$$

One can prove this by reducing to the case in which $i$ is an inclusion $\partial\Delta^n \hookrightarrow \Delta^n$ and verifying this case explicitly.

4. Finally, the functor $X \mapsto \widetilde{\mathbb{Z}}\{X\}$ is *of finite presentation*, in that it preserves filtered colimits.

These four properties actually identify the functor $X \mapsto \widetilde{\mathbb{Z}}\{X\}$ uniquely, up to canonical natural equivalence. This is the *uniqueness of homology.*

We may regard $X \mapsto \widetilde{\mathbb{Z}}\{X\}$ as a kind of categorified version of *a line of slope* 1. The first two conditions describe the values of the functor on two objects — the one-point space $*$, which (as the unit for $\vee$) is our analogue of 0, and $S^0$, which (as the unit for $\wedge$) is our analogue of 1. Under our analogy, we have insisted that $f(0) = 0$ and $f(1) = 1$. The other two axioms declare that $X \mapsto \widetilde{\mathbb{Z}}\{X\}$ is *linear*. This linearity now determines the values of this functor on all other objects.

## Spectra

If we merely eliminate the "slope 1" condition (unitality), we arrive at the notion of a *spectrum*. Here is the definition.

Definition 4.1.1. Write $\mathcal{S}_*$ for the $\infty$-category of pointed spaces (the full subcategory of $\mathrm{Fun}(\Delta^1, \mathcal{S})$ spanned by those objects $X \to Y$ in which $X$ is contractible).

Then a functor $E \colon \mathcal{S}_* \to \mathcal{S}_*$ is called a *linear functor* or a *spectrum* if $E$ is reduced, excisive, and of finite presentation. That is:

1. The functor $E$ is *reduced*: for any contractible pointed space $P$, the pointed space $E(P)$ is also contractible.

2. The functor $E$ is *excisive*: for any pushout square

$$
\begin{array}{ccc}
U & \longrightarrow & V \\
\downarrow & & \downarrow \\
W & \longrightarrow & X
\end{array}
$$

in $\mathcal{S}_*$, the induced square

$$
\begin{array}{ccc}
E(U) & \longrightarrow & E(V) \\
\downarrow & & \downarrow \\
E(W) & \longrightarrow & E(X)
\end{array}
$$

is a pullback in $\mathcal{S}_*$.

3. The functor $E$ is *of finite presentation*: for any filtered diagram $\alpha \mapsto X_\alpha$ in $\mathcal{S}_*$, the natural map

$$
\mathrm{colim}_\alpha E(X_\alpha) \to E(\mathrm{colim}_\alpha X_\alpha)
$$

is an equivalence.

This makes precise the sense in which spectra are said to "be" generalised homology theories. But in order to come to grips with this definition, we must do some work to unpack the axioms in turn.

## Reduced functors

Reducedness is nothing profound. If one has a functor $F \colon \mathcal{S}_* \to \mathcal{S}_*$ that isn't reduced, one may "repair" it by passing to the reduction $F^{red}$, which carries a pointed space $X$ to the cofiber of the map $F(*) \to F(X)$.

## Finitely presented functors

Finite presentability is also relatively straightforward. We say that a pointed space $X$ is *finite* if it can be expressed as a finite colimit of contractible pointed spaces. Every pointed space is the filtered colimit of the finite spaces that map to it, so a functor $F \colon \mathcal{S}_* \to \mathcal{S}_*$ is of finite presentation if and only if it is left Kan extended from its restriction to the $\infty$-category $\mathcal{S}_*^{fin}$ of finite spaces.

So a finitely presented functor $\mathcal{S}_* \to \mathcal{S}_*$ is uniquely determined by its restriction to $\mathcal{S}_*^{fin}$. That is, the $\infty$-category of finitely presented functors $\mathcal{S}_* \to \mathcal{S}_*$ is equivalent to the $\infty$-category of (arbitrary) functors $\mathcal{S}_*^{fin} \to \mathcal{S}_*$.

## Excisive functors

The excision condition is where the rubber meets the road. An important special case of a square in $\mathcal{S}_*$ is when the corners are contractible spaces:

$$
\begin{array}{ccc}
Y & \longrightarrow & * \\
\downarrow & & \downarrow \\
* & \longrightarrow & X
\end{array}
\tag{4.1.1}
$$

If (4.1.1) is a pushout square, then $X$ is the *suspension* $\Sigma Y$; since the forgetful functor $\mathcal{S}_* \to \mathcal{S}$ preserves pushouts, the pointing of $Y$ is irrelevant. Dually, if (4.1.1) is a pullback square in $\mathcal{S}_*$, then $Y$ is the *loopspace* $\Omega X$; here, the pointing of $X$ is important.

4.1.2. Suspension is left adjoint to loopspace:

$$
\Sigma \colon \mathcal{S}_* \rightleftarrows \mathcal{S}_* \colon \Omega.
$$

In particular, we have the unit $\varepsilon \colon id \to \Omega\Sigma$ and the counit $\eta \colon \Sigma\Omega \to id$.

Now if $F \colon \mathcal{S}_* \to \mathcal{S}_*$ is a reduced functor, then we may apply it to the pushout square

$$
\begin{array}{ccc}
Y & \longrightarrow & * \\
\downarrow & & \downarrow \\
* & \longrightarrow & \Sigma Y
\end{array}
$$

to obtain a canonical map

$$
\sigma_Y \colon FY \to \Omega F\Sigma Y.
$$

We may also write $\sigma_Y^F$ whenever disambiguation is called for. It is clear that if $F$ is excisive, then the natural transformation $\sigma_Y$ is an equivalence. However, it is relatively surprising that this condition *suffices* to ensure the excisiveness of $F$.

Lemma 4.1.3. *Let $F \colon \mathcal{S}_* \to \mathcal{S}_*$ be a reduced functor. Then $F$ is excisive if and only if, for any pointed space $Y$, the map*

$$
\sigma_Y \colon FY \to \Omega F\Sigma Y
$$

*is an equivalence.*

*Proof.* The "only if" direction is trivial, so we focus on the "if" direction. For this, suppose

$$
\begin{array}{ccc}
U & \longrightarrow & V \\
\downarrow & & \downarrow \\
W & \longrightarrow & X
\end{array}
$$

is a pushout square in $\mathcal{S}_*$. We expand this square into a diagram

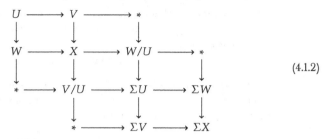

$$(4.1.2)$$

of $\mathcal{S}_*$ in which every square is a pushout. When we apply $F$ to this diagram, we obtain a solid arrow square

$$
\begin{array}{ccc}
F(U) & \xrightarrow{\hspace{3cm}} & \Omega F(\Sigma U) \\
\downarrow & \overset{\lambda}{\dashrightarrow} & \downarrow \\
F(W) \times_{F(X)} F(V) & \longrightarrow & \Omega F(\Sigma V) \times_{\Omega F(\Sigma X)} \Omega F(\Sigma W)
\end{array}
$$

$$(4.1.3)$$

in which both horizontal maps are, by assumption, equivalences. The universal property of $\Omega F(\Sigma U)$ supplies us with a dotted lift $\lambda$, and it follows that every map in this square is an equivalence. $\qquad\square$

**Notation 4.1.4.** We write $\mathrm{Fun}^{fp,red}(\mathcal{S}_*, \mathcal{S}_*)$ for the full subcategory of the $\infty$-category $\mathrm{Fun}(\mathcal{S}_*, \mathcal{S}_*)$ spanned by the reduced functors $F$ of finite presentation, and we write $\mathbf{Sp} \subset \mathrm{Fun}^{fp,red}(\mathcal{S}_*, \mathcal{S}_*)$ for the full subcategory spanned by the spectra.

**4.1.5.** Let $E$ be a spectrum. Then we obtain a sequence of spaces

$$\{X_n = E(S^n)\}_{n \geq 0}$$

along with equivalences $\{X_n \xrightarrow{\sim} \Omega X_{n+1}\}_{n \geq 0}$. Thus a spectrum gives rise to what we might call a *sequential spectrum*. We will show that these are in fact equivalent homotopy theories.

*Exercise 4.1.6.* Show that $\mathrm{Fun}^{fp,red}(\mathcal{S}_*, \mathcal{S}_*)$ is a presentable $\infty$-category:

For any finite pointed space $X$, denote by $h^X \colon \mathcal{S}_* \to \mathcal{S}_*$ the functor corepresented by $X$, so that $h^X(Y) \simeq \mathrm{Map}_{\mathcal{S}_*}(X, Y)$. Observe that $h^X$ is a reduced functor of finite presentation. For any pointed space $Y$, the counit $\Sigma\Omega \to id$ induces a map

$$d_X(Y) \colon \Sigma\,\mathrm{Map}_{\mathcal{S}_*}(\Sigma X, Y) \simeq \Sigma\Omega\,\mathrm{Map}_{\mathcal{S}_*}(X, Y) \to \mathrm{Map}_{\mathcal{S}_*}(X, Y);$$

this is functorial in $Y$, whence we obtain a natural transformation $d_X \colon \Sigma \circ h^{\Sigma X} \to h^X$. Show that a reduced finitely presented functor $F$ is a spectrum if and only if it is local with respect to the set of maps $\{d_X : X \in \mathcal{S}_*^{fin}\}$. Thus $\mathbf{Sp}$ is the accessible localisation of $\mathrm{Fun}^{fp,red}(\mathcal{S}_*, \mathcal{S}_*)$, and the class of morphisms that are inverted by the localisation is exactly the saturated class generated by $\{d_X : X \in \mathcal{S}_*^{fin}\}$.

Deduce that $\mathbf{Sp}$ is a presentable $\infty$-category, and the fully faithful inclusion functor $\mathbf{Sp} \hookrightarrow \mathrm{Fun}^{fp,red}(\mathcal{S}_*, \mathcal{S}_*)$ preserves limits and filtered colimits.

### Shifting

There is a nontrivial auto-equivalence of the loopspace or suspension of any space. If $X$ is a pointed space, then the universal property of the kernel product provides an endomorphism

$$(-1)\colon \Omega X = {*} \times_X {*} \to {*} \times_X {*} = \Omega X$$

obtained by exchanging the roles of the two points. The map $(-1)$ is clearly an auto-equivalence, and it is a nontrivial one, because it *reverses* the direction of the (implicit) homotopy in the square

$$
\begin{array}{ccc}
\Omega X & \longrightarrow & * \\
\downarrow & & \downarrow \\
* & \longrightarrow & X
\end{array}
$$

The map $(-1)$ is a natural auto-equivalence on the functor $\Omega$. It is not homotopic to *id*, but it is an involution in the sense that its composition $(-1)^2$ with itself is homotopic (in a canonical fashion) to *id*. This goes some way to justifying the notation.

Geometrically, each point of $\Omega X$ corresponds to a parametrised loop, and $(-1)$ takes each point to the point representing the same loop, parametrised in the reverse direction. On $\pi_1 X = \pi_0 \Omega X$, the auto-equivalence $(-1)$ induces the assignment $\gamma \mapsto \gamma^{-1}$.

In precisely the same manner, we obtain an involution

$$(-1)\colon \Sigma \to \Sigma$$

of the suspension functor.

These two involutions are compatible under the adjunction between suspension and loopspace. Indeed, in a square

$$
\begin{array}{ccc}
Y & \longrightarrow & * \\
\downarrow & & \downarrow \\
* & \longrightarrow & X
\end{array}
$$

reversing the direction of the implicit homotopy is at once tantamount to the composition of the map $Y \to \Omega X$ with $(-1)$ and to the composition of $(-1)$ with the map $\Sigma Y \to X$. (This point, silly as it is, is the origin of virtually *all* the signs throughout stable homotopy theory and homological algebra.)

**Warning 4.1.7.** There are two ways to iterate the suspension maps $\sigma_Y$, and they are not homotopic; they differ by a sign. For any reduced functor $F\colon \mathcal{S}_* \to \mathcal{S}_*$, one has a natural homotopy

$$-\Omega \sigma^F \Sigma \simeq \sigma^{\Omega F \Sigma} \tag{4.1.4}$$

between the two functors $\Omega F \Sigma \to \Omega^2 F \Sigma^2$.

*Exercise 4.1.8.* Construct the homotopy (4.1.4) by contemplating the diagrams (4.1.2) and (4.1.3) in the case in which both $V$ and $W$ are contractible.

Notation 4.1.9.   Let $E\colon \mathcal{S}_* \to \mathcal{S}_*$ be a reduced excisive functor. Then for any natural numbers $a \le b$, the natural transformation

$$\sigma^{\Omega^{b-1} E\Sigma^{b-1}} \cdots \sigma^{\Omega^{a+1} E\Sigma^{a+1}} \sigma^{\Omega^a E\Sigma^a} \colon \Omega^a E\Sigma^a \to \Omega^b E\Sigma^b$$

is an equivalence, which induces, for any pointed space $X$, a natural isomorphism

$$\pi_{n+a} E\Sigma^a X \cong \pi_{n+b} E\Sigma^b X$$

for any integer $n$ such that $n \ge -a$. Consequently, we may define, for any integer $n$, an abelian group

$$E_n X = \pi_{n+a} E\Sigma^a X$$

for some $a$ such that $a \ge \max\{2, -n\}$, secure in our knowledge that this abelian group is canonically independent of the choice of $a$. We thus obtain a functor

$$E_* \colon \mathcal{S}_* \to \mathbf{Ab}^{\mathbb{Z}},$$

where the target is the 1-category of $\mathbb{Z}$-graded abelian groups. This is the $E$-*homology* functor.

If $E$ is a spectrum, then we may define the *suspension* or *shift by* 1 of $E$ as the spectrum $E[1] = E \circ \Sigma$. In the other direction, we may define the *loop* or *shift by* $-1$ of $E$ as the spectrum $E[-1] = \Omega \circ E$. Iterating these, we obtain shifts $E[m]$ for any $m \in \mathbb{Z}$, and we note that on homology theories,

$$E[m]_n \cong E_{n-m}.$$

It is quite common in the literature to see $E_n$ as a shorthand for the *group* $E_n(S^0)$.

## Homology and cohomology

Let $X$ be a pointed space, and let $E$ be a spectrum. Then we define the $E$-*homology and $E$-cohomology* of $X$ as the groups

$$E_n(X) = \pi_n E(X) \qquad \text{and} \qquad E^n(X) = \pi_{-n} \mathrm{Map}(X, E(S^0)).$$

## 4.2    Examples

### Eilenberg–Mac Lane spectra

Our motivation for the definition of a spectrum was our contemplation of ordinary homology. We therefore already have one class of examples in hand:

Example 4.2.1.   The functor $X \mapsto \widetilde{\mathbb{Z}}\{X\}$ is a spectrum $H\mathbb{Z}\colon \mathcal{S}_* \to \mathcal{S}_*$. This is the *Eilenberg–Mac Lane spectrum of* $\mathbb{Z}$. The groups $(H\mathbb{Z})_*(X)$ are zero in negative degrees, and in nonnegative degrees, they are the reduced homology groups $\widetilde{H}_*(X, \mathbb{Z})$ of $X$.

More generally, for any abelian group $A$, let us contemplate the functor $X \mapsto \tilde{A}\{X\}$, which as a functor on pointed simplicial sets carries $X$ to the pointed simplicial set $\tilde{A}\{X\}_* = \widetilde{\mathbb{Z}}\{X\}_* \otimes A$. This is a spectrum $HA\colon \mathcal{S}_* \to \mathcal{S}_*$, called the *Eilenberg–Mac Lane* *spectrum of $A$*. The groups $(HA)_*(X)$ are zero in negative degrees, and in nonnegative degrees, they are the reduced homology groups $\widetilde{H}_*(X, A)$ of $X$.

## The derivative

Just as one may often find a best linear approximation of a general (differentiable) function by forming the derivative, we can construct the best linear approximation of a general (reduced and finitely presented) functor. This provides us with a few more useful examples.

Indeed, we have already seen that the full subcategory $\mathbf{Sp} \subseteq \mathrm{Fun}^{fp,red}(\mathcal{S}_*, \mathcal{S}_*)$ is a localisation; that is, the inclusion admits a left adjoint $D$, which we call the *derivative*. The bonus good news is that we can write a convenient formula for this $D$.

**Construction 4.2.2.** Let $F\colon \mathcal{S}_* \to \mathcal{S}_*$ be a reduced functor of finite presentation. Then we may look at the sequence of reduced functors of finite presentation

$$F \xrightarrow{\ \sigma^F\ } \Omega F\Sigma \xrightarrow{\ \sigma^{\Omega F\Sigma}\ } \Omega^2 F\Sigma^2 \xrightarrow{\ \sigma^{\Omega^2 F\Sigma^2}\ } \cdots,$$

which is indexed on the natural numbers $\mathbb{N}$. We write $DF = \mathrm{colim}_n \Omega^n F\Sigma^n$ for the colimit of this diagram of functors.

The assignment $F \mapsto DF$ comes equipped with a natural transformation $\alpha\colon id \to D$.

**Lemma 4.2.3.** *For any reduced functor*

$$F\colon \mathcal{S}_* \to \mathcal{S}_*$$

*of finite presentation, the functor $DF\colon \mathcal{S}_* \to \mathcal{S}_*$ is a spectrum.*

*Proof.* Let $Y$ be a pointed space, and consider the morphism

$$\sigma_Y^{DF}\colon (DF)Y \to \Omega(DF)\Sigma Y.$$

We represent $\sigma_Y^{DF}$ as the filtered colimit of the solid arrow sequence of morphisms, shown on the diagram to the side. The dotted arrows are all equivalences that make this diagram commute, and thus in the colimit they define an inverse to $\sigma_Y^{DF}$.

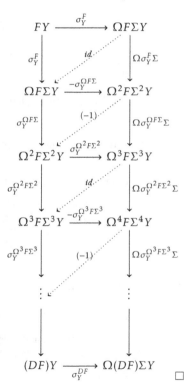

We note that if $F$ is already a spectrum, then $\alpha_F\colon F \to DF$ is in fact already an equivalence. In fact, we now show that $DF$ is the universal linear approximation to $F$.

**Proposition 4.2.4.** *The natural transformation $\alpha$ exhibits $D$ as a localisation functor on $\mathrm{Fun}^{fp,red}(\mathcal{S}_*, \mathcal{S}_*)$ whose essential image is precisely $\mathbf{Sp}$.*

*Proof.* If $E$ is a spectrum, then $\alpha_E \colon E \to DE$ is an equivalence. Thus **Sp** is the essential image of $D$, and for any reduced functor $F$ of finite presentation, $\alpha_{DF} \colon DF \to DDF$ is an equivalence. We must also check that $D\alpha_F \colon DF \to DDF$ is an equivalence; for this, it suffices to note that $D\sigma^F \colon DF \to D(\Omega F\Sigma)$ is an equivalence.  $\square$

## Suspension spectra

Now let us use this construction to define some interesting examples of spectra.

**Construction 4.2.5.** Let $\mathbb{S}^0 = \Sigma^\infty S^0$ be the spectrum $D\,\mathrm{id}$. That is, for any point space $Y$, we have

$$(\Sigma^\infty S^0)Y = \mathrm{colim}_n \Omega^n \Sigma^n Y.$$

This is the *sphere spectrum*, which represents stable homotopy:

$$\pi^s_m(Y) = (\Sigma^\infty S^0)_m Y \cong \mathrm{colim}_n \pi_{m+n}\Sigma^n Y;$$

by Freudenthal, one has $\pi^s_m(Y) \cong \pi_{2m+2}(\Sigma^{m+2}Y)$.

More generally, for any space $X$, consider the reduced, finitely presented functor $s_X \colon \mathbb{S}_* \to \mathbb{S}_*$ given by the assignment $Y \mapsto X \wedge Y$. We define

$$\Sigma^\infty X = Ds_X;$$

this is the *suspension spectrum of $X$*. We therefore obtain

$$(\Sigma^\infty X)Y = \mathrm{colim}_n \Omega^n(X \wedge \Sigma^n Y) \simeq \mathrm{colim}_n \Omega^n \Sigma^n(X \wedge Y) \simeq (\mathbb{S}^0)(X \wedge Y).$$

When $X$ is a sphere, we write $\mathbb{S}^n = \Sigma^\infty S^n$, and we observe that

$$\mathbb{S}^n \simeq \mathbb{S}^0[n].$$

The suspension spectrum is a functor $\Sigma^\infty \colon \mathbb{S}_* \to$ **Sp**. In the other direction, we have a functor $\Omega^\infty \colon$ **Sp** $\to \mathbb{S}_*$ that carries a spectrum $E$ to the value $E(S^0)$. They are related in the following manner:

**Proposition 4.2.6.** *The functor $\Sigma^\infty$ is left adjoint to the functor $\Omega^\infty$.*

*Exercise 4.2.7.* Verify this.

4.2.8. With the suspension functor in hand, we may define the $E$-*cohomology* of a pointed space $X$ as

$$E^n(X) = \pi_{-n}\mathrm{Map}_{\mathbf{Sp}}(\Sigma^\infty X, E)$$

## Spanier–Whitehead duals

It is also possible to generalise the sphere spectrum $\mathbb{S}^0$ in a dual manner. We will study the phenomenon of Spanier–Whitehead duality in a structured manner soon.

Example 4.2.9. Let $X$ be a finite pointed space, and let $h^X \colon \mathcal{S}_* \to \mathcal{S}_*$ be the functor corepresented by $X$; that is, $h^X(Y) = \operatorname{Map}_{\mathcal{S}_*}(X, Y)$. Since $X$ is finite, $h^X$ is finitely presented, and we obtain

$$(Dh^X)Y = \operatorname{colim}_n \Omega^n \operatorname{Map}_{\mathcal{S}_*}(X, \Sigma^n Y) \simeq \operatorname{Map}_{\mathcal{S}_*}(X, \operatorname{colim}_n \Omega^n \Sigma^n Y)$$
$$\simeq \operatorname{Map}_{\mathcal{S}_*}(X, (\Sigma^\infty S^0)(Y)).$$

The spectrum $(\Sigma^\infty X)^\vee = Dh^X$ is the *Spanier–Whitehead dual of* $X$.

The assignment $X \mapsto (\Sigma^\infty X)^\vee$ is a *contravariant* functor from pointed finite spaces to spectra.

Also, the Spanier–Whitehead dual of a finite pointed space may well have negative homotopy groups. For example, when $X$ is a sphere, we obtain an identification

$$(\Sigma^\infty S^n)^\vee \simeq \mathcal{S}^0[-n],$$

whence we are compelled to define $\mathcal{S}^{-n} = (\Sigma^\infty S^n)^\vee$.

*Exercise 4.2.10.* For any spectrum $E$ and any finite pointed space $X$, exhibit a homotopy equivalence

$$\operatorname{Map}_{\mathbf{Sp}}((\Sigma^\infty X)^\vee, E) \simeq E(X).$$

## Thom spectra

Definition 4.2.11. Let $X$ be a space. Then a *local system of spectra on* $X$ is a functor $X^{op} \to \mathbf{Sp}$; we write $\mathbf{Sp}_X = \operatorname{Fun}(X^{op}, \mathbf{Sp})$ for the ∞-category of local systems of spectra.

4.2.12. If we unpack the definitions a bit, a local system of spectra on a space $X$ is a functor $X^{op} \times \mathcal{S}_* \to \mathcal{S}_*$, written $(x, T) \mapsto \zeta(x)(T)$, such that for any point $x \in X$, the functor $\zeta(x) \colon \mathcal{S}_* \to \mathcal{S}_*$ is a spectrum.

Example 4.2.13. For any spectrum $E$, we have a *constant local system* $E_X$ at $E$.

Example 4.2.14 (The *J* homomorphism). To any finite-dimensional real vector space $V$ we can attach the one-point compactification $S^V$. This is a topologically enriched functor from finite-dimensional real vector spaces and isomorphisms to topological spaces. After passing to the attached ∞-categories, we may compose this functor with the suspension functor to obtain a local system

$$\coprod_{n \geq 0} BO(n) \to \mathbf{Sp}.$$

This functor factors through the group completion $\mathbb{Z} \times BO \to \mathbf{Sp}$. The *J homomorphism* is then the restricted map

$$J_O \colon BO \simeq \{0\} \times BO \subset \mathbb{Z} \times BO \to \mathbf{Sp},$$

which is a local system over $BO$. If $X$ is a topological space with a real vector bundle $\nu \colon X \to BO$, one obtains a local system of spectra by composition with $J_O$.

In the same manner, we obtain a local system

$$J_U \colon BU \to \mathbf{Sp},$$

and if $X$ is a topological space with a complex vector bundle $\nu \colon X \to BU$, one obtains a local system of spectra by composition with $J_U$.

Definition 4.2.15. Let $X$ be a space. A *stable spherical fibration over* $X$ is a local system of spectra $\zeta \colon X^{op} \to \mathbf{Sp}$ such that, for each point $x \in X$, the spectrum $\zeta(x)$ is (abstractly) equivalent to the sphere spectrum $\mathbb{S}^0$.

The *Thom spectrum* $X^\zeta$ of a stable spherical fibration $\zeta$ is the colimit of the diagram

$$\zeta \colon X^{op} \to \mathbf{Sp}.$$

4.2.16. For any stable spherical fibration $\zeta$ over $X$, the Thom spectrum enjoys the following universal property: for any spectrum $E$, we have a natural weak homotopy equivalence

$$\mathrm{Map}_{\mathbf{Sp}}(X^\zeta, E) \simeq \mathrm{Map}_{\mathbf{Sp}_X}(\zeta, E_X).$$

As a functor $\mathcal{S}_* \to \mathcal{S}_*$, the Thom spectrum $X^\zeta$ carries a space $T$ to the space

$$\mathrm{colim}_{n \to +\infty} \Omega^n (\mathrm{colim}_{x \in X} \Sigma^n \zeta(x)(T)).$$

Example 4.2.17. If $\zeta \colon X^{op} \to \mathbf{Sp}$ is a constant spherical fibration, then the Thom spectrum $X^\zeta$ is nothing more than $\Sigma^\infty X_+$.

Example 4.2.18 (Cobordism). By taking the Thom spectra attached to the $J$ homomorphism, we obtain

$$MO = (BO)^{J_O} \qquad \text{and} \qquad MU = (BU)^{J_U}.$$

These spectra are the *real* and *complex cobordism spectra*, respectively.

The homotopy of $MO$ and $MU$ are known — the former by Thom and the latter by Milnor:

$$\pi_* MO \cong \mathbb{F}_2[x_n : n \geq 2, n \neq 2^j - 1, |x_n| = n];$$
$$\pi_* MU \cong \mathbb{Z}[z_n : n \geq 1, |z_n| = 2n].$$

Example 4.2.19. If $X$ is a topological space with a real vector bundle $\nu$, then we may abuse notation slightly and write $X^\nu$ for the Thom spectrum $X^{J_O \circ \nu}$. We may define the Thom spectrum of a complex vector bundle in the same manner.

Example 4.2.20 (Atiyah duality). Let $X$ be a compact manifold. For a sufficiently general embedding of $X$ into $\mathbb{R}^n$, the Spanier–Whitehead dual $(\Sigma^\infty X_+)^\vee$ is naturally equivalent to $\Sigma^\infty(\mathbb{R}^n/(\mathbb{R}^n - X))[-n]$, which in turn can be identified with the Thom spectrum of the stable normal bundle of $X$.

## 4.3        Smash products

One of the most important aspects of the theory of spectra is the presence of the smash product, which provides $\mathbf{Sp}$ with a symmetric monoidal structure. We won't dive headlong into the details of the theory of symmetric monoidal structures on $\infty$-categories, but the setup of higher categories makes it possible to characterize the smash product of spectra with a homotopy-coherent universal property.

### Day convolution

Let $E_1, \ldots, E_n$ be a finite collection of reduced functors of finite presentation. Since these can be regarded as functors $\mathcal{S}_*^{fin} \to \mathcal{S}_*$, and since both source and target are endowed with the smash product symmetric monoidal structure, we may form their *Day convolution*: this is the functor

$$E_1 \star \cdots \star E_n \colon \mathcal{S}_*^{fin} \to \mathcal{S}_*$$

defined as the left Kan extension of the functor $(K_1, \ldots, K_n) \mapsto E_1 K_1 \wedge \cdots \wedge E_n K_n$ along the functor $(K_1, \ldots, K_n) \mapsto K_1 \wedge \cdots \wedge K_n$. In other words, we have, for any finite pointed space $Y$, the formula

$$(E_1 \star \cdots \star E_n) Y = \operatorname{colim}_{K_1 \wedge \cdots \wedge K_n \to Y} E_1 K_1 \wedge \cdots \wedge E_n K_n,$$

where the colimit is taken over the $\infty$-category

$$(\mathcal{S}_*^{fin} \times \cdots \times \mathcal{S}_*^{fin}) \times_{\mathcal{S}_*^{fin}} (\mathcal{S}_*^{fin})_{/Y}.$$

It is immediate from this formula that $E_1 \star \cdots \star E_n$ is a reduced functor. When $n = 0$, it's immediate that the unit is the inclusion functor $\mathcal{S}_*^{fin} \hookrightarrow \mathcal{S}_*$.

4.3.1.   The Day convolution actually defines a symmetric monoidal structure on the $\infty$-category $\mathrm{Fun}^{fp,red}(\mathcal{S}_*, \mathcal{S}_*)$, but we won't concern ourselves with that now. For now, we simply observe that $\star$ is associative and symmetric up to homotopy in the most naïve sense possible.

Let us note that, for any finite collection $X_1, \ldots, X_n$ of finite pointed spaces, the natural morphism

$$h^{X_1} \star \cdots \star h^{X_n} \to h^{X_1 \wedge \cdots \wedge X_n}$$

is an equivalence, and the Day convolution $(E_1, \ldots, E_n) \mapsto E_1 \star \cdots \star E_n$ preserves colimits separately in each variable.

The point here is that, since every reduced functor of finite presentation is a colimit of corepresentables, the Day convolution is controlled by its behaviour on the corepresentables, where it mirrors the smash product of pointed spaces.

Smash product

Now if $E_1, \ldots, E_n$ are spectra, we may form the *smash product*

$$E_1 \wedge \cdots \wedge E_n = D(E_1 \star \cdots \star E_n).$$

This gives us the explicit (but, in all honesty, not tremendously useful) formula

$$(E_1 \wedge \cdots \wedge E_n)Y = \mathrm{colim}_m \, \Omega^m (\mathrm{colim}_{K_1 \wedge \cdots \wedge K_n \to \Sigma^m Y} E_1 K_1 \wedge \cdots \wedge E_n K_n).$$

This is only a reasonable definition because of the following technical lemma, which expresses a compatibility of the Day convolution with the derivative $D$. Here recall the collection of natural transformations $\{d_X : X \in \mathbb{S}_*^{fin}\}$ from 4.1.6.

**Lemma 4.3.2.** *For any finite space $X$ and any finitely presented reduced functor $F$, the natural transformation $d_X : \Sigma h^{\Sigma X} \to h^X$ induces a morphism*

$$d_X * id : \Sigma h^{\Sigma X} * F \to h^X * F$$

*that lies in the strongly saturated class of morphisms of $\mathrm{Fun}^{fp, red}(\mathbb{S}_*, \mathbb{S}_*)$ generated by $\{d_X : X \in \mathbb{S}_*^{fin}\}$.*

*Proof.* Any finitely presented reduced functor $F$ is a colimit of functors of the form $h^Y$ for $Y$ a finite pointed space, so it suffices to assume that $F = h^Y$. In that case, $d_X * id$ is homotopic to the natural transformation

$$d_{X \wedge Y} : \Sigma h^{\Sigma X \wedge Y} \to h^{X \wedge Y}. \qquad \square$$

This lemma will actually imply that **Sp** is symmetric monoidal under the smash product, and the derivative $D$ is symmetric monoidal. For now, we will make do with the following less structured assertion:

**Proposition 4.3.3.** *The smash product preserves colimits separately in each variable. Additionally, $D$ carries the convolution product to the smash product in the sense that if $E_1, \ldots, E_n$ are reduced functors $\mathbb{S}_*^{fin} \to \mathbb{S}_*$, then the canonical natural transformation on Day convolutions $\alpha_{E_1} \star \cdots \star \alpha_{E_n} : E_1 \star \cdots \star E_n \to DE_1 \star \cdots \star DE_n$ induces an equivalence*

$$D(E_1 \star \cdots \star E_n) \simeq DE_1 \wedge \cdots \wedge DE_n.$$

*In particular, the sphere spectrum $\mathbb{S}^0$ is a unit for the smash product.*

*Proof.* The first claim is formal. For the second, we observe that by the previous lemma, $\alpha_{E_1} \star \cdots \star \alpha_{E_n}$ lies in the saturated class generated by $\{d_X : X \in \mathbb{S}_*^{fin}\}$. $\qquad \square$

Many of the spectra we've been contemplating so far are obtained via the derivative. This result shows that when we are smashing derivatives, we may delay the application of the derivative to the last possible moment. We deduce the following pleasant corollary:

**Corollary 4.3.4.** *If $X_1, \ldots, X_n$ are pointed spaces, the natural map is an equivalence:*

$$(\Sigma^\infty X_1) \wedge \cdots \wedge (\Sigma^\infty X_n) \simeq \Sigma^\infty (X_1 \wedge \cdots \wedge X_n).$$

4.3.5.   The smash product also appears in the classical formula for the value of an excisive functor. If $E: \mathcal{S}_* \to \mathcal{S}_*$ is a spectrum, then any map $K_1 \wedge K_2 \to S^0$ induces a map $E(K_1) \wedge K_2 \wedge X \to E(K_1 \wedge K_2 \wedge X) \to E(X)$, natural in $X$; together, these define an equivalence

$$\Omega^\infty(\Sigma^\infty X \wedge E) \xrightarrow{\sim} E(X),$$

natural in $X$.

## Function spectra and duality

For any spectrum $E$, the functor $E'' \mapsto E'' \wedge E$ preserves colimits, and since **Sp** is presentable, it follows that there exists a right adjoint $E' \mapsto \mathbb{F}(E, E')$ thereto. This is the *function spectrum* from $E$ to $E'$. As a functor on pointed spaces, it is given by the assignment

$$X \mapsto \mathrm{Map}_{\mathbf{Sp}}((\Sigma^\infty X)^\vee \wedge E, E').$$

For any pointed finite space $X$ and for any map $f: K_1 \wedge K_2 \to Y$ of pointed finite spaces, evaluation defines a map

$$f \circ (ev \wedge id): \mathrm{Map}(X, K_1) \wedge X \wedge K_2 \to K_1 \wedge K_2 \to Y.$$

Letting $f$ and $Y$ vary, we obtain a natural transformation $h^X \star s_X \to h^{S^0}$. Applying $D$, we obtain a morphism of spectra

$$(\Sigma^\infty X)^\vee \wedge \Sigma^\infty X \to \mathbb{S}^0,$$

which in turn specifies a map

$$\delta_X: (\Sigma^\infty X)^\vee \to \mathbb{F}(\Sigma^\infty X, \mathbb{S}^0),$$

which turns out to be an equivalence. We therefore take this as motivation for the following definition.

**Definition 4.3.6.**  For any spectrum $E$, the *dual* of $E$ is the spectrum

$$E^\vee = \mathbb{F}(E, \mathbb{S}^0).$$

4.3.7.   If $E$ is a spectrum, then there is a morphism of spectra

$$E \wedge E^\vee \simeq E^\vee \wedge E \to \mathbb{S}^0,$$

which corresponds to a morphism $E \to E^{\vee\vee}$.

Let's classify the finite objects of **Sp**. It turns out that finiteness in **Sp** is a far simpler matter than in $\mathcal{S}$; in effect, problems that the Wall finiteness obstruction catches in $\mathcal{S}$ are finessed in **Sp**:

**Theorem 4.3.8.**  *Let $E$ be a spectrum. The following are equivalent.*

1. *There exists a finite pointed space $X$, an integer $n \in \mathbb{Z}$, and an equivalence $E \simeq (\Sigma^\infty X)[n]$.*

2. *The spectrum E can be expressed as a finite colimit of spectra of the form $\mathbb{S}^n$ for $n \in \mathbb{Z}$.*

3. *The spectrum E is compact as an object of* **Sp**.

4. *The natural morphism $E \to E^{\vee\vee}$ is an equivalence.*

Definition 4.3.9. A spectrum is said to be *finite* if it satisfies the conditions of 4.3.8. We write $\mathbf{Sp}^{fin} \subset \mathbf{Sp}$ for the full subcategory spanned by the finite spectra.

## 4.4    Stable ∞-categories

A stable ∞-category is much like an abelian category, except that what is asked of monomorphisms or epimorphisms in an abelian category is asked of *all* morphisms of a stable ∞-category. In an abelian category, every monomorphism is the kernel of its cokernel, and every epimorphism is the cokernel of its kernel. The definition of stable ∞-category is rigged so that *every morphism* of a stable ∞-category is both the kernel of its cokernel and the cokernel of its kernel.

Another way of thinking about stable ∞-categories is in relation to triangulated categories. A central theme in modern mathematics is the idea of encoding geometric structure in terms of a triangulated category of modules of some sort, such as the derived category of a scheme, the stable module category of a finite group, or the Fukaya category of a symplectic manifold. A lot of work (e.g., see [214]) permits the use of triangulated categories as a setting for abstract stable homotopy theory. This is explained by the connection to stable ∞-categories. The structure of a triangulated category is, in a precise sense, the shadow of the structure of a stable ∞-category: the homotopy category of a stable ∞-category is a triangulated category. However, stable ∞-categories are much easier to work with. For one thing, the definition is considerably more concise, as the axioms of a triangulated category immediately become basic computations with kernels and cokernels. For another, a variety of problems go away — notably, the formation of cokernels is functorial in a stable ∞-category, but it is almost never so in a triangulated category. As a result, there are important invariants that require functorial cokernels, like algebraic $K$-theory, that really only make sense for an ∞-category: they are capable of distinguishing two stable ∞-categories with triangulated-equivalent homotopy categories (e.g., see [258]).

Definition 4.4.1. An ∞-category $A$ is said to be *stable* if the following conditions obtain.

1. There is a *zero object* — that is, an object that is both initial and terminal — in $A$.

2. The ∞-category $A$ has all finite limits and all finite colimits.

3. A square

$$\begin{array}{ccc} U & \longrightarrow & V \\ \downarrow & & \downarrow \\ W & \longrightarrow & X \end{array}$$

is a pushout if and only if it is a pullback.

If $A$ and $B$ are stable ∞-categories, a functor $f : A \to B$ is left exact (i.e., finite-limit-preserving) if and only if it is right exact (i.e., finite-colimit-preserving). In this case, we simply call $f$ *exact*. The subcategory of $\mathbf{Cat}_\infty$ whose objects are stable ∞-categories and whose morphisms are exact functors is denoted $\mathbf{Stab}_\infty$.

**Example 4.4.2.** Naturally, $\mathbf{Sp}$ is stable, as is $\mathbf{Sp}^{fin}$. On the other hand, although $\mathcal{S}_*$ and $\mathcal{S}_*^{fin}$ have zero objects, finite limits, and finite colimits, they certainly aren't stable.

**Example 4.4.3.** For any small ∞-category $C$ and any stable ∞-category $A$, the ∞-category $\mathrm{Fun}(C, A)$ is stable. In particular, for any space $X$, the ∞-category $\mathbf{Sp}_X$ of local systems of spectra on $X$ is stable.

**Example 4.4.4.** If $A$ is a stable ∞-category, then so is $A^{op}$.

*Exercise 4.4.5.* Show that if $A$ is a small stable ∞-category, then so is $\mathrm{Ind}(E)$.

## Kernels and cokernels

Let $A$ be an ∞-category with a zero object 0, and let $f : X \to Y$ be a morphism thereof. We can form the *kernel*[1] or *fibre* or *cocone* $i : K \to X$ of $f$, which is the pullback

$$
\begin{array}{ccc}
K & \longrightarrow & 0 \\
{\scriptstyle i}\downarrow & & \downarrow \\
X & \xrightarrow{\ f\ } & Y
\end{array}
$$

and the *cokernel* or *cofibre* or *cone* $p : Y \to C$ of $f$, which is the pushout

$$
\begin{array}{ccc}
X & \xrightarrow{\ f\ } & Y \\
\downarrow & & \downarrow{\scriptstyle p} \\
0 & \longrightarrow & C
\end{array}
$$

In a stable ∞-category, pullback squares and pushout squares coincide, so $f$ is both the cokernel of $i$ and the kernel of $p$. We can keep pushing and pulling with the aid of the loopspace and the suspension:

**Construction 4.4.6.** If $A$ has all finite colimits, then we have the endofunctors

$$X \mapsto \Sigma X = 0 \cup^X 0 \quad \text{and} \quad X \mapsto \Omega X = 0 \times_X 0.$$

Note that the functors $\Sigma$ and $\Omega$ each admit an involution $-1$ given by swapping the zero objects.

These functors are adjoint, but if $A$ is stable, they are also inverse to each other;

---

[1] We have opted to keep the terms "kernel" and "cokernel" in circulation — even though this is uncommon lingo in stable ∞-category literature — because we think the parallel to abelian categories is highlighted clearly this way.

that is, the unit $id \to \Omega\Sigma$ and the counit $\Sigma\Omega \to id$ are each equivalences. In that case, we also may write

$$X[1] = \Sigma X \quad \text{and} \quad X[-1] = \Omega X,$$

and we call these the *shift* functors. In particular, for any object $X$, there is an endomorphism $-1 \colon X \to X$ that arises from thinking of $X$ as $X[1][-1]$ or $X[-1][1]$.

4.4.7.   When $A$ is stable, the kernel of the cokernel of our morphism $f$ is $f$ again, and the cokernel of the kernel of $f$ is $f$ again. The kernel of the kernel of $f$ is the morphism $-\Omega p \colon Y[-1] \to C[-1] \simeq F$, and the cokernel of the cokernel of $f$ is the morphism $-\Sigma i \colon C \simeq F[1] \to X[1]$. If we continue to form kernels and cokernels, we obtain a diagram

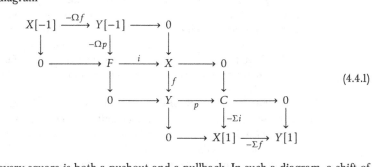

$$(4.4.1)$$

in which every square is both a pushout and a pullback. In such a diagram, a shift of a morphism changes sign precisely when it turns from horizontal to vertical or vice versa.

4.4.8.   If $f \colon X \to Y$ and $g \colon Y \to Z$ are morphisms of a stable ∞-category $A$, and if $\eta$ is a *nullhomotopy* of $gf$, that is, a homotopy between $gf$ and the *zero morphism* $0 \colon X \to Z$, which is the composite of the unique morphisms $X \to 0$ and $0 \to Z$, then we can ask whether $\eta$ exhibits $g$ is the cokernel of $f$. If it does, then there is a further morphism $h \colon Z \to X[1]$, and one calls the sequence $X \to Y \to Z \to X[1]$ a *distinguished triangle* or a *fibre/cofibre sequence*. The "triangle" here is the diagram

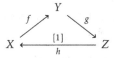

where the arrow marked $[1]$ isn't a morphism as shown but rather the morphism $h \colon Z \to X[1]$. The value of drawing it this way is that it can be rotated:

The homotopy category $hA$, with shift functor $X \mapsto X[1]$ and distinguished triangles as above, is in fact a *triangulated category*. The proof of this claim is Theorem 1.1.2.14 in [168]. However, working with a stable ∞-category is always preferable to — and usually easier than — working with a triangulated category.

## Universal property of Sp

The ∞-category **Sp** admits a universal property as an object of the ∞-category $\mathbf{Pr}^L_{st}$ of presentable stable ∞-categories and colimit-preserving functors. Precisely, **Sp** is the free presentable stable ∞-category on one generator; that is, for any presentable stable ∞-category $E$, evaluation at $S^0$ defines an equivalence $\mathrm{Fun}^L(\mathbf{Sp}, E) \simeq E$, where $\mathrm{Fun}^L$ is the category of colimit-preserving functors.

Though we won't go into detail about symmetric monoidal structures on ∞-categories, it is useful to note that $\mathbf{Pr}^L_{st}$ has such a structure: for any pair of presentable stable ∞-categories $C$ and $D$, there exists a presentable stable ∞-category $C \otimes D$ such that $\mathrm{Fun}^L(C \otimes D, A)$ is equivalent to the ∞-category of functors $C \times D \to A$ that preserve colimits separately in each variable. In this symmetric monoidal structure, the unit is **Sp**.

Since the ∞-category **Sp** is the unit for the symmetric monoidal structure $\mathbf{Pr}^L_{st}$ that we discussed above, it follows that **Sp** admits a unique symmetric monoidal structure $\mathbf{Sp} \times \mathbf{Sp} \to \mathbf{Sp}$ that preserves colimits separately in each variable. This gives a pleasant universal characterisation of the smash product.

One consequence of the universal property of the ∞-category of spectra is the following omnibus comparison result to the models of spectra considered in Chapter 3.

**Theorem 4.4.9.** *The underlying ∞-categories of the categories of orthogonal spectra, symmetric spectra, classical prespectra, and EKMM spectra with the stable equivalences are all equivalent to the ∞-category of spectra.*

## Additivity

One point that we will address carefully is the presence of *direct sums* and the *additivity* of a stable ∞-category.

**Definition 4.4.10.** If $A$ is an ∞-category, we say that $A$ *admits direct sums* if the following conditions obtain.

1. The ∞-category $A$ admits finite products and finite coproducts.
2. The natural morphism from the initial object to the terminal object is an equivalence, so that there is a zero object in $A$.
3. For any objects $X, Y \in A$, the map

$$I = \begin{pmatrix} id & 0 \\ 0 & id \end{pmatrix} : X \sqcup Y \to X \times Y$$

is an equivalence.

In this case, we write $X \oplus Y$ for the identified product and coproduct.

If $A$ admits direct sums, then the homotopy category $hA$ acquires an enrichment in the category of commutative monoids: for any morphisms $f, g \colon X \to Y$, one defines

$$f + g = (id \;\; id)\begin{pmatrix} f & 0 \\ 0 & g \end{pmatrix}\begin{pmatrix} id \\ id \end{pmatrix} \colon X \to X \oplus X \to Y \oplus Y \to Y$$

One says moreover that $A$ is *additive* if $\mathrm{Map}_{hA}(X, Y)$ is an abelian group.

The homotopy category $hA$ of a stable ∞-category $A$ is automatically enriched in abelian groups, thanks to the natural equivalence $\mathrm{Map}_A(S, T) \simeq \Omega^2 \mathrm{Map}_A(S[-2], T)$. But in fact even more is true:

**Proposition 4.4.11.** *Any stable ∞-category $A$ is additive.*

*Proof.* $A$ contains a zero object, and it admits finite products and finite coproducts. To see that these coincide, we claim that $id \times 0 \colon X \to X \times Y$ and $0 \times id \colon Y \to X \times Y$ together exhibit $X \times Y$ as the coproduct $X \sqcup Y$. For any object $Z$, the induced map

$$\mathrm{Map}_A(X \times Y, Z) \to \mathrm{Map}_A(X, Z) \times \mathrm{Map}_A(Y, Z)$$

admits a homotopy inverse given by the formula $(f, g) \mapsto (f \times 0) + (0 \times g)$ (using the enrichment of $hA$ in abelian groups). Finally, the Eckmann–Hilton argument shows that the commutative monoid enrichment of $hA$ arising from the presence of direct sums coincides with the abelian group enrichment of $hA$ arising from the stability of $A$. □

Notation 4.4.12. If $A$ is a stable ∞-category and if $X, Y \in A$, we obtain abelian groups

$$\mathrm{Ext}_A^n(X, Y) = \mathrm{Mor}_{hA}(X[-n], Y) \simeq \mathrm{Mor}_{hA}(X, Y[n]).$$

When $n \leq 0$, we have

$$\mathrm{Ext}_A^n(X, Y) \cong \pi_{-n} \mathrm{Map}_A(X, Y).$$

These abelian groups are the homotopy groups of *mapping spectra* associated to objects $X, Y \in A$. In fact, in a precise sense the category of stable ∞-categories is equivalent to the category of spectral categories (where equivalences are the "Morita equivalences" of spectral categories, defined in terms of equivalences on associated module categories). See for example [46, 4.23] for a discussion of this.

## Loopspace and suspension

The argument of 4.1.3 works in general here, and it implies that, in order to verify stability, it is enough to check that $\Sigma$ and $\Omega$ are inverse:

**Theorem 4.4.13.** *Let $A$ be an ∞-category with a zero object, all finite limits, and all finite colimits. If the functors $\Sigma$ and $\Omega$ on $A$ are inverse, then $A$ is stable.*

Example 4.4.14. If $A$ is stable, then a *stable subcategory* is a full subcategory that is stable under equivalences, contains the zero object, and is stable under finite limits and colimits.

*Exercise 4.4.15.* Let $C$ be an ∞-category with a zero object and all finite limits. Show that the limit $\mathbf{Sp}^{seq}(C)$ in $\mathbf{Cat}_\infty$ of the sequence

$$\cdots \xrightarrow{\;\Omega\;} C \xrightarrow{\;\Omega\;} C \xrightarrow{\;\Omega\;} C$$

is stable. (Hint: the tricky point is to confirm that $\mathbf{Sp}^{seq}(C)$ admits all finite colimits.)

The "cartesian section" point of view on $\mathbf{Sp}^{seq}(C)$ is that its objects are sequences $\{X_n\}_{n\geq 0}$ of objects of $C$ along with sequences of equivalences $\{X_n \to \Omega X_{n+1}\}_{n\geq 0}$. This goes some way to explaining the notation. The equivalence with "true" spectra will be addressed in the next section.

## 4.5     Generalisations

One may ask what happens when one has only part of the axioms of a stable ∞-category. These ∞-categories often appear as subcategories of stable ∞-categories, but they also arise directly from applications.

### Prestable ∞-categories

For example, the subcategory of *connective spectra* — those whose homotopy is confined to nonnegative degrees — is only closed under suspension, but not loopspace. More generally, we have the following subcategories of spectra:

Example 4.5.1. For any integer $k$, write $\mathbf{Sp}_{\geq k} \subset \mathbf{Sp}$ for the full subcategory spanned by the *$k$-connective spectra*, i.e., those spectra $E$ such that $E_n = 0$ for $n < k$. Dually, write $\mathbf{Sp}_{\leq k} \subset \mathbf{Sp}$ for the full subcategory spanned by the *$k$-coconnective spectra*, i.e., those spectra $E$ such that $E_n = 0$ for $n > k$.

The objects of the ∞-category $\mathbf{Sp}_{\geq k}$ are called the *$k$-connective spectra*, and the objects of $\mathbf{Sp}_{\leq k}$ are called the *$k$-truncated spectra*.

We will study systems of subcategories like this in detail in Section 4.7. Here, we are more interested in the intrinsic properties of the ∞-category $\mathbf{Sp}_{\geq k}$. Right away, we notice that the suspension of a $k$-connective spectrum remains $k$-connective, but the loopspace of a $k$-connective spectrum is in general no longer $k$-connective. Consequently, we are interested in situations in which we have "half" of our stability conditions:

Definition 4.5.2. An ∞-category $A$ is said to be *prestable* if and only if the following conditions obtain.

1. The ∞-category $A$ admits a zero object.
2. The ∞-category $A$ has all finite colimits.

3. Every morphism $f \colon X \to \Sigma Y$ of $A$ admits a kernel $i \colon F \to X$, and the square

$$
\begin{array}{ccc}
F & \xrightarrow{\ i\ } & X \\
\downarrow & & \downarrow f \\
0 & \longrightarrow & \Sigma Y
\end{array}
$$

exhibits $f$ as the cokernel of $i$.

4.5.3. By the same argument as 4.1.3, we see that an $\infty$-category that contains a zero object and all finite limits and colimits is prestable if and only if the suspension $\Sigma$ is fully faithful.

4.5.4. As in Proposition 4.4.11, a prestable $\infty$-category $A$ is automatically additive. To see this, note that we have the abelian group enrichment, thanks to the equivalence $\mathrm{Map}_A(S,T) \simeq \Omega^2 \mathrm{Map}_A(S, T[2])$. The rest of the argument is as in 4.4.11.

Example 4.5.5.   For any $k \in \mathbb{Z}$, the $\infty$-category $\mathbf{Sp}_{\geq k}$ of $k$-connective spectra is prestable.

## Derived $\infty$-categories

The triangulated derived category $D(R)$ of the category of $R$-modules has some disadvantages:

- The formation of cones is not functorial; they are generally not unique, but rather they are unique up to a noncanonical isomorphism in the derived category. This is because diagrams in $D(R)$ commute up to homotopy, but the data of such a homotopy is not part of the data of such a diagram.
- In a similar vein, there is not a good theory of sheaves valued in $D(R)$. For instance, if $\{U, V, W\}$ is an open cover of a topological space $X$, and if $F$ is a sheaf on $X$ valued in $D(R)$, then the sheaf condition ensures that global sections can be recovered from local sections that agree up to homotopy on double overlaps, but this is true even without any compatibility for these homotopies on the triple overlap.

Consequently, it is often more convenient to work with the derived $\infty$-category of $R$. Here is the construction:

Construction 4.5.6.   Let $E$ be an abelian category, which we shall regard as an $\infty$-category. Assume that $E$ has enough projective objects. Write $E_{proj} \subseteq E$ for the full subcategory spanned by the projective objects.

We will construct the *nonnegative derived $\infty$-category of* $E$. It's actually convenient to start by defining the nonnegative derived $\infty$-category of $\mathrm{Ind}(E)$.

We write $D_{\geq 0}(\mathrm{Ind}(E)) \subseteq \mathrm{Fun}(E_{proj}^{op}, \mathcal{S})$ for the full subcategory spanned by those functors $E_{proj}^{op} \to \mathcal{S}_*$ that carry finite direct sums to products.

One has the Yoneda embedding $j \colon E_{proj} \hookrightarrow D_{\geq 0}(\mathrm{Ind}(E))$, which can be thought of *either* as freely generating $D_{\geq 0}(\mathrm{Ind}(E))$ under sifted colimits (that is, filtered colimits and geometric realisations) *or* as generating $D_{\geq 0}(\mathrm{Ind}(E))$ under *all* colimits, subject to the condition that $j$ preserve finite coproducts. That is:

1. For any $\infty$-category $C$ that admits all sifted colimits, the functor

$$j^*\colon \operatorname{Fun}(D_{\geq 0}(\operatorname{Ind}(E)), C) \to \operatorname{Fun}(E_{proj}, C)$$

   restricts to an equivalence between the full subcategory of $\operatorname{Fun}(D_{\geq 0}(\operatorname{Ind}(E)), C)$ spanned by those functors $D_{\geq 0}(\operatorname{Ind}(E)) \to C$ that preserve sifted colimits and $\operatorname{Fun}(E_{proj}, C)$.

2. If $E$ admits *all* colimits, then $j^*$ restricts to an equivalence between the full subcategory of $\operatorname{Fun}(D_{\geq 0}(\operatorname{Ind}(E)), C)$ spanned by those functors $D_{\geq 0}(\operatorname{Ind}(E)) \to C$ that preserve all colimits and the full subcategory of $\operatorname{Fun}(E_{proj}, C)$ spanned by those functors $A \to E$ that preserve finite coproducts.

The $\infty$-category $D_{\geq 0}(\operatorname{Ind}(E))$ is called the *nonnegative derived $\infty$-category* of $\operatorname{Ind}(E)$.

We write $D_{\geq 0}(E) \subseteq D_{\geq 0}(\operatorname{Ind}(E))$ for the smallest full subcategory that contains $E_{proj}$ and is closed under geometric realisations. Thus $D_{\geq 0}(E)$ is obtained from $A$ by freely adding geometric realisations; that is, for any $\infty$-category $E$ that admits all geometric realisations, the functor

$$j^*\colon \operatorname{Fun}(D_{\geq 0}(E), E) \to \operatorname{Fun}(E_{proj}, E)$$

restricts to an equivalence between the full subcategory of $\operatorname{Fun}(D_{\geq 0}(\operatorname{Ind}(E)), E)$ spanned by those functors $D_{\geq 0}(E) \to E$ that preserve geometric realisations and $\operatorname{Fun}(E_{proj}, E)$.

The $\infty$-category $D_{\geq 0}(E)$ is called the *nonnegative derived $\infty$-category* of $E$. A functor $F\colon D_{\geq 0}(E) \to E$ that preserves geometric realisations will be said to be the *left derived functor* of $j^*F$.

4.5.7. There is no ambiguity in our notation. If $E$ is an abelian category with enough projectives and $E' = \operatorname{Ind}(E)$, then $E'$ also has enough projectives, and our definition of $D_{\geq 0}(E')$ agrees with our definition of $D_{\geq 0}(\operatorname{Ind}(E))$: each freely adds sifted colimits to $E_{proj}$.

Example 4.5.8. For any abelian category $E$ with enough projectives, the $\infty$-categories $D_{\geq 0}(\operatorname{Ind}(E))$ and $D_{\geq 0}(E)$ are prestable. Indeed, the second universal property makes it clear that $D_{\geq 0}(E)$ admits direct sums. To prove that the suspension on $D_{\geq 0}(\operatorname{Ind}(E))$ is fully faithful, let $C\colon E_{proj}^{op} \to \mathcal{S}_*$ be an object; then since products and geometric realisations are computed objectwise in $D_{\geq 0}(\operatorname{Ind}(E))$, we may write $\Sigma C$ as the functor that carries an object $X \in A$ to the geometric realisation of the bar construction

$$B_*(0, C(X), 0)\colon n \mapsto C(X)^n.$$

This simplicial space is a grouplike Segal space, and so we have an equivalence $C(X) \simeq \Omega|B_*(0, C(X), 0)| \simeq \Omega\Sigma C(X)$.

Construction 4.5.9. Let $E$ be an abelian category with enough projectives, and let $C\colon E_{proj}^{op} \to \mathcal{S}_*$ be an object of $D_{\geq 0}(E)$. Then we obtain, for any integer $n \geq 0$, a functor

$$H_n(C) = \pi_n C\colon E_{proj}^{op} \to \mathbf{Set}_*$$

that carries direct sums to products. In other words, $H_n(C) \in \text{Ind}(E)$.

Let us quickly check that $H_n(C)$ actually lies in $E$. For any object $M \in E$, one has $H_0(j(M)) = M$, and for $n \geq 1$, $H_n(j(M)) = 0$. Furthermore, if $C_*$ is a simplicial object of $D_{\geq 0}(E)$ with the property that $H_n(C_k) \in E$ for every $k, n \geq 0$, then the obvious spectral sequence argument ensures that for every $n \geq 0$, one has $H_n|C_*| \in A$.

We have thus defined the *homology functors* $H_n \colon D_{\geq 0}(E) \to E$.

If $E$ is an abelian category with enough projective objects, the homotopy category $hD_{\geq 0}(E)$ can be shown to be the derived category of nonnegatively graded complexes in $E_{proj}$. Under this equivalence, the homology functors above agree with the classically defined functors, and the left derived functor of a functor $E \to E'$ in our sense coincides with the left derived functor in the classical sense.

**Definition 4.5.10.** For any abelian category $E$ with enough projectives, we write $\text{Fun}^\oplus(E_{proj}^{op}, \mathbf{Sp})$ for the (stable) $\infty$-category of functors $E_{proj}^{op} \to \mathbf{Sp}$ that preserve direct sums. We then define $D^-(E)$ as the smallest stable full subcategory of $\text{Fun}^\oplus(E_{proj}^{op}, \mathbf{Sp})$ that contains the essential image of $\Sigma^\infty j \colon E_{proj} \to \text{Fun}^\oplus(E_{proj}^{op}, \mathbf{Sp})$ and is closed under geometric realisations. This is the *right bounded derived $\infty$-category* of $E$.

*Exercise 4.5.11.* Verify that the functor $\Sigma^\infty \colon D_{\geq 0}(E) \to D^-(E)$ is fully faithful.

**4.5.12.** Let $E$ be an abelian category with enough projectives. If $C \colon E_{proj}^{op} \to \mathbf{Sp}$ is an object of $D_{\geq 0}(E)$, then as in 4.5.9 we obtain, for any integer $n \in \mathbb{Z}$, a functor

$$H_n(C) = \pi_n C \colon E_{proj}^{op} \to \mathbf{Set}_*$$

that carries direct sums to products, so that $H_n(C) \in \text{Ind}(E)$, and once again it turns out that $H_n(C)$ lies in $E$ itself.

We have thus defined the *homology functors* $H_n \colon D^-(E) \to A$.

**Construction 4.5.13.** If $E$ is an abelian category with enough *injective* objects, then we can define

$$D_{\leq 0}(E) = D_{\geq 0}(E^{op})^{op} \qquad \text{and} \qquad D^+(E) = D^-(E^{op})^{op};$$

we call $D^+(E)$ the *left bounded derived $\infty$-category* of $E$. We can also define the *cohomology functors* $H^{-n} = H_n$.

**4.5.14.** An even more dramatic generalisation of the stable $\infty$-categories is the notion of an *exact $\infty$-category*. These were introduced in [24] as a simultaneous generalisation of the exact categories of Quillen and stable $\infty$-categories. Exact $\infty$-categories are a natural setting for algebraic $K$-theory and Quillen's $Q$ construction.

## 4.6    Stabilisation

In this section, we give a machine for printing examples of stable $\infty$-categories. This machine is really nothing more than a formal extension of our definition of spectra.

**Definition 4.6.1.** Let $C$ and $D$ be ∞-categories. Assume that $C$ admits all finite colimits. We say that a functor $F\colon C \to D$ is *reduced* if it carries the initial object of $C$ to a terminal object of $D$, and we say that $F$ is *excisive* if it carries any pushout square in $C$ to a pullback square in $D$.

We write $\mathrm{Fun}_*(C,D) \subseteq \mathrm{Fun}(C,D)$ for the full subcategory spanned by the reduced functors; $\mathrm{Exc}(C,D) \subseteq \mathrm{Fun}(C,D)$ for the full subcategory spanned by the excisive functors; and $\mathrm{Exc}_*(C,D) \subseteq \mathrm{Fun}(C,D)$ for the full subcategory spanned by the reduced excisive functors.

**4.6.2.** Let $C$, $D$, and $F$ be as above. If $C$ is stable, then $F$ is reduced excisive if and only if $F$ is left exact. If $D$ is stable, then $F$ is reduced excisive if and only if $F$ is right exact.

If $D$ admits all finite limits, then the argument of 4.1.3 applies again to ensure that $F$ is excisive if and only if, for any object $X \in C$, the natural map $FX \to \Omega F\Sigma X$ is an equivalence.

*Exercise 4.6.3.* Check that, if $C$ is an ∞-category $C$ with all finite colimits and $D$ is an ∞-category $D$ with all finite limits, the ∞-category $\mathrm{Exc}_*(C,D)$ is stable.

**Definition 4.6.4.** For any ∞-category $D$ with all finite limits, a *spectrum in $D$* is a reduced excisive functor $\mathcal{S}_*^{fin} \to D$. We write $\mathbf{Sp}(D) = \mathrm{Exc}_*(\mathcal{S}_*^{fin}, D)$, and we call this ∞-category the *stabilisation* of $D$.

Evaluation at $S^0$ defines a functor $\Omega^\infty\colon \mathbf{Sp}(D) \to D$.

**Example 4.6.5.** Of course $\mathbf{Sp} \simeq \mathbf{Sp}(\mathcal{S})$.

We haven't got much of an excuse for the notation $\Omega^\infty$ at the moment, but we will explain it soon.

## Universal property of stabilisation

*Exercise 4.6.6.* For any ∞-category $D$ with all finite limits, show that $\Omega^\infty\colon \mathbf{Sp}(D) \to D$ is an equivalence if and only if $D$ is stable.

**4.6.7.** Let $C$ be an ∞-category with all finite colimits, and let $D$ be an ∞-category with all finite limits. Then a reduced excisive functor $C \to \mathbf{Sp}(D)$ is the same thing as a functor $C \times \mathcal{S}_*^{fin} \to D$ that is reduced and excisive separately in each variable. This, in turn, is the same thing as a spectrum in the ∞-category $\mathrm{Exc}_*(C,D)$.

**Proposition 4.6.8.** *Let $C$ be an ∞-category with all finite colimits, and let $D$ be an ∞-category with all finite limits. The functor $\Omega^\infty\colon \mathbf{Sp}(D) \to D$ induces an equivalence*

$$\mathrm{Exc}_*(C, \mathbf{Sp}(D)) \simeq \mathrm{Exc}_*(C,D).$$

*Proof.* The induced functor $\mathbf{Sp}(\mathrm{Exc}_*(C,D)) \simeq \mathrm{Exc}_*(C, \mathbf{Sp}(D)) \to \mathrm{Exc}_*(C,D)$ is $\Omega^\infty$, which is an equivalence since $\mathrm{Exc}_*(C,D)$ is stable. $\square$

This result reveals a universal property of the stabilisation: if we look at the subcategory $\mathbf{Cat}_\infty^{lex}$ whose objects are ∞-categories with all finite limits and whose morphisms are left exact functors, then the $\infty$-category $\mathbf{Stab}_\infty$ is a full subcategory of $\mathbf{Cat}^{lex}_\infty$. Now the previous proposition reveals that the stabilisation is in fact the right adjoint to the inclusion $\mathbf{Stab}_\infty \hookrightarrow \mathbf{Cat}^{lex}_\infty$. In other words, $\mathbf{Stab}_\infty$ is a colocalisation of $\mathbf{Cat}^{lex}_\infty$.

### Spectra and sequential spectra

Here is another perspective, which explains the notation $\Omega^\infty$ and refers to the construction of $\mathbf{Sp}^{seq}$ of 4.4.15:

**Proposition 4.6.9.** *Let $D$ be an $\infty$-category with a zero object and all finite limits. Then the functor*

$$\mathbf{Sp}^{seq}(D) \to D$$

*given informally by $\{X_n\}_{n \geq 0} \mapsto X_0$ exhibits $\mathbf{Sp}^{seq}(D)$ as the stabilisation of $D$:*

$$\mathbf{Sp}(D) \simeq \mathbf{Sp}^{seq}(D).$$

*Proof.* One knows from (4.4.15) that $\mathbf{Sp}^{seq}(D)$ is stable. Therefore it suffices to prove that for any stable $\infty$-category $A$, the induced functor

$$\mathrm{Exc}_*(A, \mathbf{Sp}^{seq}(D)) \to \mathrm{Exc}_*(A, D)$$

is an equivalence. But this functor is the limit of the sequence

$$\cdots \xrightarrow{\Omega} \mathrm{Exc}_*(A, D) \xrightarrow{\Omega} \mathrm{Exc}_*(A, D) \xrightarrow{\Omega} \mathrm{Exc}_*(A, D),$$

which is a diagram of equivalences over a weakly contractible $\infty$-category.  $\square$

### Complete derived $\infty$-categories

We can apply the stabilisation process to the nonnegative derived $\infty$-category:

**Definition 4.6.10.** Let $E$ be an abelian category with enough projective objects. We write $D^{-,\wedge}(E)$ for the stabilisation $\mathbf{Sp}(D_{\geq 0}(E))$. This is the *right complete derived $\infty$-category* of $E$.

Dually, if $E$ is an abelian category with enough injective objects, we write $D^{+,\wedge}(E)$ for the stabilisation $\mathbf{Sp}(D_{\geq 0}(E^{op}))^{op}$. This is the *left complete derived $\infty$-category* of $E$.

In the next section, we will be able to characterise these $\infty$-categories in an intrinsic manner.

## 4.7    *t*-structures

The most basic examples of triangulated categories possess additional structure given by shift and truncation functors. For example, for the derived category of a commutative ring, there are inverse auto-equivalences given by shifting complexes up and down, and it is often useful to study about truncated complexes that live entirely in positive or negative degrees. This structure is axiomatized in terms of additional data referred to as a *t*-structure, which specifies positive and negative subcategories whose intersection is an abelian category known as the heart of the *t*-structure. There is a natural generalisation of this theory to the setting of stable ∞-categories.

**Definition 4.7.1.** Let $A$ be a stable ∞-category, and let $A_{\geq 0}, A_{\leq 0} \subseteq A$ be a pair of full subcategories. We may shift these subcategories about:

$$A_{\geq n} = (A_{\geq 0})[n] \quad \text{and} \quad A_{\leq n} = (A_{\leq 0})[n].$$

We say that the pair $(A_{\geq 0}, A_{\leq 0})$ constitute a *t-structure* on $A$ if it enjoys the following properties.

1. If $X \in A_{\geq 0}$ and $Y \in A_{\leq -1}$, then the space $\mathrm{Map}_A(X, Y)$ is contractible.
2. The subcategory $A_{\geq 0}$ is closed under positive shifts, and the subcategory $A_{\leq 0}$ is closed under negative shifts. So $A_{\geq 1} \subseteq A_{\geq 0}$, and, dually, $A_{\leq -1} \subseteq A_{\leq 0}$ as well.
3. For every $X \in A$, there is a distinguished triangle

$$\tau_{\geq 0} X \to X \to \tau_{\leq -1} X \to (\tau_{\geq 0} X)[1],$$

where $\tau_{\geq 0} X \in A_{\geq 0}$ and $(\tau_{\leq -1} X)[1] \in A_{\leq 0}$.

**4.7.2.** For any object $X$, a distinguished triangle $\tau_{\geq 0} X \to X \to \tau_{\leq -1} X \to (\tau_{\geq 0} X)[1]$ exhibits $\tau_{\leq -1} X \in A_{\leq -1}$ as a $(A_{\leq -1})$-localisation of $X$. Consequently, $\tau_{\leq -1}$ organises itself into a left adjoint to the inclusion $A_{\leq -1} \hookrightarrow A$. One may shift to find that for any $n \in \mathbb{Z}$, the functor $\tau_{\leq n}$ defined by

$$\tau_{\leq n} X = (\tau_{\leq -1}(X[-n]))[n]$$

exhibits the subcategory $A_{\leq n} \subseteq A$ as a localisation.

Dually, the distinguished triangle $\tau_{\geq 0} X \to X \to \tau_{\leq -1} X \to (\tau_{\geq 0} X)[1]$ exhibits $\tau_{\geq 0} X \in A_{\geq 0}$ as a $(A_{\geq 0})$-colocalisation of $X$, and for any $n \in \mathbb{Z}$, the functor $\tau_{\geq n}$ defined by

$$\tau_{\geq n} X = (\tau_{\geq 0}(X[n]))[-n]$$

exhibits the subcategory $A_{\geq n} \subseteq A$ as a colocalisation.

**Example 4.7.3.** We have already encountered the *t*-structure on the ∞-category **Sp** of spectra. The spectra that lie in $\mathbf{Sp}_{\geq 0}$ are called *connective*.

**Example 4.7.4.** Let $E$ be an abelian category with enough projectives. We have also encountered the *t*-structure on right bounded derived ∞-category. Then $D_{\leq 0}(E)$, regarded as a full subcategory of $D^-(E)$, is a *t*-structure.

This notion is compatible with the classical notion; if $A$ is a stable ∞-category with a *t*-structure in the sense above, then the homotopy category of $A$ is a triangulated category with a *t*-structure.

Warning 4.7.5. Here we are following the homotopy theory convention of homological indexing. This is mostly for the sake of compatibility with Lurie's text.

However, one may expect to encounter cohomological indexing in the literature. These *should* be written with superscripts rather than subscripts:

$$A^{\leq n} = A_{\geq -n}, \qquad A^{\geq n} = A_{\leq -n}, \qquad \tau^{\leq n} = \tau_{\geq -n}, \qquad \tau^{\geq n} = \tau_{\leq -n}.$$

Unfortunately, even when cohomological indexing is being employed, the truncation functors are sometimes written with subscripts (notably, in [34]), so one must remain vigilant.

We emphasise that the meaning of the shift functor $X \mapsto X[1]$ is *always* suspension. So one has the formulas

$$A^{\leq n} = (A^{\leq 0})[-n] \qquad \text{and} \qquad A^{\geq n} = (A^{\geq 0})[-n].$$

4.7.6. A *t*-structure on a stable $\infty$-category $A$ is uniquely specified by giving, for some $n \in \mathbb{Z}$, any one of the following pieces of data:

1. the full subcategory $A_{\geq n} \subseteq A$;
2. the full subcategory $A_{\leq n} \subseteq A$;
3. the functor $\tau_{\geq n} \colon A \to A$; or
4. the functor $\tau_{\leq n} \colon A \to A$.

4.7.7. For any $n \in \mathbb{Z}$, the $\infty$-category $A_{\geq n}$ is an exact $\infty$-category in which every morphism is ingressive, and $A_{\leq n}$ is an exact $\infty$-category in which every morphism is egressive.

4.7.8. For integers $a \leq b$, one may define $A_{[a,b]} = A_{\geq a} \cap A_{\leq b}$. The restriction of $\tau_{\leq b}$ to $A_{\geq a}$ is a left adjoint $A_{\geq a} \to A_{[a,b]}$, and the restriction of $\tau_{\geq a}$ to $A_{\leq b}$ is a left adjoint $A_{\leq b} \to A_{[a,b]}$. A simple "five lemma" argument furnishes us with a natural equivalence

$$\tau_{\leq b} \tau_{\geq a} \simeq \tau_{\geq a} \tau_{\leq b} \colon A \to A_{[a,b]},$$

and we shall write $\tau_{[a,b]}$ for this functor.

The $\infty$-category $A_{[a,b]}$ is an exact $\infty$-category in which the ingressive morphisms are those morphisms that are ingressive in $A_{\geq a}$, and the egressive morphisms are those morphisms that are egressive in $A_{\leq b}$.

As a special case, we write $A^{\heartsuit} = A_{[0,0]}$; this is called the *heart* of the *t*-structure. Note that the shift functor restricts to a *specified* equivalence $A^{\heartsuit} \simeq A_{[n,n]}$ for any $n \in \mathbb{Z}$; we now define the *homological functors* attached to the *t*-structure:

$$\pi_n = \tau_{[n,n]} \colon A \to A_{[n,n]} \simeq A^{\heartsuit}.$$

We have chosen this notation again for the sake of compatibility with Lurie. Other authors may write $H_n$ for this functor, and those who use cohomological indexing are liable to write $H^n = \tau^{[n,n]}$.

Proposition 4.7.9. *Let $A$ be a stable $\infty$-category endowed with a t-structure. Then the heart $A^{\heartsuit}$ is (equivalent to the $\infty$-category corresponding to) an ordinary abelian category.*

*Proof.* If $X, Y \in A^\heartsuit$, then for any $n \geq 1$, one has

$$\pi_n \mathrm{Map}_A(X, Y) \cong \mathrm{Ext}_A^{-n}(X, Y) \cong 0,$$

whence $A^\heartsuit$ is (equivalent to) a 1-category. We have already seen that $A^\heartsuit$ is an exact ∞-category, whence it is an exact category in Quillen's sense. To show that it is abelian, one just has to note that the ingressives are precisely the monomorphisms, and the egressives are precisely the epimorphisms.    □

**Example 4.7.10.** The heart $\mathbf{Sp}^\heartsuit$ is the category $\mathbf{Ab}$ of abelian groups. The homological functors attached to this $t$-structure are precisely the usual stable homotopy group functors $\pi_n$.

**Example 4.7.11.** Let $E$ be an abelian category with enough projectives. Then the heart $D^-(E)^\heartsuit$ is again $E$. The homological functors attached to this $t$-structure are precisely the homology functors $H_n$.

## Boundedness and completeness

The previous examples show that stable ∞-categories with $t$-structures are not determined by their hearts. There is, however, a special class of stable ∞-categories with $t$-structures that *are* determined by their hearts. These are the *derived* ∞-categories of abelian categories. To describe them, we must discuss some different kinds of $t$-structures.

**Definition 4.7.12.** Let $A$ be a stable ∞-category equipped with a $t$-structure. Define

$$A^- = \bigcup_{m \in \mathbb{Z}} A_{\geq m}, \qquad A^+ = \bigcup_{n \in \mathbb{Z}} A_{\leq n}, \qquad A^b = A^+ \cap A^- = \bigcup_{m,n \in \mathbb{Z}} A_{[m,n]}.$$

We call

1. the objects of $A^-$ *bounded below*,
2. the objects of $A^+$ *bounded above*, and
3. the objects of $A^b$ *bounded*.

We say that the $t$-structure is

4. *right bounded* if $A = A^-$,
5. *left bounded* if $A = A^+$, and
6. *bounded* if $A = A^b$.

**Example 4.7.13.** The $t$-structure on $\mathbf{Sp}^{fin}$ is right bounded.

**Definition 4.7.14.** Let $A$ be a stable ∞-category equipped with a $t$-structure. We define $A^{\wedge,R} \subseteq \mathrm{Fun}(\mathbb{Z}^{op}, A)$ as the full subcategory spanned by those sequences $X$ such that $X(m) \in A_{\geq m}$ for any $m \in \mathbb{Z}$, and that the induced morphism $X(n) \to \tau_{\geq n} X(m)$ is an equivalence for any $m \leq n$. Dually, we define $A^{\wedge,L} \subseteq \mathrm{Fun}(\mathbb{Z}^{op}, A)$ as the full subcategory spanned by those objects $X$ such that $X(m) \in A_{\leq m}$ for any $m \in \mathbb{Z}$, and that the induced morphism $\tau_{\leq m} X(n) \to X(m)$ is an equivalence for any $m \leq n$.

We call

1. $A^{\wedge,R}$ the *right completion* of $A$ with respect to its *t*-structure, and
2. $A^{\wedge,L}$ the *left completion* of $A$ with respect to its *t*-structure.

We say that the *t*-structure is

3. *right complete* if the natural map $A \to A^{\wedge,R}$ is an equivalence,
4. *left complete* if the natural map $A \to A^{\wedge,L}$ is an equivalence, and
5. *complete* if it is both left and right complete.

4.7.15. If $A$ is a stable $\infty$-category equipped with a *t*-structure, the right completion of $A^-$ coincides with the right completion of $A$ itself, and the bounded below objects of the right completion $A^{\wedge,R}$ coincide with the bounded below objects of $A$ itself. It follows that there is an equivalence between the $\infty$-category of right bounded *t*-structures and that of right complete *t*-structures.

Example 4.7.16. Let $E$ be an abelian category. If $E$ has enough projectives, then the *t*-structure on $D^-(E)$ is right bounded (whence the notation!) and left complete. Dually, if $E$ has enough injectives, then the *t*-structure on $D^+(E)$ is left bounded and right complete.

In the same vein, if $E$ has enough projectives, then $D^{-,\wedge}(E)$ is complete, and if $E$ has enough injectives, then $D^{+,\wedge}(E)$ is complete. In fact, $D^{-,\wedge}(E)$ is the right completion of $D^-(E)$, and $D^{+,\wedge}(E)$ is the left completion of $D^+(E)$.

It is *a priori* difficult to determine whether a *t*-structure is right or left complete. Fortunately, there is a reasonable criterion for this.

Definition 4.7.17. Let $A$ be a stable $\infty$-category equipped with a *t*-structure. We define

$$A_{-\infty} = \bigcap_{n \in \mathbb{Z}} A_{\leq n} \quad \text{and} \quad A_{+\infty} = \bigcap_{n \in \mathbb{Z}} A_{\geq n}.$$

We say that the *t*-structure is

1. *right separated* if $A_{-\infty} = 0$,
2. *left separated* if $A_{+\infty} = 0$, and
3. *separated* if it is both left and right separated.

Proposition 4.7.18 ([168, Proposition 1.2.1.19]). *Let $A$ be a stable $\infty$-category with countable coproducts. Let $\tau$ be a t-structure on $A$ with the property that $A_{\geq 0}$ is stable under countable coproducts. Then $\tau$ is right complete if and only if it is right separated.*

*Exercise 4.7.19.* Use this criterion to check that the *t*-structure on **Sp** is complete.

### Derived $\infty$-categories

Roughly speaking, the constructions $E \mapsto D^-(E)$ and $E \mapsto D^+(E)$ are left adjoint to the construction $A \mapsto A^\heartsuit$. To make this precise, we must specify which $\infty$-category of stable $\infty$-categories we will to use:

**Definition 4.7.20.** Let $A$ and $B$ be stable $\infty$-categories equipped with $t$-structures. An exact functor $f \colon A \to B$ is said to be *right $t$-exact* if $f(A_{\geq 0}) \subseteq B_{\geq 0}$. Dually, an exact functor $f \colon A \to B$ is said to be *left $t$-exact* if $f(A_{\leq 0}) \subseteq B_{\leq 0}$. An exact functor $f \colon A \to B$ is said to be *$t$-exact* if it is both left and right $t$-exact.

Let us say that a right $t$-exact functor $A \to B$ is *left derived* if it carries the projective objects of $A^\heartsuit$ into $B^\heartsuit$. We write $\mathrm{Fun}^{lder}(A, B) \subseteq \mathrm{Fun}(A, B)$ for the full subcategory spanned by the left derived right $t$-exact functors $A \to B$. Dually, let us say that a left $t$-exact functor $A \to B$ is *right derived* if it carries the injective objects of $A^\heartsuit$ into $B^\heartsuit$. We write $\mathrm{Fun}^{rder}(A, B) \subseteq \mathrm{Fun}(A, B)$ for the full subcategory spanned by the right derived right $t$-exact functors $A \to B$.

**Theorem 4.7.21.** *Let $E$ be an abelian category with enough projectives, and let $B$ be a stable $\infty$-category equipped with a left complete $t$-structure. The construction $F \mapsto \tau_{\leq 0} F|_E$ is an equivalence of $\infty$-categories*

$$\mathrm{Fun}^{lder}(D^-(E), B) \to \mathrm{Fun}^{rex}(E, B^\heartsuit),$$

*where $\mathrm{Fun}^{rex}(E, B^\heartsuit) \subseteq \mathrm{Fun}(E, B^\heartsuit)$ is the full subcategory spanned by the right exact functors $E \to B^\heartsuit$.*

*Dually, let $E$ be an abelian category with enough injectives, and let $B$ be a stable $\infty$-category equipped with a right complete $t$-structure. The construction $G \mapsto \tau_{\geq 0} G|_E$ is an equivalence of $\infty$-categories*

$$\mathrm{Fun}^{rder}(D^+(E), B) \to \mathrm{Fun}^{lex}(E, B^\heartsuit),$$

*where $\mathrm{Fun}^{lex}(E, B^\heartsuit) \subseteq \mathrm{Fun}(E, B^\heartsuit)$ is the full subcategory spanned by the left exact functors $E \to B^\heartsuit$.*

**4.7.22.** If $E$ has enough projectives and $B$ is a stable $\infty$-category equipped with a left complete $t$-structure, then we call $F \colon D^-(E) \to B$ the *left derived functor* of $f = \tau_{\leq 0} F|_A$, and we write $\mathbb{L}f = F$.

Dually, if $E$ has enough injectives and $B$ is a stable $\infty$-category equipped with a right complete $t$-structure, then we call $G \colon D^+(E) \to B$ the *right derived functor* of $g = \tau_{\geq 0} G|_E$, and we write $\mathbb{R}g = G$.

**Example 4.7.23.** Since $\mathbf{Sp}^\heartsuit \simeq \mathbf{Ab}$, we obtain a $t$-exact functor $H \colon D^-(\mathbf{Ab}) \to \mathbf{Sp}$, which carries a chain complex $C$ to the *generalised Eilenberg–Mac Lane spectrum $HC$.*

This result also allows us to recognise derived $\infty$-categories.

**Corollary 4.7.24.** *Let $A$ be a stable $\infty$-category equipped with a left complete $t$-structure. Assume that $A^\heartsuit$ has enough projectives. The unique $t$-exact functor $K \colon D^-(A^\heartsuit) \to A$ is fully faithful if and only if, for any projective object $M \in A^\heartsuit$ and any object $N \in A^\heartsuit$, the groups $\mathrm{Ext}^n(M, N) = 0$ for any $n \geq 1$. In this case, the essential image of $K$ is $A^- \subseteq A$.*

4.7.25.   As a final comment, we observe that if $E$ is a Grothendieck abelian category, then there is also an *unbounded derived* $\infty$-*category* $D(E)$ equipped with a $t$-structure. Since $E$ has enough injectives, we obtain a stable $\infty$-category $D^+(E)$. The previous corollary ensures that the unique $t$-exact functor $D^+(E) \to D(E)$ is fully faithful, and it identifies $D^+(E)$ with the bounded above objects of $D(E)$.

One might be therefore tempted to believe that $D(E)$ coincides with the left complete derived $\infty$-category $D^{+,\wedge}(E)$. We emphasise, however, that this is not generally true: there are abelian categories $E$, such as the category of representations of $\mathbb{G}_a$ over a field of positive characteristic, for which $D(E)$ is not left complete.

If, however, countable products in $E$ are exact, the criterion of 4.7.18 works to ensure that $D(E)$ is left complete. Then we can identify $D(E)$ with $D^{+,\wedge}(E)$, and so 4.7.21 furnishes us with a universal characterisation of $D(E)$ in this case. The author does not know a universal characterisation of $D(E)$ for a general Grothendieck abelian category.

# 5    Operads and operadic algebras in homotopy theory

by Michael A. Mandell

## 5.1    Introduction

Operads first appeared in the book *Geometry of iterated loop spaces* by J. P. May [194], though Boardman and Vogt had earlier implicitly defined a mathematically equivalent notion as a "PROP in standard form" [49, §2]. In those works, operads and operadic algebra structures provide a recognition principle and a delooping machine for $n$-fold loop spaces and infinite loop spaces. The basic idea is that an operad should encode the operations in some kind of homotopical algebraic structure. For example, an $n$-fold loop space $\Omega^n X$ comes with $n$ different multiplications $(\Omega^n X)^2 \to \Omega^n X$, which can be iterated and generalized to a space of $m$-ary maps $\mathcal{C}_n(m)$ (from $(\Omega^n X)^m$ to $\Omega^n X$); here $\mathcal{C}_n$ is the Boardman–Vogt little $n$-cubes operad (see Construction 5.3.5 and Section 5.11 below). The content of the recognition theorem is that $\mathcal{C}_n$ specifies a structure that is essentially equivalent to the structure of an $n$-fold loop space for connected spaces. It was clear even at the time of introduction that operads were a big idea and in the almost 50 years since then, operads have found a wide range of other uses in a variety of areas of mathematics: a quick MathSciNet search for papers since 2015 with "operad" in the title comes up with papers in combinatorics, algebraic geometry, nonassociative algebra, geometric group theory, free probability, mathematical modeling, and physics, as well as in algebraic topology and homological algebra.

Even the topic of operads in algebraic topology is too broad to cover or even summarize in a single article. This expository article concentrates on what I view as the basic topics in the homotopy theory of operadic algebras: the definition of operads, the definition of algebras over operads, structural aspects of categories of algebras over operads, model structures on algebra categories, and comparison of algebra categories when changing operad or underlying category. In addition, it includes two applications of the theory: the original application to $n$-fold loop spaces, and an application to algebraic models of homotopy types (chosen purely on the basis of personal bias). This leaves out a long list of other topics that could also fit in this chapter, such as model structures on operads, Koszul duality, deformation theory and Quillen (co)homology, multiplicative structures in stable homotopy theory (for example, on Thom spectra, $K$-theory spectra, etc.), Deligne and Kontsevich conjectures, string topology, factorization homology, construction of moduli spaces, and Goodwillie calculus, just to name a few areas.

## Notation and conventions

Although we concentrate on operads and operadic algebras in topology, much of the background applies very generally. Because of this and because we will want to discuss both the case of spaces and the case of spectra, we will use neutral notation: let $\mathcal{M}$ denote a symmetric monoidal category [145, §1.4], writing $\square$ for the monoidal product and $\mathbf{1}$ for the unit. (We will uniformly omit notation for associativity isomorphisms and typically omit notation for commutativity isomorphisms, but when necessary, we will write $c_\sigma$ for the commutativity isomorphism associated to a permutation $\sigma$.) Usually, we will want $\mathcal{M}$ to have coproducts and sometimes more general colimits, which we will expect to commute with $\square$ on each side (keeping the other side fixed). This exactness of $\square$ is automatic if the monoidal structure is closed [145, §1.5], i.e., if for each fixed object $X$ of $\mathcal{M}$, the functor $(-)\square X$ has a right adjoint; this is often convenient to assume, and when we do, we will use $F(X,-)$ for the right adjoint. The three basic classes of examples to keep in mind are:

   (i) Convenient categories of topological spaces, including compactly generated weak Hausdorff spaces [206]; then $\square$ is the categorical product, $\mathbf{1}$ is the final object (one point space), and $F(X,Y)$ is the function space, often written $Y^X$.
  (ii) Modern categories of spectra, including EKMM $S$-modules [94], symmetric spectra [133], and orthogonal spectra [178]; then $\square$ is the smash product, $\mathbf{1}$ is the sphere spectrum, and $F(-,-)$ is the function spectrum.
 (iii) The category of chain complexes of modules over a commutative ring $R$; then $\square$ is the tensor product over $R$, $\mathbf{1}$ is the complex $R$ concentrated in degree zero, and $F(-,-)$ is the Hom-complex $\mathrm{Hom}_R(-,-)$.

We now fix a convenient category of spaces and just call it "the category of spaces" and the objects in it "spaces", ignoring the classical category of topological spaces.

In the context of operadic algebras in spectra (i.e., (ii) above), it is often technically convenient to use operads of spaces. However, for uniformity of exposition, we have written this article in terms of operads internally in $\mathcal{M}$. The unreduced suspension functor $\Sigma_+^\infty(-)$ converts operads in spaces to operads in the given category of spectra.

## Outline

The basic idea of an operad is that the pieces of it should parametrize a class of $m$-ary operations. From this perspective, the fundamental example of an operad is the *endomorphism operad* of an object $X$,

$$\mathcal{E}\mathrm{nd}_X(m) := F(X^{(m)}, X), \qquad X^{(m)} := \underbrace{X \,\square \cdots \square\, X}_{m \text{ factors}},$$

which parametrizes all $m$-ary maps from $X$ to itself. Abstracting the symmetry and composition properties leads to the definition of operad in [194]. We review this definition in Section 5.2.

Section 5.3 presents some basic examples of operads important in topology, including some $A_\infty$ operads, $E_\infty$ operads, and $E_n$ operads.

May chose the term "operad" to match the term "monad" (see [191]), to show their close connection. Basically, a monad is an abstract way of defining some kind of structure on objects in a category, and an operad gives a very manageable kind of monad. Section 5.4 reviews the monad associated to an operad and defines algebras over an operad.

Section 5.5 gives the basic definition of a module over an operadic algebra and reviews the basics of the homotopy theory of module categories.

Section 5.6 discusses limits and colimits in categories of operadic algebras. It includes a general filtration construction that often provides the key tool to study pushouts of operadic algebras homotopically in terms of colimits in the underlying category. Section 5.7 discusses when categories of operadic algebras are enriched, and in the case of categories of algebras enriched over spaces, discusses the geometric realization of simplicial and cosimplicial algebras. Although these sections may seem less basic and more technical than the previous sections, the ideas here provide the tools necessary for further work with operadic algebras using the modern methods of homotopy theory.

Model structures on categories of operadic algebras provide a framework for proving comparison theorems and rectification theorems. Section 5.8 reviews some aspects of model category theory for categories of operadic algebras. In the terminology of this article, a *comparison theorem* is an equivalence of homotopy theories between categories of algebras over different operads that are equivalent in some sense (for example, between categories of algebras over different $E_\infty$ operads) or between categories of algebras over equivalent base categories (for example, $E_\infty$ algebras in spaces versus $E_\infty$ algebras in simplicial sets). A *rectification theorem* is a comparison theorem where one of the operads is discrete in some sense: a comparison theorem for the category of algebras over an $A_\infty$ operad and the category of associative algebras is an example of a rectification theorem, as is the comparison theorem for $E_\infty$ algebras and commutative algebras in modern categories of spectra. Section 5.9 discusses these and other examples of comparison and rectification theorems. In both Sections 5.8 and 5.9, instead of stating theorems of maximal generality, we have chosen to provide "Example Theorems" that capture some examples of particular interest in homotopy theory and stable homotopy theory. Both the statements and the arguments provide examples: the arguments apply or can be adapted to apply in a wide range of generality.

The Moore space is an early rectification technique (predating operads and $A_\infty$ monoids) for producing a genuine associative monoid version of the loop space; the construction applies generally to a little 1-cubes algebra to produce an associative algebra that we call the *Moore algebra*. The concept of modules over an operadic algebra leads to another way of producing an associative algebra, called the *enveloping algebra*. Section 5.10 compares these constructions and the rectification of $A_\infty$ algebras constructed in Section 5.9.

Sections 5.11 and 5.12 review two significant applications of the theory of operadic algebras. Section 5.11 reviews the original application: the theory of iterated loop spaces and the recognition principle in terms of $E_n$ algebras. Section 5.12 reviews the equivalence between the rational and $p$-adic homotopy theory of spaces with the homotopy theory of $E_\infty$ algebras.

### Acknowledgments

The author benefited from conversations and advice from Clark Barwick, Agnès Beaudry, Julie Bergner, Myungsin Cho, Bjørn Dundas, Tyler Lawson, Andrey Lazarev, Amnon Neeman, Brooke Shipley, and Michael Shulman while working on this chapter. The author thanks Peter May for his mentorship in the 1990s (and beyond) while learning these topics and for help straightening out some of the history described here. The author thanks the Isaac Newton Institute for Mathematical Sciences for support and hospitality during the program "Homotopy harnessing higher structures" (HHH) when work on this chapter was undertaken; this work was supported by: EPSRC Grant Number EP/R014604/1. The author was supported in part by NSF grants DMS-1505579 and DMS-1811820 while working on this project. Finally, the author thanks Andrew Blumberg for extensive editorial advice.

## 5.2     Operads and endomorphisms

We start with the definition of an operad. The collection of $m$-ary endomorphism objects $\mathcal{E}nd_X(m) = F(X^{(m)}, X)$ provides the prototype for the definition, and we use its intrinsic structure to motivate and explain it. Although the endomorphism objects only make sense when the symmetric monoidal category is "closed" (which means that function objects exist), the definition of operad will not require or assume function objects, nor will the definition of operadic algebra in Section 5.4. To take in the picture, it might be best just to take $\mathcal{M}$ to be the category of spaces, the category of vector spaces over a field, or the category of sets on first introduction to this material.

In our basic classes of examples, and more generally as a principle of enriched category theory, function objects behave like sets of morphisms: the counit of the defining adjunction

$$F(X, Y) \,\square\, X \to Y$$

is often called the *evaluation map* (and denoted *ev*). It allows "element-free" definition and study of composition: iterating evaluation maps

$$F(Y, Z) \,\square\, F(X, Y) \,\square\, X \to F(Y, Z) \,\square\, Y \to Z$$

induces (by adjunction) a *composition map*

$$\circ\colon F(Y, Z) \,\square\, F(X, Y) \to F(X, Z).$$

One can check just using the basic properties of adjunctions that this composition is associative in the obvious sense. It is also unital: the identity element of $\mathscr{M}(X,X)$ specifies a map $1_X \colon \mathbf{1} \to F(X,X)$,

$$\mathrm{id}_X \in \mathscr{M}(X,X) \cong \mathscr{M}(\mathbf{1} \square X, X) \cong \mathscr{M}(\mathbf{1}, F(X,X)),$$

where the first isomorphism is induced by the unit isomorphism; essentially by construction, the composite

$$\mathbf{1} \square X \xrightarrow{1_X \square \mathrm{id}_X} F(X,X) \square X \xrightarrow{ev} X$$

is the unit isomorphism. It follows that the diagram

$$\begin{array}{ccccc}
\mathbf{1} \square F(X,Y) & \xrightarrow{\;\cong\;} & F(X,Y) & \xleftarrow{\;\cong\;} & F(X,Y) \square \mathbf{1} \\
{\scriptstyle 1_Y \square \mathrm{id}_{F(X,Y)}} \downarrow & & \| & & \downarrow {\scriptstyle \mathrm{id}_{F(X,Y)} \square 1_X} \\
F(Y,Y) \square F(X,Y) & \xrightarrow[\circ]{} & F(X,Y) & \xleftarrow[\circ]{} & F(X,Y) \square F(X,X)
\end{array}$$

commutes, where the top-level isomorphisms are the unit isomorphisms. More is true: the function objects enrich the category $\mathscr{M}$ over itself, and the $\square, F$ parametrized adjunction is itself enriched [145, §1.5–6].

In the case when $\mathscr{M}$ is the category of spaces, the evaluation map is just the map that evaluates functions on their arguments; thinking in these terms will make the formulas and checks clearer for the reader not used to working with adjunctions. Since in the category of spaces $\mathbf{1}$ is the one-point space, a map out of $\mathbf{1}$ just picks out an element of the target space and the map $\mathbf{1} \to F(X,X)$ is just the map that picks out the identity map of $X$.

The basic compositions above generalize to associative and unital $m$-ary compositions; now for simplicity and because it is the main case of interest here, we restrict to considering a fixed object $X$. The $m$-ary composition takes the form

$$F(X^{(m)},X) \square (F(X^{(j_1)},X) \square \cdots \square F(X^{(j_m)},X)) \to F(X^{(j)},X),$$

where $j = j_1 + \cdots + j_m$ and (as in the introduction) $X^{(m)}$ denotes the $m$-th $\square$ power of $X$; we think of the $m$-ary composition as plugging in the $m$ $j_i$-ary maps into the first $m$-ary map; it is adjoint to the map

$$F(X^{(m)},X) \square F(X^{(j_1)},X) \square \cdots \square F(X^{(j_m)},X) \square X^{(j)} \cong$$
$$F(X^{(m)},X) \square F(X^{(j_1)},X) \square \cdots \square F(X^{(j_m)},X) \square X^{(j_1)} \square \cdots \square X^{(j_m)} \to X$$

that does the evaluation map

$$F(X^{(j_i)},X) \square X^{(j_i)} \to X,$$

then collects the resulting $m$ factors of $X$ and does the evaluation map

$$F(X^{(m)},X) \square X^{(m)} \to X.$$

In this double evaluation, implicitly we have shuffled some of the factors of $X$ past some of the endomorphism objects, but we take care not to permute factors of $X$

among themselves or the endomorphism objects among themselves. This defines a composition map

$$\Gamma^m_{j_1,\dots,j_m} : \mathcal{E}nd_X(m) \,\square\, \mathcal{E}nd_X(j_1) \,\square\, \cdots \,\square\, \mathcal{E}nd_X(j_m) \to \mathcal{E}nd_X(j).$$

The composition is associative and unital in the obvious sense (which we write out in the definition of an operad, Definition 5.2.1, below).

We now begin systematically writing $\mathcal{E}nd_X(m)$ for $F(X^{(m)}, X)$. We observe that $\mathcal{E}nd_X(m) = F(X^{(m)}, X)$ has a right action by the symmetric group $\Sigma_m$ induced by the left action of $\Sigma_m$ on $X^{(m)}$ corresponding to permuting the $\square$-factors. In general, for a permutation $\sigma$, we write $c_\sigma$ for the map that permutes $\square$-factors and $a_\sigma$ for the action of $\sigma$ on $\mathcal{E}nd_X(m)$, i.e., the map that does $c_\sigma$ on the domain of $\mathcal{E}nd_X(m) = F(X^{(m)}, X)$. We now study what happens when we permute the various factors in the formula for $\Gamma$ above. (As these are a bit tricky, we do the formulas out here and repeat them below in the definition of an operad, Definition 5.2.1.)

First consider what happens when we permute the factors of $X$. We have nothing to say for an arbitrary permutation of the factors of $X$, but in the composition $\Gamma^m_{j_1,\dots,j_m}$, we can say something for a permutation that permutes the factors only within their given blocks of size $j_1,\dots,j_m$, i.e., when the overall permutation $\sigma$ of all $j$ factors is the block sum of permutations $\sigma_1 \oplus \cdots \oplus \sigma_m$ with $\sigma_i$ in $\Sigma_{j_i}$. By extranaturality, performing the right action of $\sigma_i$ on $\mathcal{E}nd_X(j_i)$ and evaluating is the same as applying the left action of $\sigma_i$ on $X^{(j_i)}$ and evaluating. It follows that the composition $\Gamma^m_{j_1,\dots,j_m}$ is $(\Sigma_{j_1} \times \cdots \times \Sigma_{j_m})$-equivariant where we use the $\Sigma_{j_i}$-actions on the $\mathcal{E}nd_X(j_i)$'s in the source and block sum with the $\Sigma_j$-action on $\mathcal{E}nd_X(j)$ on the target.

Permuting the endomorphism object factors is easier to understand when we also permute the corresponding factors of $X$. In the context of $\Gamma^m_{j_1,\dots,j_m}$, for $\sigma$ in $\Sigma_m$, let $\sigma_{j_1,\dots,j_m} \in \Sigma_j$ permute the blocks $X^{(j_1)},\dots,X^{(j_m)}$ as $\sigma$ permutes $1,\dots,m$. So, for example, if $m = 3$, $j_1 = 1$, $j_2 = 3$, $j_3 = 2$, and $\sigma = (23)$, then $\sigma_{1,3,2}$ is the permutation

$$(23)_{1,3,2} = \left\{ \begin{matrix} 1 & 2 & 3 & 4 & 5 & 6 \\ \downarrow & \downarrow & \downarrow & \downarrow & \downarrow & \downarrow \\ 1 & 5 & 6 & 2 & 3 & 4 \end{matrix} \right\} = (25364).$$

In $\mathcal{E}nd_X(j_1) \,\square\, \cdots \,\square\, \mathcal{E}nd_X(j_m) \,\square\, X^{(j)}$, if we apply $\sigma$ to permute the endomorphism object factors and $\sigma_{j_1,\dots,j_m}$ to permute the $X$ factors, then evaluation pairs the same factors as with no permutation and the diagram

$$
\begin{array}{ccc}
(\mathcal{E}nd_X(j_1) \,\square\, \cdots \,\square\, \mathcal{E}nd_X(j_m)) \,\square\, X^{(j)} & \xrightarrow{\;ev\;} & X^{(m)} \\
{\scriptstyle c_\sigma \square c_{\sigma_{j_1,\dots,j_m}}} \downarrow & & \downarrow {\scriptstyle c_\sigma} \\
(\mathcal{E}nd_X(j_{\sigma^{-1}(1)}) \,\square\, \cdots \,\square\, \mathcal{E}nd_X(j_{\sigma^{-1}(m)})) \,\square\, X^{(j)} & \xrightarrow[\;ev\;]{} & X^{(m)}
\end{array}
$$

commutes. This now tells us what happens with $\Gamma^m_{j_1,\dots,j_m}$ and the permutation action on $\mathcal{E}nd_X(n)$: the composite of the right action of $\sigma$ on $\mathcal{E}nd_X(m)$ with $\Gamma^m_{j_1,\dots,j_m}$,

$$\mathcal{E}nd_X(m) \,\square\, (\mathcal{E}nd_X(j_1) \,\square\, \cdots \,\square\, \mathcal{E}nd_X(j_m))$$

$$\xrightarrow{\;a_\sigma \square \mathrm{id}\;} \mathcal{E}nd_X(m) \,\square\, (\mathcal{E}nd_X(j_1) \,\square\, \cdots \,\square\, \mathcal{E}nd_X(j_m)) \xrightarrow{\;\Gamma^m_{j_1,\dots,j_m}\;} \mathcal{E}nd_X(j),$$

is equal to the composite of the $\square$-permutation $c_\sigma$ on the $\mathcal{E}nd(j_i)$'s, the composition map $\Gamma^m_{j_{\sigma^{-1}(1)},\ldots,j_{\sigma^{-1}(m)}}$, and the right action of $\sigma_{j_1,\ldots,j_m}$ on $\mathcal{E}nd_X(j)$:

$$\mathcal{E}nd_X(m) \square (\mathcal{E}nd_X(j_1) \square \cdots \square \mathcal{E}nd_X(j_m))$$

$$\xrightarrow{\mathrm{id}\,\square\,c_\sigma} \mathcal{E}nd_X(m) \square (\mathcal{E}nd_X(j_{\sigma^{-1}(1)}) \square \cdots \square \mathcal{E}nd_X(j_{\sigma^{-1}(m)}))$$

$$\xrightarrow{\Gamma^m_{j_{\sigma^{-1}(1)},\ldots,j_{\sigma^{-1}(m)}}} \mathcal{E}nd_X(j) \xrightarrow{a_{\sigma_{j_1,\ldots,j_m}}} \mathcal{E}nd_X(j).$$

See Figure 5.2 on p. 191 for this equation written as a diagram.

Although we did not emphasize this above, we need to allow any of $m$, $j_1, \ldots, j_m$, or $j$ to be zero, where we understand empty $\square$-products to be the unit $\mathbf{1}$. The formulations above still work with this extension, using the unit isomorphism where necessary. The purpose of allowing these "zero-ary" operations is that it allows us to encode a unit object into the structure: For example, in the context of spaces $\mathbf{1}$ is the one point space $*$ and to describe the structure of a topological monoid, not only do we need the binary operation $X \times X \to X$, but we also need the zero-ary operation $* \to X$ for the unit.

Rewriting the properties of $\mathcal{E}nd_X$ above as a definition, we get an element-free version of the definition of operad of May [194, 1.2].[1]

**Definition 5.2.1.** An operad in a symmetric monoidal category $\mathcal{M}$ consists of a sequence of objects $\mathcal{O}(m)$, $m = 0, 1, 2, 3, \ldots$, together with

(a) a right action of the symmetric group $\Sigma_m$ on $\mathcal{O}(m)$ for all $m$,
(b) a *unit map* $1: \mathbf{1} \to \mathcal{O}(1)$, and
(c) a *composition rule*

$$\Gamma^m_{j_1,\ldots,j_m}: \mathcal{O}(m) \square \mathcal{O}(j_1) \square \cdots \square \mathcal{O}(j_m) \to \mathcal{O}(j)$$

for every $m$, $j_1, \ldots, j_m$, where $j = j_1 + \cdots + j_m$, typically written $\Gamma$ when $m$ and $j_1, \ldots, j_m$ are understood or irrelevant,

satisfying the following conditions:

(i) The composition rule $\Gamma$ is associative in the sense that for any $m$, $j_1, \ldots, j_m$ and $k_1, \ldots, k_j$, letting $j = j_1 + \cdots + j_m$, $k = k_1 + \cdots + k_j$, $t_i = j_1 + \cdots + j_{i-1}$ (with $t_1 = 0$), and $s_i = k_{t_i+1} + \cdots + k_{t_i+j_i}$, the equation

$$\Gamma^j_{k_1,\ldots,k_j} \circ (\Gamma^m_{j_1,\ldots,j_m} \square \mathrm{id}_{\mathcal{O}(k_1)} \square \cdots \square \mathrm{id}_{\mathcal{O}(k_j)})$$
$$= \Gamma^m_{s_1,\ldots,s_m} \circ (\mathrm{id}_{\mathcal{O}(m)} \square \Gamma^{j_1}_{k_1,\ldots,k_{j_1}} \square \cdots \square \Gamma^{j_m}_{k_{t_m+1},\ldots,k_j}) \circ c$$

holds in the set of maps

$$\mathcal{O}(m) \square \mathcal{O}(j_1) \square \cdots \square \mathcal{O}(j_m) \square \mathcal{O}(k_1) \square \cdots \square \mathcal{O}(k_j) \to \mathcal{O}(k),$$

---

[1] In the original definition, May required $\mathcal{O}(0) = \mathbf{1}$ in order to provide $\mathcal{O}$-algebras with units, which was desirable in the iterated loop space context, but standard convention has since dropped this requirement to allow non-unital algebras and other unit variants.

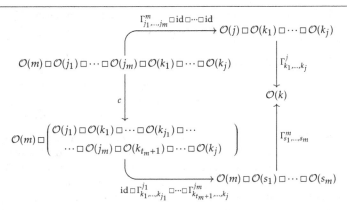

**Figure 5.1** The diagram for 5.2.1(i). Here $c$ is the $\square$-permutation that shuffles $\mathcal{O}(k_\ell)$'s past $\mathcal{O}(j_i)$'s as displayed, $j = j_1 + \cdots + j_m$, $t_i = j_1 + \cdots + j_{i-1}$ (with $t_1 = 0$), $s_i = k_{t_i+1} + \cdots + k_{t_i+j_i}$, and $k = k_1 + \cdots + k_j = s_1 + \cdots + s_m$.

where $c$ is the $\square$-permutation

$$\mathcal{O}(m) \square \mathcal{O}(j_1) \square \cdots \square \mathcal{O}(j_m) \square \mathcal{O}(k_1) \square \cdots \square \mathcal{O}(k_j) \rightarrow$$
$$\mathcal{O}(m) \square (\mathcal{O}(j_1) \square \mathcal{O}(k_1) \square \cdots \square \mathcal{O}(k_{j_1})) \square \cdots \square (\mathcal{O}(j_m) \square \mathcal{O}(k_{t_m+1}) \square \cdots \square \mathcal{O}(k_j)).$$

that shuffles the $\mathcal{O}(k_\ell)$'s and $\mathcal{O}(j_i)$'s as displayed (see Figure 5.1 for the diagram).

(ii) The unit map 1 is a left and right unit for the composition rule $\Gamma$ in the sense that

$$\Gamma^1_m \circ (1 \square \mathrm{id}): \quad 1 \square \mathcal{O}(m) \xrightarrow{1 \square \mathrm{id}} \mathcal{O}(1) \square \mathcal{O}(m) \xrightarrow{\Gamma^1_m} \mathcal{O}(m)$$

is the unit isomorphism and

$$\Gamma^m_{1,\ldots,1} \circ (\mathrm{id} \square 1^{(m)}): \quad \mathcal{O}(m) \square 1^{(m)} \xrightarrow{\mathrm{id} \square 1^{(m)}} \mathcal{O}(m) \square \mathcal{O}(1)^{(m)} \xrightarrow{\Gamma^m_{1,\ldots,1}} \mathcal{O}(m)$$

is the iterated unit isomorphism for $\mathcal{O}(m)$ for all $m$.

(iii) The map $\Gamma^m_{j_1,\ldots,j_m}$ is $(\Sigma_{j_1} \times \cdots \times \Sigma_{j_m})$-equivariant for the block sum inclusion of $\Sigma_{j_1} \times \cdots \times \Sigma_{j_m}$ in $\Sigma_j$.

(iv) For any $m, j_1, \ldots, j_m$ and any $\sigma \in \Sigma_m$, the equation

$$\Gamma^m_{j_1,\ldots,j_m} \circ (a_\sigma \square \mathrm{id}_{\mathcal{O}(j_1)} \square \cdots \square \mathrm{id}_{\mathcal{O}(j_m)}) = a_{\sigma_{j_1,\ldots,j_m}} \circ \Gamma^m_{j_{\sigma^{-1}(1)},\ldots,j_{\sigma^{-1}(m)}} \circ (\mathrm{id}_{\mathcal{O}(m)} \square c_\sigma)$$

holds in the set of maps

$$\mathcal{O}(m) \square \mathcal{O}(j_1) \square \cdots \square \mathcal{O}(j_m) \rightarrow \mathcal{O}(j),$$

where $\sigma_{j_1,\ldots,j_m}$ denotes the block permutation in $\Sigma_j$ corresponding to $\sigma$ on the blocks of size $j_1, \ldots, j_m$, $a$ denotes the right action of (a), and $c_\sigma$ denotes the $\square$-permutation corresponding to $\sigma$ (see Figure 5.2 for the diagram).

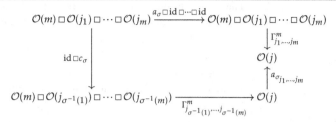

**Figure 5.2** The diagram for 5.2.1 (iv). Here $\sigma \in \Sigma_m$, $c_\sigma$ is the $\square$-permutation corresponding to $\sigma$, $\sigma_{j_1,\dots,j_m} \in \Sigma_j$ is the block permutation performing $\sigma$ on blocks of sizes $j_1,\dots,j_m$, $j = j_1 + \cdots + j_m$, and $a$ denotes the $\Sigma_m$ action on $\mathcal{O}(m)$ and the $\Sigma_j$-action on $\mathcal{O}(j)$.

A map of operads consists of a map of each object that commutes with the structure:

**Definition 5.2.2.** A map of operads $(\{\mathcal{O}(m)\}, 1, \Gamma) \to (\{\mathcal{O}'(m)\}, 1', \Gamma')$ consists of $\Sigma_m$-equivariant maps $\phi_m \colon \mathcal{O}(m) \to \mathcal{O}'(m)$ for all $m$ such that

$$\Gamma'^m_{j_1,\dots,j_m} \circ (\phi_m \,\square\, \phi_{j_1} \,\square\, \cdots \,\square\, \phi_{j_m}) = \phi_j \circ \Gamma^m_{j_1,\dots,j_m}$$

for all $m, j_1,\dots,j_m$ and $1' = \phi_1 \circ 1$; in commuting diagrams:

$$
\begin{array}{ccc}
\mathcal{O}(m) \,\square\, \mathcal{O}(j_1) \,\square\, \cdots \,\square\, \mathcal{O}(j_m) & \xrightarrow{\;\Gamma^m_{j_1,\dots,j_m}\;} & \mathcal{O}(j) \\
{\scriptstyle \phi_m \square \phi_{j_1} \square \cdots \square \phi_{j_m}}\downarrow & & \downarrow{\scriptstyle \phi_j} \\
\mathcal{O}'(m) \,\square\, \mathcal{O}'(j_1) \,\square\, \cdots \,\square\, \mathcal{O}'(j_m) & \xrightarrow[\;\Gamma'^m_{j_1,\dots,j_m}\;]{} & \mathcal{O}'(j)
\end{array}
\qquad
\begin{array}{ccc}
 & \mathbf{1} & \\
{\scriptstyle 1}\swarrow & & \searrow{\scriptstyle 1'} \\
\mathcal{O}(1) & \xrightarrow[\;\phi_1\;]{} & \mathcal{O}'(1).
\end{array}
$$

The endomorphism operad $\mathcal{E}nd_X$ gives an example of an operad in any closed symmetric monoidal category (for any object $X$). Here are some additional important examples.

**Example 5.2.3 (The identity operad).** Assume the symmetric monoidal category $\mathcal{M}$ has an initial object $\emptyset$. If $\square$ preserves the initial object in each variable, $\emptyset \,\square\, (-) \cong \emptyset \cong (-) \,\square\, \emptyset$ (which is automatic in the closed case, i.e., when function objects exist), we also have the example of the *identity operad* $\mathcal{I}$, which has $\mathcal{I}(1) = \mathbf{1}$ (with $\mathbf{1}$ the identity) and $\mathcal{I}(m)$ the initial object for $m \neq 1$; this is the initial object in the category of operads.

**Example 5.2.4 (The commutative algebra operad).** The operad $\mathcal{C}om$ exists in any symmetric monoidal category:

$$\mathcal{C}om(m) = \mathbf{1}$$

for all $m$ with the trivial symmetric group actions and composition law $\Gamma$ given by the unit isomorphism; its category of algebras (see the next section) is isomorphic to the category of commutative monoids for $\square$ in $\mathcal{M}$ (defined in terms of the usual diagrams, i.e., [174, VII§3] plus commutativity); see Example 5.4.3.

Example 5.2.5 (The associative algebra operad). If $\mathscr{M}$ has finite coproducts and $\square$ preserves finite coproducts in each variable, then we also have the operad $\mathcal{A}ss$:

$$\mathcal{A}ss(m) = \coprod_{\Sigma_m} \mathbf{1}$$

with symmetric group action induced by the natural (right) action of $\Sigma_m$ on $\Sigma_m$ and composition law $\Gamma$ induced by block permutation and block sum of permutations,

$$\sigma \in \Sigma_m, \tau_1 \in \Sigma_{j_1}, \ldots, \tau_m \in \Sigma_{j_m} \mapsto \sigma_{j_1,\ldots,j_m} \circ (\tau_1 \oplus \cdots \oplus \tau_m) \in \Sigma_j.$$

Its category of algebras is isomorphic to the category of monoids for $\square$ in $\mathscr{M}$; see Example 5.4.4.

For operads like $\mathcal{A}ss$, it is often useful to work in terms of *non-symmetric operads*, which come without the permutation action.

Definition 5.2.6. A *non-symmetric operad* consists of a sequence of objects $\mathcal{O}(m)$, $m = 0, 1, 2, 3, \ldots$, together with a unit map and composition rule as in 5.2.1(b) and (c) satisfying the associativity and unit rules of 5.2.1(i) and (ii). A map of non-symmetric operads consists of a map of their object sequences that commutes with the unit map and the composition rule.

Forgetting the permutation action on $\mathcal{C}om$ gives a non-unital operad called $\overline{\mathcal{A}ss}$ that is the non-symmetric version of the operad $\mathcal{A}ss$. In general, under the finite coproduct assumption in Example 5.2.5, given a non-symmetric operad $\overline{\mathcal{O}}$, the product $\overline{\mathcal{O}} \square \mathcal{A}ss$ has the canonical structure of an operad; it is the *operad associated to $\overline{\mathcal{O}}$*. In the category of spaces (or sets, but not in the category of abelian groups, the category of chain complexes, or the various categories of spectra), an operad $\mathcal{O}$ comes from a non-symmetric operad exactly when it admits a map to $\mathcal{A}ss$: the corresponding non-symmetric operad $\overline{\mathcal{O}}$ has $\overline{\mathcal{O}}(n)$ the subobject that maps to the identity permutation summand of $\mathcal{A}ss$, and there is a canonical isomorphism $\mathcal{O} \cong \overline{\mathcal{O}} \square \mathcal{A}ss$ (which depends only on the original choice of map $\mathcal{O} \to \mathcal{A}ss$).

## 5.3     $A_\infty$, $E_\infty$, and $E_n$ operads

This section reviews some of the most important classes of examples of operads in homotopy theory, the $A_\infty$, $E_\infty$, and $E_n$ operads. We concentrate on the case of (unbased) spaces, with notes about the appropriate definition of such operads in other contexts. For example, in stable homotopy theory, the unbased suspension spectrum functor $\Sigma_+^\infty$ converts model $E_n$ operads into operads in the various modern categories of spectra. The universal role played by spaces in homotopy theory typically allows for reasonable definitions of these classes of operads in any homotopy theoretic setting.

The terminology of $A_\infty$ space and the basic model of an $A_\infty$ operad, due to Stasheff [282], preceded the definition of operad by several years.

Definition 5.3.1. An $A_\infty$ operad in spaces is a non-symmetric operad whose $m$-th space is contractible for all $m$.

Informally, an operad (with symmetries) is $A_\infty$ when there is an understood isomorphism to the operad associated to some $A_\infty$ operad. The definition of $A_\infty$ operad usually has a straightforward generalization to other symmetric monoidal categories with a notion of homotopy theory: contractibility corresponds to a weak equivalence with the unit $\mathbf{1}$ of the symmetric monoidal structure, and we should add the requirement that the non-symmetric operad composition rule should be a weak equivalence for all indexes (which is automatic in spaces). One wrinkle is that a flatness condition may be needed and should be imposed to ensure that the functor $\overline{\mathcal{O}}(m) \square X^{(m)}$ is weakly equivalent to $X^{(m)}$ (cf. Section 5.9); in the case of spaces, contractibility implicitly includes such a condition. In symmetric spectra and orthogonal spectra, a good flatness condition is to be homotopy equivalent to a cofibrant object; in EKMM $S$-modules, a good flatness condition is to be homotopy equivalent to a semi-cofibrant object (see [157, §6]).

We have already seen an example of an $A_\infty$ operad: the operad $\overline{\mathcal{A}ss}$. The associahedra $\mathcal{K}(m)$ of Stasheff [282, I.§6] have the structure of a non-symmetric operad using the insertion maps [*ibid.*] for the composition rule, and this is an example of an $A_\infty$ operad. The Boardman–Vogt little 1-cubes (non-symmetric) operad $\overline{\mathcal{C}}_1$ described below gives a third example.

Next we discuss $E_\infty$ operads. Recall that a free $\Sigma_m$-cell complex is a space built by cells of the form $(\Sigma_m \times D^n, \Sigma_m \times S^{n-1})$, where $D^n$ denotes the unit disk in $\mathbb{R}^n$. The definition of $E_\infty$ operad asks for the constituent spaces to have the $\Sigma_m$-equivariant homotopy type of a free $\Sigma_m$-cell complex and the non-equivariant homotopy type of a point.

Definition 5.3.2.    An operad $\mathcal{E}$ in spaces is an $E_\infty$ operad when for each $m$, its $m$-th space is a universal $\Sigma_m$ space: $\mathcal{E}(m)$ has the $\Sigma_m$-equivariant homotopy type of a free $\Sigma_m$-cell complex and is non-equivariantly contractible.

Unlike the $A_\infty$ case, the operad $\mathcal{C}om$ is not $E_\infty$ as its spaces do not have free actions. The Barratt–Eccles operad $\mathcal{E}\Sigma$ provides an example:

Example 5.3.3 (The Barratt–Eccles operad).    Let $\mathcal{E}\Sigma(m)$ denote the nerve of the category $E\Sigma_m$ whose set of objects is $\Sigma_m$ and which has a unique map between any two objects. The symmetric group $\Sigma_m$ acts strictly on the category and the nerve $\mathcal{E}\Sigma(m)$ inherits a $\Sigma_m$-action; moreover, as the action of $\Sigma_m$ on the simplices is free, the simplicial triangulation of $\mathcal{E}\Sigma(m)$ has the structure of a free $\Sigma_m$-cell complex. It is non-equivariantly contractible because every object of $E\Sigma_m$ is a zero object. The multiplication is induced by an operad structure on the sequence of categories using block sums of permutations as in the operad structure on $\mathcal{A}ss$. The resulting operad is called the *Barratt–Eccles operad*.

Boardman and Vogt [49, §2] defined another $E_\infty$ operad, built out of linear isometries.

Example 5.3.4 (The linear isometries operad).    The Boardman–Vogt linear isometries operad $\mathcal{L}$ has its $m$-th space the space of linear isometries

$$(\mathbb{R}^\infty)^m = \mathbb{R}^\infty \oplus \cdots \oplus \mathbb{R}^\infty \to \mathbb{R}^\infty$$

(where $\mathbb{R}^\infty = \bigcup \mathbb{R}^n$), with operad structure defined as in the example of an endo-morphism operad. The topology comes from the identification

$$\mathcal{L}(m) = \lim_k \operatorname{colim}_n \mathcal{I}((\mathbb{R}^k)^m, \mathbb{R}^n)$$

for $\mathcal{I}((\mathbb{R}^k)^m, \mathbb{R}^n)$ the space of linear isometries $(\mathbb{R}^k)^m \to \mathbb{R}^n$ (with the usual manifold topology). The $\Sigma_m$-action induced by the action on the direct sum $(\mathbb{R}^\infty)^m$ is clearly free; each $\mathcal{I}((\mathbb{R}^k)^m, \mathbb{R}^n)$ is a $\Sigma_m$-manifold, and $\mathcal{L}(m)$ is homotopy equivalent to a free $\Sigma_m$-cell complex. Since $\mathcal{I}((\mathbb{R}^k)^m, \mathbb{R}^n)$ is $(n - km - 1)$-connected, it follows that $\mathcal{L}(m)$ is non-equivariantly contractible.

The Boardman–Vogt little $\infty$-cubes operad $\mathcal{C}_\infty$ described below gives a third example of an $E_\infty$ operad.

The requirement for freeness derives from infinite loop space theory. As we review in Section 5.11, infinite loop spaces are algebras for the little $\infty$-cubes operad $\mathcal{C}_\infty$. As we review in Section 5.9, for any $E_\infty$ operad $\mathcal{E}$ in spaces, the category of $\mathcal{E}$-algebras has an equivalent homotopy theory to the category of $\mathcal{C}_\infty$-algebras. On the other hand, any algebra in spaces for the operad $\mathcal{C}om$ must be a generalized Eilenberg–Mac Lane space, and the category of $\mathcal{C}om$-algebras does not have an equivalent homotopy theory to the category of $\mathcal{C}_\infty$-algebras. In generalizing the notion of $E_\infty$ to other categories, getting the right category of algebras is key. For symmetric spectra, orthogonal spectra, and EKMM $S$-modules and for chain complexes of modules over a ring containing the rational numbers, it is harmless to allow $\mathcal{C}om$ to fit the definition of $E_\infty$ operad (cf. Examples 5.9.3, 5.9.4); in spaces and chain complexes of modules over a finite field, some freeness condition is required. In general, the condition should be a flatness condition on $\mathcal{O}(m)$ for $(\mathcal{O}(m) \square X^{(m)})/\Sigma_m$ as a functor of $X$ (for suitable $X$) (cf. Definition 5.9.1).

Unlike the definition of $E_\infty$ or $A_\infty$ operad, which are defined in terms of homotopical conditions on the constituent spaces, the definition of $E_n$ operads for other $n$ depends on specific model operads first defined by Boardman–Vogt [49] and called the *little n-cubes operads* $\mathcal{C}_n$.

**Construction 5.3.5** (The little $n$-cubes operad). The $m$-th space $\mathcal{C}_n(m)$ of the little $n$-cubes operad is the space of $m$ ordered almost disjoint parallel axis affine embeddings of the unit $n$-cube $[0, 1]^n$ in itself. So $\mathcal{C}_n(0)$ is a single point representing the unique way to embed $0$ unit $n$-cubes in the unit $n$-cube. A parallel axis affine embedding of the unit cube in itself is a map of the form

$$(t_1, \ldots, t_n) \in [0, 1]^n \mapsto (x_1 + a_1 t_1, \ldots, x_n + a_n t_n) \in [0, 1]^n$$

for some fixed $(x_1, \ldots, x_n)$ and $(a_1, \ldots, a_n)$ with each $a_i > 0$, $x_i \geq 0$, and $x_i + a_i \leq 1$; it is determined by the point $(x_1, \ldots, x_n)$ where it sends $(0, \ldots, 0)$ and the point

$$(y_1, \ldots, y_n) = (x_1 + a_1, \ldots, x_n + a_n)$$

where it sends $(1,\ldots,1)$. So $\mathcal{C}_n(1)$ is homeomorphic to the subspace

$$\{((x_1,\ldots,x_n),(y_1,\ldots,y_n)) \in [0,1]^n \times [0,1]^n \mid x_1 < y_1, x_2 < y_2,\ldots,x_n < y_n\}$$

of $[0,1]^n \times [0,1]^n$. For $m \geq 2$, almost disjoint means that the images of the open subcubes are disjoint (the embedded cubes only intersect on their boundaries), and $\mathcal{C}_n(m)$ is homeomorphic to a subset of $\mathcal{C}_n(1)^m$. The map 1 is specified by the element of $\mathcal{C}_n(1)$ that gives the identity embedding of the unit $n$-cube. The action of the symmetric group is to re-order the embeddings. The composition law $\Gamma^m_{j_1,\ldots,j_m}$ composes the $j_1$ embeddings in $\mathcal{C}_n(j_1)$ with the first embedding in $\mathcal{C}_n(m)$, the $j_2$ embeddings in $\mathcal{C}_n(j_2)$ with the second embedding in $\mathcal{C}_n(m)$, etc., to give $j = j_1 + \cdots + j_m$ total embeddings. See Figure 5.3 for a picture in the case $n = 2$. Taking cartesian product with the identity map on $[0,1]$ takes a self-embedding of the unit $n$-cube to a self-embedding of the unit $(n+1)$-cube and induces maps of operads $\mathcal{C}_n \to \mathcal{C}_{n+1}$ that are closed inclusions of the underlying spaces. Let $\mathcal{C}_\infty(m) = \bigcup \mathcal{C}_n(m)$; the operad structures on the $\mathcal{C}_n$ fit together to define an operad structure on $\mathcal{C}_\infty$.

The space $\mathcal{C}_n(m)$ has the $\Sigma_m$-equivariant homotopy type of the configuration space $C(m, \mathbb{R}^n)$ of $m$ (ordered) points in $\mathbb{R}^n$, or equivalently, $C(m, (0,1)^n)$ of $m$ points in $(0,1)^n$. To see this, since both spaces are free $\Sigma_m$-manifolds (non-compact, and with boundary in the case of $\mathcal{C}_n(m)$), it is enough to show that they are non-equivariantly weakly equivalent, but it is in fact no harder to produce a $\Sigma_m$-equivariant homotopy equivalence explicitly. We have a $\Sigma_m$-equivariant map $\mathcal{C}_n(m) \to C(m, (0,1)^n)$ by taking the center point of each embedded subcube. It is easy to define a $\Sigma_m$-equivariant section of this map by continuously choosing cubes centered on the given configuration; one way to do this is to make them all have the same equal side length of $1/2$ of the minimum of the distance between each of the points and the distance from each point to the boundary of $[0,1]^n$. A $\Sigma_m$-equivariant homotopy from the composite map on $\mathcal{C}_n(m)$ to the identity could (for example) first linearly shrink all sides that are bigger than their original length and then linearly expand all remaining sides to their original length. In particular, $\mathcal{C}_n(1)$ is always contractible and $\mathcal{C}_n(2)$ is $\Sigma_2$-equivariantly homotopy equivalent to the sphere $S^{n-1}$ with the antipodal action. For

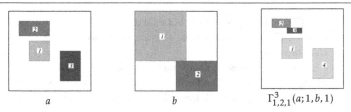

$a$ $\qquad\qquad\qquad$ $b$ $\qquad\qquad\qquad$ $\Gamma^3_{1,2,1}(a;1,b,1)$

**Figure 5.3** Composition of little 2-cubes. Shown is the composition
$$\Gamma^3_{1,2,1}: \mathcal{C}_2(3) \times \mathcal{C}_2(1) \times \mathcal{C}_2(2) \times \mathcal{C}_2(1) \to \mathcal{C}_2(4)$$
applied to elements $a \in \mathcal{C}_2(3)$, $1 \in \mathcal{C}_2(1)$, $b \in \mathcal{C}_2(2)$, $1 \in \mathcal{C}_2(1)$, with $a$ and $b$ as pictured.

$m > 2$, the configuration spaces can be described in terms of iterated fibrations, and their Borel homology was calculated by Cohen in [70] and [71, IV].

We can say more about the homotopy types in the cases $n = 1$, $n = 2$, and $n = \infty$. For $n = 1$, the natural order of the interval $[0, 1]$, gives a natural order to the embedded sub-intervals (1-cubes); let $\bar{C}_1(m)$ denote the subspace of $C_1(m)$ where the sub-intervals are numbered in their natural order. The spaces $\bar{C}_1(m)$ are contractible and form a non-symmetric operad with $C_1$ (canonically) isomorphic to the associated operad. In other words, the map of operads $C_1 \to \mathcal{A}ss$ that takes a sequence of embeddings and just remembers the order they come in is a $\Sigma_m$-equivariant homotopy equivalence at each level. In particular $C_1$ is an $A_\infty$ operad.

For $n = 2$, the configuration space $C(m, \mathbb{R}^2)$ is easily seen to be an Eilenberg–Mac Lane space $K(A_m, 1)$, where $A_m$ is the pure braid group (of braids with fixed endpoints) on $m$ strands (see, for example, [194, §4]).

For $n = \infty$, $C_\infty$ is an $E_\infty$ operad; each $C_\infty(m)$ is a universal $\Sigma_m$-space. To see this, it is easier to work with

$$C(m, \mathbb{R}^\infty) := \bigcup C(m, \mathbb{R}^n).$$

Choosing a homeomorphism $(0, 1) \cong \mathbb{R}$ that sends $1/2$ to $0$, the induced homeomorphisms $C_n(m) \to C(m, \mathbb{R}^n)$ are compatible with the inclusions $C_n(m) \to C_{n+1}(m)$ and $C(m, \mathbb{R}^n) \to C(m, \mathbb{R}^{n+1})$; as these inclusions are embeddings of closed submanifolds (with boundary in the case of $C_n(m)$), the induced map

$$C_\infty(m) = \bigcup C_n(m) \to \bigcup C(m, \mathbb{R}^n) = C(m, \mathbb{R}^\infty)$$

remains a homotopy equivalence. One way to see that $C(m, \mathbb{R}^\infty)$ is non-equivariantly contractible is to start by choosing a homotopy though injective linear maps from the identity on $\mathbb{R}^\infty$ to the shift map that on basis elements sends $e_i$ to $e_{i+m}$. We then homotope the configuration (which now starts with the first $m$ coordinates all zero) so that the $i$-th point has $i$-th coordinate 1 and the remainder of the first $m$ coordinates zero. Finally, we homotope the configuration to the configuration with $i$-th point at $e_i$.

We use the operads $C_n$ to define $E_n$ operads:

**Definition 5.3.6.** An operad $\mathcal{E}$ in spaces is an $E_n$ operad when there is a zigzag of maps of operads relating it to $C_n$, each of which is a $\Sigma_m$-equivariant homotopy equivalence on $m$-th spaces for all $m$.

This definition is standard, but a bit awkward, because it defines a property, whereas a better definition would define a structure and ask for at least a preferred equivalence class of zigzag.

As we review in Section 5.9, such maps induce equivalences of homotopy categories of algebras (indeed, Quillen equivalences). We have implicitly given two different definitions of $E_\infty$ operad; the following proposition justifies this.

**Proposition 5.3.7.** *An operad $\mathcal{E}$ of spaces is $E_\infty$ in the sense of Definition 5.3.2 if and only if it is $E_\infty$ in the sense of Definition 5.3.6.*

Before reviewing the proof, we state a closely related proposition.

**Proposition 5.3.8.**  *An operad $\mathcal{E}$ of spaces is $E_1$ if and only if it is isomorphic to the associated operad of an $A_\infty$ operad.*

The previous two propositions (and their common proof) are the gist of the second half of §3 of May [194]. In each case one direction is clear, since $\mathcal{C}_1$ and $\mathcal{C}_\infty$ are $A_\infty$ and $E_\infty$ (respectively), and the conditions of Definitions 5.3.1 and 5.3.2 are preserved by the zigzags considered in Definition 5.3.6. The proof of the other direction is to exhibit an explicit zigzag:

*Proof.* Let $\mathcal{E}$ be the operad in question and assume it is either $E_\infty$ in the sense of Definition 5.3.2 (for the first proposition) or $A_\infty$ in the sense of Definition 5.3.1 ff. (for the second proposition). In the case of the first proposition, consider the product in the category of operads $\mathcal{C}_\infty \times \mathcal{E}$; it satisfies

$$(\mathcal{C}_\infty \times \mathcal{E})(m) = \mathcal{C}_\infty(m) \times \mathcal{E}(m)$$

with the diagonal $\Sigma_m$-action and the unit and composition maps the product of those for $\mathcal{C}_\infty$ and $\mathcal{E}$. The projections

$$\mathcal{C}_\infty \leftarrow \mathcal{C}_\infty \times \mathcal{E} \rightarrow \mathcal{E}$$

give a zigzag as required by Definition 5.3.6. For the second proposition, do the same trick with the non-symmetric operads $\bar{\mathcal{E}}$ and $\bar{\mathcal{C}}_1$ and then pass to the associated operads.                    □

Definitions 5.3.1 and 5.3.2 mean that identifying $A_\infty$ and $E_\infty$ operads is pretty straightforward. In unpublished work, Fiedorowicz [98] defines the notion of a *braided operad*, which provides a good criterion for identifying $E_2$ operads. For $n > 2$ (finite), the spaces $\mathcal{C}_n(m)$ are not Eilenberg–Mac Lane spaces (for $m > 1$), and that makes identification of such operads much harder; however, Berger [36, 1.16] proves a theorem (which he attributes to Fiedorowicz) that gives a method to identify $E_n$ operads that seems to work well in practice; see [205, §14], [37, §1.6].

The work of Dunn [85] and Fiedorowicz–Vogt [97] is the start of an abstract identification of $E_n$ operads: The derived tensor product of $n$ $E_1$ operads is an $E_n$ operad. Here "tensor product" refers to the Boardman–Vogt tensor product of operads (or PROPs) in [48, 2§3], which is the universal pairing subject to "interchange", meaning that an $\mathcal{O} \otimes \mathcal{P}$-algebra structure consists of an $\mathcal{O}$-algebra and a $\mathcal{P}$-algebra structure on a space where the $\mathcal{O}$- and $\mathcal{P}$-structure maps commute (see *ibid.* for more details on the construction of the tensor product). This still essentially defines $E_n$ operads in terms of reference models, though in principle, it gives a wide range of additional models. (I do not know an example where this is actually put to use, but [62] comes close.) The concept of interchange makes sense in any cartesian symmetric monoidal structure, so this also in principle tells how to extend the notion of $E_n$ to other cartesian symmetric monoidal categories with a homotopy theory of operads for which the Boardman–Vogt tensor product is reasonably well-behaved. (Again, I know no examples where this is put to use, but perhaps work by Barwick (unpublished), Gepner (unpublished), and Lurie [164] on $E_n$ structures is in a similar spirit.)

In categories suitably related to spaces, $E_n$ algebras are defined by a reference model suitably related to $C_n$. For example, in the context of simplicial sets, the total singular complex of the little $n$-cubes operad has the canonical structure of an operad of simplicial sets, and we define $E_n$ operads in terms of this reference model. In symmetric spectra and orthogonal spectra, we have the reference model given by the unbased suspension spectrum functor: an operad is an $E_n$ operad when it is related to $\Sigma_+^\infty C_n$ by a zigzag of operad maps that are (non-equivariant) weak equivalences on $m$-th objects for all $m$. For categories of chain complexes, we use the singular chain complex of the little $n$-cubes operad to define the reference model. To make the singular chains an operad, we use the Eilenberg–Mac Lane shuffle map to relate tensor product of chains to chains on the cartesian product; the shuffle map is a lax symmetric monoidal natural transformation

$$C_*(X) \otimes C_*(Y) \to C_*(X \times Y),$$

meaning that it commutes strictly with the symmetry isomorphisms

$$C_*(X) \otimes C_*(Y) \cong C_*(Y) \otimes C_*(X) \quad \text{and} \quad C_*(X \times Y) \cong C_*(Y \times X)$$

and makes the following associativity diagram commute:

$$
\begin{array}{ccc}
C_*(X) \otimes C_*(Y) \otimes C_*(Z) & \longrightarrow & C_*(X \times Y) \otimes C_*(Z) \\
\downarrow & & \downarrow \\
C_*(X) \otimes C_*(Y \times Z) & \longrightarrow & C_*(X \times Y \times Z)
\end{array}
$$

See, for example, [200, §29].

The fact that $E_n$ operads need to be defined in terms of a reference model is not entirely satisfactory, especially in homotopical contexts that are not topological. Nevertheless, the definition for spaces, simplicial sets, or chain complexes seems to suffice to cover all other contexts that arise in practice.[2]

## 5.4    Operadic algebras and monads

In the original context of iterated loop spaces and in many current contexts in homotopy theory and beyond, the main purpose of operads is to parametrize operations, which is to say, to define operadic algebras. For a closed symmetric monoidal category, there are three equivalent definitions, one in terms of operations, one in terms of endomorphism operads, and one in terms of monads. This section reviews the three definitions.

Viewing $\mathcal{O}(m)$ as parametrizing some $m$-ary operations on an object $X$ means that we have an *action map*

$$\mathcal{O}(m) \,\square\, X^{(m)} \to X.$$

---

[2]  In theory, the definition for simplicial sets should suffice for all homotopical contexts, but this may require changing models, which for a particular problem may be inconvenient or more complicated, or make it less concrete.

Since the right action of $\Sigma_m$ on $\mathcal{O}(m)$ corresponds to reordering the arguments of the operations, applying $\sigma \in \Sigma_m$ to $\mathcal{O}(m)$ (and then performing the action map) should have the same effect as applying $\sigma$ to permute the factors in $X^{(m)}$. A concise way of saying this is to say that the map is equivariant for the diagonal (left) action on the source $\mathcal{O}(m) \square X^{(m)}$ and the trivial action on the target $X$ (using the standard convention that the left action $\sigma$ on $\mathcal{O}(m)$ is given by the right action of $\sigma^{-1}$). The action map should also respect the composition law $\Gamma$, making $\Gamma$ correspond to composition of operations, and respect the identity 1, making 1 act by the identity operation. The following gives the precise definition:

**Definition 5.4.1.** Let $\mathcal{M}$ be a symmetric monoidal category and $\mathcal{O} = (\{\mathcal{O}(m)\}, \Gamma, 1)$ an operad in $\mathcal{M}$. An $\mathcal{O}$-algebra (in $\mathcal{M}$) consists of an object $A$ in $\mathcal{M}$ together with *action maps*

$$\xi_m \colon \mathcal{O}(m) \square A^{(m)} \to A$$

that are equivariant for the diagonal (left) $\Sigma_m$-action on the source and the trivial $\Sigma_m$-action on the target and that satisfy the following associativity and unit conditions:

(i) For all $m$, $j_1, \ldots, j_m$,

$$\xi_m \circ (\mathrm{id}_{\mathcal{O}(m)} \square \xi_{j_1} \square \cdots \square \xi_{j_m}) = \xi_j \circ (\Gamma^m_{j_1,\ldots,j_m} \square \mathrm{id}_A^{(j)}),$$

i.e., the diagram

$$
\begin{array}{ccc}
\mathcal{O}(m) \square \mathcal{O}(j_1) \square \cdots \square \mathcal{O}(j_m) \square A^{(j)} & \xrightarrow{\Gamma^m_{j_1,\ldots,j_m} \square \mathrm{id}_A^{(j)}} & \mathcal{O}(j) \square A^{(j)} \\
{\scriptstyle \mathrm{id}_{\mathcal{O}(m)} \square \xi_{j_1} \square \cdots \square \xi_{j_m}} \downarrow & & \downarrow {\scriptstyle \xi_j} \\
\mathcal{O}(m) \square A^{(m)} & \xrightarrow{\xi_m} & A
\end{array}
$$

commutes.

(ii) The map $\xi_1 \circ (1 \square \mathrm{id}_A) \colon 1 \square A \to A$ is the unit isomorphism for $\square$.

A map of $\mathcal{O}$-algebras from $(A, \{\xi_m\})$ to $(A', \{\xi'_m\})$ consists of a map $f \colon A \to A'$ in $\mathcal{M}$ that commutes with the action maps, i.e., that make the diagrams

$$
\begin{array}{ccc}
\mathcal{O}(m) \square A^{(m)} & \xrightarrow{\xi_m} & A \\
{\scriptstyle \mathrm{id}_{\mathcal{O}(m)} \square f^{(m)}} \downarrow & & \downarrow {\scriptstyle f} \\
\mathcal{O}(m) \square A'^{(m)} & \xrightarrow{\xi'_m} & A'
\end{array}
$$

commute for all $m$. We write $\mathcal{M}[\mathcal{O}]$ for the category of $\mathcal{O}$-algebras.

**Example 5.4.2.** When $\mathcal{M}$ has an initial object and $\square$ preserves the initial object in each variable, the structure of an algebra over the identity operad $\mathcal{I}$ is no extra structure on an object of $\mathcal{M}$.

Per (ii) above and as illustrated in the previous example, the 1 in the structure of the operad corresponds to the identity operation. In some contexts algebras have units; when that happens, the unit is encoded in $\mathcal{O}(0)$ as in the examples of monoids and commutative monoids. Recall that a *monoid object for* $\square$ *in* $\mathcal{M}$ (or $\square$-*monoid* for short) consists of an object $M$ together with a *multiplication map* $\mu\colon M \square M \to M$ and *unit map* $\eta\colon 1 \to M$ satisfying the following associativity and unit diagrams

$$
\begin{array}{ccc}
M \square M \square M & \xrightarrow{\mu\square\mathrm{id}} & M \square M \\
\mathrm{id}\square\mu \downarrow & & \downarrow \mu \\
M \square M & \xrightarrow{\mu} & M
\end{array}
\qquad
\begin{array}{ccccc}
1 \square M & \xrightarrow{\eta\square\mathrm{id}} & M \square M & \xleftarrow{\mathrm{id}\square\eta} & M \square 1 \\
& {}_{\cong}\searrow & \downarrow \mu & \swarrow_{\cong} & \\
& & M & &
\end{array}
$$

(where the diagonal maps are the unit isomorphisms in $\mathcal{M}$). The *opposite multiplication* is the composite of the symmetry morphism $c\colon M \square M \to M \square M$ with $\mu$, and a $\square$-monoid is *commutative* when $\mu = \mu \circ c$.

**Example 5.4.3.** Given a $\mathcal{C}$om-algebra $A$, defining $\eta$ to be the action map $\xi_0$

$$\eta\colon 1 = \mathcal{C}\mathrm{om}(0) \xrightarrow{\xi_0} A$$

and $\mu$ to be the composite of the (inverse) unit isomorphism and the action map $\xi_2$

$$\mu\colon A \square A \cong \mathcal{C}\mathrm{om}(2) \square A \square A \xrightarrow{\xi_2} A$$

endows $A$ with the structure of a commutative $\square$-monoid: associativity follows from the fact that the maps $\Gamma^2_{1,2}$ and $\Gamma^2_{2,1}$ are both unit maps for $\square$ so under the canonical isomorphisms

$$A \square A \square A \cong \mathcal{C}\mathrm{om}(2) \square (\mathcal{C}\mathrm{om}(1) \square \mathcal{C}\mathrm{om}(2)) \square (A \square A \square A),$$
$$A \square A \square A \cong \mathcal{C}\mathrm{om}(2) \square (\mathcal{C}\mathrm{om}(2) \square \mathcal{C}\mathrm{om}(1)) \square (A \square A \square A),$$

both maps induce the same map $A \square A \square A \to A$. Likewise, the unit condition follows from the fact that

$$\Gamma^2_{0,1}\colon \mathcal{C}\mathrm{om}(2) \square (\mathcal{C}\mathrm{om}(0) \square \mathcal{C}\mathrm{om}(1)) \to \mathcal{C}\mathrm{om}(1),$$
$$\Gamma^2_{1,0}\colon \mathcal{C}\mathrm{om}(2) \square (\mathcal{C}\mathrm{om}(1) \square \mathcal{C}\mathrm{om}(0)) \to \mathcal{C}\mathrm{om}(1)$$

are both unit maps. The multiplication is commutative because the action of the symmetry map on $1 = \mathcal{C}\mathrm{om}(2)$ is trivial. Conversely, we can convert a commutative $\square$-monoid to a $\mathcal{C}$om-algebra by taking $\xi_0$ to be the unit $\eta$, $\xi_1$ to be the unit isomorphism for $\square$, $\xi_2$ to be induced by the unit isomorphism for $\square$ and the multiplication, and all higher $\xi_m$'s induced by the unit isomorphism for $\square$ and (any) iterated multiplication. This defines a bijective correspondence between the set of commutative $\square$-monoid structures and the set of $\mathcal{C}$om-algebra structures on a fixed object and an isomorphism between the category of commutative $\square$-monoids and the category of $\mathcal{C}$om-algebras.

For a non-symmetric operad, defining an algebra in terms of the associated operad or in terms of the analogue of Definition 5.4.1 without the equivariance requirement produce the same structure.

**Example 5.4.4.** The constructions of Example 5.4.3 applied to the non-symmetric operad $\overline{\mathcal{A}ss}$ give a bijective correspondence between the set of $\square$-monoid structures and the set of $\mathcal{A}ss$-algebra structures on a fixed object and an isomorphism between the category of $\square$-monoids and the category of $\mathcal{A}ss$-algebras.

The monoid and commutative monoid objects in the category of sets (with the usual symmetric monoidal structure given by cartesian product) are just the monoids and commutative monoids in the usual sense. Likewise, in spaces, they are the topological monoids and topological commutative monoids. In the category of abelian groups (with the usual symmetric monoidal structure given by the tensor product), the monoid objects are the rings and the commutative monoid objects are the commutative rings. In the category of chain complexes of $R$-modules for a commutative ring $R$ (with the usual symmetric monoidal structure given by tensor product over $R$), the monoid objects are the differential graded $R$-algebras and the commutative monoid objects are the commutative differential graded $R$-algebras. In a modern category of spectra, the monoid objects are called $\mathbb{S}$-*algebras* or sometimes *strictly associative ring spectra*. Some authors take the term "ring spectrum" to be synonymous with $\mathbb{S}$-algebra, but others take it to mean the weaker notion of monoid object in the stable category (or even weaker notions). Work of Schwede–Shipley [265, 3.12.(3)] shows that the homotopy category of monoid objects in any modern category of spectra is equivalent to an appropriate full subcategory of the (mutually equivalent) homotopy category of monoid objects in EKMM $S$-modules, symmetric spectra, or orthogonal spectra (at least when "modern category of spectra" is used as a technical term to mean a model category with a preferred equivalence class of symmetric monoidal Quillen equivalence to the currently known modern categories of spectra); cf. Example Theorem 5.9.6 below. The analogous result does not hold for commutative monoid objects; see [151]. The term "commutative $\mathbb{S}$-algebra" is typically reserved for examples where the homotopy category of commutative monoid objects is equivalent to an appropriate full subcategory of the (mutually equivalent) homotopy category of commutative monoid objects in EKMM $S$-modules, symmetric spectra, or orthogonal spectra. See Chapter 6 of this volume for more on commutative ring spectra.

Returning to the discussion of operadic algebras, in the case when $\mathcal{M}$ is a closed symmetric monoidal category, adjoint to the action map

$$\xi_m \colon \mathcal{O}(m) \,\square\, A^{(m)} \to A$$

is a map

$$\phi_m \colon \mathcal{O}(m) \to F(A^{(m)}, A) = \mathcal{E}\mathrm{nd}_A(m).$$

Equivariance for $\xi_m$ is equivalent to equivariance for $\phi_m$. Similarly, conditions (i) and (ii) in the definition of $\mathcal{O}$-algebra (Definition 5.4.1) are adjoint to the diagrams in the definition of map of operads (Definition 5.2.2). This proves the following proposition, which gives an alternative definition of $\mathcal{O}$-algebra.

**Proposition 5.4.5.** *Let $\mathcal{M}$ be a closed symmetric monoidal category, let $\mathcal{O}$ be an operad in $\mathcal{M}$, and let $X$ be an object in $\mathcal{M}$. The adjunction rule $\xi_m \leftrightarrow \phi_m$ above defines a bijection between the set of $\mathcal{O}$-algebra structures on $X$ and the set of maps of operads $\mathcal{O} \to \mathcal{E}\mathrm{nd}_X$.*

In the case when $\mathcal{M}$ is (countably) cocomplete (has (countable) colimits) and $\square$ preserves (countable) colimits in each variable (which includes the case when it is closed), algebras can also be defined in terms of a monad. The idea for the underlying functor is to gather the domains of all the action maps into a coproduct; since the action maps are equivariant with target having the trivial action, they factor through the coinvariants (quotient by the symmetric group action), and this goes into the definition.

**Notation 5.4.6.** Let $\mathcal{M}$ be a symmetric monoidal category with countable colimits, and let $\mathcal{O}$ be an operad in $\mathcal{M}$. Define the endofunctor $\mathbb{O}$ of $\mathcal{M}$ (i.e., a functor $\mathbb{O} \colon \mathcal{M} \to \mathcal{M}$) by

$$\mathbb{O}X = \coprod_{m=0}^{\infty} \mathcal{O}(m) \,\square_{\Sigma_m}\, X^{(m)},$$

where $\mathcal{O}(m) \,\square_{\Sigma_m}\, X^{(m)} := (\mathcal{O}(m) \,\square\, X^{(m)})/\Sigma_m$.

(When we use other letters for operads, we typically use the corresponding letters for the associated monad; for example, we write $\mathbb{A}$ for the monad associated to an operad $\mathcal{A}$, $\mathbb{B}$ for the monad associated to an operad $\mathcal{B}$, etc.)

The action maps for an $\mathcal{O}$-algebra $A$ then specify a map $\xi \colon \mathbb{O}A \to A$; the conditions for defining an $\mathcal{O}$-structure also admit a formulation in terms of this map. The basic observation is that we have a canonical isomorphism

$$(\mathbb{O}X)^{(m)} \cong \coprod_{j_1=0}^{\infty} \cdots \coprod_{j_m=0}^{\infty} (\mathcal{O}(j_1) \,\square_{\Sigma_{j_1}}\, X^{(j_1)}) \,\square \cdots \square\, (\mathcal{O}(j_m) \,\square_{\Sigma_{j_m}}\, X^{(j_m)})$$

$$\cong \coprod_{j=0}^{\infty} \coprod_{\substack{j_1,\dots,j_m \\ \Sigma j_i = j}} (\mathcal{O}(j_1) \,\square \cdots \mathcal{O}(j_m)) \,\square_{\Sigma_{j_1} \times \cdots \times \Sigma_{j_m}}\, X^{(j)},$$

using the symmetry isomorphism to shuffle like factors without permuting them. We can use this isomorphism to give $\mathbb{O}X$ the canonical structure of an $\mathcal{O}$-algebra, defining the action map

$$\mu_m \colon \mathcal{O}(m) \,\square\, (\mathbb{O}X)^{(m)} \to \mathbb{O}X$$

by commuting the coproduct past $\square$, using the operad composition law, and passing to the quotient by the full permutation group:

$$\mathcal{O}(m) \,\square\, (\mathbb{O}X)^{(m)} \cong \coprod_{j=0}^{\infty} \coprod_{\substack{j_1,\dots,j_m \\ \Sigma j_i = j}} \mathcal{O}(m) \,\square\, (\mathcal{O}(j_1) \,\square \cdots \mathcal{O}(j_m)) \,\square_{\Sigma_{j_1} \times \cdots \times \Sigma_{j_m}}\, X^{(j)}$$

$$\xrightarrow{\;\coprod\coprod \Gamma^m_{j_1,\dots,j_m} \,\square\, \mathrm{id}_X^{(j)}\;} \coprod_{j=0}^{\infty} \mathcal{O}(j) \,\square_{\Sigma_{j_1} \times \cdots \times \Sigma_{j_m}}\, X^{(j)} \longrightarrow \coprod_{j=0}^{\infty} \mathcal{O}(j) \,\square_{\Sigma_j}\, X^{(j)} = \mathbb{O}X.$$

The pictured map is well-defined because of the $(\Sigma_{j_1} \times \cdots \times \Sigma_{j_m})$-equivariance of $\Gamma^m_{j_1,\ldots,j_m}$ (5.2.1(iii)). The other permutation rule (5.2.1(iv)) implies that $\mu_m$ is $\Sigma_m$-equivariant. The remaining two parts of the definition of operad show that the $\mu_m$ define an $\mathcal{O}$-algebra structure map: 5.2.1(i)–(ii) imply 5.4.1(i)–(ii), respectively. This $\mathcal{O}$-algebra structure then defines a map

$$\mu \colon \mathbb{OO}X \to \mathbb{O}X$$

as above, which is natural in $X$. The map $1 \,\square\, \mathrm{id}_X$ also induces a natural transformation

$$\eta \colon X \to \mathbb{O}X.$$

These two maps together give $\mathbb{O}$ the structure of a monad.

**Proposition 5.4.7.** *Let $\mathcal{M}$ be a symmetric monoidal category with countable colimits and assume that $\square$ commutes with countable colimits in each variable. For an operad $\mathcal{O}$, the functor $\mathbb{O}$ and natural transformations $\mu$, $\eta$ form a monad: the diagrams*

$$
\begin{array}{ccc}
\mathbb{OOO}X & \xrightarrow{\;\mu\;} & \mathbb{OO}X \\
{\scriptstyle \mathbb{O}\mu}\big\downarrow & & \big\downarrow{\scriptstyle \mu} \\
\mathbb{OO}X & \xrightarrow[\;\mu\;]{} & \mathbb{O}X
\end{array}
\qquad\qquad
\begin{array}{ccc}
\mathbb{O}X & \xrightarrow{\;\eta\;} & \mathbb{OO}X \\
 & \searrow & \big\downarrow{\scriptstyle \mu} \\
 & & \mathbb{O}X
\end{array}
$$

*commute (where the top map in the left-hand diagram is the map $\mu$ for the object $\mathbb{O}X$).*

The proof is applying 5.4.1(i)–(ii) for $\mathbb{O}X$.

**Example 5.4.8.** Under the hypotheses of the previous proposition, the monad associated to the identity operad $\mathcal{I}$ is canonically isomorphic (via the unit isomorphism) to the identity monad Id. The monad associated to the operad $\mathcal{C}$om is canonically isomorphic to the free commutative monoid monad

$$\mathbb{P}X = \coprod_{j=0}^{\infty} X^{(j)}/\Sigma_j.$$

The monad associated to the algebra $\mathcal{A}$ss is canonically isomorphic to the free monoid monad

$$\mathbb{T}X = \coprod_{j=0}^{\infty} X^{(j)}.$$

An algebra over the monad $\mathbb{O}$ consists of an object $A$ and a map $\xi \colon \mathbb{O}A \to A$ such that the diagrams

$$
\begin{array}{ccc}
\mathbb{OO}A & \xrightarrow{\;\mu\;} & \mathbb{O}A \\
{\scriptstyle \mathbb{O}\xi}\big\downarrow & & \big\downarrow{\scriptstyle \xi} \\
\mathbb{O}A & \xrightarrow[\;\xi\;]{} & A
\end{array}
\qquad\qquad
\begin{array}{ccc}
A & \xrightarrow{\;\eta\;} & \mathbb{O}A \\
 & \searrow & \big\downarrow{\scriptstyle \xi} \\
 & & A
\end{array}
$$

commute. Given an $\mathcal{O}$-algebra $(A, \{\xi_m\})$, the map $\xi \colon \mathbb{O}A \to A$ constructed as the coproduct of the induced maps on coinvariants then is an $\mathcal{O}$-algebra action map. Conversely, given an $\mathcal{O}$-algebra $(A, \xi)$, defining $\xi_m$ to be the composite

$$\mathcal{O}(m) \,\square\, A^{(m)} \to \mathcal{O}A \xrightarrow{\xi} A,$$

the maps $\xi_m$ make $A$ an $\mathcal{O}$-algebra. This gives a second alternative definition of $\mathcal{O}$-algebra.

**Proposition 5.4.9.** *Under the hypotheses of Proposition 5.4.7, for $X$ an object of $\mathcal{M}$, the correspondence $\{\xi_m\} \leftrightarrow \xi$ above defines a bijection between the set of $\mathcal{O}$-algebra structures on $X$ and the set of $\mathbb{O}$-algebra structures on $X$ and an isomorphism between the category of $\mathcal{O}$-algebras and the category of $\mathbb{O}$-algebras.*

## 5.5     Modules over operadic algebras

Just as an operad defines a category of algebras, an algebra defines a category of modules. Because this chapter concentrates on the theory of operadic algebras, we will only touch on the theory of modules. A complete discussion could fill a book and many of the aspects of the theory of operads we omit in this chapter (including Koszul duality, Quillen (co)homology, Deligne and Kontsevich conjectures) correspond to statements about categories of modules; even an overview could form its own chapter. We will just give a brief review of the definitions and the homotopy theory.

The original definition of modules over an operadic algebra seems to be due to Ginzburg and Kapranov [104, §1.6].

**Definition 5.5.1.** For an operad $\mathcal{O}$ and an $\mathcal{O}$-algebra $A$, an $(\mathcal{O}, A)$-module (or just $A$-module when $\mathcal{O}$ is understood) consists of an object $M$ of $\mathcal{M}$ and structure maps

$$\zeta_m \colon \mathcal{O}(m+1) \,\square\, (A^{(m)} \,\square\, M) \to M$$

for $m \geq 0$ such that the associativity, symmetry, and unit diagrams in Figure 5.4 commute. A map of $A$-modules is a map of the underlying objects of $\mathcal{M}$ that commutes with the structure maps.

Although the definition appears to favor $A$ on the left, we obtain analogous right-hand structure maps

$$\mathcal{O}(m+1) \,\square\, (M \,\square\, A^{(m)}) \to M$$

satisfying the analogous right-hand version of the diagrams in Figure 5.4 by applying an appropriate permutation. Thus, an $A$-module structure can equally be regarded as either a left or right module structure. The following example illustrates this point.

**Example 5.5.2.** When $\mathcal{O} = \mathcal{A}$ss, the (symmetric) operad for associative algebras, and $A$ is an $\mathcal{O}$-algebra (i.e., $\square$-monoid), an $(\mathcal{O}, A)$-module in the sense of the previous definition is precisely an $A$-bimodule in the usual sense: it has structure maps

$$\lambda \colon A \,\square\, M \to M \qquad \text{and} \qquad \rho \colon M \,\square\, A \to M$$

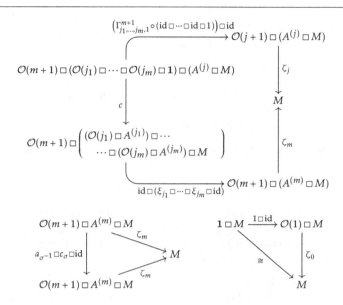

Figure 5.4 The diagrams for Definition 5.5.1. In the first diagram, $j = j_1 + \cdots + j_m$ and $c$ is the $\square$-permutation that shuffles the $\mathcal{O}(j_i)$'s past the $M$ and $A$'s as displayed composed with the unit isomorphism for $\square$; $\xi_i$ denote the $\mathcal{O}$-algebra structure maps for $A$. In the second diagram, $\sigma$ is a permutation of $\{1, \ldots, m\}$, permuting the factors of $A$, viewed as an element of $\Sigma_{m+1}$ for permutation action on $\mathcal{O}(m + 1)$. In the third diagram, the diagonal isomorphism is the unit isomorphism for $\square$.

satisfying the following associativity, unity, and interchange diagrams:

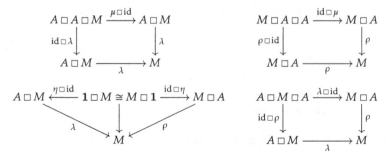

where $\mu$ denotes the multiplication and $\eta$ the unit for $A$ and the unlabeled arrow is the unit isomorphism for $\square$.

Obtaining a theory of modules closer to the idea of a left module (or right module) over an associative algebra requires working with non-symmetric operads.

**Definition 5.5.3.** Let $\overline{\mathcal{O}}$ be a non-symmetric operad and let $A$ be an $\overline{\mathcal{O}}$-algebra. A left $(\overline{\mathcal{O}}, A)$-module (or just left $A$-module when $\overline{\mathcal{O}}$ is understood) consists of an object $M$ of $\mathcal{M}$ and structure maps

$$\zeta_m \colon \overline{\mathcal{O}}(m+1) \,\square\, (A^{(m)} \,\square\, M) \to M$$

for $m \geq 0$ such that the associativity and unit diagrams in Figure 5.4 commute (with $\overline{\mathcal{O}}$ in place of $\mathcal{O}$). A map of left $A$-modules is a map of the underlying objects of $\mathcal{M}$ that commutes with the structure maps.

We also have the evident notion of a right $A$-module defined in terms of structure maps

$$\zeta_m \colon \overline{\mathcal{O}}(m+1) \,\square\, (M \,\square\, A^{(m)}) \to M$$

and the analogous right-hand associativity and unit diagrams.

Unlike in the case of operadic algebras, where working with a non-symmetric operad and its corresponding symmetric operad results in the same theory, in the case of modules, the results are very different.

**Example 5.5.4.** When $\overline{\mathcal{O}} = \overline{\mathcal{A}ss}$, the non-symmetric operad for associative algebras, and $A$ is an $\overline{\mathcal{O}}$-algebra (i.e., a $\square$-monoid), a left $(\overline{\mathcal{A}ss}, A)$-module in the sense of the previous definition is precisely a left $A$-module in the usual sense defined in terms of an associative and unital left action map $A \,\square\, M \to M$. Likewise, a right $(\overline{\mathcal{A}ss}, A)$-module is precisely a right $A$-module in the usual sense.

Under mild hypotheses, the category of $(\mathcal{O}, A)$-modules is a category of modules for a $\square$-monoid called the enveloping algebra of $A$.

**Construction 5.5.5** (The enveloping algebra). Assume that $\mathcal{M}$ admits countable colimits and $\square$ preserves countable colimits in each variable. For an operad $\mathcal{O}$ and an $\mathcal{O}$-algebra $A$, let $U^{\mathcal{O}}A$ (or $UA$ when $\mathcal{O}$ is understood) be the coequalizer

$$\coprod_{m=0}^{\infty} \mathcal{O}(m+1) \,\square_{\Sigma_m}\, (\mathcal{O}A)^{(m)} \rightrightarrows \coprod_{m=0}^{\infty} \mathcal{O}(m+1) \,\square_{\Sigma_m}\, A^{(m)} \longrightarrow U^{\mathcal{O}}A,$$

where we regard $\Sigma_m$ as the usual subgroup of $\Sigma_{m+1}$ of permutations that fix $m+1$. Here one map is induced by the action map $\mathcal{O}A \to A$ and the other is induced by the operadic multiplication

$$\mathcal{O}(m+1) \,\square\, (\overline{\mathcal{O}}A)^{(m)} \cong \coprod_{j_1, \dots, j_m} \mathcal{O}(m+1) \,\square\, (\mathcal{O}(j_1) \,\square\, A^{(j_1)}) \,\square\, \cdots \,\square\, (\mathcal{O}(j_m) \,\square\, A^{(j_m)})$$

$$\cong \coprod_{j_1, \dots, j_m} \left( \mathcal{O}(m+1) \,\square\, (\mathcal{O}(j_1) \,\square\, \cdots \,\square\, \mathcal{O}(j_m) \,\square\, \mathbf{1}) \right) \,\square\, A^{(j)}$$

$$\xrightarrow{\coprod \Gamma_{j_1, \dots, j_m, 1}^{m+1} \,\square\, \mathrm{id}} \mathcal{O}(j+1) \,\square\, A^{(j)}$$

(where we have omitted writing $1 \colon \mathbf{1} \to \mathcal{O}(1)$ and as always $j = j_1 + \cdots + j_m$). Let

$\eta\colon \mathbf{1} \to UA$ be the map induced by $1\colon \mathbf{1} \to \mathcal{O}(1)$ and the inclusion of the $m = 0$ summand and let $\mu\colon UA \,\square\, UA \to UA$ be the map induced from the maps

$$(\mathcal{O}(m+1) \,\square\, A^{(m)}) \,\square\, (\mathcal{O}(n+1) \,\square\, A^{(n)}) \to \mathcal{O}(m+n+1) \,\square\, A^{(m+n)}$$

obtained from the map $\circ_{m+1}\colon \mathcal{O}(m+1) \,\square\, \mathcal{O}(n+1) \to \mathcal{O}(m+n+1)$ defined as the composite

$$\mathcal{O}(m+1) \,\square\, \mathcal{O}(n+1) \cong \mathcal{O}(m+1) \,\square\, (\mathbf{1} \,\square\, \cdots \,\square\, \mathbf{1} \,\square\, \mathcal{O}(n+1)) \xrightarrow{\Gamma^{m+1}_{1,\ldots,1,n+1}} \mathcal{O}(m+n+1)$$

(where again we have omitted writing $1\colon \mathbf{1} \to \mathcal{O}(1)$). Associativity of the operad multiplication implies that $\eta$ and $\mu$ give $UA$ the structure of an associative monoid for $\square$ and the resulting object is called the *enveloping algebra* of $A$ over $\mathcal{O}$.

An easy argument from the definitions and universal property of the coequalizer proves the following proposition.

**Proposition 5.5.6.** *Assume $\mathcal{M}$ admits countable coproducts and $\square$ preserves them in each variable. Let $\mathcal{O}$ be an operad and let $A$ be an $\mathcal{O}$-algebra. For an object $X$ of $\mathcal{M}$, $(\mathcal{O}, A)$-module structures on $X$ are in bijective correspondence with left $U^{\mathcal{O}}A$-module structures. In particular, the category of $(\mathcal{O}, A)$-modules is isomorphic to the category of left $U^{\mathcal{O}}A$-modules.*

Similarly, in the case of non-symmetric operads, we can construct a left module enveloping algebra $\overline{U}^{\overline{\mathcal{O}}}A$ (denoted $\overline{U}A$ when $\overline{\mathcal{O}}$ is understood) as the coequalizer

$$\coprod_{m=0}^{\infty} \overline{\mathcal{O}}(m+1) \,\square\, (\overline{\mathcal{O}}A)^{(m)} \rightrightarrows \coprod_{m=0}^{\infty} \overline{\mathcal{O}}(m+1) \,\square\, A^{(m)} \longrightarrow \overline{U}^{\overline{\mathcal{O}}}A \qquad (5.5.1)$$

with maps defined as in Construction 5.5.5. The analogous identification of module categories holds.

**Proposition 5.5.7.** *Assume $\mathcal{M}$ admits countable coproducts and $\square$ preserves them in each variable. Let $\overline{\mathcal{O}}$ be a non-symmetric operad and let $A$ be an $\overline{\mathcal{O}}$-algebra. For an object $X$ of $\mathcal{M}$, left $(\overline{\mathcal{O}}, A)$-module structures on $X$ are in bijective correspondence with left $\overline{U}A$-module structures. In particular, the category of left $(\overline{\mathcal{O}}, A)$-modules is isomorphic to the category of left $\overline{U}A$-modules.*

We develop some tools to study enveloping algebras in the next section. In the meantime, we can identify the enveloping algebra in some specific examples.

**Example 5.5.8.** For $\mathcal{O} = \mathcal{A}ss$ and $A$ an $\mathcal{A}ss$-algebra (a $\square$-monoid), $U^{\mathcal{A}ss}A$ is $A \,\square\, A^{\mathrm{op}}$, the usual enveloping algebra for a $\square$-monoid. Viewing $A$ as an $\overline{\mathcal{A}ss}$-algebra, $\overline{U}^{\overline{\mathcal{A}ss}}A$ is the $\square$-monoid $A$. If $A$ is a $\mathcal{C}om$-algebra (a commutative $\square$-monoid), then $U^{\mathcal{C}om}A$ makes sense and is also the $\square$-monoid $A$.

**Example 5.5.9.** Let $\mathcal{L}$ denote the Boardman–Vogt linear isometries operad of

Example 5.3.4. For an $\mathcal{L}$-algebra, the underlying space of $U^{\mathcal{L}}A$ is the pushout

$$
\begin{array}{ccc}
\mathcal{L}(2) \times_{\mathcal{L}(1)} \mathcal{L}(0) & \xrightarrow{\mathrm{id} \times \xi_0} & \mathcal{L}(2) \times_{\mathcal{L}(1)} A \\
{\scriptstyle \circ_1} \downarrow & & \downarrow \\
\mathcal{L}(1) & \longrightarrow & U^{\mathcal{L}}A
\end{array}
$$

where $\circ_1$ is the map induced by $1 \colon * \to \mathcal{L}(1)$ and $\Gamma_{0,1}^2$ (as in Construction 5.5.5) and the right action on $\mathcal{L}(2)$ of $\mathcal{L}(1) \cong \mathcal{L}(1) \times *$ is via $\Gamma_{1,1}^2 \circ (\mathrm{id} \times 1)$. The inclusions of the $m = 0$ and $m = 1$ summands induce the map from the pushout above to the coequalizer defining $U^{\mathcal{L}}A$; the inverse isomorphism uses the "Hopkins' Lemma" [94, I.5.4] isomorphism

$$
\mathcal{L}(2) \times_{\mathcal{L}(1) \times \mathcal{L}(1)} (\mathcal{L}(i) \times \mathcal{L}(j)) \cong \mathcal{L}(i+j) \tag{HL}
$$

for $i, j \geq 1$. The pushout explicitly admits maps in from the $m = 0$ and $m = 1$ summands of the coequalizer, and for $m > 1$ we have the map

$$
\mathcal{L}(m+1) \times_{\Sigma_m} A^{(m)} \cong \mathcal{L}(m+1) \times_{\Sigma_m \times \mathcal{L}(1)} (A^{(m)} \times \mathcal{L}(1))
$$
$$
\underset{(\mathrm{HL})}{\cong} \mathcal{L}(2) \times_{\mathcal{L}(1) \times \mathcal{L}(1)} ((\mathcal{L}(m) \times_{\Sigma_m} A^{(m)}) \times \mathcal{L}(1))
$$
$$
\xrightarrow{\mathrm{id} \times (\xi_m \times \mathrm{id})} \mathcal{L}(2) \times_{\mathcal{L}(1) \times \mathcal{L}(1)} (A \times \mathcal{L}(1)) \cong \mathcal{L}(2) \times_{\mathcal{L}(1)} A.
$$

The previous display also indicates how the multiplication of $U^{\mathcal{L}}A$ works in the pushout description: it is induced by the map

$$
(\mathcal{L}(2) \times A) \times (\mathcal{L}(2) \times A) \cong (\mathcal{L}(2) \times \mathcal{L}(2)) \times A^{(2)}
$$
$$
\xrightarrow{\circ_2 \times \mathrm{id}} \mathcal{L}(3) \times A^{(2)} \to \mathcal{L}(3) \times_{\Sigma_2} A^{(2)} \to \mathcal{L}(2) \times_{\mathcal{L}(1)} A,
$$

where the last map is the $m = 2$ case of the map above. It turns out that the map $U^{\mathcal{L}}A \to A$ induced by the operadic algebra action maps is always a weak equivalence. (The proof is not obvious but uses the ideas from EKMM, especially [94, I.8.5, XI.3.1] in the context of the theory of $\mathcal{L}(1)$-spaces, as in for example [28, §6], [44, §4.6], or [45, §4.3].) If we forget the symmetries in $\mathcal{L}$ to create a non-symmetric operad $\mathcal{L}_{\Sigma}$, then $\overline{U}^{\mathcal{L}_{\Sigma}}A \cong U^{\mathcal{L}}A$. Even when $A$ is just an $\mathcal{L}_{\Sigma}$-algebra, $\overline{U}^{\mathcal{L}_{\Sigma}}A$ can still be identified as the same pushout construction pictured above using the analogous comparison isomorphisms with the coequalizer definition (5.5.1). Analogous formulations also hold in the context of orthogonal spectra, symmetric spectra, and EKMM $S$-modules using the operad $\Sigma_+^{\infty} \mathcal{L}$ in the respective categories. In the context of Lewis–May spectra, these observations are closely related to the foundations of EKMM $S$-modules and the properties of the smash product ($\wedge_{\mathcal{L}}$, $\wedge$, and $\wedge_A$); this is the start of a much longer story on monoidal products and balanced products for $A_{\infty}$ module categories (see, for example, [184] and [47, §17-18]).

Although in both previous examples we had an isomorphism of enveloping algebras for symmetric and non-symmetric constructions, this is not a general phenomenon, as can be seen, for example, by comparing $U^{\mathcal{A}ss}$ and $\overline{U}^{\mathcal{A}ss_{\underline{\Sigma}}}$, where $\mathcal{A}ss_{\underline{\Sigma}}$ is the non-symmetric operad formed from $\mathcal{A}ss$ by forgetting the symmetry. (In this style of notation, $\overline{\mathcal{A}ss} = \mathcal{C}om_{\underline{\Sigma}}$.)

The left module enveloping algebra construction for the non-symmetric little 1-cubes operad, $\overline{U}^{\overline{C}_1}(-)$, also admits a concrete description [184, §2], which we review in Section 5.10. It shares the feature with the previous two examples that for any $\overline{C}_1$-algebra $A$, $\overline{U}^{\overline{C}_1} A$ is weakly equivalent to $A$ (see [184, 1.1] or Proposition 5.10.3).

Given Propositions 5.5.6 and 5.5.7, the homotopy theory of modules over operadic algebras reduces to (1) the homotopy theory of modules over $\square$-monoids and (2) the homotopy theory of $U^{\mathcal{O}} A$ (or $\overline{U}^{\overline{\mathcal{O}}} A$) as a functor of $\mathcal{O}$ (or $\overline{\mathcal{O}}$) and $A$. The latter first requires a study of the homotopy theory of operadic algebras, which we review (in part) in the next few sections, before returning to this question in Corollary 5.9.7. On the other hand the homotopy theory of modules over $\square$-monoids is very straightforward, and we give a short review of the main results in the remainder of this section. We discuss this in terms of closed model categories. (For an overview of closed model categories as a setting for homotopy theory, we refer the reader to [91]. See also Chapter 2 of this volume.) The following theorem gives a comprehensive result in some categories of primary interest.

**Theorem 5.5.10.** *Let $(\mathcal{M}, \square, 1)$ be the category of simplicial sets, spaces, symmetric spectra, orthogonal spectra, EKMM $S$-modules, simplicial abelian groups, chain complexes, or any category of modules over a commutative monoid object in one of these categories, with the usual monoidal product and one of the standard cofibrantly generated model structures. Let $A$ be a monoid object in $\mathcal{M}$. The category of $A$-modules is a closed model category with weak equivalences and fibrations created in $\mathcal{M}$.*

The proof in all cases is much like the argument in [94, VII§4] or [267, 2.3]. Heuristically, whenever the small object argument applies and $\square$ behaves well with respect to weak equivalences, pushouts, and sequential or filtered colimits, a version of the previous theorem should hold. For an example of a more general statement, see [267, 4.1].

A map of monoid objects $A \rightarrow B$ induces an obvious *restriction of scalars* functor from the category of $B$-modules to the category of $A$-modules. When $\mathcal{M}$ admits coequalizers and $\square$ preserves coequalizers in each variable (as is the case in the examples in the previous theorem), the restriction of scalars functor admits a left adjoint *extension of scalars* functor $B \square_A (-)$ which on the underlying objects is constructed as the coequalizer

$$B \square A \square M \rightrightarrows B \square M \longrightarrow B \square_A M,$$

where one map is induced by the $A$-action on $M$ and the other by the $A$ action on $B$ (induced by the map of monoid objects). When the categories of modules have closed model structures with weak equivalences and fibrations created in the underlying category $\mathcal{M}$, this adjunction is automatically a Quillen adjunction, which implies a derived adjunction on homotopy categories. When the map $A \rightarrow B$ is a weak equivalence, we can often expect the Quillen adjunction to be a Quillen equivalence and induce an equivalence of homotopy categories; this is the case in the setting of the previous theorem.

**Theorem 5.5.11.** *Let $\mathcal{M}$ be one of the symmetric monoidal model categories of Theorem 5.5.10. A weak equivalence of monoid objects induces a Quillen equivalence on categories of modules.*

Again, significantly more general results hold; see, for example, [157], especially Theorem 8.3 and the subsection of Section 1 entitled "Extension of scalars".

## 5.6     Limits and colimits in categories of operadic algebras

Before going on to the homotopy theory of categories of operadic algebras, we say a few words about certain constructions, limits and colimits in this section, and geometric realization in the next section. While limits of operadic algebras are pretty straightforward (as explained below), colimits tend to be more complicated and we take some space to describe in detail what certain colimits look like.

We start with limits. Let $D: \mathcal{D} \to \mathcal{M}[\mathcal{O}]$ be a diagram, i.e., a functor from a small category $\mathcal{D}$, where $\mathcal{M}$ is a symmetric monoidal category and $\mathcal{O}$ is an operad in $\mathcal{M}$. By neglect of structure, we can regard $D$ as a diagram in $\mathcal{M}$, and suppose the limit $L$ exists in $\mathcal{M}$. Then for each $d \in \mathcal{D}$, we have the canonical map $L \to D(d)$, and using the $\mathcal{O}$-algebra structure map for $D(d)$, we get a map

$$\mathcal{O}(m) \,\square\, L^{(m)} \to \mathcal{O}(m) \,\square\, D(d)^{(m)} \to D(d).$$

These maps satisfy the required compatibility to define a map

$$\mathcal{O}(m) \,\square\, L^{(m)} \to L,$$

which together are easily verified to provide structure maps for an $\mathcal{O}$-algebra structure on $L$. This $\mathcal{O}$-algebra structure has the universal property for the limit of $D$ in $\mathcal{M}[\mathcal{O}]$.

**Proposition 5.6.1.** *For any symmetric monoidal category $\mathcal{M}$, any operad $\mathcal{O}$ in $\mathcal{M}$, and any diagram of $\mathcal{O}$-algebras, if the limit exists in $\mathcal{M}$, then it has a canonical $\mathcal{O}$-algebra structure that gives the limit in $\mathcal{M}[\mathcal{O}]$.*

We cannot expect general colimits of operadic algebras to be formed in the underlying category, as can be seen from the examples of coproducts of $\square$-monoids ($\mathcal{A}ss$-algebras) or of commutative $\square$-monoids ($\mathcal{C}om$-algebras). The discussion of colimits simplifies if we assume that $\mathcal{M}$ has countable colimits and that $\square$ preserves countable colimits in each variable, so that Proposition 5.4.9 holds and the category of $\mathcal{O}$-algebras is the category of algebras over the monad $\mathbb{O}$. The main technical tool in this case is the following proposition; because we have assumed in particular that $\square$ preserves coequalizers in each variable, it follows that the $m$-th $\square$-power functor preserves reflexive coequalizers (see [94, II.7.2] for a proof) and the filtered colimits that exist (by an easy cofinality argument).

**Proposition 5.6.2.** *If $\mathcal{M}$ satisfies the hypotheses of Proposition 5.4.7, then for any operad $\mathcal{O}$, the monad $\mathbb{O}$ preserves reflexive coequalizers in $\mathcal{M}$ and the filtered colimits that exist in $\mathcal{M}$.*

Recall that a reflexive coequalizer is a coequalizer

$$X \overset{a}{\underset{b}{\rightrightarrows}} Y \overset{c}{\rightarrow} C$$

where there exists a map $r\colon Y \to X$ such that $a \circ r = \mathrm{id}_Y$ and $b \circ r = \mathrm{id}_Y$; $r$ is called a *reflexion*. The proposition says that if the above coequalizer exists in $\mathcal{M}$ and is reflexive then the diagram obtained by applying $\mathbb{O}$,

$$\mathbb{O}X \overset{\mathbb{O}a}{\underset{\mathbb{O}b}{\rightrightarrows}} \mathbb{O}Y \overset{\mathbb{O}c}{\rightarrow} \mathbb{O}C,$$

is also a (reflexive) coequalizer diagram in $\mathcal{M}$. Now suppose that $a$ and $b$ are maps of $\mathcal{O}$-algebras. Then the diagrams

$$
\begin{array}{ccc}
\mathbb{O}X & \overset{\mathbb{O}a}{\longrightarrow} & \mathbb{O}Y \\
\downarrow & & \downarrow \\
X & \underset{a}{\longrightarrow} & Y
\end{array}
\qquad \text{and} \qquad
\begin{array}{ccc}
\mathbb{O}X & \overset{\mathbb{O}b}{\longrightarrow} & \mathbb{O}Y \\
\downarrow & & \downarrow \\
X & \underset{b}{\longrightarrow} & Y
\end{array}
$$

commute (where the vertical maps are the $\mathcal{O}$-algebra structure maps) and we get an induced map

$$\mathbb{O}C \to C.$$

Repeating this for $\mathbb{O}X \rightrightarrows \mathbb{O}Y$ and the two maps $\mathbb{O}\mathbb{O}X \rightrightarrows \mathbb{O}\mathbb{O}Y$ to $\mathbb{O}X \rightrightarrows \mathbb{O}Y$, we see that the map $\mathbb{O}C \to C$ constructed above is an $\mathcal{O}$-algebra structure map and an easy check of universal properties shows that $C$ with this $\mathcal{O}$-algebra structure is the coequalizer in $\mathcal{M}[\mathcal{O}]$. This shows that if a pair of parallel arrows in $\mathcal{M}[\mathcal{O}]$ has a reflexion in $\mathcal{M}$, then the coequalizer in $\mathcal{M}$ has the canonical structure of an $\mathcal{O}$-algebra and is the coequalizer in $\mathcal{M}[\mathcal{O}]$.

We can turn the observation in the previous paragraph into a construction of colimits of arbitrary shapes in $\mathcal{M}[\mathcal{O}]$. Given a diagram $D\colon \mathcal{D} \to \mathcal{M}[\mathcal{O}]$, assume that the colimit of the underlying functor to $\mathcal{M}$ exists, and denote it by $\mathrm{colim}^{\mathcal{M}} D$. If $\mathrm{colim}^{\mathcal{M}} \mathbb{O}D$ also exits, then we get a pair of parallel arrows

$$\mathbb{O}(\mathrm{colim}^{\mathcal{M}} \mathbb{O}D) \rightrightarrows \mathbb{O}(\mathrm{colim}^{\mathcal{M}} D)\,, \tag{5.6.1}$$

where one arrow is induced by the $\mathcal{O}$-algebra structure maps $\mathbb{O}D(d) \to D(d)$ and the other is the composite

$$\mathbb{O}(\mathrm{colim}^{\mathcal{M}} \mathbb{O}D) \overset{\mathbb{O}i}{\longrightarrow} \mathbb{O}\mathbb{O}(\mathrm{colim}^{\mathcal{M}} D) \overset{\mu}{\longrightarrow} \mathbb{O}(\mathrm{colim}^{\mathcal{M}} D),$$

where $\mu$ is the monadic multiplication $\mathbb{O}\mathbb{O} \to \mathbb{O}$ and

$$i\colon \mathrm{colim}^{\mathcal{M}} \mathbb{O}D \to \mathbb{O}(\mathrm{colim}^{\mathcal{M}} D)$$

is the map assembled from the maps $\mathbb{O}D(d) \to \mathbb{O}(\mathrm{colim}^{\mathcal{M}} D)$ induced by applying $\mathbb{O}$ to the canonical maps $D(d) \to \mathrm{colim}^{\mathcal{M}} D$. We also have a reflexion

$$\mathbb{O}(\mathrm{colim}^{\mathcal{M}} D) \to \mathbb{O}(\mathrm{colim}^{\mathcal{M}} \mathbb{O}D)$$

induced by the unit map $D(d) \to \mathbb{O}D(d)$. Thus, the coequalizer of (5.6.1) in $\mathcal{M}$ has the canonical structure of an $\mathcal{O}$-algebra which provides the coequalizer in $\mathcal{M}[\mathcal{O}]$; a check of universal properties shows that the coequalizer is the colimit in $\mathcal{M}[\mathcal{O}]$ of $D$.

**Proposition 5.6.3.** *Assume $\mathcal{M}$ satisfies the hypotheses of Proposition 5.4.7. For any operad $\mathcal{O}$ and any diagram $D: \mathcal{D} \to \mathcal{M}[\mathcal{O}]$, if the colimit of $D$ and the colimit of $\mathbb{O}D$ exist in $\mathcal{M}$, then the colimit of $D$ exists in $\mathcal{M}[\mathcal{O}]$ and is given by the coequalizer of the reflexive pair displayed in (5.6.1).*

For example, the coproduct $A \amalg^{\mathcal{M}[\mathcal{O}]} B$ in $\mathcal{M}[\mathcal{O}]$ can be constructed as the coequalizer

$$\mathbb{O}(\mathbb{O}A \amalg \mathbb{O}B) \rightrightarrows \mathbb{O}(A \amalg B) \longrightarrow A \amalg^{\mathcal{M}[\mathcal{O}]} B.$$

When $B = \mathbb{O}X$ for some $X$ in $\mathcal{M}$, we can say more by recognizing that the category of $\mathcal{O}$-algebras under $A$ is itself the category of algebras over an operad.

**Construction 5.6.4** (The enveloping operad). For $m \geq 0$, define $\mathcal{U}_A^{\mathcal{O}}(m)$ by the coequalizer diagram

$$\coprod_{\ell=0}^{\infty} \mathcal{O}(\ell + m) \, \Box_{\Sigma_\ell} \, (\mathbb{O}A)^{(\ell)} \rightrightarrows \coprod_{\ell=0}^{\infty} \mathcal{O}(\ell + m) \, \Box_{\Sigma_\ell} \, A^{(\ell)} \longrightarrow \mathcal{U}_A^{\mathcal{O}}(m),$$

where one arrow is induced by the operadic multiplication

$$\Gamma_{j_1,\ldots,j_\ell,1,\ldots,1}^{\ell+m}: \mathcal{O}(\ell + m) \,\Box\, \mathcal{O}(j_1) \,\Box\, \cdots \,\Box\, \mathcal{O}(j_\ell) \,\Box\, \mathbf{1} \,\Box\, \cdots \,\Box\, \mathbf{1} \to \mathcal{O}(j + m)$$

and the other by the $\mathcal{O}$-algebra action map $\mathbb{O}A \to A$. We think of the $\ell$ factors of $A$ (or $\mathbb{O}A$) as being associated with the first $\ell$ inputs of $\mathcal{O}(\ell + m)$, leaving the last $m$ inputs open. We then have a $\Sigma_m$-action induced from the $\Sigma_m$-action on $\mathcal{O}(\ell + m)$ on the open inputs, unit map $1: \mathbf{1} \to \mathcal{U}_A^{\mathcal{O}}(1)$ induced by the unit map of $\mathcal{O}$ (on the summand $\ell = 0$), and operadic composition $\Gamma$ induced by applying the operadic multiplication of $\mathcal{O}$ using the open inputs.

This operad is called the *enveloping operad* of $A$ and generalizes the enveloping algebra $U^{\mathcal{O}}A$ of Construction 5.5.5: for $m = 1$, $\mathcal{U}_A^{\mathcal{O}}(1)$ is precisely the coequalizer defining $U^{\mathcal{O}}A$ and the operad unit and multiplication $\Gamma_1^1$ coincide with the $\Box$-monoid unit and multiplication.

To return to the discussion of the category of $\mathcal{O}$-algebras under $A$, we note that for $m = 0$, the coequalizer in Construction 5.6.4 is

$$\mathbb{O}\mathbb{O}A \rightrightarrows \mathbb{O}A \longrightarrow \mathcal{U}_A^{\mathcal{O}}(0),$$

giving a canonical isomorphism $A \to \mathcal{U}_A^{\mathcal{O}}(0)$, and so a $\mathcal{U}_A^{\mathcal{O}}$-algebra $T$ comes with a structure map $A \to T$. Looking at the summands with $\ell = 0$ above, we get a map of operads $\mathcal{O} \to \mathcal{U}_A^{\mathcal{O}}$, giving $T$ an $\mathcal{O}$-algebra structure; the map $A \to T$ is a map of $\mathcal{O}$-algebras. On the other hand, given an $\mathcal{O}$-algebra $B$ and a map of $\mathcal{O}$-algebras $A \to B$, we have maps

$$\mathcal{O}(\ell + m) \,\square\, A^{(\ell)} \,\square\, B^{(m)} \to \mathcal{O}(\ell + m) \,\square\, B^{(\ell)} \,\square\, B^{(m)} \to B$$

which together induce maps $\mathcal{U}_A^{\mathcal{O}}(m) \square B^{(m)} \to B$ that are easily checked to provide $\mathcal{U}_A^{\mathcal{O}}$-algebra structure maps. This gives a bijection between the structure of an $\mathcal{O}$-algebra under $A$ and the structure of a $\mathcal{U}_A^{\mathcal{O}}$-algebra.

**Proposition 5.6.5.** *Let $\mathcal{M}$ satisfy the hypotheses of Proposition 5.4.7. For an object $X$ of $\mathcal{M}$, the set of $\mathcal{U}_A^{\mathcal{O}}$-algebra structures on $X$ is in bijective correspondence with the set of ordered pairs consisting of an $\mathcal{O}$-algebra structure on $X$ and a map of $\mathcal{O}$-algebras $A \to X$ for that structure.*

As a consequence we have the following description of the coproduct of $\mathcal{O}$-algebras $A \amalg^{\mathcal{M}[\mathcal{O}]} \mathbb{O}X$, since $A \amalg^{\mathcal{M}[\mathcal{O}]} \mathbb{O}(-)$ is the left adjoint of the forgetful functor from $\mathcal{O}$-algebras under $A$ to $\mathcal{M}$.

**Proposition 5.6.6.** *When $\mathcal{M}$ satisfies the hypotheses of Proposition 5.4.7,*

$$A \amalg^{\mathcal{M}[\mathcal{O}]} \mathbb{O}X \cong \mathbb{U}_A^{\mathcal{O}}X = \coprod_{m=0}^{\infty} \mathcal{U}_A^{\mathcal{O}}(m) \,\square_{\Sigma_m}\, X^{(m)}$$

*(where the coproduct symbol undecorated by a category denotes coproduct in $\mathcal{M}$).*

The decomposition above can be useful even without further information on $\mathcal{U}_A^{\mathcal{O}}$, but in fact we can be more concrete about what $\mathcal{U}_A^{\mathcal{O}}$ looks like in the case when $A$ is built up iteratively from pushouts of free objects in $\mathcal{M}[\mathcal{O}]$. As a base case, an easy calculation gives

$$\mathcal{U}_{\mathbb{O}X}^{\mathcal{O}}(m) = \coprod_{\ell=0}^{\infty} \mathcal{O}(\ell + m) \,\square_{\Sigma_\ell}\, X^{(\ell)}.$$

Now suppose $A' = A \amalg_{\mathbb{O}X}^{\mathcal{M}[\mathcal{O}]} \mathbb{O}Y$ for some maps $X \to A$ and $X \to Y$ in $\mathcal{M}$; we can then describe $\mathcal{U}_{A'}^{\mathcal{O}}$ in terms of $\mathcal{U}_A^{\mathcal{O}}$ and pushouts in $\mathcal{M}[\mathcal{O}]$ as follows. (In particular, the calculation of $\mathcal{U}_{A'}^{\mathcal{O}}(0)$ describes $A'$ in these terms and the calculation of $\mathcal{U}_{A'}^{\mathcal{O}}(1)$ describes $UA'$ in these terms.) First, using the observations on colimits above, a little work shows that the coequalizer defining $\mathcal{U}_{A'}^{\mathcal{O}}$ simplifies in this case to

$$\coprod_{\ell=0}^{\infty} \mathcal{U}_A^{\mathcal{O}}(\ell + m) \,\square_{\Sigma_\ell}\, (X \amalg Y)^{(\ell)} \rightrightarrows \coprod_{\ell=0}^{\infty} \mathcal{U}_A^{\mathcal{O}}(\ell + m) \,\square_{\Sigma_\ell}\, Y^{(\ell)} \longrightarrow \mathcal{U}_{A'}^{\mathcal{O}}(m)$$

where one map is induced by the map $X \to A$ $(= \mathcal{U}_A^{\mathcal{O}}(0))$ and the other is induced by the map $X \to Y$. We then have a filtration on $\mathcal{U}_{A'}^{\mathcal{O}}(m)$ by powers of $Y$; specifically, define $F^k \mathcal{U}_{A'}^{\mathcal{O}}(m)$ by the coequalizer

$$\coprod_{\ell=0}^{k} \mathcal{U}_A^{\mathcal{O}}(\ell + m) \,\square_{\Sigma_\ell}\, (X \amalg Y)^{(\ell)} \rightrightarrows \coprod_{\ell=0}^{k} \mathcal{U}_A^{\mathcal{O}}(\ell + m) \,\square_{\Sigma_\ell}\, Y^{(\ell)} \longrightarrow F^k \mathcal{U}_{A'}^{\mathcal{O}}(m)$$

Then $\operatorname{colim}_k F^k \mathcal{U}^{\mathcal{O}}_{A'}(m) = \mathcal{U}^{\mathcal{O}}_{A'}(m)$. Comparing the universal properties for $F^{k-1}\mathcal{U}^{\mathcal{O}}_{A'}(m)$ and $F^k \mathcal{U}^{\mathcal{O}}_{A'}(m)$, we see that the following diagram is a pushout (in $\mathcal{M}$):

$$
\begin{array}{ccc}
\mathcal{U}^{\mathcal{O}}_{A}(k+m) \square_{\Sigma_{k-1}} (X \square Y^{(k-1)}) & \longrightarrow & \mathcal{U}^{\mathcal{O}}_{A}(k+m) \square_{\Sigma_k} Y^{(k)} \\
\downarrow & & \downarrow \\
F^{k-1}\mathcal{U}^{\mathcal{O}}_{A'}(m) & \longrightarrow & F^k \mathcal{U}^{\mathcal{O}}_{A'}(m)
\end{array}
$$

This describes $\mathcal{U}^{\mathcal{O}}_{A'}$ in terms of iterated pushouts in $\mathcal{M}$, but we can do somewhat better, as can be seen in the example where $\mathcal{M}$ is the category of spaces and $X \to Y$ is a closed inclusion. In the pushout above, the top horizontal map comes from the map

$$\Sigma_k \times_{\Sigma_{k-1}} (X \times Y^{k-1}) \to Y^k$$

which fails to be an inclusion for $k > 1$ except in trivial cases; however, the image of this map can be described as an iterated pushout, starting with $X^k$ and gluing in higher powers of $Y$. This works as well in the general case (which we now return to). Let $Q^k_0(X \to Y) = X^{(k)}$, an object of $\mathcal{M}$ with a $\Sigma_k$-action and a $\Sigma_k$-equivariant map to $Y^{(k)}$. Inductively, for $i > 0$, define $Q^k_i(X \to Y)$ as the pushout

$$
\begin{array}{ccc}
\Sigma_k \times_{\Sigma_{k-i} \times \Sigma_i} (X^{(k-i)} \square Q^i_{i-1}(X \to Y)) & \longrightarrow & \Sigma_k \times_{\Sigma_{k-i} \times \Sigma_i} (X^{(k-i)} \square Y^{(i)}) \\
\downarrow & & \downarrow \\
Q^k_{i-1}(X \to Y) & \longrightarrow & Q^k_i(X \to Y)
\end{array}
\qquad (5.6.2)
$$

with the evident $\Sigma_k$-action and $\Sigma_k$-equivariant map

$$Q^k_i(X \to Y) \to Y^{(k)}.$$

Then for all $j > 0$, we have a $(\Sigma_j \times \Sigma_k)$-equivariant map

$$X^{(j)} \square Q^k_i(X \to Y) \to Q^{j+k}_i(X \to Y)$$

induced by the map

$$X^{(j)} \square X^{(k-i)} \square Y^{(i)} \cong X^{(j+k-i)} \square Y^{(i)} \to Q^{j+k}_i(X \to Y)$$

and the compatible (inductively defined) map

$$X^{(j)} \square Q^k_{i-1}(X \to Y) \to Q^{j+k}_{i-1}(X \to Y) \to Q^{j+k}_i(X \to Y),$$

which allows us to continue the induction. In the case when $\mathcal{M}$ is the category of topological spaces and $X \to Y$ is a closed inclusion, the maps

$$Q^k_0(X \to Y) \to \cdots \to Q^k_{k-1}(X \to Y) \to Y^{(k)}$$

are closed inclusions with $Q^k_i(X \to Y)$ the subspace of $Y^k$ where at most $i$ coordinates are in $Y \setminus X$. In the general case, an inductive argument shows that the map

$$\Sigma_k \times_{\Sigma_{k-i} \times \Sigma_i} (X^{(k-i)} \square Y^{(i)}) \to Q^k_i(X \to Y)$$

is a categorical epimorphism and that the map

$$\mathcal{U}_A^O(k+m) \,\square_{\Sigma_{k-1}} (X \,\square\, Y^{(k-1)}) \to \mathcal{U}_A^O(k+m) \,\square_{\Sigma_k} Q_{k-1}^k(X \to Y)$$

is a categorical epimorphism. Since this factors the map

$$\mathcal{U}_A^O(k+m) \,\square_{\Sigma_{k-1}} (X \,\square\, Y^{(k-1)}) \to \mathcal{U}_A^O(k+m) \,\square_{\Sigma_k} Y^{(k)},$$

we get the following more sophisticated identification of $F^k \mathcal{U}_{A'}^O(m)$ as a pushout:

$$
\begin{array}{ccc}
\mathcal{U}_A^O(k+m) \,\square_{\Sigma_k} Q_{k-1}^k(X \to Y) & \longrightarrow & \mathcal{U}_A^O(k+m) \,\square_{\Sigma_k} Y^{(k)} \\
\downarrow & & \downarrow \\
F^{k-1} \mathcal{U}_{A'}^O(m) & \longrightarrow & F^k \mathcal{U}_{A'}^O(m)
\end{array}
\tag{5.6.3}
$$

In practice, the map $Q_{k-1}^k(X \to Y) \to Y^{(k)}$ is some kind of cofibration when $X \to Y$ is nice enough; the above formulation is then useful for deducing homotopical information in the presence of cofibrantly generated model category structures, as discussed in Section 5.8.

## 5.7  Enrichment and geometric realization

Categories of operadic algebras in spaces or spectra come with a canonical enrichment in spaces, i.e., they have mapping spaces and an intrinsic notion of homotopy. While more abstract notions of homotopy, for example, in terms of model structures, now play a more significant role in homotopy theory, the topological enrichment provides some powerful tools, including and especially geometric realization of simplicial objects.

We begin with a general discussion of enrichment of operadic algebra categories. When $\mathcal{M}$ satisfies the hypotheses of Proposition 5.4.7, Proposition 5.4.9 describes the maps in the category of $\mathcal{O}$-algebras as an equalizer

$$\mathcal{M}[\mathcal{O}](A,B) \longrightarrow \mathcal{M}(A,B) \rightrightarrows \mathcal{M}(\mathcal{O}A,B),$$

where one arrow $\mathcal{M}(A,B) \to \mathcal{M}(\mathcal{O}A,B)$ is induced by the action map $\mathcal{O}A \to A$ and the other is induced by applying the functor $\mathcal{O}: \mathcal{M}(A,B) \to \mathcal{M}(\mathcal{O}A, \mathcal{O}B)$ and then using the action map $\mathcal{O}B \to B$. When $\mathcal{M}$ is enriched over a complete symmetric monoidal category (for example, when the mapping sets of $\mathcal{M}$ are topologized or simplicial), then $\mathcal{M}[\mathcal{O}]$ becomes enriched exactly when $\mathcal{O}$ has the structure of an enriched functor, defining the enrichment of $\mathcal{M}[\mathcal{O}]$ by the equalizer above. Clearly it is not always possible for $\mathcal{O}$ to be enriched: if $\mathcal{M}$ is the category of abelian groups and $\mathcal{O} = \mathcal{A}ss$ or $\mathcal{C}om$, then $\mathcal{O}$ is not an additive functor so cannot be enriched over abelian groups; this corresponds to the fact that the categories of rings and commutative rings are not enriched over abelian groups. On the other hand, enrichments over spaces and simplicial sets are usually inherited by algebra categories; the reason, as we now explain, derives from the fact that spaces and simplicial sets are cartesian.

For convenience, consider the case when $\mathcal{M}$ is a closed symmetric monoidal category. Let $\mathscr{E}, \times, *$ be a symmetric monoidal category (which we will eventually assume to be cartesian), and let $L : \mathscr{E} \to \mathcal{M}$ be a strong symmetric monoidal functor that is a left adjoint; let $R$ denote its right adjoint. For formal reasons $R$ is then lax symmetric monoidal and in particular $RF$ provides an $\mathscr{E}$-enrichment of $\mathcal{M}$ (where, as always, $F$ denotes the mapping object in $\mathcal{M}$). These hypotheses are not all necessary but avoid some review of enriched category theory and concisely state a lot of coherence data that more minimal hypotheses would force us to spell out. The iterated symmetric monoidal product in $\mathcal{M}$ then gives a multivariable enriched functor

$$RF(A_1, B_1) \times \cdots \times RF(A_m, B_m) \to RF(A_1 \,\square\, \cdots \,\square\, A_m, B_1 \,\square\, \cdots \,\square\, B_m).$$

Now assume that $\times$ is a cartesian monoidal product, meaning that it is the categorical product, the unit is the final object, and the symmetry and unit isomorphisms are the universal ones. With this assumption, we have a natural diagonal map $E \to E \times E$, which we can apply in particular to the object $RF(A, B)$ to get a natural map

$$RF(A, B) \to RF(A, B) \times \cdots \times RF(A, B) \to RF(A^{(m)}, B^{(m)}). \qquad (5.7.1)$$

This makes the $m$-th $\square$-power into an $\mathscr{E}$-enriched functor for $m > 0$. In the case $m = 0$, we have the final map

$$RF(A, B) \to * \to R1 \xrightarrow{\cong} RF(A^{(0)}, B^{(0)}).$$

From here the rest is easy: the $\square, F$ adjunction gives a natural (and $\mathscr{E}$-natural) map

$$RF(A^{(m)}, B^{(m)}) \to RF(\mathcal{O}(m) \,\square\, A^{(m)}, \mathcal{O}(m) \,\square\, B^{(m)})$$

and the composite to $RF(\mathcal{O}(m) \,\square\, A^{(m)}, \mathcal{O}(m) \,\square_{\Sigma_m} B^{(m)})$ admits a canonical factorization

$$RF(A, B) \to RF(\mathcal{O}(m) \,\square_{\Sigma_m} A^{(m)}, \mathcal{O}(m) \,\square_{\Sigma_m} B^{(m)}),$$

since the target is a limit (in $\mathscr{E}$) that exists by right adjoint considerations when the quotient $\mathcal{O}(m) \,\square_{\Sigma_m} B^{(m)} = (\mathcal{O}(m) \,\square\, B^{(m)})/\Sigma_m$ in $\mathcal{M}$ exists. When we assume that $\mathcal{M}$ has countable coproducts, composing further into

$$RF(\mathcal{O}(m) \,\square_{\Sigma_m} A^{(m)}, \mathcal{O}B),$$

the countable categorical product over $m$ exists, giving an $\mathscr{E}$-natural map

$$RF(A, B) \to RF(\mathcal{O}A, \mathcal{O}B)$$

which provides the $\mathscr{E}$-enrichment of $\mathcal{O}$. We state this as a theorem:

**Theorem 5.7.1.** *Let $\mathcal{M}$ be a closed symmetric monoidal category with countable colimits, and let $\mathcal{O}$ be an operad in $\mathcal{M}$. Let $\mathscr{E}$ be a cartesian monoidal category and let $\mathscr{E} \to \mathcal{M}$ be a strong symmetric monoidal functor with a right adjoint. Regarding $\mathcal{M}$ as $\mathscr{E}$-enriched over the right adjoint, the category $\mathcal{M}[\mathcal{O}]$ of $\mathcal{O}$-algebras has a canonical $\mathscr{E}$-enrichment for which the forgetful functor $\mathcal{M}[\mathcal{O}] \to \mathcal{M}$ is $\mathscr{E}$-enriched.*

We apply this now in the discussion of geometric realizations of (co)simplicial objects. Let $\mathscr{S}$ denote either the category of spaces or of simplicial sets, and write $C(-,-)$ for the internal mapping objects in $\mathscr{S}$. To avoid awkward circumlocutions, we will refer to objects of $\mathscr{S}$ as spaces in either case for the rest of the section. We now assume that $\mathscr{M}$ is closed symmetric monoidal and has countable coproducts and that we have a left adjoint symmetric monoidal functor $L$ from $\mathscr{S}$ to $\mathscr{M}$, as above, so that Theorem 5.7.1 applies. We write $R$ for the right adjoint to $L$ as above, so that $RF(-,-)$ provides mapping spaces for $\mathscr{M}$. The category $\mathscr{M}$ then has *tensors* $X \otimes T$ and *cotensors* $T \pitchfork Y$, defined by the natural isomorphisms

$$RF(X \otimes T, -) \cong C(T, RF(X, -)) \qquad \text{(tensor)},$$
$$RF(-, T \pitchfork Y)) \cong C(T, RF(-, Y)) \qquad \text{(cotensor)},$$

for spaces $T$ and objects $X$ and $Y$ of $\mathscr{M}$, constructed as follows.

**Proposition 5.7.2.** *In the context of Theorem 5.7.1, tensors and cotensors with spaces exist and are given by $X \otimes T = X \square LT$ and $T \pitchfork Y = F(LT, Y)$ for a space $T$ and objects $X, Y$ in $\mathscr{M}$.*

The proposition is an easy consequence of the formal isomorphism

$$RF(LT, X) \cong C(T, RX), \qquad (5.7.2)$$

natural in spaces $T$ and objects $X$ of $\mathscr{M}$; the isomorphism in the forward direction is adjoint to the map

$$RF(LT, X) \times T \to RF(LT, X) \times RLT \to R(F(LT, X) \square LT) \to RX$$

and the isomorphism in the backwards direction is adjoint to the map $LC(T, RX) \to F(LT, X)$ adjoint to the map

$$LC(T, RX) \square LT \cong L(C(T, RX) \times T) \to LRX \to X.$$

Let $RF^{\mathscr{M}[\mathcal{O}]}(-,-)$ denote the mapping spaces constructed above for the category of $\mathcal{O}$-algebras; despite the suggestion of the notation, this is not typically a composite functor. For an $\mathcal{O}$-algebra $A$, $F(-, A)$ does not typically carry a canonical $\mathcal{O}$-algebra structure, but for a space $T$, $F(LT, A) = T \pitchfork A$ does: the structure map

$$\mathcal{O}(n) \square (T \pitchfork A)^{(n)} \to T \pitchfork A$$

is adjoint to the map

$$\mathcal{O}(n) \square (T \pitchfork A)^{(n)} \square LT = \mathcal{O}(n) \square (F(LT, A))^{(n)} \square LT \to A$$

constructed as the composite

$$\mathcal{O}(n) \square (F(LT, A))^{(n)} \square LT \to \mathcal{O}(n) \square (F(LT, A))^{(n)} \square (LT)^{(n)} \to \mathcal{O}(n) \square A^{(n)} \to A$$

using the diagonal map on the space $T$ and the structure map on $A$. A check of universal properties then shows that $T \pitchfork A$ is the cotensor of $A$ with $T$ in the category of $\mathcal{O}$-algebras. Tensors in $\mathcal{M}[\mathcal{O}]$ can be constructed as reflexive coequalizers

$$\mathbb{O}(\mathbb{O}A \otimes T) \rightrightarrows \mathbb{O}(A \otimes T) \longrightarrow A \otimes^{\mathcal{M}[\mathcal{O}]} T.$$

Writing $\Delta[n]$ for the standard $n$-simplex, we then have the standard definition of geometric realization of simplicial objects in $\mathcal{M}$ and $\mathcal{M}[\mathcal{O}]$ (without additional assumptions) and geometric realization (often called "Tot") of cosimplicial objects in $\mathcal{M}$ and $\mathcal{M}[\mathcal{O}]$ when certain limits exist. Given a simplicial object $X_\bullet$ or a cosimplicial object $Y^\bullet$, the degeneracy subobject $sX_n$ of $X_n$ is defined as the colimit of the degeneracy maps and the degeneracy quotient object $sY^n$ of $Y^n$ is defined as the limit (if it exists) of the degeneracy maps. (In some literature, $sX_n$ is called the "latching object" and $sY^n$ the "matching object"; see [124, §15.2].) The geometric realization of $X_\bullet$ in $\mathcal{M}$ or $\mathcal{M}[\mathcal{O}]$ is then the sequential colimit of $|X_\bullet|_n$, where $|X_\bullet|_0 = X_0$ and $|X_\bullet|_n$ is defined inductively as the pushout

$$\begin{array}{ccc}
(sX_n \otimes \Delta[n]) \cup_{(sX_n \otimes \partial\Delta[n])} (X_n \otimes \partial\Delta[n]) & \longrightarrow & X_n \otimes \Delta[n] \\
\downarrow & & \downarrow \\
|X_\bullet|_{n-1} & \longrightarrow & |X_\bullet|_n
\end{array}$$

with both the tensor and the pushouts performed in $\mathcal{M}$ to define the geometric realization in $\mathcal{M}$ or performed in $\mathcal{M}[\mathcal{O}]$ to define the geometric realization in $\mathcal{M}[\mathcal{O}]$. The analogous, opposite construction defines the geometric realization of $Y^\bullet$ when all the limits exist. Because cotensors and limits (when they exist) coincide in $\mathcal{M}$ and $\mathcal{M}[\mathcal{O}]$, geometric realization of cosimplicial objects (when it exists) also coincides in $\mathcal{M}$ and $\mathcal{M}[\mathcal{O}]$. Because pushouts generally look very different in $\mathcal{M}$ than in $\mathcal{M}[\mathcal{O}]$, one might expect that geometric realization of simplicial objects in $\mathcal{M}$ and in $\mathcal{M}[\mathcal{O}]$ would also look very different; this turns out not to be the case.

**Theorem 5.7.3.** *Assume $\mathcal{M}$ satisfies the hypotheses of Theorem 5.7.1 for $\mathcal{E}$ either the category of spaces or the category of simplicial sets.*

(i) *Let $A^\bullet$ be a cosimplicial object in $\mathcal{M}[\mathcal{O}]$. If the limits defining the geometric realization (Tot) exist in $\mathcal{M}$, then that geometric realization has the canonical structure of an $\mathcal{O}$-algebra and is isomorphic to the geometric realization Tot in $\mathcal{M}[\mathcal{O}]$.*

(ii) *Let $A_\bullet$ be a simplicial object in $\mathcal{M}[\mathcal{O}]$. Then the geometric realization of $A_\bullet$ in $\mathcal{M}$ has the canonical structure of an $\mathcal{O}$-algebra and is isomorphic to the geometric realization of $A_\bullet$ in $\mathcal{M}$.*

As discussed above, only (ii) requires additional argument. For clarity in the argument for the theorem, we will write $|\cdot|$ for geometric realization in $\mathcal{M}$ and $|\cdot|^{\mathcal{M}[\mathcal{O}]}$ for geometric realization in $\mathcal{M}[\mathcal{O}]$. Here is the key fact:

**Lemma 5.7.4.** *For $\mathcal{M}$ as in the previous theorem, geometric realization in $\mathcal{M}$ is strong symmetric monoidal.*

*Proof.* Although we wrote a more constructive definition of geometric realization above, it can also be described as a coend

$$|X_\bullet| = \int^{\mathbf{\Delta}^{\mathrm{op}}} X_\bullet \otimes \Delta[\bullet],$$

where $\mathbf{\Delta}$ denotes the category of simplexes (the category with objects $[n] = \{0, \ldots, n\}$ for $n = 0, 1, 2, \ldots$, and maps the non-decreasing functions) and $\Delta[n]$ denotes the standard $n$-simplex in spaces or simplicial sets. Because the symmetric monoidal product $\square$ for $\mathscr{M}$ is assumed to commute with colimits in each variable, we can identify the product of geometric realizations also as a coend

$$|X_\bullet| \square |Y_\bullet| \cong \int^{\mathbf{\Delta}^{\mathrm{op}} \times \mathbf{\Delta}^{\mathrm{op}}} (X_\bullet \square Y_\bullet) \otimes (\Delta[\bullet] \times \Delta[\bullet]).$$

On the other hand,

$$|X_\bullet \square Y_\bullet| = \int^{\mathbf{\Delta}^{\mathrm{op}}} \mathrm{diag}(X_\bullet \square Y_\bullet) \otimes \Delta[\bullet].$$

Next, we need a purely formal observation, which is an adjoint form of the Yoneda lemma: if coproducts of appropriate cardinality exist in $\mathscr{C}$, then given a functor $F: \mathscr{C} \to \mathscr{D}$, functoriality of $F$ induces a natural isomorphism

$$\int^{c \in \mathscr{C}} F(c) \times \mathscr{C}(c, -) \xrightarrow{\cong} F(-)$$

(where $\times$ denotes coproduct over the given set; this coend exists and the identification holds with no further hypotheses on $\mathscr{C}$ or $\mathscr{D}$). Applying this to

$$F((\bullet, \bullet)) = X_\bullet \square Y_\bullet : \mathbf{\Delta}^{\mathrm{op}} \times \mathbf{\Delta}^{\mathrm{op}} \to \mathscr{M}$$

and pre-composing with diag, we get an isomorphism

$$X_p \square Y_p \cong \int^{(m,n) \in \mathbf{\Delta}^{\mathrm{op}} \times \mathbf{\Delta}^{\mathrm{op}}} (X_m \square Y_n) \times (\mathbf{\Delta}^{\mathrm{op}}(m, p) \times \mathbf{\Delta}(n, p))$$

of functors $p \in \mathbf{\Delta}^{\mathrm{op}} \to \mathscr{M}$. Commuting coends, we can reorganize the double coend

$$|X_\bullet \square Y_\bullet| \cong \int^{p \in \mathbf{\Delta}^{\mathrm{op}}} \left( \int^{(m,n) \in \mathbf{\Delta}^{\mathrm{op}} \times \mathbf{\Delta}^{\mathrm{op}}} (X_m \square Y_n) \times (\mathbf{\Delta}^{\mathrm{op}}(m, p) \times \mathbf{\Delta}^{\mathrm{op}}(n, p)) \right) \otimes \Delta[p]$$

as

$$\int^{(m,n) \in \mathbf{\Delta}^{\mathrm{op}} \times \mathbf{\Delta}^{\mathrm{op}}} (X_m \square Y_n) \otimes \left( \int^{p \in \mathbf{\Delta}^{\mathrm{op}}} (\mathbf{\Delta}^{\mathrm{op}}(m, p) \times \mathbf{\Delta}^{\mathrm{op}}(n, p)) \times \Delta[p] \right).$$

In the latter formula, the expression in parentheses is the coend formula for the geometric realization (in spaces) of the product of standard simplices (in simplicial sets) $\Delta[m]_\bullet \times \Delta[n]_\bullet$, which is $\Delta[m] \times \Delta[n]$ by the classic version of the lemma for geometric realization in spaces. This then constructs the natural isomorphism $|X_\bullet| \square |Y_\bullet| \cong |X_\bullet \square Y_\bullet|$, and a little more fiddling with coends shows that this natural transformation is symmetric monoidal. $\square$

Because of the previous lemma, we have a natural isomorphism $\mathcal{O}|X_\bullet| \cong |\mathcal{O}X_\bullet|$ that makes the appropriate diagrams commute, so that the geometric realization (in $\mathcal{M}$) of a simplicial object $A_\bullet$ in $\mathcal{M}[\mathcal{O}]$ acquires the natural structure of an $\mathcal{O}$-algebra. Moreover, the canonical maps $A_n \otimes \Delta[n] \to |A_\bullet|$ induce maps of $\mathcal{O}$-algebras $A_n \otimes^{\mathcal{M}[\mathcal{O}]} \Delta[n] \to |A_\bullet|$ that assemble into a natural map of $\mathcal{O}$-algebras

$$|A_\bullet|^{\mathcal{M}[\mathcal{O}]} \to |A_\bullet|.$$

In the case when $A_\bullet = \mathcal{O}X_\bullet$, under the identification of colimits $|\mathcal{O}X_\bullet|^{\mathcal{M}[\mathcal{O}]} = \mathcal{O}|X_\bullet|$, this map is the isomorphism $\mathcal{O}|X_\bullet| \to |\mathcal{O}X_\bullet|$ above. To see that it is an isomorphism for arbitrary $A_\bullet$, write $A_\bullet$ as a (reflexive) coequalizer

$$\mathcal{O}\mathcal{O}A_\bullet \rightrightarrows \mathcal{O}A_\bullet \to A_\bullet,$$

apply the functors, and compare diagrams.

## 5.8     Model structures for operadic algebras

The purpose of this section is to review the construction of model structures on some of the categories of operadic algebras that are of interest in homotopy theory; we use these in the next section in comparison theorems giving Quillen equivalences between some of these categories. Constructing model structures for algebras over operads is a special case of constructing model structures for algebras over monads; chapter VII of EKMM [94] seems to be an early reference for this kind of result, but it concentrates on the category of LMS-spectra and related categories. Schwede–Shipley [267] studies the general case of monads in cofibrantly generated monoidal model categories, which Spitzweck [280] specializes to the case of operads. Because less sharp results hold in the general case than in the special cases of interest, we state the results on model structures as a list of examples. This is an "example theorem" both in the sense that it gives a list of examples, but also in the sense that it fits into the general rubric of the kind of theorem that should hold very generally under appropriate technical hypotheses with essentially the same proof outline. Some terminology and notation is explained after the statement.

Example Theorem 5.8.1.   *Let $\mathcal{M}$ be a symmetric monoidal category with a cofibrantly generated model structure and let $\mathcal{O}$ be an operad in $\mathcal{M}$ from one of the examples listed below. Then the category of $\mathcal{O}$-algebras in $\mathcal{M}$ is a closed model category with*

(i)   *weak equivalences the underlying weak equivalences in $\mathcal{M}$,*
(ii)  *fibrations the underlying fibrations in $\mathcal{M}$, and*
(iii) *cofibrations the retracts of regular $\mathcal{O}I$-cofibrations.*

*This theorem holds in particular in the examples:*

(a)  *$\mathcal{M}$ is the category of symmetric spectra (of spaces or simplicial sets) with its positive stable model structure or orthogonal spectra with its positive stable model structure or* *the category of EKMM S-modules with its standard model structure (with $\square$ the smash product, $1$ the sphere spectrum) and $\mathcal{O}$ is any operad in $\mathcal{M}$.* [68, 8.1]

(b) $\mathcal{M}$ *is the category of spaces or simplicial sets (with $\square = \times$, $1 = *$), or simplicial R-modules for some simplicial commutative ring R (with $\square = \otimes_R$, $1 = R$) and $\mathcal{O}$ is any operad.*

(c) $\mathcal{M}$ *is the category of (unbounded) chain complexes in R-modules for a commutative ring R (with $\square = \otimes_R$, $1 = R$) and either $R \supset \mathbb{Q}$ or $\mathcal{O}$ admits a map of operads $\mathcal{O} \to \mathcal{O} \otimes \mathcal{E}$ which is a section for the map $\mathcal{O} \otimes \mathcal{E} \to \mathcal{O} \otimes \mathcal{C}om \cong \mathcal{O}$, where $\mathcal{E}$ is any $E_\infty$ operad that naturally acts on the normalized cochains of simplicial sets.* [37, 3.1.3]

(d) $\mathcal{M}$ *is a monoidal model category in the sense of* [267, 3.1] *that satisfies the Monoid Axiom of* [267, 3.3] *and $\mathcal{O}$ is a cofibrant operad in the sense of* [280, §3]. [280, §4, Theorem 4]

The category of EKMM $\mathbb{L}$-spectra [94, I§4] also fits into example (a) if we allow $\mathcal{M}$ to be a "weak" symmetric monoidal category in the sense of [94, II.7.1]; the theorem then covers categories of operadic algebras in LMS spectra for operads over the linear isometries operad that have the form $\mathcal{O} \times \mathcal{L} \to \mathcal{L}$; see [68, 3.5].

In part (c), we note that for an operad that satisfies the section condition (or when $R \supset \mathbb{Q}$), the functor $\mathcal{O}(n) \times_{R[\Sigma_n]} (-)$ preserves preserve exactness of (homologically) bounded-below exact sequences of $R$-free $R[\Sigma_n]$-modules (for all $n$). For operads that satisfy this more general condition but not necessarily the section condition, the algebra category still has a theory of cofibrant objects and a good homotopy theory for those objects; see, for example, [181, §2].

It is beyond the scope of this chapter to do a full review of closed model category theory terminology, but we recall that a "cofibrantly generated model category" has a set $I$ of "generating cofibrations" and a set $J$ of "generating acyclic cofibrations" for which the Quillen small object argument can be done (perhaps transfinitely, but in the examples of (a), (b), and (c), sequences suffice). Then

$$\mathbb{O}I = \{\mathbb{O}f \mid f \in I\}$$

is the set of maps of $\mathcal{O}$-algebras obtained by applying $\mathbb{O}$ to each of the maps in $I$. The point of $\mathbb{O}I$ is that a map of $\mathcal{O}$-algebras has the left lifting property with respect to $\mathbb{O}I$ in $\mathcal{O}$-algebras exactly when the underlying map in $\mathcal{M}$ has the left lifting property with respect to $I$. The same definition and observations apply replacing $I$ with $J$. The strategy for proving the previous theorem is to define the fibrations and weak equivalences of $\mathcal{O}$-algebras as in (i),(ii), and define cofibrations in terms of the left lifting property (obtaining the characterization in (iii) as a theorem). The advantage of this approach is that fibrations and acyclic fibrations are also characterized by lifting properties: a map of $\mathcal{O}$-algebras is a fibration if and only if it has the right lifting property with respect to $\mathbb{O}J$ and a map of $\mathcal{O}$-algebras is an acyclic fibration if and only if it has the right lifting property with respect to $\mathbb{O}I$. For these lifting properties, we can attempt the small object argument. We now outline the remaining steps in this approach.

Recall that a *regular $\mathbb{O}I$-cofibration* is a map formed as a (transfinite) composite of pushouts along coproducts of maps in $\mathcal{O}I$. This is the generalization of the notion of a *relative $\mathcal{O}I$-cell complex*, which is the colimit of a sequence of pushouts of coproducts of maps in $\mathcal{O}I$; in the case of examples (a), (b), and (c), in a regular $\mathcal{O}I$-cofibration the transfinite composite can always be replaced simply by a sequential composite and so a regular $\mathcal{O}I$-cofibration is a relative $\mathcal{O}I$-cell complex. The small object argument for $I$ and $J$ in $\mathcal{M}$ implies the small object argument for $\mathcal{O}I$ and $\mathcal{O}J$, which gives factorization in $\mathcal{O}$-algebras of a map as either a regular $\mathcal{O}I$-cofibration followed by an acyclic fibration or a regular $\mathcal{O}J$-cofibration followed by a fibration. (A small wrinkle comes up in going from the small object argument in $\mathcal{M}$ to the small object argument in $\mathcal{M}[\mathcal{O}]$ in the topological examples of (a) and (b): we need to check that regular $\mathcal{O}I$-cofibrations are nice maps, for example, closed inclusions on the constituent spaces; see the "Cofibration Hypothesis" of [94, VII§4] or [178, 5.3].)

This gets us most of the way to a model structure. Having defined a cofibration of $\mathcal{O}$-algebras as a map that has the left lifting property with respect to the acyclic fibrations, the free-forgetful adjunction shows that regular $\mathcal{O}I$-cofibrations are cofibrations; moreover, it follows formally that any cofibration is the retract of a regular $\mathcal{O}I$-cofibration: given a cofibration $f\colon A \to B$, factor it as $p \circ i$ for $i\colon A \to B'$ a regular $\mathcal{O}I$-cofibration and $p\colon B' \to B$ an acyclic fibration, then solving the lifting problem

$$\begin{array}{ccc} A & \xrightarrow{i} & B' \\ {\scriptstyle f}\downarrow & {\scriptstyle g}\nearrow & \downarrow{\scriptstyle p} \\ B & \xrightarrow[\text{id}]{} & B \end{array}$$

to produce a map $g\colon B \to B'$ exhibits $f$ as a retract of $i$.

$$\begin{array}{ccccc} A & \xrightarrow{\text{id}} & A & \xrightarrow{\text{id}} & A \\ {\scriptstyle f}\downarrow & & {\scriptstyle i}\downarrow & & \downarrow{\scriptstyle f} \\ B & \xrightarrow{g} & B' & \xrightarrow{p} & B \end{array}$$

We can try the same thing with regular $\mathcal{O}J$-cofibrations; they have the left lifting property with respect to all fibrations so are in particular cofibrations, but are they weak equivalences? This is the big question and what keeps us from having a fully general result for Theorem 5.8.1, especially in (c). If regular $\mathcal{O}J$-cofibrations are weak equivalences, then the trick in the previous argument shows that every acyclic cofibration is a retract of a regular $\mathcal{O}J$-cofibration, and the lifting property for acyclic cofibrations follows as does the other factorization, proving the model structure. (Conversely, if the model structure exists, because regular $\mathcal{O}J$-cofibrations have the left lifting property for all fibrations, it follows that they are weak equivalences.)

In many examples, including examples (a) and (b) in the theorem above, the homogeneous filtration on the pushout that we studied in Section 5.6 can be used to prove that regular $\mathcal{O}J$-cofibrations are weak equivalences. Specifically, for $X \to Y$ a map in $J$, taking $A' = A \amalg_{\mathcal{O}X}^{\mathcal{M}[\mathcal{O}]} \mathcal{O}Y$, the case $m = 0$ of the filtration on the enveloping operad for $A$ gives a filtration on $A'$ by objects of $\mathcal{M}$ starting from $A$. Now from the inductive definition of $Q_{k-1}^k(X \to Y)$ in (5.6.2), it can be checked in examples (a) and (b) that the map $Q_{k-1}^k(X \to Y) \to Y^{(k)}$ is an equivariant Hurewicz cofibration of the underlying spaces or a monomorphism of the underlying simplicial sets as well as being a weak equivalence. The pushout (5.6.3) then identifies the maps in the filtration of $A'$ as weak equivalences as well. (This approach can also be used to prove versions of the "Cofibration Hypothesis" of [94, VII§4] or [178, 5.3] that regular $\mathcal{O}I$-cofibrations are closed inclusions on the constituent spaces.)

Example (d) is similar, except that it uses a filtration argument on the construction of a cofibrant operad; see [280, §4].

Example (c) fits into the case of the general theorem of Schwede–Shipley [267, 2.3], where every object is fibrant and has a path object. To complete the argument here, we need to show that every map $f : A \to B$ factors as a weak equivalence followed by a fibration:

$$A \xrightarrow{\simeq} A' \twoheadrightarrow B.$$

We then get the factorization of an acyclic cofibration followed by a fibration by using the factorization already established:

$$A \xrightarrowtail{\simeq} A'' \xrightarrow{\simeq} A' \twoheadrightarrow B.$$

In the case of (c) where we hypothesize a map of operads $\mathcal{O} \to \mathcal{O} \otimes \mathcal{E}$, this map gives a natural $\mathcal{O}$-algebra structure on $B \otimes C^*(-)$; the hypothesis that the composite map on $\mathcal{O}$ is the identity implies that the canonical isomorphism

$$B \cong B \otimes C^*(\Delta[0])$$

is an $\mathcal{O}$-algebra map. Looking at the maps between $\Delta[0]$ and $\Delta[1]$, we get maps of $\mathcal{O}$-algebras

$$B \to B \otimes C^*(\Delta[1]) \to B \times B$$

and the usual mapping path object construction

$$A \xrightarrow{\simeq} A \times_B (B \otimes C^*(\Delta[1])) \twoheadrightarrow B$$

consists of maps of $\mathcal{O}$-algebras and gives the factorization. In the case when $R \supset \mathbb{Q}$, the polynomial de Rham functor $A^*$ reviewed in Section 5.12 is a functor from simplicial sets to commutative differential graded $\mathbb{Q}$-algebras, which can be used in the same way to construct a factorization

$$A \xrightarrow{\simeq} A \times_B (B \otimes_{\mathbb{Q}} A^*(\Delta[1])) \twoheadrightarrow B.$$

In the case of operadic algebras in spaces in example (b) and EKMM $S$-modules in example (a), we have another argument taking advantage of the topological enrichment. In these examples, the maps in $J$ are deformation retractions, and so the maps in $\mathcal{O}J$ are deformation retractions in the category of $\mathcal{O}$-algebras. It follows that regular $\mathcal{O}J$-cofibrations are also deformation retractions and in particular homotopy equivalences. Since homotopy equivalences are weak equivalences, regular $\mathcal{O}J$-cofibrations are weak equivalences in examples where this argument can be made. The specific examples again fit into the case of [267, 2.3] where every object is fibrant and has a path object.

## 5.9        Comparison and rectification theorems for operadic algebras

This section discusses Quillen equivalences and Quillen adjunctions between the model categories in Example Theorem 5.8.1. When we change from simplicial sets to spaces or when we change the underlying symmetric monoidal category between the Quillen equivalent modern categories of spectra, we get Quillen equivalences of categories of operadic algebras under only mild technical hypotheses on the operad; this gives several comparison theorems. We also consider Quillen adjunctions and Quillen equivalences obtained by change of operads. In wide generality, the augmentation map $A \to \mathcal{A}ss$ for an $A_\infty$ operad induces a Quillen equivalence between categories of algebras. Likewise, in the case of modern categories of spectra, the augmentation map $\mathcal{E} \to \mathcal{C}om$ for an $E_\infty$ operad induces a Quillen equivalence between categories of algebras. These comparison theorems are rectification theorems in that they show that a homotopical algebraic structure can be replaced up to weak equivalence with a strict algebraic structure.

We begin by reviewing the change of operad adjunction. Let $f : A \to B$ be a map of operads in a symmetric monoidal category $\mathcal{M}$. Such a map certainly gives a restriction functor $U_f$ from $B$-algebras to $A$-algebras, and under mild hypothesis, this functor has a left adjoint. As in the discussion of colimits in Section 5.6, if we assume that $\mathcal{M}$ satisfies the hypotheses of Proposition 5.4.7 then we can define $P_f : \mathcal{M}[A] \to \mathcal{M}[B]$ by the reflexive coequalizer

$$\mathbb{B}(\mathbb{A}A) \rightrightarrows \mathbb{B}A \to P_f(A),$$

where $\mathbb{A}$ and $\mathbb{B}$ denote the monads associated to $A$ and $B$, one arrow is induced by the $A$-algebra structure on $A$, and the other arrow is the composite $\mathbb{B}\mathbb{A} \to \mathbb{B}\mathbb{B} \to \mathbb{B}$ induced by the map of operads $f$ and the monadic product on $\mathbb{B}$. As a side remark, not related to the rest of this section, we note that in this situation the category $B$-algebras can be identified as the category of algebras for the monad $U_f P_f$ in $\mathcal{M}[A]$ (for a general formal proof, see [94, II.6.6.1]).

Now suppose that $\mathcal{M}$ has a closed model structure and $\mathcal{M}[A]$ and $\mathcal{M}[B]$ are closed model categories with fibrations and weak equivalences created in $\mathcal{M}$. For a map of operads $f : A \to B$, we then get a Quillen adjunction

$$P_f : \mathcal{M}[A] \rightleftarrows \mathcal{M}[B] : U_f.$$

When can we expect it to be a Quillen equivalence? It is tempting to define an equivalence of operads in $\mathcal{M}$ to be a map $f$ such that derived adjunction induces an equivalence of homotopy categories; then we have a tautological result that an equivalence of operads induces a Quillen equivalence of model structures. Instead we propose the following definition, which leads to a theorem with some substance

(Example Theorem 5.9.5). It is the condition used in practice in proving comparison and rectification theorems.

**Definition 5.9.1.** Let $\mathcal{M}$ be a closed model category with countable coproducts and with a symmetric monoidal product that preserves countable colimits in each variable. We say that a map $f \colon \mathcal{A} \to \mathcal{B}$ of operads in $\mathcal{M}$ is a *derived monad equivalence* if the induced map $\mathbb{A}Z \to \mathbb{B}Z$ is a weak equivalence for every cofibrant object $Z$ in $\mathcal{M}$.

Though we have not put enough hypotheses on $\mathcal{M}$ to ensure it, in practice countable coproducts of reasonable objects in $\mathcal{M}$ will preserve and reflect weak equivalences and then $f$ will be a derived monad equivalence if and only if each of the maps

$$\mathcal{A}(m) \,\square_{\Sigma_m}\, Z^{(m)} \to \mathcal{B}(m) \,\square_{\Sigma_m}\, Z^{(m)}$$

is a weak equivalence. In our examples of main interest, we have more intrinsic sufficient conditions for a map of operads to be a derived monad equivalence.

**Example 5.9.2.** In the category of spaces (or more generally, any topological or simplicial model category), a map of operads $f \colon \mathcal{A} \to \mathcal{B}$ that induces an equivariant homotopy equivalence $\mathcal{A}(m) \to \mathcal{B}(m)$ for all $m$ is a derived monad equivalence. Indeed, the map $\mathbb{A}Z \to \mathbb{B}Z$ is a homotopy equivalence for all $Z$, and a homotopy equivalence in a topological or simplicial model category is a weak equivalence. As a special case, when $\overline{\mathcal{A}}$ is a non-symmetric operad with $\overline{\mathcal{A}}(m)$ contractible for all $m$, the map of operads $\mathcal{A} \to \mathcal{A}$ss is a derived monad equivalence.

**Example 5.9.3.** In the category of symmetric spectra (of spaces or simplicial sets) with its positive stable model structure or the category of orthogonal spectra with its positive model structure, a map of operads $f \colon \mathcal{A} \to \mathcal{B}$ that induces a (non-equivariant) weak equivalence $\mathcal{A}(n) \to \mathcal{B}(n)$ is a derived monad equivalence. This can be proved by generalizing the argument of [178, 15.5] (see [68, 8.3.(i)] for slightly more details). In the case of EKMM $S$-modules, if $f \colon \mathcal{A} \to \mathcal{B}$ is a map of operads of spaces that is a (non-equivariant) homotopy equivalence $\mathcal{A}(n) \to \mathcal{B}(n)$ for all $n$, then $\Sigma_+^\infty f$ is a derived monad equivalence. This can be proved by generalizing the argument of [94, III.5.1]. (See [68, 8.3.(ii)] for a more general statement.) In particular, in these categories, the augmentation map $\mathcal{E} \to \mathcal{C}$om for an $E_\infty$ operad (assumed to come from spaces in the EKMM $S$-module case) is a derived monad equivalence.

**Example 5.9.4.** In the context of chain complexes of $R$-modules, a map of operads $\mathcal{A} \to \mathcal{B}$ that is an $R[\Sigma_n]$-module chain homotopy equivalence $\mathcal{A}(n) \to \mathcal{B}(n)$ for all $n$ is a derived monad equivalence. If the functors $\mathcal{A}(n) \otimes_{R[\Sigma_n]} (-)$ and $\mathcal{B}(n) \otimes_{R[\Sigma_n]} (-)$ preserve exactness of (homologically) bounded-below exact sequences of $R$-free $R[\Sigma_n]$-modules (for all $n$), then a weak equivalence $\mathcal{A} \to \mathcal{B}$ is a derived monad equivalence. This occurs in particular for part (c) of Example Theorem 5.8.1 when $\mathcal{A}$ and $\mathcal{B}$ both satisfy the stated operad hypotheses.

To go with these examples, we have the following example theorem.

Example Theorem 5.9.5.  *Let $\mathcal{M}$ be a symmetric monoidal category and $f: A \to B$ a map of operads in $\mathcal{M}$, where $\mathcal{M}$, $A$, and $B$ fall into one of the examples of Example Theorem 5.8.1(a)-(c). If $f$ is a derived monad equivalence then the Quillen adjunction $P_f: \mathcal{M}[A] \rightleftarrows \mathcal{M}[B] : U_f$ is a Quillen equivalence.*

Again, as in the previous section, this is an "example theorem" in that it gives an example of the kind of theorem that holds much more generally with a proof that can also be adapted to work much more generally. We outline the proof after the change of categories theorem below, as the arguments for both are quite similar.

In terms of change of categories, one should expect comparison theorems of the following form to hold quite generally:

*Let $L: \mathcal{M} \rightleftarrows \mathcal{M}' : R$ be a Quillen equivalence between monoidal model categories with $L$ strong symmetric monoidal, and let $\mathcal{O}$ be an operad in $\mathcal{M}$. With some technical hypotheses, the adjunction*

$$L: \mathcal{M}[\mathcal{O}] \rightleftarrows \mathcal{M}'[L\mathcal{O}] : R$$

*on operadic algebra categories is also a Quillen equivalence.*

A minimal technical hypothesis is that $L\mathcal{O}$ be "the right thing" and an easy way to ensure this is to put some kind of cofibrancy condition on the objects $\mathcal{O}(n)$. In our cases of interest, we could certainly state such a theorem, but it would not cover the example in modern categories of spectra when $\mathcal{O}$ is the suspension spectrum functor applied to an operad of spaces; for such an operad, the spectra $\mathcal{O}(n)$ will not be cofibrant. On the other hand, in these examples the right adjoint preserves all weak equivalences and not just weak equivalences between fibrant objects; in this setup it seems more convenient to consider an operad $\mathcal{O}'$ in $\mathcal{M}'$ and a map of operads $\mathcal{O} \to R\mathcal{O}'$ (or equivalently, $L\mathcal{O} \to \mathcal{O}'$) that induces a weak equivalence

$$\mathcal{O}Z \to R(\mathcal{O}'LZ)$$

for all cofibrant objects $Z$ of $\mathcal{M}$. We state such a theorem for our examples of interest.

Example Theorem 5.9.6.  *Let $L: \mathcal{M} \rightleftarrows \mathcal{M}' : R$ be one of the Quillen adjunctions of symmetric monoidal categories listed below. Let $A$ be an operad in $\mathcal{M}$, let $B$ be an operad in $\mathcal{M}'$, and let $f: A \to RB$ be a map of operads that induces a weak equivalence*

$$AZ \to R(BLZ)$$

*for all cofibrant objects $Z$ of $\mathcal{M}$. Then the induced Quillen adjunction*

$$P_{L,f}: \mathcal{M}[A] \rightleftarrows \mathcal{M}'[B] : U_{R,f}$$

*is a Quillen equivalence. This theorem holds in particular in the examples:*

(a) *$\mathcal{M}$ is the category of simplicial sets (with the usual model structure) or the category of symmetric spectra of simplicial sets, $\mathcal{M}'$ is the category of spaces or the category of symmetric spectra in spaces (respectively), and $L, R$ is the geometric realization, singular simplicial set adjunction.*

(b) $\mathcal{M}$ *is the category of symmetric spectra,* $\mathcal{M}'$ *is the category of orthogonal spectra and* $L, R$ *is the prolongation, restriction adjunction of* [178, p. 442].

(c) $\mathcal{M}$ *is the category of symmetric spectra or orthogonal spectra,* $\mathcal{M}'$ *is the category of EKMM S-modules, and* $L, R$ *is the adjunction of* [263] *or* [177, I.1.1].

As indicated in the paragraph above the statement, the statement takes advantage of the fact that in the examples being considered in this section, the right adjoint preserves all weak equivalences; a general statement for other examples should use a fibrant replacement for $\mathbb{B}LZ$ in place of $\mathbb{B}LZ$. The proof sketch below also takes advantage of this property of the right adjoint. In generalizing the argument to the case when fibrant replacement is required in the statement, the fibrant replacement of the filtration can be performed in $\mathcal{M}'$.

The proof of the theorems above uses the homogeneous filtration on a pushout of the form $A' = A \sqcup_{\mathbb{O}X}^{\mathcal{M}[\mathcal{O}]} \mathbb{O}Y$ studied in Section 5.6. This is the $m = 0$ case of the filtration on the enveloping operad $\mathcal{U}_{A'}^{\mathcal{O}}$, and we will need to use the filtration on the whole operad for an inductive argument even though we are only interested in the $m = 0$ case in the end. We will use uniform notation in the sketch proof that follows, taking $\mathcal{M}' = \mathcal{M}$ with adjoint functors $L$ and $R$ to be the identity in the case of Example Theorem 5.9.5. We use the notation $I$ for the preferred set of generators for the cofibrations of $\mathcal{M}$ (as in Section 5.8).

Because fibrations and weak equivalences in the algebra categories are created in the underlying symmetric monoidal categories, the adjunction $P_{L,f}, U_{R,f}$ is automatically a Quillen adjunction (as indicated already in the statements), and we just have to prove that the unit of the adjunction

$$A \rightarrow U_{R,f}(P_{L,f}A) \tag{5.9.1}$$

is a weak equivalence for any cofibrant $\mathcal{A}$-algebra $A$. Every cofibrant $\mathcal{A}$-algebra is the retract of an $\mathbb{A}I$-cell $\mathcal{A}$-algebra, and so it suffices to consider the case when $A$ is an $\mathbb{A}I$-cell $\mathcal{A}$-algebra; then $A = \operatorname{colim} A_n$ where $A_0 = \mathcal{A}(0)$ and each $A_{n+1}$ is formed from $A_n$ by cell attachment (of possibly an infinite coproduct of cells). As we shall see below, the underlying maps $A_n \rightarrow A_{n+1}$ are nice enough that $A$ is the homotopy colimit (in $\mathcal{M}$ or $\mathcal{M}[\mathcal{A}]$) of the system of the finite stages $A_n$ (this is the subject of the "Cofibration Hypothesis" of [94, VII§4] mentioned parenthetically in the previous section). Analogous observations apply for $P_{L,f}A$, which is a cell $\mathbb{B}LI$-algebra with stages $P_{L,f}A_n$. Thus, it will be enough to see that (5.9.1) is a weak equivalence for each $A_n$. By the hypothesis of the theorem, we know that this holds for $A_0$ (which is the free $\mathcal{A}$-algebra on the initial object of $\mathcal{M}$); moreover, as the enveloping operad of $A_0$ is $\mathcal{A}$ and the enveloping operad of $P_{L,f}A_0$ is $\mathcal{B}$, we can assume as an inductive hypothesis that

$$\mathbb{U}_{A_n}^{\mathcal{A}} Z \rightarrow \mathbb{U}_{P_{L,f}A_n}^{\mathcal{B}} LZ$$

is a weak equivalence for all cofibrant $Z$; in other words, we can assume by induction that the hypothesis of the theorem holds for the map of enveloping operads $\mathcal{U}_{A_n}^{\mathcal{A}} \rightarrow R(\mathcal{U}_{P_{L,f}A_n}^{\mathcal{B}})$. It then suffices to prove that the hypothesis of the theorem holds for the map of enveloping operads $\mathcal{U}^{\mathcal{A}}_{A_{n+1}} \to R(\mathcal{U}^{\mathcal{B}}_{P_{L,f}A_{n+1}})$; this is because in the categories $\mathcal{M}$ and $\mathcal{M}'$ of the examples, countable coproducts preserve and reflect weak equivalences and the unit map $A_{n+1} \to U_{R,f}(P_{L,f}A_{n+1})$ is the restriction of the map of monads to the homogeneous degree zero summand (at least in the homotopy category of $\mathcal{M}$).

To prove this, let $X \to Y$ be the coproduct of maps in $I$ such that $A_{n+1} = A_n \amalg^{\mathcal{M}[\mathcal{A}]}_{\mathbb{A}X} \mathbb{A}Y$ and consider the filtration on $\mathcal{U}^{\mathcal{A}}_{A_{n+1}}(m)$ and $\mathcal{U}^{\mathcal{B}}_{P_{L,f}A_{n+1}}(m)$ studied in Section 5.6. We note that the induction hypothesis on $A_n$ also implies that the map

$$\mathcal{U}^{\mathcal{A}}_{A_n}(m) \square_{\Sigma_{m_1} \times \cdots \times \Sigma_{m_i}} (Z_1^{(m_1)} \square \cdots \square Z_i^{(m_i)})$$
$$\to R(\mathcal{U}^{\mathcal{B}}_{P_{L,f}A_n}(m) \square_{\Sigma_{m_1} \times \cdots \times \Sigma_{m_i}} (LZ_1^{(m_1)} \square \cdots \square LZ_i^{(m_i)}))$$

is a weak equivalence for all cofibrant objects $Z_1, \ldots, Z_i$ (where $m = m_1 + \cdots + m_i$) as this is a summand of the map

$$\mathcal{U}^{\mathcal{A}}_{A_n}(m) \square_{\Sigma_m} (Z_1 \amalg \cdots \amalg Z_i)^{(m)} \to R(\mathcal{U}^{\mathcal{B}}_{P_{L,f}A_n}(m) \square_{\Sigma_m} L(Z_1 \amalg \cdots \amalg Z_i)^{(m)}).$$

Looking at the pushout square (5.6.2) that inductively defines $Q_i^k(X \to Y)$, a bit of analysis shows that in our example categories the maps $Q_{i-1}^k \to Q_i^k$ are $\Sigma_k$-equivariant Hurewicz cofibrations (or in the simplicial categories, maps that geometrically realize to such). It follows that for any cofibrant object $Z$, the maps

$$\mathcal{U}^{\mathcal{A}}_{A_n}(k+m) \square_{\Sigma_k \times \Sigma_m} (Q_{i-1}^k(X \to Y) \square Z^{(m)})$$
$$\to \mathcal{U}^{\mathcal{A}}_{A_n}(k+m) \square_{\Sigma_k \times \Sigma_m} (Q_i^k(X \to Y) \square Z^{(m)})$$

are (or geometrically realize to) Hurewicz cofibrations (likewise in $\mathcal{M}'$) and that the maps

$$\mathcal{U}^{\mathcal{A}}_{A_n}(k+m) \square_{\Sigma_k \times \Sigma_m} (Q_i^k(X \to Y) \square Z^{(m)})$$
$$\to R(\mathcal{U}^{\mathcal{B}}_{P_{L,f}A_n}(k+m) \square_{\Sigma_k \times \Sigma_m} (Q_i^k(LX \to LY) \square LZ^{(m)}))$$

are weak equivalences. Now the pushout square (5.6.3) shows that for any cofibrant object $Z$, at each filtration level $k$, the map

$$F^{k-1}\mathcal{U}^{\mathcal{A}}_{A_{n+1}}(m) \square_{\Sigma_m} Z^{(m)} \to F^k\mathcal{U}^{\mathcal{A}}_{A_{n+1}}(m) \square_{\Sigma_m} Z^{(m)}$$

is (or geometrically realizes to) a Hurewicz cofibration (likewise in $\mathcal{M}'$) and that the maps

$$F^k\mathcal{U}^{\mathcal{A}}_{A_{n+1}}(m) \square_{\Sigma_m} Z^{(m)} \to R(F^k\mathcal{U}^{\mathcal{B}}_{P_{L,f}A_{n+1}}(m) \square_{\Sigma_m} LZ^{(m)})$$

are weak equivalences. The colimit is then weakly equivalent to the homotopy colimit and we get a weak equivalence

$$\mathcal{U}^{\mathcal{A}}_{A_{n+1}}(m) \square_{\Sigma_m} Z^{(m)} \to R(\mathcal{U}^{\mathcal{B}}_{P_{L,f}A_{n+1}}(m) \square_{\Sigma_m} LZ^{(m)}),$$

completing the induction and the sketch proof of Example Theorems 5.9.5 and 5.9.6.

The argument above proved the comparison theorems by proving equivalences of enveloping operads. Since the unary part of the enveloping operad is the enveloping algebra, we also get module category comparison results. We state this as the following corollary, which says that as long as the algebras are cofibrant, changing categories by Quillen equivalences and the algebras by derived monad equivalences results in Quillen equivalent categories of modules.

**Corollary 5.9.7.** *Let* $L\colon \mathcal{M} \rightleftarrows \mathcal{M}'\colon R$ *be one of the Quillen adjunctions of symmetric monoidal categories in Example Theorem 5.9.6 or the identity functor adjunction on one of the categories in Example Theorem 5.9.5. Let* $f\colon A \to R\mathcal{B}$ *be a map of operads that induces a weak equivalence* $AZ \to R(\mathbb{B}LZ)$ *for all cofibrant objects* $Z$, *and let* $g\colon A \to RB$ *be a weak equivalence of* $A$-*algebras for an* $A$-*algebra* $A$ *and a* $\mathcal{B}$-*algebra* $B$. *If* $A$ *and* $B$ *are cofibrant (in* $\mathcal{M}[A]$ *and* $\mathcal{M}'[\mathcal{B}]$, *respectively), then* $f$ *and* $g$ *induce a Quillen equivalence of the category of* $(A, A)$-*modules and the category of* $(\mathcal{B}, B)$-*modules.*

*Sketch proof.* The argument above shows that under the given hypotheses, the map of $\square$-monoids $U^A A \to R(U^B B)$ is a weak equivalence. The left and right adjoint functors in the Quillen adjunction on module categories are given by $U^B B \square_{L(U^A A)} L(-)$ and $R$, respectively. These both preserve coproducts, homotopy cofiber sequences, and sequential homotopy colimits up to weak equivalence. It follows that the unit of the adjunction $X \to R(U^B B \square_{L(U^A A)} LX)$ is a weak equivalence for every cofibrant $A$-module $X$.                                                    □

The analogous result also holds for modules over algebras on non-symmetric operads, proved by essentially the same filtration argument: we have a non-symmetric version $\overline{U}^{\mathcal{O}}_A(m)$ of Construction 5.6.4. In this case, the resulting objects do not assemble into an operad; nevertheless, $\overline{U}^{\mathcal{O}}_A(1)$ still has the structure of a $\square$-monoid and coincides with the (non-symmetric) enveloping algebra $\overline{U}^{\mathcal{O}}A$. The non-symmetric analogue of (5.6.3) holds, and the filtration argument (under the hypotheses of the previous corollary) proves that the map $\overline{U}^A A \to R(\overline{U}^B B)$ is a weak equivalence of $\square$-monoids. We conclude that the unit map $X \to R(\overline{U}^B B \square_{L\overline{U}^A A} LX)$ is a weak equivalence for every cofibrant $A$-module $X$.

## 5.10  Enveloping algebras, Moore algebras, and rectification

In the special case of Example 5.9.2, Example Theorem 5.9.5 gives an equivalence of the homotopy category of $A_\infty$ algebras (over a given $A_\infty$ operad) with the homotopy category of associative algebras, in particular constructing an associative algebra rectification of an $A_\infty$ algebra. We know another way to construct an associative algebra from an $A_\infty$ algebra, namely the (non-symmetric) enveloping algebra. In the case when the $A_\infty$ operad is the operad of little $1$-cubes $\overline{\mathcal{C}}_1$, there is also a classical rectification called the Moore algebra. The purpose of this section is to compare these constructions.

We first consider the rectification of Example Theorem 5.9.5 and the non-symmetric enveloping algebra. Let $\overline{\mathcal{O}}$ be a non-symmetric operad and $\epsilon\colon \overline{\mathcal{O}} \to \overline{\mathcal{A}ss}$ a weak equivalence. Under the hypotheses of Example Theorem 5.9.5, the rectification (change of operads) functor $P_\epsilon$ associated to $\epsilon$ gives a $\square$-monoid $P_\epsilon A$ and a map of $\bar{\mathcal{O}}$-algebras $A \to P_\epsilon A$ that is a weak equivalence when $A$ is cofibrant. As part of the proof of Example Theorem 5.9.5, we get a weak equivalence of enveloping operads

$$\mathcal{U}^{\mathcal{O}}_A \to \mathcal{U}^{\overline{Ass}}_{P_\epsilon A}.$$

As mentioned at the end of the previous section, the non-symmetric version of this argument also works to give a weak equivalence of $\square$-monoids

$$\bar{U}^{\bar{\mathcal{O}}} A \to \bar{U}^{\overline{Ass}}(P_\epsilon A).$$

Moreover, in the case of the associative algebra operad $\overline{Ass}$, we have a natural isomorphism of $\square$-monoids $\bar{U}^{\overline{Ass}} M \to M$ for any $\square$-monoid $M$. Putting this together, we get:

**Theorem 5.10.1.** *Let $\mathcal{M}$ be a symmetric monoidal category and $\mathcal{O}$ an $A_\infty$ operad that fall into one of the examples of Theorem 5.8.1(a)–(c). Write $\epsilon \colon \bar{\mathcal{O}} \to \overline{Ass}$ for the weak equivalence identifying $\mathcal{O}$ as an $A_\infty$ operad. If $A$ is a cofibrant $\mathcal{O}$-algebra then the natural maps*

$$A \to P_\epsilon A \cong \bar{U}^{\overline{Ass}} P_\epsilon A \leftarrow \bar{U}^{\bar{\mathcal{O}}} A$$

*are weak equivalences of $\mathcal{O}$-algebras.*

We now focus on $A_\infty$ algebras for the little $1$-cubes operad $\bar{\mathcal{C}}_1$, where we can describe results both more concretely and in much greater generality. For the rest of the section we work in the context of a symmetric monoidal category enriched over topological spaces as in Section 5.7: Let $\mathcal{M}$ be a closed symmetric monoidal category with countable colimits, and let $L \colon \mathcal{S} \to \mathcal{M}$ be strong symmetric monoidal left adjoint functor (whose right adjoint we denote as $R$). Then, by Theorem 5.7.1, $\mathcal{M}$ becomes enriched over topological spaces and we have a notion of homotopies and homotopy equivalences in $\mathcal{M}$, defined in terms of mapping spaces or equivalently in terms of tensor with the unit interval. We also have $L\bar{\mathcal{C}}_1$ as a non-symmetric operad in $\mathcal{M}$; for an $L\bar{\mathcal{C}}_1$-algebra $A$, we give a concrete construction of the enveloping algebra $\bar{U}A$, mostly following [184, §2]. We first write the formulas and then explain where they come from.

**Construction 5.10.2.** [184, §2] Let $\bar{D}$ be the space of subintervals of $[0,1]$ and let $D$ be the subspace of $\bar{D}$ of those intervals that do not start at $0$. We have a canonical isomorphism $\bar{D} \cong \bar{\mathcal{C}}_1(1)$ (sending a subinterval to the $1$-tuple containing it) that we elide notation for. Under this isomorphism, the composition law $\Gamma^1_1$ defines a pairing $\gamma \colon \bar{D} \times \bar{D} \to \bar{D}$ that satisfies the formula

$$\gamma([x,y],[x',y']) = [x + (y-x)x', \, x + (y-x)y'].$$

We note that $\gamma$ restricts to a pairing $D \times D \to D$, and that for formal reasons $\gamma$ is associative:

$$\gamma(\gamma([x,y],[x',y']),[x'',y'']) = [x+(y-x)x'+(y-x)(y'-x')x'', x+(y-x)x'+(y-x)(y'-x')y'']$$
$$= \gamma([x,y],\gamma([x',y'],[x'',y''])),$$

and unital:

$$\gamma([0,1],[x,y]) = [x,y] = \gamma([x,y],[0,1]),$$

making $\bar{D}$ a topological monoid and $D$ a subsemigroup. Define $\alpha\colon D \times D \to \bar{\mathcal{C}}_1(2)$ by

$$\alpha([x,y],[x',y']) = \left(\left[0, \frac{x}{x+(y-x)x'}\right], \left[\frac{x}{x+(y-x)x'}, 1\right]\right).$$

Let $DA$ be the object of $\mathcal{M}$ defined by the pushout diagram

$$
\begin{array}{ccc}
LD \square 1 & \longrightarrow & LD \square A \\
\downarrow & & \downarrow \\
L\bar{D} \square 1 & \longrightarrow & DA
\end{array}
$$

where the top map is induced by the composite of the isomorphism $1 \cong L\bar{\mathcal{C}}_1(0)$ (from the strong symmetric monoidal structure on $L$) and the $L\bar{\mathcal{C}}_1$-action map $L\bar{\mathcal{C}}_1(0) \to A$. We use $\gamma$ and $\alpha$ to define a multiplication on $DA$ as follows. We use the map

$$(LD \square A) \square (LD \square A) \to LD \square A \to DA$$

coming from the map

$$(LD \square A) \square (LD \square A) \cong L(D \times D) \square (A \square A) \to$$
$$L(D \times \bar{\mathcal{C}}_1(2)) \square (A \square A) \cong LD \square (L\bar{\mathcal{C}}_1(2) \square (A \square A)) \to LD \square A$$

induced by the map $(\gamma, \alpha)\colon D \times D \to D \times \bar{\mathcal{C}}_1(2)$ and the $L\bar{\mathcal{C}}_1$-action map on $A$. We note that both associations

$$(LD \square A) \square (LD \square A) \square (LD \square A) \to LD \square A$$

coincide: both factor through the map

$$(LD \square A) \square (LD \square A) \square (LD \square A) \cong L(D \times D \times D) \square A^{(3)} \to L(D \times \bar{\mathcal{C}}_1(3)) \square A^{(3)}$$

induced by the map $D \times D \times D \to D \times \bar{\mathcal{C}}_1(3)$ given on the $D$ factor as $\gamma \circ (\gamma \times \mathrm{id}) = \gamma \circ (1 \times \gamma)$ and on the $\bar{\mathcal{C}}_1(3)$ factor by the formula

$$[x,y],[x',y'],[x'',y''] \mapsto ([0,a],[a,b],[b,1]),$$

where

$$a = \frac{x}{x+(y-x)(x'+(y'-x')x'')}, \qquad b = \frac{x+(y-x)x'}{x+(y-x)(x'+(y'-x')x'')}.$$

When restricted to maps

$$(LD \square 1) \square (LD \square A), (LD \square A) \square (LD \square 1) \to DA,$$

this map coincides with the map induced by just $\gamma$ and the unit isomorphism of $\mathcal{M}$ and so extends to compatible maps

$$(L\bar{D} \square 1) \square (L\bar{D} \square 1) \to DA,$$
$$(L\bar{D} \square 1) \square (LD \square A) \to DA,$$
$$(LD \square A) \square (L\bar{D} \square 1) \to DA,$$

and defines an associative multiplication on $DA$. The map $1 \to DA$ induced by the inclusion of the element $[0,1]$ of $\bar{D}$ is a unit for this multiplication.

To understand the construction, it is useful to think of $D$ as a subspace of $\bar{C}_1(2)$ rather than a subspace of $\bar{C}_1(1)$, via the embedding

$$[x,y] \mapsto ([0,x],[x,y]).$$

Then we have a map $DA \to \bar{U}A$ sending $L\bar{D}\square 1$ and $LD\square A$ to the 0 and 1 summands

$$L\bar{D} \square 1 \cong L\bar{C}_1 \square A^{(0)} \quad \text{and} \quad LD \square A \to L\bar{C}_1(2) \square A$$

in the coequalizer (5.5.1) for $\bar{U}A$. We also have a map back that sends the summand $L\bar{C}_1(n+1)\square A^{(n)}$ (for $n \geq 1$) to $LD\square A$ by remembering just the last interval and using the rest to do the multiplication on $A$; specifically, for $[x_1,y_1],\ldots,[x_{n+1},y_{n+1}]$, we use the element of $\bar{C}_1(n)$ corresponding to

$$\left[\frac{x_1}{x_{n+1}}, \frac{y_1}{x_{n+1}}\right], \ldots, \left[\frac{x_n}{x_{n+1}}, \frac{y_n}{x_{n+1}}\right]$$

for the map $A^{(n)} \to A$. It is straightforward to check that these maps give inverse isomorphisms of objects of $\mathcal{M}$; see [184, 2.5].

The isomorphism of the previous paragraph then forces the formula for the multiplication. Intuitively speaking, the first box in $D$ (viewed as a subset of $\bar{C}_1(2)$) holds the algebra (from the tensor) and the second box is a placeholder to plug in the module variable; the complement $\bar{D} \setminus D$ corresponds to the first box having length zero and then only the unit of the algebra can go there. For the composition, the right copy gets plugged into the second box of the left copy to give an element of $\bar{C}_1(3)$ (i.e., the operadic composition $\ell \circ_2 r = \Gamma^2_{1,2}(\ell;1,r)$, where $\ell$ is the element of the left copy of $D$ and $r$ is the element of the right copy of $D$); the first and second boxes are on the one hand rescaled to an element of $\bar{C}_1(2)$ that does the multiplication on the copies of $A$ and on the other hand joined to give with the third box the new element of $D$, viewed as a subspace of $\bar{C}_1(2)$. The associativity is straightforward to visualize in terms of plugging in boxes when written down on paper. (See Section 2 of [184].) When one of the elements comes from $\bar{D} \setminus D$, the corresponding copy of $A$ is restricted to the unit 1 and the first box of zero length also works like a unit.

Using the isomorphism of $\square$-monoids $\bar{U}A \cong DA$, we have the following comparison theorem for the underlying objects of $\bar{U}A$ and $A$.

**Proposition 5.10.3** ([184, 1.1]). *The map of $\bar{U}A$-modules $\bar{U}A \cong 1 \square \bar{U}A \to A$ induced by the map $1 \cong L\bar{C}_1(0) \to A$ is a homotopy equivalence of objects of $\mathcal{M}$.*

*Proof.* In concrete terms, the map in the statement is induced by the map

$$LD \,\square\, A \to L\bar{C}_1(1) \,\square\, A \to A$$

for the map $D \to \bar{C}_1(1)$ that sends $[x, y]$ to $([0, x])$, which is compatible with the map

$$L\bar{D} \,\square\, 1 \to 1 \to A.$$

We can use any element of $D$ to produce a map (in $\mathcal{M}$) from $A$ to $\bar{U}A$; a path to the operad identity element 1 in $\bar{C}_1(1)$ (which corresponds to $[0,1] \subseteq [0,1]$) then induces a homotopy of the composite map $A \to A$ to the identity map of $A$. We can construct a homotopy from the composite to the identity on $\bar{U}A$ using a homotopy of self-maps of $\bar{C}_1(1)$ from the identity to the constant map on 1 (combined with the $\bar{C}_1(1)$ action map on $A$) and a homotopy of self-maps of the pair $(\bar{D}, D)$ from the constant map (on the chosen element of $D$) to the identity map. For example, if the chosen element of $D$ corresponds to the subinterval $[a, b]$ (with $a \neq 0$) then the linear homotopy

$$[x, y], t \mapsto [xt + a(1 - t), yt + b(1 - t)]$$

is such a homotopy of self-maps of the pair.    $\square$

In the context of spaces, J. C. Moore invented an associative version of the based loop space by parametrizing loops with arbitrary length intervals. This idea extends to the current context to give another even simpler construction of a $\square$-monoid equivalent (in $\mathcal{M}$) to an $L\bar{C}_1$-algebra $A$.

**Construction 5.10.4.** Define $MA$ to be the object of $\mathcal{M}$ defined by the pushout diagram

$$\begin{array}{ccc} L\mathbb{R}^{>0} \,\square\, 1 & \longrightarrow & L\mathbb{R}^{>0} \,\square\, A \\ \downarrow & & \downarrow \\ L\mathbb{R}^{\geq 0} \,\square\, 1 & \longrightarrow & MA \end{array}$$

(where $\mathbb{R}^{>0} \subset \mathbb{R}^{\geq 0}$ are the usual subspaces of positive and non-negative real numbers, respectively). We give this the structure of a $\square$-monoid with the unit $1 \to MA$ induced by the inclusion of 0 in $\mathbb{R}^{\geq 0}$ and multiplication $MA \,\square\, MA \to MA$ induced by the map

$$(L\mathbb{R}^{>0} \,\square\, A) \,\square\, (L\mathbb{R}^{>0} \,\square\, A) \cong L(\mathbb{R}^{>0} \times \mathbb{R}^{>0}) \,\square\, (A \,\square\, A)$$

$$\to L(\mathbb{R}^{>0} \times \bar{C}_1(2)) \,\square\, (A \,\square\, A) \cong L\mathbb{R}^{>0} \,\square\, (L\bar{C}_1(2) \,\square\, (A \,\square\, A)) \to L\mathbb{R}^{>0} \,\square\, A$$

induced by the $\bar{C}_1$-action on $A$ and the map

$$c \colon (r, s) \in \mathbb{R}^{>0} \times \mathbb{R}^{>0} \mapsto (r + s, ([0, \tfrac{r}{r+s}], [\tfrac{r}{r+s}, 1])) \in \mathbb{R}^{>0} \times \bar{C}_1(2).$$

The idea is that the element of $\mathbb{R}^{>0}$ specifies a length (with the zero length only available for the unit) and the multiplication uses the proportionality of the two lengths to choose an element of $\bar{C}_1(2)$ for the multiplication on $A$; the two lengths add to give the length in the result. In the case when $\mathcal{M}$ is the category of spaces and $A = \Omega X$ is the based loop space of a space $X$, $MA$ is the Moore loop space. An element is specified by an element $r$ of $\mathbb{R}^{\geq 0}$ together with an element of $\Omega X$ (which must be the basepoint when $r = 0$) but can be visualized as a based loop parametrized by $[0, r]$ (or for $r = 0$ the constant length zero loop at the basepoint). The multiplication concatenates loops by concatenating the parametrizations, an operation that is strictly associative and unital.

We can compare the $\square$-monoids $MA$ and $\bar{U}A$ through a third $\square$-monoid $NA$ constructed as follows. Let $N = \mathbb{R}^{>0} \times \mathbb{R}^{>0} \times \mathbb{R}^{\geq 0}$, let $\bar{N} = \mathbb{R}^{\geq 0} \times \mathbb{R}^{>0} \times \mathbb{R}^{\geq 0}$, and define $NA$ by the pushout diagram

$$
\begin{array}{ccc}
LN \square \mathbf{1} & \longrightarrow & LN \square A \\
\downarrow & & \downarrow \\
L\bar{N} \square \mathbf{1} & \longrightarrow & NA
\end{array}
$$

We have maps $\bar{N} \times \bar{N} \to \bar{N}$ and $N \times N \to \bar{C}_1(2)$ defined by

$$((r, s, t), (r', s', t')) \in \bar{N} \times \bar{N} \mapsto (r + sr', ss', st' + t) \in \bar{N},$$

$$((r, s, t), (r', s', t')) \in N \times N \mapsto c(t, st') = \left( \left[ 0, \tfrac{r}{r+sr'} \right], \left[ \tfrac{r}{r+sr'}, 1 \right] \right) \in \bar{C}_1(2),$$

which we use to construct the multiplication on $NA$ by the same scheme as above

$$(LN \square A) \square (LN \square A) \cong L(N \times N) \square (A \square A) \to L(N \times \bar{C}_1(2)) \square (A \square A) \to LN \square A.$$

The unit is the map $\mathbf{1} \to NA$ induced by the inclusion of $(0, 1, 0)$ in $\bar{N}$.

The parametrizing space $N = \{(r, s, t)\}$ generalizes $D$ by allowing $[r, s]$ to be a subinterval of $[0, r+s+t]$ instead of $[0, 1]$, or from another perspective, generalizes lengths in the definition on the Moore algebra by incorporating a scaling factor $s$ and padding of length $t$. In other words, we have maps

$$[x, y] \in \bar{D} \mapsto (x, y - x, 1 - y) \in \bar{N},$$

$$r \in \mathbb{R}^{\geq 0} \mapsto (r, 1, 0) \in \bar{N}.$$

These maps induce maps of $\square$-monoids $\bar{U}A \cong DA \to NA$ and $MA \to NA$, respectively, and the argument of Proposition 5.10.3 shows that these maps are homotopy equivalences in $\mathcal{M}$. We state this as a theorem, repeating the conventions of this part of the section for easy reference.

**Theorem 5.10.5.** *Let $\mathcal{M}$ be a closed symmetric monoidal category admitting countable colimits and enriched over spaces via a strong symmetric monoidal left adjoint functor $L$. Then for algebras over the little 1-cubes operad ($L\bar{C}_1$-algebras) the non-symmetric enveloping algebra $\bar{U}A$ and the Moore algebra $MA$ fit in a natural zigzag of $\square$-monoids*

$$\bar{U}A \to NA \leftarrow MA,$$

*where the maps are homotopy equivalences in $\mathcal{M}$. Moreover, the canonical maps $\bar{U}A \to A$ and $MA \to A$ are homotopy equivalences in $\mathcal{M}$.*

To compare $MA$ and $A$ as $A_\infty$ algebras, we use a new $A_\infty$ operad $\bar{C}^\ell$ defined as follows.

Construction 5.10.6. Let $\overline{C}^\ell(0) = \mathbb{R}^{\geq 0}$ and for $m > 0$, let $\overline{C}^\ell(m)$ be the set of ordered pairs $(S, r)$ with $r$ a positive real number and $S$ a list of $m$ almost non-overlapping closed subintervals of $[0, r]$ in their natural order, topologized analogously as in the definition of $\overline{C}_1$ (as a semilinear submanifold of $\mathbb{R}^{2m+1}$). The operadic composition is defined by scaling and replacement of the subintervals: the basic composition

$$\Gamma_j^1((([x, y]), r), (([x'_1, y'_1], \ldots, [x'_j, y'_j]), r')) =$$
$$(([x + ax'_1, x + ay'_1], \ldots, [x + ax'_j, x + ay'_j]), r + a(r' - 1))$$

(with $a := y - x$) scales the interval $[0, r']$ to length $ar'$ and inserts that in place of $[x, y] \subset [0, r]$; the resulting final interval then has size $r - a + ar'$. The general composition $\Gamma_{j_1, \ldots, j_m}^m$ does this operation on each of the $m$ subintervals:

$$\Gamma_{j_1, \ldots, j_m}^m : (([x_1^0, y_1^0], \ldots, [x_m^0, y_m^0]), r_1),$$
$$(([x_1^1, y_1^1], \ldots, [x_{j_1}^1, y_{j_1}^1]), r_1), \ldots, (([x_1^m, y_1^m], \ldots, [x_{j_m}^m, y_{j_m}^m]), r_m),$$
$$\longmapsto$$
$$(([x_1^0 + a_1 x_1^1, x_1^0 + a_1 y_1^1], \ldots, [s_{m-1} + x_m^m + a_m x_{j_m}^m, s_{m-1} + x_m^m + a_m y_{j_m}^m]), r_0 + s_m),$$

where $a_i := y_i^0 - x_i^0$ and $s_i = a_1(r_1 - 1) + \cdots + a_i(r_i - 1)$. When one of the $j_i$ is zero, that $j_i$ contributes no subintervals but still scales the original subinterval $[x_i^0, y_i^0]$ to length $a_i r_i$ (or removes it when $r_i = 0$). The operad identity element is the element $(([0, 1]), 1) \in \overline{C}^\ell(1)$.

The maps $\overline{C}_1(m) \to \overline{C}^\ell(m)$ that include $\overline{C}_1(m)$ as the length 1 subspace assemble to a map of operads $i : \overline{C}_1 \to \overline{C}^\ell$. We also have a map of operads $j : \overline{Ass} \to \overline{C}^\ell$ induced by sending the unique element of $\overline{Ass}(m)$ to the element

$$(([0, 1], [1, 2], \ldots, [m - 1, m]), m)$$

of $\overline{C}^\ell(m)$. Using the map $j$, an $L\overline{C}^\ell$-algebra has the underlying structure of a $\square$-monoid. A straightforward check of universal properties proves the following proposition.

Proposition 5.10.7. *The functor that takes a $\overline{C}_1$-algebra $A$ to its Moore algebra $MA$ is naturally isomorphic to the functor that takes $A$ to the underlying $\square$-monoid of the pushforward $P_{Li}A$ for the map of operads $Li : L\overline{C}_1 \to L\overline{C}^\ell$.*

The $\overline{C}^\ell$-action map $L\overline{C}^\ell(m) \square (MA)^{(m)} \to MA$ is induced by the map

$$\overline{C}^\ell(m) \times (\mathbb{R}^{>0})^n \to \overline{C}^\ell(m) \times \overline{C}^\ell(1)^n \xrightarrow{\Gamma_{1, \ldots, 1}^m} \overline{C}^\ell(m) \cong \mathbb{R}^{>0} \times \overline{C}_1(m)$$

that includes $\mathbb{R}^{>0}$ in $\overline{C}^\ell(1)$ by $r \mapsto (([0, r]), r)$, where the isomorphism is the map that takes an element $(([x_1, y_1], \ldots, [x_m, y_m]), r)$ of $\overline{C}^\ell(m)$ to the element

$$\left(r, \left(\left[\tfrac{x_1}{r}, \tfrac{y_1}{r}\right], \ldots, \left[\tfrac{x_m}{r}, \tfrac{y_m}{r}\right]\right)\right)$$

of $\mathbb{R}^{>0} \times \overline{C}_1(m)$.

The map of $\overline{C}_1$-algebras that is the unit of the change of operads adjunction $A \to P_{Li}A$ is induced by the inclusion of 1 in $\mathbb{R}^{>0}$ and is a homotopy equivalence by a (simpler) version of the homotopy argument of Proposition 5.10.3. I do not see how to do a similar argument for the pushforward $P_{Lj}$ from □-monoids to $\bar{C}^\ell$-algebras, so we do not get a direct comparison of $\bar{C}_1$-algebras between $A$ (or $P_{Li}A$) and $MA$ with the $\bar{C}_1$-algebra structure inherited from its □-monoid structure without some kind of rectification result (such as Example Theorem 5.9.5) comparing the category of $L\bar{C}^\ell$-algebras with the category of $\overline{A}$ss-algebras.

The argument in [184, 2.5] that identifies $\bar{U}^{\bar{C}_1}A$ as $DA$ generalizes to identify $\bar{U}^{\bar{C}^\ell}P_{Li}A$ as $NA$; the maps in Theorem 5.10.5 can then be viewed as the natural maps on enveloping algebras induced by maps of operads and maps of algebras.

## 5.11    $E_n$ spaces and iterated loop space theory

The recognition principle for iterated loop spaces provided the first application for operads. Although the summary here has been spiced up with model category notions and terminology (in the adjoint functor formulation of [196, §8]), the mathematics has not changed significantly from the original treatment by May in [194], except for the improvements noted in the appendix to [71], which extend the results from connected to grouplike $E_n$ spaces. ($E_n$ spaces = $E_n$ algebras in spaces.)

The original idea for the little $n$-cubes operads $C_n$ and the start of the relationship between $E_n$ spaces and $n$-fold loop spaces is the Boardman–Vogt observation that every $n$-fold loop space comes with the natural structure of a $C_n$-algebra. The action map

$$C_n(m) \times \Omega^n X \times \cdots \times \Omega^n X \to \Omega^n X$$

is defined as follows. We view $S^n$ as $[0,1]^n/\partial$. Given an element $c \in C_n(m)$, and elements $f_1, \ldots, f_m \colon S^n \to X$ of $\Omega^n X$, let $f_{c; f_1, \ldots, f_n} \colon S^n \to X$ be the function that sends a point $x$ in $S^n$ to the basepoint if $x$ is not in one of the embedded cubes; the $i$-th embedded cube gets sent to $X$ using the inverse of the embedding and the quotient map $[0,1]^n \to S^n$ followed by the map $f_i \colon S^n \to X$. This is a continuous based map $S^n \to X$ since the boundary of each embedded cube gets sent to the basepoint. Phrased another way, $c$ defines a based map

$$S^n \to S^n \vee \cdots \vee S^n$$

with the $i$-th embedded cube mapping to the $i$-th wedge summand of $S^n$ by collapsing all points not in an open cube to the basepoint and rescaling; we then apply $f_i \colon S^n \to X$ to the $i$-th summand to get a composite map $S^n \to X$.

The construction of the previous paragraph factors $\Omega^n$ as a functor from based spaces to $C_n$-spaces (= $C_n$-algebras in spaces). It is clear that not every $C_n$-space arises as $\Omega^n X$ because $\pi_0 \Omega^n X$ is a group (for its canonical multiplication), whereas for the free $C_n$-space $\mathbb{C}_n X$, $\pi_0 \mathbb{C}_n X$ is not a group unless $X$ is the empty set; for example, $\pi_0 \mathbb{C}_n X \cong \mathbb{N}$ when $X$ is path connected. We say that a $C_n$-space $A$ is *grouplike* when $\pi_0 A$ is a group (for its canonical multiplication). The following is the fundamental theorem of iterated loop space theory; it gives an equivalence of homotopy theories between $n$-fold loop spaces and grouplike $C_n$-spaces.

Theorem 5.11.1 (May [194], Boardman–Vogt [48, §6]).  *The functor $\Omega^n$ from based spaces to $C_n$-spaces is a Quillen right adjoint. The unit of the derived adjunction*

$$A \to \Omega^n B^n A$$

*is an isomorphism in the homotopy category of $C_n$-spaces if (and only if) $A$ is grouplike. The counit of the derived adjunction*

$$B^n \Omega^n X \to X$$

*is an isomorphism in the homotopy category of spaces if (and only if) $X$ is $(n-1)$-connected; in general it is an $(n-1)$-connected cover.*

We have written the derived functor of the left adjoint in Theorem 5.11.1 as $B^n$, suggesting an iterated bar construction. Although neither the point-set adjoint functor nor the model for its derived functor used in the argument of Theorem 5.11.1 is constructed iteratively, Dunn [86] shows that the derived functor is naturally equivalent to an iterated bar construction.

As a consequence of the statement of the theorem, the unit of the derived adjunction $A \to \Omega^n B^n A$ is the initial map in the homotopy category of $C_n$-spaces from $A$ to a grouplike $C_n$-space and so deserves to be called "group completion". Group completion has various characterizations and for the purposes of sketching the ideas behind the proof of the theorem, it works best to choose one of them as the definition and state the property of the unit map as a theorem. One such characterization uses the classifying space construction, which we understand as the Eilenberg–Mac Lane bar construction (after converting the underlying $C_1$-spaces to topological monoids) or the Stasheff bar construction (choosing compatible maps from the Stasheff associahedra into the spaces $C_n(m)$).

Definition 5.11.2.   A map $f : A \to G$ of $C_n$-spaces is a *group completion* if $G$ is grouplike and $f$ induces a weak equivalence of classifying spaces.

In the case $n > 1$ (and under some hypotheses if $n = 1$), Quillen [227] gives a homological criterion for a map to be group completion: if $G$ is grouplike, then a map $A \to G$ of $C_n$-spaces is group completion if and only if

$$H_*(A)[(\pi_0 A)^{-1}] \to H_*(G)$$

is an isomorphism. Counterexamples exist in the case $n = 1$ (indeed, McDuff [208] gives a counterexample for every loop space homotopy type), but recent work of Braun, Chuang, and Lazarev [59] gives an analogous derived category criterion in terms of derived localization at the multiplicative set $\pi_0 A$. Using Definition 5.11.2 or any equivalent independent characterization of group completion, we have the following addendum to Theorem 5.11.1.

Addendum 5.11.3.   *The unit of the derived adjunction in Theorem 5.11.1 is group completion.*

The homotopical heart of the proof of Theorem 5.11.1 is the May–Cohen–Segal Approximation Theorem ([194, §6–7], [70], [270]), which we now review. This theorem studies a version of the free $C_n$-algebra functor $\tilde{C}_n$ whose domain is the category of based spaces, where the basepoint becomes the identity element in the $C_n$-algebra structure. This version of the free functor has the advantage that for a connected space $X$, $\tilde{C}X$ is also a connected space; May's Approximation Theorem identifies $\tilde{C}X$ in this case as a model for $\Omega^n\Sigma^nX$. Cohen (following conjectures of May) and Segal (working independently) then extended this to non-connected spaces: the group completion of $\tilde{C}X$ is a model for $\Omega^n\Sigma^nX$.

For a based space $X$, $\tilde{C}_nX$ is formed as a quotient of

$$\mathbb{C}X = \coprod C_n(m) \times_{\Sigma_m} X^m$$

by the equivalence relation that identifies $(c,(x_1,\ldots,x_i,*,\ldots,*)) \in C_n(m) \times X^m$ with $(c',(x_1,\ldots,x_i)) \in C_n(i) \times X^i$ for $c' = \Gamma(c;1,\ldots,1,0,\ldots,0)$ where 1 denotes the identity element in $C_n(1)$ and 0 denotes the unique element in $C_n(0)$. This is actually an instance of the operad pushforward construction: let $\mathcal{I}_{\mathrm{dbp}}$ be the operad with $\mathcal{I}_{\mathrm{dbp}}(0) = \mathcal{I}_{\mathrm{dbp}}(1) = *$ and $\mathcal{I}_{\mathrm{dbp}}(m) = \emptyset$ for $m > 1$. The functor associated to $\mathcal{I}_{\mathrm{dbp}}$ is the functor $(-)_+$ that adds a disjoint basepoint with the monad structure $((-)_+)_+ \to (-)_+$ that identifies the two disjoint basepoints; the category of algebras for this monad is the category of based spaces. The functor $\tilde{C}_n$ from based spaces to $C_n$-algebras is the pushforward $P_f$ for $f$ the unique map of operads $\mathcal{I}_{\mathrm{dbp}} \to C_n$: formally $P_f$ is the coequalizer described in Section 5.9, that in this case takes the form

$$\mathbb{C}_n(X_+) \rightrightarrows \mathbb{C}_nX \longrightarrow \tilde{C}_nX.$$

As mentioned in an aside in that section (or as can be seen concretely here using the operad multiplication on $C_n$ directly), the endofunctor $\tilde{C}_n$ on based spaces (i.e., $U_fP_f$) has the structure of a monad, and we can identify the category of $C_n$-spaces as the category of algebras over the monad $\tilde{C}_n$.

The factorization of the functor $\Omega^n$ through $C_n$-spaces has the formal consequence of producing a map of monads (in based spaces)

$$\tilde{C}_n \to \Omega^n\Sigma^n.$$

Formally the map is induced by the composite

$$\tilde{C}_nX \xrightarrow{\tilde{C}_n\eta} \tilde{C}_n\Omega^n\Sigma^nX \xrightarrow{\xi} \Omega^n\Sigma^nX,$$

where $\eta$ is the unit of the $\Sigma^n,\Omega^n$-adjunction and $\xi$ is the $C_n$-action map. This map has the following concrete description: an element $(c,(x_1,\ldots,x_m)) \in C_n(m) \times X^m$ maps to the element $\gamma: S^n \to \Sigma^nX$ of $\Omega^n\Sigma^nX$ given by the composite of the map

$$S^n \to S^n \vee \cdots \vee S^n$$

associated to $c$ (as described above) and the map

$$S^n \cong \Sigma^n\{x_i\}_+ \subset \Sigma^nX$$

on the $i$-th factor of $S^n$. Either using this concrete description, or following diagrams in a formal categorical argument, it is straightforward to check that this defines a map of monads. We can now state the May–Cohen–Segal Approximation Theorem.

**Theorem 5.11.4** (May–Cohen–Segal Approximation Theorem [194, 6.1], [70, 3.3], [270, Theorem 2]).
*For any non-degenerately based space $X$, the map of $C_n$-spaces $\tilde{\mathbb{C}}_n X \to \Omega^n \Sigma^n X$ is group completion.*

("Non-degenerately based" means that the inclusion of the basepoint is a cofibration. Both $\tilde{\mathbb{C}}_n$ and $\Omega^n \Sigma^n$ preserve weak equivalences in non-degenerately based spaces, but for other spaces, either or both may have the wrong weak homotopy type.)

From here a sketch of the proof of Theorem 5.11.1 goes as follows. Since $\Omega^n$ as a functor from based spaces to based spaces has left adjoint $\Sigma^n$, a check of universal properties shows that the functor from $C_n$-spaces to based spaces defined by the coequalizer

$$\Sigma^n \tilde{\mathbb{C}}_n A \rightrightarrows \Sigma^n A \longrightarrow \Sigma^n \otimes_{\mathbb{C}_n} A$$

is the left adjoint to $\Omega^n$ viewed as a functor from based spaces to $C_n$-spaces. (In the coequalizer, one map is induced by the $C_n$-action map on $A$ and the other is adjoint to the map of monads $\tilde{\mathbb{C}} \to \Omega^n \Sigma^n$.) Because $\Omega^n$ preserves fibrations and weak equivalences, this is a Quillen adjunction.

The main tool to study the $\Sigma^n \otimes_{\mathbb{C}_n}(-), \Omega^n$-adjunction is the two-sided monadic bar construction, invented in [194, §9] for this purpose. Given a monad $\mathbb{T}$ and a right action of $\mathbb{T}$ on a functor $F$ (say, to based spaces), the two-sided monadic bar construction is the functor on $\mathbb{T}$-algebras $B(F, \mathbb{T}, -)$ defined as the geometric realization of the simplicial object

$$B_m(F, \mathbb{T}, A) = F \underbrace{\mathbb{T} \cdots \mathbb{T}}_{m} A,$$

with face maps induced by the action map $F\mathbb{T} \to F$, the multiplication map $\mathbb{T}\mathbb{T} \to \mathbb{T}$ and the action map $\mathbb{T}A \to A$, and degeneracy maps induced by the unit map $\mathrm{Id} \to \mathbb{T}$. In the case when $F = \mathbb{T}$, the simplicial object $B_\bullet(\mathbb{T}, \mathbb{T}, A)$ has an extra degeneracy and the map from $B_\bullet(\mathbb{T}, \mathbb{T}, A)$ to the constant simplicial object on $A$ is a simplicial homotopy equivalence (in the underlying category for $\mathbb{T}$, though not generally in the category of $\mathbb{T}$-algebras).

Because geometric realization commutes with colimits and finite cartesian products, we have a canonical isomorphism

$$\tilde{\mathbb{C}}_n B(\tilde{\mathbb{C}}_n, \tilde{\mathbb{C}}_n, A) \to B(\tilde{\mathbb{C}}_n \tilde{\mathbb{C}}_n, \tilde{\mathbb{C}}_n, A)$$

and the multiplication map $\tilde{\mathbb{C}}_n \tilde{\mathbb{C}}_n \to \tilde{\mathbb{C}}_n$ then gives $B(\tilde{\mathbb{C}}_n, \tilde{\mathbb{C}}_n, A)$ the natural structure of a $C_n$-algebra. (See Section 5.7 for a more general discussion.) For the same reason, the canonical map

$$\Sigma^n \otimes_{\mathbb{C}_n} B(\tilde{\mathbb{C}}_n, \tilde{\mathbb{C}}_n, A) \to B(\Sigma^n \otimes_{\mathbb{C}_n} \tilde{\mathbb{C}}_n, \tilde{\mathbb{C}}_n, A) = B(\Sigma^n, \tilde{\mathbb{C}}_n, A)$$

is an isomorphism. The latter functor clearly[3] preserves weak equivalences of $C_n$-spaces $A$ whose underlying based spaces are non-degenerately based. (Besides being a hypothesis of the May–Cohen–Segal Approximation Theorem, non-degenerately based here also ensures that the inclusion of the degenerate subspace (or latching object) is a cofibration.) As a consequence of Theorem 5.7.3 it follows that when the underlying based space of $A$ is cofibrant (which is the case in particular when $A$ is cofibrant as a $C_n$-space), then $B(\tilde{\mathbb{C}}_n, \tilde{\mathbb{C}}_n, A)$ is a cofibrant $C_n$-space. Because $\Sigma^n \otimes_{\mathbb{C}_n}(-)$ is a Quillen left adjoint, it preserves weak equivalences between cofibrant objects, and looking at a cofibrant approximation $A' \xrightarrow{\sim} A$, we see from the weak equivalences

$$B(\Sigma^n, \tilde{\mathbb{C}}_n, A) \xleftarrow{\sim} B(\Sigma^n, \tilde{\mathbb{C}}_n, A') \cong \Sigma^n \otimes_{\mathbb{C}_n} B(\tilde{\mathbb{C}}_n, \tilde{\mathbb{C}}_n, A') \xrightarrow{\sim} \Sigma^n \otimes_{\mathbb{C}_n} A'$$

that $B(\Sigma^n, \tilde{\mathbb{C}}_n, A)$ models the derived functor $B^n A$ of $\Sigma^n \otimes_{\mathbb{C}_n}(-)$ whenever $A$ is non-degenerately based.

To complete the argument, we need the theorem of [194, §12] that $\Omega^n$ commutes up to weak equivalence with geometric realization of (proper) simplicial spaces that are $(n-1)$-connected in each level. Then for $A$ non-degenerately based, we have that the vertical maps are weak equivalences of $C_n$-spaces

$$
\begin{array}{ccc}
B(\tilde{\mathbb{C}}_n, \tilde{\mathbb{C}}_n, A) & \longrightarrow & B(\Omega^n \Sigma^n, \tilde{\mathbb{C}}_n, A) \\
\downarrow & & \downarrow \\
A & & \Omega^n B(\Sigma^n, \tilde{\mathbb{C}}_n, A)
\end{array}
$$

while by the May–Cohen–Segal Approximation Theorem, the horizontal map is group completion. This proves that the unit of the derived adjunction is group completion.

For the counit of the derived adjunction, we have from the model above that $B^n$ is always $(n-1)$-connected and the unit

$$\Omega^n X \to \Omega^n B^n \Omega^n X$$

on $\Omega^n X$ is a weak equivalence. Looking at $\Omega^n$ of the counit,

$$\Omega^n B^n \Omega^n X \to \Omega^n X,$$

the composite with the unit is the identity on $\Omega^n X$, and so it follows that $\Omega^n$ of the counit is a weak equivalence. Thus, the counit of the derived adjunction is an $(n-1)$-connected cover map.

## 5.12 $E_\infty$ algebras in rational and $p$-adic homotopy theory

In the 1960's and 1970's, Quillen [228] and Sullivan [284, 286] showed that the rational homotopy theory of simply connected spaces (or simplicial sets) has an algebraic model

---

[3] At the time when May wrote the argument, this was far from clear: some of the first observations about when geometric realization of simplicial spaces preserves levelwise weak equivalences were developed in [194, §11] precisely for this argument.

in terms of rational differential graded commutative algebras or coalgebras. In the 1990's, I proved a mostly analogous theorem relating $E_\infty$ differential graded algebras and $p$-adic homotopy theory and a bit later some results for using $E_\infty$ differential graded algebras or $E_\infty$ ring spectra to identify integral homotopy types. In this section, we summarize this theory following mostly the memoir of Bousfield–Gugenheim [57], and the papers [181][4] and [180]. In what follows $k$ denotes a commutative ring, which is often further restricted to be a field.

In both the rational commutative differential graded algebra case and the $E_\infty$ $k$-algebra case, the theory simplifies by working with simplicial sets instead of spaces, and the functor is some variant of the cochain complex. Sullivan's approach to rational homotopy theory uses a rational version of the de Rham complex, originally due to Thom (unpublished), consisting of forms that are polynomial on simplices and piecewise matched on faces:

Definition 5.12.1. The algebra $\nabla^*[n]$ of polynomial forms on the standard simplex $\Delta[n]$ is the rational commutative differential graded algebra free on generators $t_0, \ldots, t_n$ (of degree zero), $dt_0, \ldots, dt_n$ (of degree one) subject to the relations $t_0 + \cdots + t_n = 1$ and $dt_0 + \cdots + dt_n = 0$ (as well as the differential relation implicit in the notation).

Viewing $t_0, \ldots, t_n$ as the barycentric coordinate functions on $\Delta[n]$ determines their behavior under face and degeneracy maps, making $\nabla^*[\bullet]$ a simplicial rational commutative differential graded algebra.

Definition 5.12.2. For a simplicial set $X$, the rational de Rham complex $A^*(X)$ is the rational graded commutative algebra of maps of simplicial sets from $X$ to $\nabla^*[\bullet]$, or equivalently, the end over the simplex category

$$A^*(X) := \Delta^{\mathrm{op}} \mathcal{S}\mathrm{et}(X, \nabla^*[\bullet]) = \int_{\Delta^{\mathrm{op}}} \mathcal{S}\mathrm{et}(X_n, \nabla^*[n]) = \int_{\Delta^{\mathrm{op}}} \prod_{X_n} \nabla^*[n]$$

(the last formula indicating how to regard $A^*(X)$ as a rational commutative differential graded algebra).

More concretely, $A^*(X)$ is the rational commutative differential graded algebra where an element of degree $q$ consists of a choice of element of $\nabla^q[n]$ for each non-degenerate $n$-simplex of $X$ (for all $n$) which agree under restriction by face maps, with multiplication and differential done on each simplex. (When $X$ is a finite simplicial complex $A^*(X)$ also has a Stanley–Reisner ring style description; see [284, G.i)].) The simplicial differential graded $\mathbb{Q}$-module $\nabla^q[n]$ is a contractible Kan complex for each fixed $q$ (the extension lemma [57, 1.1]) and is acyclic in the sense that the inclusion of the unit $\mathbb{Q} \to \nabla^*[n]$ is a chain homotopy equivalence for each fixed $n$ (the Poincaré lemma [57, 1.3]). These formal properties imply that the cohomology of $A^*(X)$ is canonically naturally isomorphic to $H^*(X; \mathbb{Q})$, the rational cohomology of $X$ (even

---

[4] In the published version, in addition to several other unauthorized changes, the copy editors changed the typefaces with the result that the same symbols are used for multiple different objects or concepts; the preprint version available at https://pages.iu.edu/~mmandell/papers/einffinal.pdf does not have these changes and should be much more readable.

uniquely naturally isomorphic, relative to the canonical isomorphism $\mathbb{Q} \cong A^*(\Delta[0])$).
The canonical isomorphism can be realized as a chain map to the normalized cochain
complex $C^*(X; \mathbb{Q})$ defined in terms of integrating differential forms; see [57, 1.4,2.1,2.2].

In the $p$-adic case, we can use the normalized cochain complex $C^*(X; k)$ directly
as it is naturally an $E_\infty$ $k$-algebra. In the discussion below, we use the $E_\infty$ $k$-algebra
structure constructed by Berger–Fresse [37, §2.2] for the Barratt–Eccles operad $\mathcal{E}$
(the normalized chains of the Barratt–Eccles operad of categories or simplicial sets
described in Example 5.3.3). Hinich–Schechtmann [123] and (independently) Smirnov
[279] appear to have been the first to explicitly describe a natural operadic algebra
structure on cochains; McClure–Smith [205] describes a natural $E_\infty$ structure that
generalizes classical $\cup_i$ product and bracket operations. The "cochain theory" theory
of [179] shows that all these structures are equivalent in the sense that they give
naturally quasi-isomorphic functors into a common category of $E_\infty$ $k$-algebras, as
does the polynomial de Rham complex functor $A^*$ when $k = \mathbb{Q}$.

Both $A^*(X)$ and $C^*(X; k)$ fit into adjunctions of the contravariant type that send
colimits to limits. Concretely, for a rational commutative differential graded algebra $A$
and an $E_\infty$ $k$-algebra $E$, define simplicial sets by the formulas

$$T(A) := \mathcal{C}_{\mathbb{Q}}(A, \nabla^*[\bullet]), \qquad U(E) := \mathcal{E}_k(E, C^*(\Delta[\bullet])),$$

where $\mathcal{C}_{\mathbb{Q}}$ denotes the category of rational commutative differential graded algebras
and $\mathcal{E}_k$ denotes the category of $E_\infty$ $k$-algebras (over the Barratt–Eccles operad). An
easy formal argument shows that

$$A^*: \Delta^{\mathrm{op}}\mathcal{S}et \rightleftarrows \mathcal{C}_{\mathbb{Q}}^{\mathrm{op}} :T, \qquad C^*: \Delta^{\mathrm{op}}\mathcal{S}et \rightleftarrows \mathcal{E}_k^{\mathrm{op}} :U,$$

are adjunctions. As discussed in Section 5.8, both $\mathcal{C}_{\mathbb{Q}}$ and $\mathcal{E}_k$ have closed model structures with weak equivalences the quasi-isomorphisms and fibrations the surjections.
Because both $A^*$ and $C^*$ preserve homology isomorphisms and convert injections to
surjections, these are Quillen adjunctions. The main theorems of [57] and [181] then
identify subcategories of the homotopy categories on which the adjunction restricts to
an equivalence.

Before stating the theorems, first recall the $H_*(-; k)$-local model structure on simplicial sets: this has cofibrations the inclusions and weak equivalences the $H_*(-; k)$
homology isomorphisms. When $k$ is a field, the weak equivalences depend only on
the characteristic, and we also call this the *rational model structure* (in the case of
characteristic zero) or the *$p$-adic model structure* (in the case of characteristic $p > 0$);
we call the associated homotopy categories, the *rational homotopy category* and *$p$-adic
homotopy category*, respectively. As with any localization, the local homotopy category
is the homotopy category of local objects (that is to say, the fibrant objects): in the case
of rational homotopy theory, the local objects are the Kan complexes of the homotopy
type of rational spaces. In $p$-adic homotopy theory, the local objects are the Kan
complexes that satisfy a $p$-completeness property described explicitly in [54, §5,7–8].

We say that a simplicial set $X$ is *finite $H_*(-; k)$-type* (or *finite rational type* when
$k$ is a field of characteristic zero or *finite $p$-type* when $k$ is a field of characteristic
$p > 0$) when $H_*(X; k)$ is finitely generated over $k$ in each degree (or, equivalently if

$k$ is a field, when $H^*(X; k)$ is finite dimensional in each degree). Similarly a rational commutative differential graded algebra or $E_\infty$ $k$-algebra $A$ is *finite type* when its homology is finitely generated over $k$ in each degree. It is *simply connected* when the inclusion of the unit induces an isomorphism $k \to H^0(A)$, $H^1(A) \cong 0$, and $H^n(A) \cong 0$ for $n < 0$ (with the usual cohomological grading convention that $H^n(A) := H_{-n}(A)$). With this terminology, the main theorem of [57] is the following:

**Theorem 5.12.3** ([57, Section 8, Theorem 9.4]). *The polynomial de Rham complex functor, $A^*\colon \Delta^{\mathrm{op}}\mathcal{S}\mathrm{et} \to \mathscr{C}_{\mathbb{Q}}^{\mathrm{op}}$, is a left Quillen adjoint for the rational model structure on simplicial sets. The left derived functor restricts to an equivalence of the full subcategory of the rational homotopy category consisting of the simply connected simplicial sets of finite rational type and the full subcategory of the homotopy category of rational commutative differential graded algebras consisting of the simply connected rational commutative differential graded algebras of finite type.*

For the $p$-adic version below, we need to take into account Steenrod operations. For $k = \mathbb{F}_p$, the Steenrod operations arise from the coherent homotopy commutativity of the $p$-fold multiplication, which is precisely encoded in the action of the $E_\infty$ operad. Specifically, the $p$-th complex $\mathcal{E}(p)$ of the operad is a $k[\Sigma_p]$-free resolution of $k$, and by neglect of structure, we can regard it as a $k[C_p]$-free resolution of $k$ where $C_p$ denotes the cyclic group of order $p$. The operad action induces a map

$$\mathcal{E}(p) \otimes_{k[C_p]} (C^*(X; k))^{(p)} \to \mathcal{E}(p) \otimes_{k[\Sigma_p]} (C^*(X; k))^{(p)} \to C^*(X; k).$$

The homology of $\mathcal{E}(p) \otimes_{k[C_p]} (C^*(X; k))^{(p)}$ is a functor of the homology of $C^*(X; k)$ and the Steenrod operations $P^s$ are precisely the images of certain classes under this map; see, for example, [198, 2.2]. This process works for any $E_\infty$ $k$-algebra, not just the cochains on spaces, to give natural operations on the homology of $\mathcal{E}$-algebras, usually called Dyer–Lashof operations. The numbering conventions for these are opposite those of the Steenrod operations: on the cohomology of $C^*(X; \mathbb{F}_p)$, the Dyer–Lashof operation $Q^s$ performs the Steenrod operation $P^{-s}$. If $k$ is of characteristic $p$ but not $\mathbb{F}_p$, the operations constructed this way are $\mathbb{F}_p$-linear but satisfy $Q^s(ax) = \phi(a)Q^s(x)$ for $a \in k$, where $\phi$ denotes the Frobenius automorphism of $k$.

The $\mathbb{F}_p$ cochain algebra of a space has the special property that the Steenrod operation $P^0 = Q^0$ is the identity operation on its cohomology; this is not true of the zeroth Dyer–Lashof operation in general. Indeed for a commutative $\mathbb{F}_p$-algebra regarded as $E_\infty$ $\mathbb{F}_p$-algebra, $Q^0$ is the Frobenius. (That $Q^0$ is the identity for the $\mathbb{F}_p$-cochain algebra of a space is related to the fact that it comes from a cosimplicial $\mathbb{F}_p$-algebra where the Frobenius in each degree is the identity.) So when $X$ is finite $p$-type, $C^*(X; k)$ in each degree has a basis that is fixed by $Q^0$. We say that a finite type $E_\infty$ $k$-algebra is *spacelike* when in each degree its homology has a basis that is fixed by $Q^0$.

**Theorem 5.12.4** ([181, Main Theorem, Theorem A.1]). *The cochain complex with coefficients in $k$, $C^*(-; k)\colon \Delta^{\mathrm{op}}\mathcal{S}\mathrm{et} \to \mathscr{E}_k^{\mathrm{op}}$, is a left Quillen adjoint for the $H_*(-; k)$-local model structure on simplicial sets. If $k = \mathbb{Q}$ or $k$ is characteristic $p$ and $1 - \phi$ is surjective* *on $k$, then the left derived functor restricts to an equivalence of the full subcategory of the $H_*(-;k)$-local homotopy category consisting of the simply connected simplicial sets of finite $H_*(-;k)$-type and the full subcategory of the homotopy category of $E_\infty$ $k$-algebras consisting of the spacelike simply connected $E_\infty$ $k$-algebras of finite type.*

Given the Quillen equivalence between rational commutative differential graded algebras and $E_\infty$ $\mathbb{Q}$-algebras (Theorem 5.9.5) and the natural quasi-isomorphism (zigzag) between $A^*(-)$ and $C^*(-;\mathbb{Q})$ [179, p. 549], the rational statement in Theorem 5.12.4 is equivalent to Theorem 5.12.3. The Sullivan theory in the latter often includes observations on *minimal models*. A simply connected finite type rational commutative differential graded algebra $A$ has a cofibrant approximation $A' \to A$ whose underlying graded commutative algebra is free and such that the differential of every element is decomposable (i.e., is a sum of terms, all of which have word length greater than 1 in the generators); $A'$ is called a minimal model and is unique up to isomorphism. As a consequence, simply connected simplicial sets of finite rational type are rationally equivalent if and only if their minimal models are isomorphic. The corresponding theory also works in the context of $E_\infty$ $\mathbb{Q}$-algebras with the analogous definitions and proofs. The corresponding theory does not work in the context of $E_\infty$ algebras in characteristic $p$ for reasons closely related to the fact that unlike the rational homotopy groups, the $p$-adic homotopy groups of a simplicial set are not vector spaces.

The equivalences in Theorems 5.12.3 and 5.12.4 also extend to the nilpotent simplicial sets of finite type, but the corresponding category of $E_\infty$ $k$-algebras does not have a known intrinsic description in the $p$-adic homotopy case; in the rational case, the corresponding algebraic category consists of the finite type algebras whose homology is zero in negative cohomological degrees and whose $H^0$ is isomorphic as a $\mathbb{Q}$-algebra to the cartesian product of copies of $\mathbb{Q}$ (cf. [182, §3]).

For other fields not addressed in the second part of Theorem 5.12.4, the adjunction does not necessarily restrict to the indicated subcategories and even when it does, it is never an equivalence. To be an equivalence, the unit of the derived adjunction would have to be an $H_*(-;k)$-isomorphism for simply connected simplicial sets of finite type. If $k \neq \mathbb{Q}$ is characteristic zero, then the right derived functor of $U$ takes $C^*(S^2;k)$ to a simplicial set with $\pi_2$ isomorphic to $k$; if $k$ is characteristic $p$, then the right derived functor of $U$ takes $C^*(S^2;k)$ to a simplicial set with $\pi_1$ isomorphic to the cokernel of $1 - \phi$. See [181, Appendix A] for more precise results. Because the algebraic closure of a field $k$ of characteristic $p$ does have $1 - \phi$ surjective, even when $C^*(-;k)$ is not an equivalence, it can be used to detect $p$-adic equivalences. This kind of observation extends to the case $k = \mathbb{Z}$:

**Theorem 5.12.5** ([180, Main Theorem]). *Finite type nilpotent spaces or simplicial sets $X$ and $Y$ are weakly equivalent if and only if $C^*(X;\mathbb{Z})$ and $C^*(Y;\mathbb{Z})$ are quasi-isomorphic as $E_\infty$ $\mathbb{Z}$-algebras.*

Using the spectral version of Theorem 5.12.4 in [181, Appendix C], the proof of the previous theorem in [180] extends to show that when $X$ and $Y$ are finite nilpotent simplicial sets then $X$ and $Y$ are weakly equivalent if and only if their Spanier–

Whitehead dual spectra are weakly equivalent as $E_\infty$ ring spectra. (This was the subject of a talk by the author at the Newton Institute in December 2002.)

We use the rest of the section to outline the argument for Theorems 5.12.3 and 5.12.4, using the notation of the latter. We fix a field $k$, which is either $\mathbb{Q}$ or is characteristic $p > 0$ and has $1 - \phi$ surjective. We write $C^*$ for $C^*(-;k)$ or when $k = \mathbb{Q}$ and we are working in the context of Theorem 5.12.3, we understand $C^*$ as $A^*$. We also use $C^*$ to denote the derived functor and write $\mathbf{U}$ for the derived functor of its adjoint. The idea of the proof, going back to Sullivan, is to work with Postnikov towers, and so the first step is to find cofibrant approximations for $C^*(K(\pi, n))$. For $k = \mathbb{Q}$, this is easy since $H^*(K(\mathbb{Q}, n); \mathbb{Q})$ is the free graded commutative algebra on a generator in degree $n$.

**Proposition 5.12.6.** *If $k = \mathbb{Q}$ then $C^*(K(\mathbb{Q}, n))$ is quasi-isomorphic to the free ($E_\infty$ or commutative differential graded) $\mathbb{Q}$-algebra on a generator in cohomological degree $n$.*

We use the notation $\mathbb{E}k[n]$ to denote the free $E_\infty$ $k$-algebra on a generator in cohomological degree $n$. When $k$ is characteristic $p$, there is a unique map in the homotopy category from $\mathbb{E}k[n] \to C^*(K(\mathbb{Z}/p, n))$ that sends the generator $x_n$ to a class $i_n$ representing the image of the tautological element of $H^n(K(\mathbb{Z}/p, n); \mathbb{Z}/p)$. Unlike the characteristic zero case, this is not a quasi-isomorphism since $Q^0[i_n] = [i_n]$ in $H^*(C^*(K(\mathbb{Z}/p, n)))$, but $Q^0[x_n] \neq [x_n]$ in $H^*(\mathbb{E}k[n])$. Let $B_n$ be the homotopy pushout of a map $\mathbb{E}k[n] \to \mathbb{E}k[n]$ sending the generator to a class representing $[x_n] - Q^0[x_n]$ and the map $\mathbb{E}k[n] \to k$ sending the generator to 0. Then the map $\mathbb{E}k[n] \to C^*(K(\mathbb{Z}/p, n))$ factors through a map $B_n \to C^*(K(\mathbb{Z}/p, n))$. (The map in the homotopy category turns out to be independent of the choices.) The following is a key result of [181], whose proof derives from a calculation of the relationship between the Dyer–Lashof algebra and the Steenrod algebra.

**Theorem 5.12.7** ([181], 6.2]). *Let $k$ be a field of characteristic $p > 0$. Then*

$$B_n \to C^*(K(\mathbb{Z}/p, n))$$

*is a cofibrant approximation.*

(As suggested by the hypothesis, we do not need $1 - \phi$ to be surjective in the previous theorem; indeed, the easiest way to proceed is to prove it in the case $k = \mathbb{F}_p$ and it then follows easily for all fields of characteristic $p$ by extension of scalars.)

The two previous results can be used to calculate $\mathbf{U}(C^*(K(\mathbb{Q}, n)))$ and $\mathbf{U}(C^*(K(\mathbb{Z}/p, n)))$. In the rational case,

$$\mathbf{U}(C^*(K(\mathbb{Q}, n))) \simeq U(\mathbb{E}\mathbb{Q}[n]) = Z(C^n(\Delta[\bullet])),$$

the simplicial set of $n$-cocycles of $C^*(\Delta[\bullet]; \mathbb{Q})$; this is the original model for $K(\mathbb{Q}, n)$, and a straightforward argument shows that the unit map $K(\mathbb{Q}, n) \to K(\mathbb{Q}, n)$ is a weak equivalence (the identity map with this model). In the context of Theorem 5.12.3, the same kind of argument is made in [57, 10.2]. In the $p$-adic case, we likewise have that $U(\mathbb{E}k[n])$ is the original model for $K(k, n)$, and so we get a fiber sequence

$$\Omega K(k, n) \to \mathbf{U}(K(\mathbb{Z}/p, n)) \to K(k, n) \to K(k, n).$$

The map $K(k,n) \to K(k,n)$ is calculated in [181, 6.3] to be the map that on $\pi_n$ induces $1 - \phi$. The kernel of $1 - \phi$ is $\mathbb{F}_p$ and the unit map $K(\mathbb{Z}/p, n) \to \mathbf{U}(C^*(K, \mathbb{Z}/p, n))$ is an isomorphism on $\pi_n$. As a consequence, when $1 - \phi$ is surjective (as we are assuming), the unit map is a weak equivalence for $K(\mathbb{Z}/p, n)$.

The game now is to show that for all finite type simply connected (or nilpotent) simplicial sets, the derived unit map $X \to \mathbf{U}C^*(X)$ is a rational or $p$-adic equivalence. The next result tells how to construct a cofibrant approximation for a homotopy pullback; it is not a formal consequence of the Quillen adjunction, but rather a version of the Eilenberg–Moore theorem.

**Proposition 5.12.8** ([57, §3], [181, §3]). *Let*

$$
\begin{array}{ccc}
W & \longrightarrow & Y \\
\downarrow & & \downarrow \\
Z & \longrightarrow & X
\end{array}
$$

*be a homotopy fiber square of simplicial sets. If $X, Y, Z$ are finite $H_*(-;k)$-type and $X$ is simply connected, then*

$$
\begin{array}{ccc}
C^*(X) & \longrightarrow & C^*(Y) \\
\downarrow & & \downarrow \\
C^*(Z) & \longrightarrow & C^*(W)
\end{array}
$$

*is a homotopy pushout square of $E_\infty$ $k$-algebras or rational commutative differential graded algebras.*

Since we can write $K(\mathbb{Z}/p^m, n)$ as the homotopy fiber of a map

$$K(\mathbb{Z}/p^{m-1}, n) \to K(\mathbb{Z}/p, n+1),$$

we see that the unit of the derived adjunction is a weak equivalence also for $K(\mathbb{Z}/p^m, n)$ (when $k$ is characteristic $p$). Likewise, since products are homotopy pullbacks, we also get that the unit of the derived adjunction is a weak equivalence for $K(A, n)$ when $A$ is a $\mathbb{Q}$ vector space (when $k = \mathbb{Q}$) or when $A$ is a finite $p$-group (when $k$ is characteristic $p$). Although also not a formal consequence of the adjunction, it is elementary to see that when a simplicial set $X$ is the homotopy limit of a sequence $X_j$ and the map $\operatorname{colim} H^*(X_j; k) \to H^*(X; k)$ is an isomorphism, then $C^*(X)$ is the homotopy colimit of $C^*(X_j)$ and $\mathbf{U}C^*(X)$ is the homotopy limit of $\mathbf{U}C^*(X_j)$. It follows that for $K(\mathbb{Z}_p^\wedge, n)$, the unit of the derived adjunction is a weak equivalence (when $k$ is characteristic $p$). For any finitely generated abelian group, the map $K(A, n) \to K(A \otimes \mathbb{Q}, n)$ is a rational equivalence and the map $K(A, n) \to K(A_p^\wedge, n)$ is a $p$-adic equivalence. Putting these results and tools all together, we see that the unit of the derived equivalence is an $H_*(-;k)$ equivalence for any $X$ that can be built as a sequential homotopy limit $\operatorname{holim} X_j$ where $X_0 = *$, the connectivity of the map $X \to X_j$ goes to infinity, and each $X_{j+1}$ is the homotopy fiber of a map $X_j \to K(\pi_{j+1}, n)$ for $\pi_{j+1}$ a finitely generated abelian group, or the rationalization (when $k = \mathbb{Q}$) or $p$-completion (when $k$ is characteristic $p$) of a finitely generated abelian group. In particular, for a simply connected simplicial set, applying this to the Postnikov tower, we get the following result.

**Theorem 5.12.9.** *Assume $k = \mathbb{Q}$ or $k$ is characteristic $p > 0$ and $1 - \phi$ is surjective. If $X$ is a simply connected simplicial set of finite $H_*(-;k)$-type, then the unit of the derived adjunction $X \to \mathbf{U}C^*(X)$ is an $H_*(-;k)$-equivalence.*

The previous theorem formally implies that $C^*$ induces an equivalence of the $H_*(-;k)$-local homotopy category of simply connected simplicial sets of finite $H_*(-;k)$-type with the full subcategory of the homotopy category $E_\infty$ $k$-algebras or rational commutative differential graded algebras of objects in its image. The remainder of Theorems 5.12.3 and 5.12.4 is identifying this image subcategory. In the case when $k = \mathbb{Q}$, it is straightforward to see that a finite type simply connected algebra has a cofibrant approximation that $\mathbf{U}$ turns into a simply connected principal rational finite type Postnikov tower. The argument for $k$ of characteristic $p$ is analogous, but more complicated; see [181, §7].

# 6 Commutative ring spectra

by Birgit Richter

## 6.1 Introduction

Since the 1990s we have had several symmetric monoidal categories of spectra at our disposal whose homotopy category is the stable homotopy category. The monoidal structure is usually denoted by $\wedge$ and is called the smash product of spectra. So since then we can talk about commutative monoids in any of these categories — these are commutative ring spectra. Even before such symmetric monoidal categories were constructed, the consequences of their existence were described. In [296, §2] Friedhelm Waldhausen outlines the role of "rings up to homotopy". He also coined the expression "brave new rings" in a 1988 talk at Northwestern University.

So what is the problem? Why don't we just write down nice commutative models of our favorite homotopy types and be done with it? Why does it make sense to have a whole chapter about this topic?

In algebra, if someone tells you to check whether a given ring is commutative, you can sit down and check the axiom for commutativity and you should be fine. In stable homotopy theory the problem is more involved, since strict commutativity may only be satisfied by some preferred point set level model of the underlying associative ring spectrum and the operadic incarnation of commutativity is an extra structure rather than a condition.

There is one class of commutative ring spectra that is easy to construct. If you take singular cohomology with coefficients in a commutative ring $R$, then this is represented by the Eilenberg–Mac Lane spectrum $HR$ and this can be represented by a commutative ring spectrum.

So it would be nice if we could have explicit models for other homotopy types that come naturally equipped with a commutative ring structure. Sometimes this is possible. If you are interested in real (or complex) vector bundles over your space, then you want to understand real (or complex) topological $K$-theory, and Michael Joachim [136, 137] for instance has produced explicit analytically flavored models for periodic real and complex topological $K$-theory with commutative ring structures.

There are a few general constructions that produce commutative ring spectra for you. For instance, the construction of Thom spectra often gives rise to commutative ring spectra. We will discuss this important class of examples in Section 6.4. A classical construction due to Graeme Segal also produces small nice models of commutative ring spectra (see Section 6.5).

Quite often, however, the spectra that we like are constructed in a synthetic way: You have some commutative ring spectrum $R$ and you kill a regular sequence of elements in its graded commutative ring of homotopy groups, $(x_1, x_2, \ldots)$, $x_i \in \pi_*(R)$, and you consider a spectrum $E$ with homotopy groups $\pi_*(E) \cong \pi_*(R)/(x_1, x_2, \ldots)$. Then it is not clear that $E$ is a commutative ring spectrum.

A notorious example is the Brown–Peterson spectrum, $BP$. Take the complex cobordism spectrum $MU$. Its homotopy groups are

$$\pi_*(MU) = \mathbb{Z}[x_1, x_2, \ldots],$$

where each $x_i$ is a generator in degree $2i$. If you fix a large even degree, then you have a lot of possible elements in that degree, so you might wish to consider a spectrum with sparser homotopy groups. Using the theory of (commutative, 1-dimensional) formal group laws you can do that: If you consider a prime $p$, then there is a spectrum, called the Brown–Peterson spectrum, that corresponds to $p$-typical formal group laws. It can be realized as the image of an idempotent on $MU$ and satisfies

$$\pi_*(BP) \cong \mathbb{Z}_{(p)}[v_1, v_2, \ldots],$$

but now the algebraic generators are spread out in an exponential manner: The degree of $v_i$ is $2p^i - 2$. You can actually choose the $v_i$ as the $x_{p^i-1}$, so you can think of $BP$ as a quotient of $MU$ in the above sense. Since its birth in 1966 [60] its multiplicative properties have been an important issue. In [29], for instance, it was shown that $BP$ has some partial coherence for homotopy commutativity, but in 2017 Tyler Lawson [152] finally showed that at the prime 2 $BP$ is *not* a commutative ring spectrum! For the non-existence of $E_\infty$-structures on $BP$ at odd primes see [271].

There are even worse examples: If you take the sphere spectrum $S$ and you try to kill the non-regular element $2 \in \pi_0(S)$ then you get the mod-2 Moore spectrum. That isn't even a ring spectrum up to homotopy. You can also kill all the generators $v_i \in \pi_*(BP)$ including $p = v_0$, leaving only one $v_n$ alive. The resulting spectrum is the connective version of Morava $K$-theory, $k(n)$. At the prime 2 this isn't even homotopy commutative. In fact, Pazhitnov, Rudyak and Würgler show more [219, 303]: If $\pi_0$ of a homotopy commutative ring spectrum has characteristic two, then it is a generalized Eilenberg–Mac Lane spectrum. Recent work of Mathew, Naumann and Noel puts severe restrictions on finite $E_\infty$-ring spectra [188].

Quite often, we end up working with ideals in the graded commutative ring of homotopy groups, but as we saw above, this is not a suitable notion of ideal. There is a notion of an ideal in the context of (commutative) ring spectra [131] due to Jeff Smith, but still several algebraic constructions do not have an analog in spectra.

So how can you determine whether a given spectrum is a commutative ring spectrum if you don't have a construction that tells you right away that it is commutative? This is where obstruction theory comes into the story.

There is an operadic notion of an $E_\infty$-ring spectrum that goes back to Boardman–Vogt and May. Comparison theorems [178, 262] then tell you whether these more complicated objects are equivalent to commutative ring spectra. In the categories of symmetric spectra, orthogonal spectra and $S$-modules they are.

Obstruction theory might help you with a decision whether a spectrum carries a commutative monoid structure: One version [27] gives obstructions for lifting the ordinary Postnikov tower to a Postnikov tower that lives within the category of commutative ring spectra. The other kind finds some obstruction classes that tell you that you cannot extend some partial bits and pieces of a nice multiplication to a fully fledged structure of an $E_\infty$-ring spectrum or that some homology or homotopy operation that you observe contradicts such a structure. This can be used for a negative result (as in [152]) or for positive statements: There are results by Robinson [246] and Goerss–Hopkins [106, 107] that tell you that you have a (sometimes even unique) $E_\infty$-ring structure on your spectrum if all the obstruction *groups* vanish. Most notably Goerss and Hopkins used obstruction theory to prove that the Morava stabilizer groups acts on the corresponding Lubin–Tate spectrum via $E_\infty$-morphisms [107].

The algebraic behavior on the level of homotopy groups can be quite deceiving: complexification turns a real vector bundle into a complex vector bundle. This induces a map $\pi_*(KO) \to \pi_*(KU)$ which can be realized as a map of commutative ring spectra $c\colon KO \to KU$. On homotopy groups we get

$$\pi_*(c)\colon \pi_*(KO) = \mathbb{Z}[\eta, y, \omega^{\pm 1}]/(2\eta, \eta^3, \eta y, y^2 - 4w) \to \mathbb{Z}[u^{\pm 1}] = \pi_*(KU). \qquad (6.1.1)$$

Here the degrees are $|\eta| = 1$, $|y| = 4$, $|w| = 8$, $|u| = 2$ and $y$ is sent to $2u^2$. So on the algebraic level $c$ is horrible. But John Rognes showed that the conjugation action on $KU$ turns the map $c\colon KO \to KU$ into a $C_2$-Galois extension of commutative ring spectra!

Even for ordinary rings, viewing a (commutative) ring $R$ as a (commutative) ring spectrum via the Eilenberg–Mac Lane spectrum functor changes the situation completely. The ring $R$ has a characteristic map $\chi\colon \mathbb{Z} \to R$ because the ring of integers is the initial ring. As a ring spectrum, $H\mathbb{Z}$ is far from being initial. The map $H\chi$ can be precomposed with the unit map of $H\mathbb{Z}$:

$$S \xrightarrow{\;\eta\;} H\mathbb{Z} \xrightarrow{\;H\chi\;} HR,$$

and the sphere spectrum $S$ is the initial ring spectrum! Now there is a lot of space between the sphere and any ring. I will discuss two consequences that this has: There is actually algebraic geometry happening between the sphere spectrum and the prime field $\mathbb{F}_p$: There is a Galois extension of commutative ring spectra (see 6.8.1) $A \to H\mathbb{F}_p$!

Another feature is that there exist differential graded algebras $A_*$ and $B_*$ that are not quasi-isomorphic, but whose associated algebra spectra over an Eilenberg–Mac Lane spectrum [275] are equivalent as ring spectra [80]. Similar phenomena happen if you consider differential graded $E_\infty$-algebras: there are non quasi-isomorphic ones whose associated commutative algebras over an Eilenberg–Mac Lane spectrum [236] are equivalent as commutative ring spectra [33].

## Content

The structure of this overview is as follows: We start with some basic features of commutative ring spectra and their model category structures in Section 6.2. The most basic way to relate classical algebra to brave new algebra is via the Eilenberg–Mac Lane spectrum functor. We study chain algebras and algebras over Eilenberg–Mac Lane ring spectra in Section 6.3. As you can study the group of units of a ring we consider units of ring spectra and Thom spectra in Section 6.4. In Section 6.5 we present a construction going back to Segal. Plugging in a bipermutative category yields a commutative ring spectrum.

In Section 6.6 we introduce topological Hochschild homology and some of its variants and topological André–Quillen homology. In Section 6.7 we discuss some versions of obstruction theory that tell you whether a given multiplicative cohomology theory can be represented by a strict commutative model.

Some concepts from algebra translate directly to spectra but some others don't. We discuss the different concepts of étale maps for commutative algebra spectra in Section 6.8. Picard and Brauer groups for commutative ring spectra are important invariants and feature in Section 6.9.

## Disclaimers

For more than 30 years, the phrase *commutative ring spectrum* meant a commutative monoid in the homotopy category of spectra. Since the 1990s this has changed. At the beginning of this new era people were careful not to use this name, in order to avoid confusion with the homotopy version. In this paper we reserve the phrase *commutative ring spectrum* for a commutative monoid in some symmetric monoidal category of spectra.

The second disclaimer is that for this paper a space is always compactly generated weak Hausdorff. I denote the corresponding category just by Top.

Last but not least: Of course, this overview is not complete. I had to omit important aspects of the field due to space constraints. Most prominently probably is the omission of topological cyclic homology and its wonderful applications to algebraic $K$-theory.

I try to give adequate references, but often it was just not feasible to describe the whole development of a topic and much worse, I probably have forgotten to cite important contributions. If you read this and it affects you, then I can only apologize.

## Acknowledgements

The author thanks the Hausdorff Research Institute for Mathematics in Bonn for its hospitality during the Trimester Program *K-Theory and Related Fields*. She thanks Akhil Mathew for clarifying one example, Jim Davis and Stefan Schwede for some corrections, Andy Baker and Peter May for pointing out omissions and Steffen Sagave for many valuable comments on a draft version of this paper.

## 6.2        Features of commutative ring spectra

### Some basics

Before we actually start with model structures, we state some basic facts about commutative ring spectra.

Let $R$ be a commutative ring spectrum. Then the category of $R$-module spectra is closed symmetric monoidal: For two such $R$-module spectra $M, N$ the smash product over $R$, $M \wedge_R N$, is again an $R$-module and the usual axioms of a symmetric monoidal category are satisfied. There is an $R$-module spectrum $F_R(M, N)$, the function spectrum of $R$-module maps from $M$ to $N$.

We denote the category of $R$-module spectra by $\mathcal{M}_R$. The category of commutative $R$-algebras is the category of commutative monoids in $\mathcal{M}_R$ and we denote it by $\mathcal{C}_R$.

By definition, every object $A$ of $\mathcal{C}_R$ receives a unit map from $R$ and hence $R$ is initial in $\mathcal{C}_R$. In particular, the sphere spectrum is the initial commutative ring spectrum. Every discrete ring is a $\mathbb{Z}$-algebra; similarly, every (commutative) ring spectrum is a (commutative) $S$-algebra. If $R$ is a commutative ring spectrum, then the category of commutative $R$-algebras is isomorphic to the category of commutative ring spectra under $R$, i.e., the category of commutative ring spectra $A$ with a distinguished map $\eta \colon R \to A$ in that category.

We allow the trivial $R$-algebra corresponding to the one-point spectrum $*$ and this spectrum is a terminal object in $\mathcal{C}_R$.

In any symmetric monoidal category $(\mathcal{C}, \otimes, 1, \tau)$ the coproduct of two commutative monoids $A$ and $B$ in $\mathcal{C}$ is $A \otimes B$. So, for two commutative $R$-algebras $A$ and $B$, their coproduct is $A \wedge_R B$.

### Model structures on commutative monoids

I will assume that you are familiar with the concept of model categories and that you have seen some examples and read Chapter 3 in this book. Good general references are Hovey's [130] and Hirschhorn's [124] books. You could also just skip this section and have in mind that there are some serious model category issues lurking in the dark.

For this section I will mainly focus on two models for spectra: symmetric spectra [133] and $S$-modules [94]. They are different concerning their model structures. In the model structure in [133] on symmetric spectra the sphere spectrum is cofibrant, whereas in the one for $S$-modules it is not, but all objects are fibrant.

The model structures on commutative monoids in either of the categories [94, 133] are special cases of a *right induced model structure*: We have a functor $P_R$ from $R$-module spectra to commutative $R$-algebra spectra assigning the free commutative $R$-algebra spectrum on $M$ to any $R$-module spectrum $M$: explicitly,

$$P_R(M) = \bigvee_{n \geqslant 0} M^{\wedge_R n}/\Sigma_n.$$

The symbol $P_R$ should remind you of a polynomial algebra. This functor has a right adjoint, the forgetful functor $U$. In a right-induced model structure one determines the fibrations and weak equivalences by the right adjoint functor. In our cases, a map of commutative $R$-algebra spectra is a fibration or a weak equivalence if it is one in the underlying category of $R$-module spectra. Note that establishing right induced model structures on commutative monoids in some model category does not always work. The standard example is the category of $\mathbb{F}_p$-chain complexes (say $p$ is an odd prime). Then the chain complex $\mathbb{D}^2$ is acyclic, having $\mathbb{F}_p$ in degrees 1 and 2 with the identity map as differential, but the free graded commutative monoid generated by it is $\Lambda_{\mathbb{F}_p}(x_1) \otimes \mathbb{F}_p[x_2]$ with $|x_i| = i$ and the induced differential is determined by $d(x_2) = x_1$ and the Leibniz rule. But then $d(x_2^p)$ is a cycle that is not a boundary, so the resulting object is not acyclic.

If $R$ is a commutative $S$-algebra in the setting of EKMM [94], then the categories of associative $R$-algebras and of commutative $R$-algebras possess a right induced model structure [94, Corollary VII.4.10]. The existence of the model structure for commutative monoids is a special case of the existence of right-induced model structures for $\mathbb{T}$-algebras [94, Theorem VII.4.9], where $\mathbb{T}$ is a continuous monad on the category of $R$-module spectra that preserves reflexive coequalizers and satisfies the cofibration hypothesis [94, VII.4]. The category of commutative $S$-algebras is identified [94, Proposition II.4.5] with the category of algebras for the monad $P_S$ as above on the category of $S$-modules.

In diagram categories such as symmetric spectra and orthogonal spectra the situation is different: In the standard model structures on these categories the sphere spectrum is cofibrant. If one were to take a right-induced model structure on the category of commutative monoids, i.e., the model structure such that a map of commutative ring spectra $f : A \to B$ is a fibration or weak equivalence if it is one in the underlying category, then the sphere would still be cofibrant. If we take a fibrant replacement of the sphere $S \to S^{\mathrm{fib}}$, then in particular $S^{\mathrm{fib}}$ would be fibrant in the model category of symmetric spectra; hence it would be an $\Omega$-spectrum and its zeroth level would be a strictly commutative model for $QS^0$. However, Moore shows [212, Theorem 3.29] that this would imply that $QS^0$ has the homotopy type of a product of Eilenberg–Mac Lane spaces — but this is false.

The usual way to avoid this problem is to consider a positive model structure on $Sp^\Sigma$ (see [178, Definition 6.1] for the general approach). Here the positive level fibrations (weak equivalences) are maps $f \in Sp^\Sigma(X, Y)$ such that $f(n)$ is a fibration (weak equivalence) for all levels $n \geqslant 1$. The positive cofibrations are then cofibrations in $Sp^\Sigma$ that are isomorphisms in level zero. The positive stable model category is then obtained by a Bousfield localization that forces the stable equivalences to be the weak equivalences and the right-induced model structure on the commutative monoids in $Sp^\Sigma$ then has the desired properties.

There is another nice model for connective spectra, given by $\Gamma$-spaces [268, 172]. This category is built out of functors from finite pointed sets to spaces, so it is a very hands-on category with explicit constructions. It is also a symmetric monoidal category with a suitable model structure. We refer to [172, 264] for background on this. Its (commutative) monoids are called (commutative) $\Gamma$-rings. Beware that commutative $\Gamma$-rings, however, do *not* model all connective commutative ring spectra. Tyler Lawson proves in [151] that commutative $\Gamma$-rings satisfy a vanishing condition for Dyer–Lashof operations of positive degree on classes in their zeroth mod-$p$-homology (for all primes $p$) and that for instance the free $E_\infty$-ring spectrum generated by $S^0$ cannot be modeled as a commutative $\Gamma$-ring.

## Behavior of the underlying modules

In the setting of EKMM it is shown that the underlying $R$-modules of cofibrant commutative $R$-algebras have a well-behaved smash product in the derived category of $R$-modules:

**Theorem 6.2.1** [94, Theorem VII.6.7]. *If $A$ and $B$ are two cofibrant commutative $R$-algebras, and if $\varphi_A\colon \Gamma A \xrightarrow{\sim} A$ and $\varphi_B\colon \Gamma B \xrightarrow{\sim} B$ are chosen cell $R$-module spectra approximations then*

$$\varphi_A \wedge_R \varphi_B\colon \Gamma A \wedge_R \Gamma B \to A \wedge_R B$$

*is a weak equivalence.*

Brooke Shipley developed a model structure for commutative symmetric ring spectra in [274] in which the underlying symmetric spectrum of a cofibrant commutative ring spectrum is also cofibrant as a symmetric spectrum [274, Corollary 4.3].

She starts with introducing a different model structure on symmetric spectra. Let $M$ denote the class of monomorphisms of symmetric sequences in pointed simplicial sets and let $S \otimes M$ denote the set $\{S \otimes f, f \in M\}$, where $\otimes$ denotes the tensor product of symmetric sequences. An $S$-*cofibration* is a morphism in $(S \otimes M)$-cof, i.e., a morphism in $Sp^\Sigma$ that has the left lifting property with respect to maps that have the right lifting property with respect to $S \otimes M$. She shows that the classes of $S$-cofibrations and stable equivalences determine a model structure with the $S$-fibrations being the class of morphisms with the right lifting property with respect to $S$-cofibrations that are also stable equivalences [274, Theorem 2.4]. This model structure was already mentioned in [133, 5.3.6]. Shipley proves that this model structure is cofibrantly generated, is monoidal and satisfies the monoid axiom [274, 2.4, 2.5].

Note that symmetric spectra are $S$-modules in symmetric sequences. This allows for a version of an $R$-*model structure* for every associative symmetric ring spectrum $R$ with $R$-cofibrations, $R$-fibrations and stable equivalences [274, Theorem 2.6]. In the positive variant of this model structure the positive $R$-cofibrations are $R$-cofibrations that are isomorphisms in level zero. Together with the stable equivalences this determines the *positive $R$-model structure*.

The corresponding right induced model structure on commutative $R$-algebra spectra for a commutative symmetric ring spectrum $R$ is then the *convenient* model structure: The weak equivalences are stable equivalences, the fibrations are positive $R$-fibrations and the cofibrations are determined by the structure.

She then shows a remarkable property of this model structure on commutative $R$-algebra spectra:

Theorem 6.2.2 [274, Corollary 4.3]. *If A is cofibrant as a commutative R-algebra then A is R-cofibrant in the R-model structure. If A is fibrant as a commutative R-algebra, then A is fibrant in the positive R-model structure on R-module spectra.*

The positive $R$-model structure ensures that $R$ is *not* cofibrant; hence cofibrant commutative $R$-algebras will not be positively $R$-cofibrant!

## Comparison, rigidification and $E_n$-structures

Stefan Schwede proves [262, Theorem 5.1] that the homotopy category of commutative $S$-algebras from [94] is equivalent to the homotopy category of commutative symmetric ring spectra by establishing a Quillen equivalence between the corresponding model categories. In [178, Theorem 0.7] the analogous comparison result is proven for commutative orthogonal ring spectra and commutative symmetric ring spectra.

Even before any symmetric monoidal category of spectra was constructed, the notion of operadically defined $E_\infty$-ring spectra [199] was available. An $E_\infty$-structure on a spectrum is a multiplication that is homotopy commutative in a coherent way. See Chapter 5 of this book for background on operads and their role in the study of spectra with additional structure.

There is an explicit comparison of the good old $E_\infty$-ring spectra and commutative ring spectra, see [94, Proposition II.4.5] or [178, Remark 0.14]; in particular, every $E_\infty$-ring spectrum $\tilde{R}$ can be rigidified to a commutative ring spectrum $R$ in such a way that the homotopy type is preserved.

There are several popular $E_\infty$-operads that will show up later: for instance the linear isometries operad (see (6.4.3)) and the Barratt–Eccles operad. The $n$-ary part of the latter is easy to describe: You take $\mathcal{O}(n) = E\Sigma_n$, a contractible space with free $\Sigma_n$-action. For compatibility reasons it is advisable to take the realization of the standard simplicial model of $E\Sigma_n$ whose set of $q$-simplices is $(\Sigma_n)^{q+1}$.

An operad with a nice geometric description is the little $m$-cubes operad, that in arity $n$ consists of the space of $n$-tuples of linearly embedded $m$-cubes in the standard $m$-cube with disjoint interiors and with axes parallel to that of the ambient cube [49, Example 5]. We call this (and every equivalent) operad in spaces $E_m$. For $m = 1$ this operad parametrizes $A_\infty$-structures and the colimit is an $E_\infty$-operad. Hence the intermediate $E_m$'s for $1 < m < \infty$ interpolate between these structures; they give $A_\infty$-structures with homotopy-commutative multiplications that are coherent up to some order.

## Power operations

The extra structure of an $E_\infty$-ring spectrum gives homology operations. The general setting allows for $H_\infty$-ring spectra [63]; for simplicity we assume that $E$ and $R$ are two $E_\infty$-ring spectra whose structure is given by the Barratt–Eccles operad, i.e., there are structure maps

$$\xi_R^n \colon (E\Sigma_n)_+ \wedge_{\Sigma_n} R^{\wedge n} \to R \tag{6.2.1}$$

for $R$ and also for $E$. McClure describes the general setting of power operations in [63, IX §1]. Fix a prime $p$ and abbreviate $(E\Sigma_p)_+ \wedge_{\Sigma_p} R^{\wedge p}$ by $D_p(R)$; one often calls $D_p R$ the $p$-*th extended power construction on R*. A *power operation* assigns to every class $[x] \in E_n R$ and every class $e \in E_m(D_p S^n)$ a class $Q^e[x] \in E_m R$; hence we can view $Q^e$ as a map

$$Q^e : E_n R \to E_m R.$$

The construction is as follows. Take a representative $x : S^n \to E \wedge R$ of $[x]$ and $e \in E_m(D_p S^n)$ and apply the following composition to $e$:

$$E_m(D_p S^n) \xrightarrow{\;E_m(D_p x)\;} E_m(D_p(E \wedge R)) \xrightarrow{\;\delta\;} E_m(D_p E \wedge D_p R) \qquad (6.2.2)$$
$$\downarrow{\scriptstyle E_m(\xi_E^p \wedge \mathrm{id})}$$
$$E_m(E \wedge D_p R)$$
$$\downarrow{\scriptstyle \mu_*}$$
$$E_m(D_p R)$$
$$\downarrow{\scriptstyle E_m(\xi_R^p)}$$
$$E_m(R).$$

Here,

$$\delta : (E\Sigma_p)_+ \wedge_{\Sigma_p} (E \wedge R)^{\wedge p} \to (E\Sigma_p)_+ \wedge_{\Sigma_p} E^{\wedge p} \wedge (E\Sigma_p)_+ \wedge_{\Sigma_p} R^{\wedge p}$$

is the canonical map induced by the diagonal on the space $E\Sigma_p$ and $\mu$ denotes the multiplication in $E$, so it induces

$$\mu_* : \pi_*(E \wedge E \wedge D_p R) \to \pi_*(E \wedge D_p R).$$

There are several important special cases of this construction:

1. For $E$ the sphere spectrum one obtains operations on the homotopy groups of an $E_\infty$-ring spectrum; see [63, IV §7].
2. For $E = H\mathbb{F}_p$ the power operations for certain classes $e_i \in H_i(\Sigma_p; \mathbb{F}_p)$ are often called (Araki–Kudo–)Dyer–Lashof operations. These are natural homomorphisms

$$Q^i : (H\mathbb{F}_p)_n(R) \to (H\mathbb{F}_p)_{n+2i(p-1)}(R) \qquad (6.2.3)$$

for odd primes and $Q^i : (H\mathbb{F}_2)_n(R) \to (H\mathbb{F}_2)_{n+i}(R)$ at the prime 2 that satisfy a list of axioms [63, Theorem III.1.1] and compatibility relations with the homology Bockstein and the dual Steenrod operations.
3. There are also important $K(n)$-local versions of such operations and we will encounter them later.

## 6.3     Chain algebras and algebras over Eilenberg–MacLane spectra

The derived category of a ring is an important object in many subjects. The initial ring is the ring of integers. Every ring $R$ has an associated Eilenberg–MacLane spectrum, $HR$.

### $HR$-module and algebra spectra

We collect some results that compare the category of chain complexes of $R$-modules with the category of module spectra over $HR$. We start with additive statements and move to comparison results for flavors of differential graded $R$-algebras. For an overview of algebraic applications of these equivalences see for instance [111].

In the 1980s, so before any strict symmetric monoidal category of spectra was constructed, Alan Robinson developed the notion of the derived category, $\mathcal{D}(E)$, of right $E$-module spectra for every $A_\infty$-ring spectrum $E$. He showed the following result.

**Theorem 6.3.1** [249, Theorem 3.1]. *For every associative ring $R$ there is an equivalence of categories between the derived category of $R$, $\mathcal{D}(R)$, and the derived category of the associated Eilenberg–MacLane spectrum, $\mathcal{D}(HR)$.*

Later, in the context of $S$-modules this corresponds to [94, IV, Theorem 2.4]. Work of Schwede and Shipley strengthened the result to a Quillen equivalence of the corresponding model categories:

**Theorem 6.3.2** [266, Theorem 5.1.6]. *The model category of unbounded chain complexes of $R$-modules is Quillen equivalent to the model category of $HR$-module spectra .*

Stefan Schwede uses the setting of $\Gamma$-spaces [264] to embed simplicial rings and modules into the stable world: He constructs a lax symmetric monoidal Eilenberg–MacLane functor $H$ from simplicial abelian groups to $\Gamma$-spaces together with a linearization functor $L$ in the opposite direction and proves the following comparison result:

**Theorem 6.3.3** [264, Theorems 4.4 and 4.5]. *If $R$ is a simplicial ring, then the adjoint functors $H$ and $L$ constitute a Quillen equivalence between the categories of simplicial $R$-modules and $HR$-module spectra. If $R$ is in addition commutative, then $H$ and $L$ induce a Quillen equivalence between the categories of simplicial $R$-algebras and $HR$-algebra spectra.*

Here, the functor $L$ is left inverse to $H$ and induces an isomorphism of $\Gamma$-spaces

$$\mathrm{Hom}(HA, HB) \cong H(\mathrm{Hom}_{\mathsf{sAb}}(A, B))$$

[264, Lemma 2.1]; thus $H$ embeds algebra into brave new algebra.

Brooke Shipley extends this equivalence to corresponding categories of monoids in the differential graded setting:

Theorem 6.3.4 [275, Theorem 1.1]. *For any commutative ring $R$, the model categories of unbounded differential graded $R$-algebras and $HR$-algebra spectra are Quillen equivalent.*

Dugger and Shipley show in [80] that there are examples of $HR$-algebras that are weakly equivalent as $S$-algebras, but that are not quasi-isomorphic. A concrete example is the differential graded ring $A_*$ which is generated by an element in degree 1, $e_1$, and has $d(e_1) = 2$ and satisfies $e_1^4 = 0$. The corresponding $H\mathbb{Z}$-algebra spectrum is equivalent as a ring spectrum to the one on the exterior algebra $B_* = \Lambda_{\mathbb{F}_2}(x_2)$ (with $|x_2| = 2$) but $A_*$ and $B_*$ are *not* quasi-isomorphic. You find more examples and proofs in [80, §§4,5].

We cannot expect that commutative $HR$-algebra spectra correspond to commutative differential graded $R$-algebras unless $R$ is of characteristic zero, because of cohomology operations, but we get the following result:

Theorem 6.3.5 [236, Corollary 8.3]. *If $R$ is a commutative ring, then there is a chain of Quillen equivalences between the model category of commutative $HR$-algebra spectra and $E_\infty$-monoids in the category of unbounded $R$-chain complexes.*

Haldun Özgür Bayındır shows [33] that one can find $E_\infty$-differential graded algebras that are not quasi-isomorphic, but whose corresponding commutative $HR$-algebra spectra are equivalent as commutative ring spectra.

## Cochain algebras

A prominent class of examples of commutative $HR$-algebra spectra consists of function spectra $F(X_+, HR)$. Here, $X$ is an arbitrary space and $R$ is a commutative ring. The diagonal $\Delta \colon X \to X \times X$ and the multiplication on $HR$, $\mu_{HR}$, induce a multiplication

$$F(X_+, HR) \wedge F(X_+, HR) \longrightarrow F(X_+ \wedge X_+, HR \wedge HR) \cong F((X \times X)_+, HR \wedge HR)$$
$$\Big\downarrow {\scriptstyle \Delta^*, \mu_{HR}}$$
$$F(X_+, HR)$$

that turns $F(X_+, HR)$ into a $HR$-algebra spectrum. As the diagonal is cocommutative and as $\mu_{HR}$ is commutative, the resulting multiplication is commutative.

These function spectra are models for the singular cochains of a space $X$ with coefficients in $R$:

$$\pi_*(F(X_+, HR)) \cong H^{-*}(X; R).$$

Beware that the homotopy groups of $F(X_+, HR)$ are concentrated in non-positive degrees — i.e., $F(X_+, HR)$ is coconnective.

Studying the singular cochains of a space $S^*(X; R)$ as a differential graded $R$-module is not enough in order to recover the homotopy type of $X$. If we work over the rational numbers, then Quillen showed that rational homotopy theory is algebraic in the sense that one can use rational differential graded Lie algebras or coalgebras as models for rational homotopy theory [228]. Sullivan [286] constructed a functor, assigning a rational differential graded commutative algebra to a space, that is closely related to the singular cochain functor with rational coefficients. He used this to classify rational homotopy types.

For a general commutative ring $R$, the singular cochains are an $E_\infty$-algebra. Mike Mandell proves [181, Main Theorem] that the singular cochain functor with coefficients in an algebraic closure of $\mathbb{F}_p$, $\overline{\mathbb{F}}_p$, induces an equivalence between the homotopy category of connected $p$-complete nilpotent spaces of finite $p$-type and a full subcategory of the homotopy category of $E_\infty$-$\overline{\mathbb{F}}_p$-algebras. He also characterizes those $E_\infty$-$\overline{\mathbb{F}}_p$-algebras that arise as cochain algebras of 1-connected $p$-complete spaces of finite $p$-type explicitly [181, Characterization Theorem]. There is also an integral version of this result, stating that finite type nilpotent spaces are weakly equivalent if and only if their $E_\infty$-algebras of integral cochains are quasi-isomorphic [180, Main Theorem].

A strictly commutative integral model of the $E_\infty$-algebra of cochains on a space is constructed in [235] using chain complexes indexed by the category of finite sets and injections.

## 6.4     Units of ring spectra and Thom spectra

One construction that can give rise to highly structured multiplications on a spectrum is the Thom spectrum construction: For instance, complex bordism, $MU$, obtains a commutative ring structure this way. Mahowald emphasized [175] early on that multiplicative properties of the structure maps for Thom spectra translate to multiplicative structures on the resulting Thom spectra. Their properties and the corresponding orientation theory is systematically studied in [199, 196]. There is the following general result by Lewis:

**Theorem 6.4.1** [155, Theorem IX.7.1 and Remark IX.7.2]. *Assume that $f$ is a map of spaces from $X$ to the classifying space for stable spherical fibrations, $BG$, that is a $C$-map for some operad $C$ over the linear isometries operad. Then the Thom spectrum $M(f)$ associated to $f$ carries a $C$-structure. In particular, infinite loop maps from $X$ to $BG$ give rise to $E_\infty$-ring spectra.*

Note that $BG$ is the classifying space of the units of the sphere spectrum, $GL_1(S)$. A seemingly naive definition of $GL_1(S)$ is given by the pullback of the diagram

$$
\begin{array}{ccc}
GL_1 S & \cdots\cdots\cdots\rightarrow & \Omega^\infty S \\
\downarrow & & \downarrow \\
\pi_0(S)^\times = \{\pm 1\} & \longrightarrow & \pi_0(S) = \mathbb{Z}
\end{array}
\tag{6.4.1}
$$

so by the components of $QS^0$ corresponding to $\pm 1 \in \mathbb{Z}$.

We next give a short overview of Thom spectra that arise in a more general context, where the target of the map is the space of units, $GL_1(R)$, for a commutative ring spectrum $R$. The first idea is to define the space $GL_1(R)$ as the space that represents the functor that sends a space $X$ to the units in $R^0(X)$. Copying the definition from (6.4.1) above with $S$ replaced by $R$ gives a valid definition of $GL_1(R)$ and it was shown in [199] that for commutative $R$ this model is an $E_\infty$-space.

In the approaches [6] and [31], the idea is to replace the above model of $GL_1(R)$ with its $E_\infty$-structure with a strictly commutative model. As spaces with an $E_\infty$-structure are not equivalent to strictly commutative spaces (that's the problem again that then $QS^0$ would be a product of Eilenberg–Mac Lane spaces [212]), one has to find a different category with the property that there is a Quillen equivalence between commutative monoids in that category and $E_\infty$-monoids in spaces and such that there are models of $\Omega^\infty(R)$ and $GL_1(R)$ in this category.

In [6] the authors work with *-modules and in [31] the authors use Schlichtkrull's model of $GL_1(R)$ in commutative $I$-spaces, where $I$ is the skeleton of the category of finite sets and injections.

The idea is to construct a spectrum version of the assembly map for discrete rings: If $R$ is a discrete ring and if $R^\times$ is its group of units, then there is a canonical map

$$\mathbb{Z}[R^\times] \to R \tag{6.4.2}$$

from the group ring $\mathbb{Z}[R^\times]$ to $R$ that takes an element $\sum_{i=1}^n a_i r_i$ of $\mathbb{Z}[R^\times]$ (with $a_i \in \mathbb{Z}$ and $r_i \in R^\times$) to the same sum, but now we use the ring structure of $R$ to convert the formal sum into an actual sum $\sum_{i=1}^n a_i r_i \in R$. Note that $R^\times$ is an abelian group if $R$ is a commutative ring.

We will sketch both constructions of Thom spectra and briefly discuss the application in [6] to the question of when a Thom spectrum allows for an $E_\infty$-map to some other $E_\infty$-ring spectrum: for instance, whether one can realize an $E_\infty$-version of the string orientation $MO\langle 8\rangle \to \mathrm{tmf}$ [7] or an $E_\infty$-version of a complex orientation [127].

The focus in [31] is on multiplicative properties of the Thom spectrum functor and on applications to topological Hochschild homology. We present the results about multiplicative structures and discuss their results on THH of Thom spectra in Section 6.6. We'll also describe how the concept of $I$-spaces can be generalized to a setting in which the units can be adapted to non-connective ring spectra.

## Thom spectra via $\mathbb{L}$-spaces and orientations

Fix a countably infinite-dimensional real vector space $U$ and consider

$$\mathbb{L} = \mathcal{L}(1) = \mathcal{L}(U, U),$$

the space of linear isometries from $U$ to itself. The notation $\mathcal{L}(1)$ is due to the fact that $\mathcal{L}(1)$ is the 1-ary part of the famous linear isometries operad [49, §1] whose term of arity $n$ is

$$\mathcal{L}(n) = \mathcal{L}(U^n, U). \tag{6.4.3}$$

See [49] or [6] for details. Note that $\mathbb{L}$ is a monoid with respect to composition.

Definition 6.4.2. The *category of* $\mathbb{L}$-*spaces*, $\mathsf{Top}[\mathbb{L}]$, is the category of spaces with a left action of the monoid $\mathbb{L}$.

Using the 2-ary part of the linear isometries operad, one can manufacture a product on $\mathsf{Top}[\mathbb{L}]$: For objects $X, Y$ of $\mathsf{Top}[\mathbb{L}]$ their product $X \times_\mathbb{L} Y$ is the coequalizer

$$\mathcal{L}(2) \times (\mathcal{L}(1) \times \mathcal{L}(1)) \times X \times Y \rightrightarrows \mathcal{L}(2) \times X \times Y \cdots\!\!\rightarrow X \times_\mathbb{L} Y.$$

Here, one map uses the $\mathcal{L}(1)$-action on the spaces $X$ and $Y$ and the other map uses the operad product $\mathcal{L}(2) \times \mathcal{L}(1) \times \mathcal{L}(1) \to \mathcal{L}(2)$.

As $\mathcal{L}(2) = \mathcal{L}(U^2, U)$ has a left $\mathcal{L}(1)$-action, $X \times_\mathbb{L} Y$ is an $\mathcal{L}(1)$-space. The product is associative and has a symmetry, but it is only weakly unital. See [45, §4] for a careful discussion.

By [45, Proposition 4.7] there is an isomorphism of categories between commutative monoids with respect to $\times_\mathbb{L}$ and $E_\infty$-spaces whose $E_\infty$-structure is parametrized by the linear isometries operad.

For strict unitality, one restricts to the full subcategory $\mathcal{M}_*$ of objects of $\mathsf{Top}[\mathbb{L}]$ for which the unit map is a homeomorphism. Such objects are called *-*modules*. The commutative monoids in $\mathcal{M}_*$ again model $E_\infty$-spaces [45, Proposition 4.11].

For an associative ring spectrum $R$, there is a strictly associative model in $\mathcal{M}_*$ of the space of units $GL_1(R)$ and the functor $GL_1$ is right adjoint to the inclusion of grouplike objects. One can form a bar construction, $B_{\times_\mathbb{L}}$, of a cofibrant replacement, $GL_1(R)^c$, of $GL_1(R)$ with respect to the monoidal product $\times_\mathbb{L}$, where $B_{\times_\mathbb{L}}(GL_1(R)^c)$ is the geometric realization of the simplicial $\mathcal{M}_*$ object

$$[n] \mapsto * \times_\mathbb{L} \underbrace{GL_1^c(R) \times_\mathbb{L} \ldots \times_\mathbb{L} GL_1^c(R)}_{n} \times_\mathbb{L} *.$$

Similarly, $E_{\times_\mathbb{L}}(GL_1(R)^c)$ is constructed out of the simplicial object

$$[n] \mapsto * \times_\mathbb{L} \underbrace{GL_1^c(R) \times_\mathbb{L} \ldots \times_\mathbb{L} GL_1^c(R)}_{n+1}.$$

Adapted to the situation there are suspension spectrum and underlying infinite loop space functors [159, Lemma 7.5]

$$\mathcal{M}_* \underset{\Omega_S^\infty}{\overset{(\Sigma_\mathbb{L}^\infty)_+}{\rightleftarrows}} \mathcal{M}_S \tag{6.4.4}$$

that are a Quillen adjoint pair of functors. Here, the suspension functor is strong symmetric monoidal and the underlying loop space functor is lax symmetric monoidal.

The spectrum version of the assembly map from (6.4.2) is

$$(\Sigma_\mathbb{L}^\infty)_+(GL_1^c(R)) \to (\Sigma_\mathbb{L}^\infty)_+(GL_1(R)) \to R,$$

where the first map comes from the cofibrant replacement of the units and the second one is the counit of an adjunction [6, (3.1)].

**Definition 6.4.3** [6, Definition 3.12]. Given a map $f: X \to B_{\times_{\mathbb{L}}}(GL_1^c(R))$, the *Thom spectrum for $f$ in $\mathcal{M}_*$* is the $R$-module spectrum (in the world of [94])

$$M(f) = (\Sigma_{\mathbb{L}}^\infty)_+ P^c \wedge_{(\Sigma_{\mathbb{L}}^\infty)_+ GL_1^c(R)} R. \tag{6.4.5}$$

Here, $P^c$ is a cofibrant replacement as a right $GL_1^c(R)$-module of the pullback

$$\begin{array}{ccc}
P & \longrightarrow & E_{\times_{\mathbb{L}}}(GL_1^c(R)) \\
\downarrow & & \downarrow \\
X & \longrightarrow & B_{\times_{\mathbb{L}}}(GL_1^c(R))
\end{array}$$

*Remark 6.4.4.* Because of the cofibrancy of $P^c$, the smash product in (6.4.5) is actually a derived smash product. See [6, §3] for the necessary background on the model structures involved.

In the commutative case, [6, §4, §5] is set in the classical framework of $E_\infty$-ring spectra and $E_\infty$-spaces as in [199]. For an $E_\infty$-ring spectrum $R$, the space $\Omega^\infty R$ is actually an $E_\infty$-*ring space* [197, Corollary 7.5]; this is a space on which a pair of $E_\infty$-operads acts: one codifying the additive structure that is present in every spectrum and one encoding the multiplicative structure [197, §1]. Actually more is true. Call an $E_\infty$-ring space ring-like if its $\pi_0$ is actually a ring and not just a *rig* — a ring without negatives. The homotopy category of ring-like $E_\infty$-ring spaces is equivalent to the homotopy category of connective $E_\infty$-ring spectra [197, Theorem 9.12].

If $R$ is a commutative ring spectrum or an $E_\infty$-ring spectrum then the space of units, $GL_1(R)$, is a group-like $E_\infty$-space and hence is an infinite loop space that has an associated connective spectrum, $gl_1(R)$, with $\Omega^\infty gl_1(R) = GL_1(R)$.

The crucial ingredient in this case is the pair of functors $(\Sigma_+^\infty \Omega^\infty, gl_1)$ that is an adjunction between the homotopy category of connective spectra and the homotopy category of $E_\infty$-ring spectra in the sense of Lewis–May–Steinberger.

In particular, one gets a version of the assembly map from (6.4.2):

$$\Sigma_+^\infty \Omega^\infty(gl_1(R)) \to R$$

for every $E_\infty$-ring spectrum. By [94] one can replace $E_\infty$-ring spectra with commutative $S$-algebras, i.e., with commutative ring spectra. This simplifies the discussion of pushouts and allows us to replace $\Sigma_+^\infty \Omega^\infty$ by $(\Sigma_{\mathbb{L}}^\infty)_+ \Omega_S^\infty$ from (6.4.4) to get

$$(\Sigma_{\mathbb{L}}^\infty)_+ \Omega_S^\infty(gl_1(R)) \to R.$$

Note that a map of infinite loop spaces $f: B \to BGL_1(R)$ encodes the same data as a map of spectra $f: b \to bgl_1(R)$, where the lowercase letters denote the associated connective spectra. As before we consider the pullback $p$:

$$\begin{array}{ccc}
p & \longrightarrow & egl_1(R) \\
\downarrow & & \downarrow \\
b & \xrightarrow{\;f\;} & bgl_1(R)
\end{array}$$

and form the corresponding derived smash product:

**Definition 6.4.5.** Let $f: b \to bgl_1(R)$ be a map of connective spectra. The *Thom spectrum associated to* $f$, $M(f)$, is the homotopy pushout in the category of commutative $S$-algebras

$$M(f) = (R \wedge (\Sigma_{\mathbb{L}}^{\infty})_+ \Omega_S^{\infty} p) \wedge_{R \wedge (\Sigma_{\mathbb{L}}^{\infty})_+ \Omega_S^{\infty} gl_1(R)}^{L} R.$$

As the (homotopy) pushout is the (derived) smash product, this resembles the construction from (6.4.5).

In the commutative ring spectrum setting the question about orientations is the following problem: Assume that there is a map of commutative ring spectra $\alpha: R \to A$, then $A$ is a commutative $R$-algebra spectrum. For a map of spectra $f: b \to bgl_1(R)$ as above we can ask whether there is a morphism of commutative $R$-algebra spectra from $M(f)$ to $A$. As $M(f)$ is defined as a (homotopy) pushout, we get a condition that says that we need maps from the ingredients of the derived smash product. As we start with a map $\alpha$ from $R$ to $A$, we get an induced map

$$gl_1(\alpha): gl_1(R) \to gl_1(A).$$

So what is missing is a map

$$(\Sigma_{\mathbb{L}}^{\infty})_+ \Omega_S^{\infty} p \to A$$

that is compatible with the map $(\Sigma_{\mathbb{L}}^{\infty})_+ \Omega_S^{\infty} gl_1(R) \to A$. With the help of the adjunction this means that we need a map

$$p \to gl_1(A)$$

whose precomposition with the map $gl_1(R) \to p$ gives $gl_1(\alpha)$. This argument can be turned into a proof for the following result:

**Theorem 6.4.6** [6, Theorem 4.6]. *The derived mapping space of commutative $R$-algebras from $M(f)$ to $A$, $Map_{C_R}(M(f), A)$, is weakly equivalent to the fiber in the map between derived mapping spaces*

$$Map_{M_S}(p, gl_1(A)) \to Map_{M_S}(gl_1(R), gl_1(A))$$

*at the basepoint $gl_1(\alpha)$ of $Map_{M_S}(gl_1(R), gl_1(A))$.*

An important example is the question of the string orientation of the spectrum of topological modular forms, tmf. For background on tmf and its variants see [74], whose Chapter 10 contains André Henriques' notes of Mike Hopkins' lecture on the string orientation. Let $BO\langle 8 \rangle$ be the 7-connected cover of $BO$ and let $bo\langle 8 \rangle$ be the associated spectrum with the canonical map $f: bo\langle 8 \rangle \to bgl_1(S)$. So we are in the situation where $R = S$ and we take $A = $ tmf. Ando, Hopkins and Rezk [7] establish the existence of an $E_{\infty}$-map

$$M\text{String} = MO\langle 8 \rangle \to \text{tmf}$$

by showing a fiber property as above.

An approach to orientations of the form $MU \to E$ is described in [127]: You start with an $E_\infty$-ring spectrum $E$ and an ordinary complex orientation of $E$ [234, §6.1] and want to know whether you can refine this to an $E_\infty$-map $MU \to E$. Hopkins and Lawson establish a filtration of $MU$ by $E_\infty$-Thom spectra

$$S \to MX_1 \to MX_2 \to \cdots \to MU$$

and for a given $E_\infty$-map $MX_n \to E$ they identify the space of extensions to an $E_\infty$-map $MX_{n+1} \to E$ [127, Theorem 1].

*Remark 6.4.7.* In [6] the authors present a different approach to Thom spectra and questions about orientations that uses $\infty$-categorical techniques. In certain cases it is unrealistic to hope for $E_\infty$-maps out of Thom spectra, for instance if one doesn't know that the target spectrum carries an $E_\infty$ structure. The space of $E_n$-maps out of Thom spectra is described in [68, Theorem 4.2] and [11, Corollary 3.18].

## Thom spectra via $I$-spaces

Let $I$ be the skeleton of the category of finite sets and injective maps. As objects we choose the sets $\mathbf{n} = \{1,\ldots,n\}$ for $n \geqslant 0$ with the convention that $\mathbf{0}$ denotes the empty set. A morphism $f \in I(\mathbf{n},\mathbf{m})$ is an injective function from $\mathbf{n}$ to $\mathbf{m}$. Hence $\mathbf{0}$ is an initial object of $I$ and the permutation group $\Sigma_n$ is the group of automorphisms of $\mathbf{n}$ in $I$. The category $I$ is symmetric monoidal with respect to the disjoint union: $\mathbf{n} \sqcup \mathbf{m} = \mathbf{n} + \mathbf{m}$ with unit $\mathbf{0}$ and non-trivial symmetry $\mathbf{n} + \mathbf{m} \to \mathbf{m} + \mathbf{n}$ given by the shuffle permutation that moves the first $n$ elements to the positions $m+1,\ldots,m+n$.

The functor category of $I$-spaces, $\mathsf{Top}^I$, i.e., the category of functors $X \colon I \to \mathsf{Top}$ together with natural transformations as morphisms, inherits a symmetric monoidal structure from $I$ and $\mathsf{Top}$ via the Day convolution product. Explicitly, one gets:

**Definition 6.4.8.** The product $X \boxtimes Y$ of two $I$-spaces $X, Y$ is the $I$-space given by

$$(X \boxtimes Y)(\mathbf{n}) = \operatorname{colim}_{\mathbf{p} \sqcup \mathbf{q} \to \mathbf{n}} X(\mathbf{p}) \times Y(\mathbf{q}).$$

The unit $1_I$ is the discrete $I$-space $\mathbf{n} \mapsto I(\mathbf{0},\mathbf{n})$.

As $\mathbf{0}$ is initial, the unit $1_I$ is the terminal object in $\mathsf{Top}^I$. Commutative monoids in $\mathsf{Top}^I$ are called *commutative $I$-space monoids* in [31] and their category is denoted by $C(\mathsf{Top}^I)$. A general fact about Day convolution products is that commutative monoids correspond to lax symmetric monoidal functors.

For an $I$-space $X$ let's denote by $X_{hI}$ the Bousfield–Kan homotopy colimit of $X$.

**Definition 6.4.9** [31, Definition 2.2]. A map of $I$-spaces $f \colon X \to Y$ is *an $I$-equivalence* if the induced map on homotopy colimits $f_{hI} \colon X_{hI} \to Y_{hI}$ is a weak homotopy equivalence in $\mathsf{Top}$.

With the corresponding $I$-model structure the category of $I$-spaces is actually Quillen equivalent to the category of spaces [256, Theorem 3.3], but there is a *positive flat model structure* on $I$-spaces (see [31, §2]) that lifts to a right-induced model structure on $C(\mathsf{Top}^I)$ that makes it Quillen equivalent to $E_\infty$-spaces.

Let $\mathsf{Sp}^\Sigma$ denote the category of symmetric spectra. There is a canonical Quillen adjoint functor pair

$$\mathsf{Top}^I \underset{\Omega^I}{\overset{\mathbb{S}^I}{\rightleftarrows}} \mathsf{Sp}^\Sigma \qquad (6.4.6)$$

modeling the suspension spectrum functor and the underlying infinite loop space functor with

$$\mathbb{S}^I X(n) = \mathbb{S}^n \wedge X(\mathbf{n}), \quad \Omega^I(E)(\mathbf{n}) = \Omega^n E_n,$$

where $\mathbb{S}^n$ is the $n$-fold smash product of the $1$-sphere with $\Sigma_n$-action given by permutation of the smash factors.

Stable equivalences in symmetric spectra do not in general agree with stable homotopy equivalences, but there is a notion of *semistable* symmetric spectra that has the feature that a map $f: E \to F$ between two semistable symmetric spectra is a stable equivalence if and only if it is a stable homotopy equivalence. See [133, §5.6] for details and other characterizations.

**Definition 6.4.10.** For a commutative semistable symmetric ring spectrum $R$ the commutative $I$-space monoid of units, $GL_1^I(R)$, has as $GL_1^I(R)(\mathbf{n})$ those components of the commutative $I$-space monoid $\Omega^I(R)(\mathbf{n}) = \Omega^n R_n$ that represent units in $\pi_0(R)$.

The adjunction from (6.4.6) gives a version of the assembly map from (6.4.2) as

$$\mathbb{S}^I(GL_1^I(R)) \to \mathbb{S}^I \Omega^I(R) \to R.$$

For technical reasons one has to work with a cofibrant replacement of $GL_1^I(R)$, $G \to GL_1^I(R)$ in the positive flat model structure on $C(\mathsf{Top}^I)$. The construction of a Thom spectrum associated to a map $f: X \to BG$ is now similar to the approach in [6]; one defines $BG$ and $EG$ via two sided-bar constructions and takes a suitable pushout:

**Definition 6.4.11** [31, Definitions 2.10, 2.12, 3.6].

- Let $BG = B_{\boxtimes}(1_I, G, 1_I)$ and let $EG$ be defined via a functorial factorization

$$B_{\boxtimes}(1_I, G, G) \overset{\sim}{\rightarrowtail} EG \twoheadrightarrow BG.$$

- For any $I$-space $X$ over $BG$ define $U(X)$ as the $I$-space with $G$-action given by the pullback

$$
\begin{array}{ccc}
U(X) & \dashrightarrow & X \\
\big\downarrow & & \big\downarrow \\
EG & \longrightarrow & BG.
\end{array}
$$

Here, $X$ and $BG$ are considered as $I$-spaces with trivial $G$-action.

- Let $R$ be a semistable commutative symmetric ring spectrum that is $S$-cofibrant. The *Thom spectrum associated with a map of $I$-spaces* $f: X \to BG$ is

$$M^I(f) = B_{\boxtimes}(\mathbb{S}^I(UX), \mathbb{S}^I G, R). \qquad (6.4.7)$$

You should think of this two-sided bar construction as

$$\mathbb{S}^I(UX) \boxtimes^L_{\mathbb{S}^I G} R$$

and then you have to admit that this looks very similar to (6.4.5). This Thom spectrum functor is homotopically meaningful (see [31, Proposition 3.8]). Concerning multiplicative structures one obtains the following result.

Proposition 6.4.12 [31, Proposition 3.10, Corollary 3.11]. *The functor $M^I(-)$ is lax symmetric monoidal and if $\mathcal{D}$ is an operad in spaces, then it sends $\mathcal{D}$-algebras in $\mathrm{Top}^I$ over $BG$ to $\mathcal{D}$-algebras in $R$-modules in symmetric spectra over $M^I GL_1(R) := B_{\boxtimes}(\mathbb{S}^I(EG), \mathbb{S}^I(G), R)$.*

If you dislike diagram categories for some reason, there is also an $I$-*spacification functor* [31, §4.1] that transforms a map of topological spaces

$$f : X \to BG_{hI} \tag{6.4.8}$$

into a map of $I$-spaces over $BG$, so you can associate a Thom spectrum to such a map as well. By abuse of notation, we will still denote this Thom spectrum by $M^I(f)$. This construction respects actions of operads augmented over the Barratt–Eccles operad and hence it also provides an $E_\infty$ Thom spectrum functor.

An important question is: Can a given ring spectrum $A$ be realized as a Thom spectrum with respect to a loop map, i.e., in the setting of [31] is $A$ equivalent to $M^I(f)$ with $f$ a loop map to $BG_{hI}$? A striking result is that one can identify certain quotients as such Thom spectra!

Theorem 6.4.13 [31, Theorem 5.6]. *Let $R$ be a commutative ring spectrum whose homotopy groups are concentrated in even degrees and let $u_i \in \pi_{2i}(R)$ be arbitrary elements with $1 \leqslant i \leqslant n-1$. Then the iterative cofiber $R/(u_1, \ldots, u_{n-1})$ of the multiplication maps by the $u_i$'s can be realized as the Thom spectrum of a loop map from $SU(n)$ to $BG_{hI}$. In particular, $R/(u_1, \ldots, u_{n-1})$ can be realized as an associative ring spectrum.*

An example of such a quotient is $R = ku \to ku/u = H\mathbb{Z}$. Note that there is no assumption on the regularity of the elements $u_i$ in the above statement. For periodic ring spectra the assumptions on the degree of the elements can be relaxed and the two-periodic version of Morava $K$-theory can be constructed as a Thom spectrum relative to $R = E_n$, the $n$-th Morava $E$-theory or Lubin–Tate spectrum [31, Corollary 5.7]. A related but different construction of quotients of Lubin–Tate spectra modeling versions of Morava $K$-theory is carried out in [128, §3].

## Graded units

There is one problem with the constructions of spaces and spectra of units as above. As they are constructed from the underlying infinite loop space of a spectrum and just take into account the units in $\pi_0$, they ignore graded units coming from periodicity elements in the homotopy groups of a spectrum. So for instance, the Bott class $u \in \pi_2(KU)$ is not represented in $GL_1(KU)$ or $GL_1^I(KU)$.

There is a construction of *graded units*. We'll sketch the construction and mention two of its applications: graded Thom spectra and logarithmic ring spectra.

**Definition 6.4.14** [256, Definition 4.2]. The category $J$ has as objects pairs of objects of $I$. A morphism in $J((\mathbf{n}_1, \mathbf{n}_2), (\mathbf{m}_1, \mathbf{m}_2))$ is a triple $(\alpha, \beta, \sigma)$ where $\alpha \in I(\mathbf{n}_1, \mathbf{m}_1)$, $\beta \in I(\mathbf{n}_2, \mathbf{m}_2)$ and $\sigma$ is a bijection

$$\sigma \colon \mathbf{m}_1 \setminus \alpha(\mathbf{n}_1) \to \mathbf{m}_2 \setminus \beta(\mathbf{n}_2).$$

For another morphism $(\gamma, \delta, \xi) \in J((\mathbf{m}_1, \mathbf{m}_2), (\mathbf{l}_1, \mathbf{l}_2))$ the composition is the morphism $(\gamma \circ \alpha, \delta \circ \beta, \tau(\xi, \sigma))$ where $\tau(\xi, \sigma)$ is the permutation

$$\tau(\xi, \sigma)(s) = \begin{cases} \xi(s) & \text{if } s \in \mathbf{l}_1 \setminus \gamma(\mathbf{m}_1), \\ \delta(\sigma(t)) & \text{if } s = \gamma(t) \in \gamma(\mathbf{m}_1 \setminus \alpha(\mathbf{n}_1)). \end{cases}$$

Note that $\mathbf{l}_1 \setminus \gamma(\alpha(\mathbf{n}_1))$ is the disjoint union of $\mathbf{l}_1 \setminus \gamma(\mathbf{m}_1)$ and $\gamma(\mathbf{m}_1 \setminus \alpha(\mathbf{n}_1))$.

With these definitions $J$ is actually a category and it inherits a symmetric monoidal structure from $I$ via componentwise disjoint union [256, Proposition 4.3]. In particular, the category of $J$-spaces, $\mathsf{Top}^J$, is symmetric monoidal with the Day convolution product. Note, however, that the unit for the monoidal structure $\boxtimes_J$ is $J((\mathbf{0}, \mathbf{0}), (-, -))$; this is not a constant functor, but $J((\mathbf{0}, \mathbf{0}), (\mathbf{n}, \mathbf{n}))$ can be identified with the symmetric group $\Sigma_n$!

**Proposition 6.4.15** [256, 4.4, 4.5]. *For every $J$-space $X$ the homotopy colimit, $X_{hJ}$, is a space over $QS^0$.*

*Proof.* It is not hard to see that $J$ is isomorphic to Quillen's category $\Sigma^{-1}\Sigma$ [256, Proposition 4.4] and its classifying space is $QS^0$ by the Barratt–Priddy–Quillen result. Therefore $BJ$ is $QS^0$. Every $J$-space has a map to the terminal $J$-space that is the constant $J$-diagram on a point and this induces a map

$$X_{hJ} \to *_{hJ} = BJ \simeq QS^0. \qquad \square$$

For any $I$-space $X$ we also get that $X_{hI}$ is a space over $BI$, but as $I$ has an initial object this just gives a map to $BI \simeq *$, the terminal object.

Let $C(\mathsf{Top}^J)$ denote the category of commutative $J$-space monoids, i.e., commutative monoids in $\mathsf{Top}^J$. The following result is crucial:

**Theorem 6.4.16** [256, Theorem 4.11]. *There is a model structure on $C(\mathsf{Top}^J)$ such that there is a Quillen equivalence between $C(\mathsf{Top}^J)$ and the category of $E_\infty$-spaces over $BJ$.*

Here, the $E_\infty$-structure is parametrized by the Barratt–Eccles operad.

For a (commutative) $J$-space monoid, one can associate units:

**Definition 6.4.17** [256, §4]. Let $A$ be a $J$-space monoid. Then let $A^\times$ be the $J$-space monoid with $A^\times(\mathbf{n}_1, \mathbf{n}_2)$ being the union of those components of $A(\mathbf{n}_1, \mathbf{n}_2)$ that represent units in $\pi_0(A_{hJ})$.

So now one has to construct a functor from spectra to $J$-spaces that sees all the homotopy groups, not just the ones in non-negative degrees:

Definition 6.4.18 [256, (4.5)].

- Let $\Omega^J$ be the functor from symmetric spectra to $J$-spaces that takes a symmetric spectrum $E$ and sends it to the $J$-space with

$$\Omega^J(E)(\mathbf{n}_1, \mathbf{n}_2) = \Omega^{n_2} E_{n_1}.$$

- If $R$ is a symmetric ring spectrum, then its $J$-*space of units* is

$$GL_1^J(R) = (\Omega^J(R))^\times.$$

Sagave and Schlichtkrull show that this is homotopically meaningful and that for a commutative symmetric ring spectrum $R$, the units $GL_1^J(R)$ are actually in $C(\mathsf{Top}^J)$ [256, §4]. Most importantly, the inclusion $GL_1^J(R) \hookrightarrow \Omega^J(R)$ realizes the inclusion of graded units $\pi_*(R)^\times$ into $\pi_*(R)$ for positively fibrant $R$.

Hence, for instance $GL_1^I(KU)$ (and any other model of the "usual" units) only detects the units $\pm 1$ in $\pi_0(KU)$ whereas $GL_1^J(KU)$ also detects the Bott class.

*Remark 6.4.19.*

1. John Rognes developed the concept of logarithmic ring spectra and in [255] and [253] this concept is fully explored with the help of graded units. The idea is that you want a spectrum that sits between a commutative ring spectrum like $ku$ and its localization $KU$, so you remember the Bott class as the extra datum of a logarithmic structure. This concept has its origin in algebraic geometry and is useful in stable homotopy theory, for instance for obtaining localization sequences in topological Hochschild homology [253].

2. In [257] Sagave and Schlichtkrull use graded units adapted to the setting of orthogonal spectra, $GL_1^W$, to construct *graded Thom spectra* associated to virtual vector bundles, i.e., associated to a map $f: X \to \mathbb{Z} \times BO$ in such a way that uses the $E_\infty$-structure on $\mathbb{Z} \times BO$. They use this for orientation theory and relate $GL_1^W$-orientations to logarithmic structures. They provide an $E_\infty$-Thom isomorphism that allows to compute the homology of spectra appearing in connection with logarithmic ring spectra [257, §§ 7,8].

## 6.5 Constructing commutative ring spectra from bipermutative categories

In section 6.4 we saw that Thom spectra give rise to commutative ring spectra. Algebraic $K$-theory is another machine that takes a commutative ring (spectrum) $R$ and produces a commutative ring spectrum $K(R)$. In this section we focus on a classical construction that takes a small bipermutative category $\mathcal{R}$ and turns it into a commutative ring spectrum. This construction goes back to Segal [268]; its multiplicative properties were investigated by May [199, 195, 192, 193], Shimada–Shimakawa [273], Woolfson [302] and Elmendorf–Mandell [95].

We sketch a simplified version of the construction, present some important examples and refer to [95] for a discussion of the multiplicative properties.

**Definition 6.5.1.** A *permutative category* $(\mathcal{C}, \oplus, 0, \tau)$ is a category $\mathcal{C}$ together with an object $0$ of $\mathcal{C}$, a functor $\oplus: \mathcal{C} \times \mathcal{C} \to \mathcal{C}$ and a natural isomorphism $\tau_{C_1, C_2}: C_1 \oplus C_2 \to C_2 \oplus C_1$ for all objects $C_1, C_2$ of $\mathcal{C}$ such that

- $\oplus$ is strictly associative, i.e., for all objects $C_1, C_2, C_3$ of $\mathcal{C}$

$$C_1 \oplus (C_2 \oplus C_3) = (C_1 \oplus C_2) \oplus C_3.$$

- $0$ is a strict unit, i.e., for all objects $C$ of $\mathcal{C}$: $C \oplus 0 = C = 0 \oplus C$.
- $\tau^2$ is the identity, i.e., for all objects $C_1, C_2$ of $\mathcal{C}$ the composite

$$C_1 \oplus C_2 \xrightarrow{\tau_{C_1, C_2}} C_2 \oplus C_1 \xrightarrow{\tau_{C_2, C_1}} C_1 \oplus C_2$$

is the identity on $C_1 \oplus C_2$.
- The diagrams

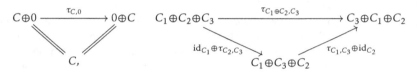

commute for all objects $C, C_1, C_2, C_3$ of $\mathcal{C}$.

We work with small permutative categories, i.e., we require that the objects of $\mathcal{C}$ form a set (and not a proper class). We recall Segal's construction from [268, §2]:

**Definition 6.5.2.** Let $\mathcal{C}$ be a small permutative category and let $X$ be a finite set with basepoint $+ \in X$. Let $\mathcal{C}(X)$ be the category whose objects are families $(C_S, \rho_{S,T})$ such that:

- $S \subset X$ and $+ \notin S$;
- $S$ and $T$ are pairs of such subsets that are disjoint;
- the $C_S$ are objects of $\mathcal{C}$ and $\rho_{S,T}$ is an isomorphism in $\mathcal{C}$:

$$\rho_{S,T}: C_S \oplus C_T \to C_{S \cup T};$$

- $C_\varnothing = 0$ and $\rho_{\varnothing, T} = \mathrm{id}_{C_T}$ for all $T$; and
- for pairwise disjoint $S, T, U$ that don't contain $+$ the following diagrams commute:

$$
\begin{array}{ccc}
C_S \oplus C_T & \xrightarrow{\rho_{S,T}} & C_{S \cup T} \\
{\scriptstyle \tau}\downarrow & & \| \\
C_T \oplus C_S & \xrightarrow{\rho_{T,S}} & C_{T \cup S},
\end{array}
\qquad
\begin{array}{ccc}
C_S \oplus C_T \oplus C_U & \xrightarrow{\rho_{S,T} \oplus \mathrm{id}_{C_U}} & C_{S \cup T} \oplus C_U \\
{\scriptstyle \mathrm{id}_{C_S} \oplus \rho_{T,U}}\downarrow & & \downarrow{\scriptstyle \rho_{S \cup T, U}} \\
C_S \oplus C_{T \cup U} & \xrightarrow{\rho_{S, T \cup U}} & C_{S \cup T \cup U}.
\end{array}
$$

Morphisms $\alpha: (C_S, \rho_{S,T}) \to (C'_S, \rho'_{S,T})$ consist of a family of morphisms $\alpha_S \in \mathcal{C}(C_S, C'_S)$

for all $S \subset X$ with $+ \notin S$ such that $\alpha_\emptyset = \mathrm{id}_0$ and for all $S, T \in X$ with $+ \notin S, T$ and $S \cap T = \emptyset$ the diagram

$$
\begin{array}{ccc}
C_S \oplus C_T & \xrightarrow{\;\rho_{S,T}\;} & C_{S \cup T} \\
{\scriptstyle a_S \oplus a_T}\downarrow & & \downarrow{\scriptstyle a_{S \cup T}} \\
C'_S \oplus C'_T & \xrightarrow{\;\rho'_{S,T}\;} & C'_{S \cup T}
\end{array}
$$

commutes.

So up to isomorphism, every object $C_S$ for $S = \{x_1, \ldots, x_n\}$ can be decomposed as

$$C_S \cong C_{\{x_1\}} \oplus \cdots \oplus C_{\{x_n\}}$$

by an iterated application of the isomorphisms $\rho$, but these isomorphisms are part of the data. Segal shows [268, Corollary 2.2] that this construction gives rise to a so-called $\Gamma$-space (see [268, Definition 1.2] for a definition) that sends a finite pointed set $X$ to the classifying space of $\mathcal{C}(X)$. Every $\Gamma$-space gives rise to a spectrum, and we denote the spectrum associated to $\mathcal{C}$ by $H\mathcal{C}$.

*Remark 6.5.3.* Segal's construction actually works for symmetric monoidal categories and it produces a spectrum whose associated infinite loop space is the group completion of the classifying space of the category $\mathcal{C}$, $B\mathcal{C}$, and the latter is the geometric realization of the nerve of $\mathcal{C}$.

**Definition 6.5.4.** A *bipermutative category* $\mathcal{R}$ is a category with two permutative category structures, $(\mathcal{R}, \oplus, 0_\mathcal{R}, \tau_\oplus)$ and $(\mathcal{R}, \otimes, 1_\mathcal{R}, \tau_\otimes)$, that are compatible in the following sense:

1. $$0_\mathcal{R} \otimes C = 0_\mathcal{R} = C \otimes 0_\mathcal{R}$$

   for all objects $C$ of $\mathcal{R}$.

2. For all objects $A, B, C$ we have an equality between $(A \oplus B) \otimes C$ and $A \otimes C \oplus B \otimes C$, and the diagram

$$
\begin{array}{ccc}
(A \oplus B) \otimes C & =\!=\!= & A \otimes C \oplus B \otimes C \\
{\scriptstyle \tau_\oplus \otimes \mathrm{id}}\downarrow & & \downarrow{\scriptstyle \tau_\oplus} \\
(B \oplus A) \otimes C & =\!=\!= & B \otimes C \oplus A \otimes C
\end{array}
$$

   commutes.

3. We define the distributivity isomorphism $d_\ell : A \otimes (B \oplus C) \to A \otimes B \oplus A \otimes C$ for all $A, B, C$ in $\mathcal{R}$ via the diagram

$$
\begin{array}{ccc}
A \otimes (B \oplus C) & \xrightarrow{\;\tau_\otimes\;} & (B \oplus C) \otimes A \\
{\scriptstyle d_\ell}\dashdownarrow & & \Vert \\
A \otimes B \oplus A \otimes C & \xleftarrow[\;\tau_\otimes \oplus \tau_\otimes\;]{} & B \otimes A \oplus C \otimes A
\end{array}
$$

Then the diagram

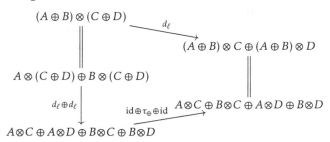

commutes.

This definition is taken from [199, Definition VI.3.3, p. 154]. The definition in [95] is less strict, but bipermutative categories in the above sense are also bipermutative in the sense of [95, Definition 3.6]. We refer to Elmendorf and Mandell for a proof that for a bipermutative category $\mathcal{R}$, one actually obtains a commutative ring spectrum $H\mathcal{R}$:

**Theorem 6.5.5** [95, Corollary 3.9]. *If $\mathcal{R}$ is a bipermutative category, then $H\mathcal{R}$ is equivalent to a strictly commutative symmetric ring spectrum.*

There is an alternative construction of an $E_\infty$-ring spectrum from a bipermutative category in [193]: May first constructs an $E_\infty$-ring space associated to a bipermutative category and then builds the corresponding $E_\infty$-ring spectrum.

Segal's construction enables us to find small and explicit models for certain connective commutative ring spectra. Famous examples of bipermutative categories and their associated commutative ring spectra are the following:

1. If $R$ is a commutative discrete ring, then the category $\mathcal{R}_R$ which has the elements of $R$ as objects and only identity morphisms is a bipermutative category with the addition in the ring being $\oplus$ and the multiplication being $\otimes$. The associated spectrum, $H\mathcal{R}_R$ is the Eilenberg–Mac Lane spectrum of the ring $R$, $HR$.

2. Let $\mathcal{E}$ denote the bipermutative category of finite sets whose objects are the finite sets $\mathbf{n} = \{1, \ldots, n\}$ for $n \in \mathbb{N}_0$. By convention $\mathbf{0}$ is the empty set. The morphisms in $\mathcal{E}$ are
$$\mathcal{E}(\mathbf{n}, \mathbf{m}) = \begin{cases} \varnothing & \text{for } n \neq m, \\ \Sigma_n & \text{for } n = m. \end{cases}$$

For the full structure see [199, VI, Example 5.1]. Here $H\mathcal{E}$ is the sphere spectrum, $S$.

3. The bipermutative category of complex vector spaces, $\mathcal{V}_\mathbb{C}$, with objects the natural numbers with zero and morphisms
$$\mathcal{V}_\mathbb{C}(n, m) = \begin{cases} \varnothing & \text{for } n \neq m, \\ U(n) & \text{for } n = m, \end{cases}$$

is bipermutative. On objects we set $n \oplus m = n + m$ and $n \otimes m = nm$ and on morphisms we use the block sum and the tensor product of matrices. The associated spectrum is $H\mathcal{V}_\mathbb{C} = ku$, the connective version of topological complex $K$-theory. Its real analog, $\mathcal{V}_\mathbb{R}$, gives a model for connective topological real $K$-theory, $ko$. You can also work with the general linear group instead of the unitary or orthogonal group.

4. If $R$ is a discrete commutative ring, then we define the category $F_R$ as the one with objects $\mathbb{N}_0$ again. As morphisms we have

$$F_R(n, m) = \begin{cases} \varnothing & \text{for } n \neq m, \\ GL_n(R) & \text{for } n = m. \end{cases}$$

This category is often called *the small category of free R-modules*. Its spectrum is the *free algebraic K-theory of R*, $K^f(R)$. Its homotopy groups agree with the algebraic $K$-groups of $R$ from degree 1 on.

## 6.6 From topological Hochschild to topological André–Quillen homology

For rings and algebras Hochschild homology contains a lot of information. For commutative rings and algebras André–Quillen homology is the adequate tool. There are spectrum level versions of these homology theories: topological Hochschild homology, THH, and topological André–Quillen homology, TAQ.

We can determine classes in the algebraic $K$-theory of a ring spectrum using the trace to topological Hochschild homology or to topological cyclic homology:

$$\text{tr}\colon K(R) \to \text{THH}(R). \tag{6.6.1}$$

For instance the trace from $K(\mathbb{Z})$ to $\text{THH}(\mathbb{Z})$ detects important classes. Bökstedt, Madsen and Rognes [50, 252] show for instance that

$$\text{tr}\colon K_{2p-1}(\mathbb{Z}) \to \text{THH}_{2p-1}(\mathbb{Z}) \cong \mathbb{Z}/p\mathbb{Z}$$

is surjective for all primes $p$.

We give a construction of topological Hochschild homology and, more generally, for commutative ring spectra $R$ we define $X \otimes R$ for $X$ a finite pointed simplicial set. We give some examples of calculations of such $X$-homology groups of $R$ and tell you about topological Hochschild cohomology as a derived center of an algebra spectrum. We define topological André–Quillen homology and we will see applications to Postnikov towers for commutative ring spectra later in Section 6.7.

### THH and friends

Let $X$ be a finite pointed simplicial set and let $R$ be a cofibrant commutative ring spectrum.

**Definition 6.6.1.** We denote by $X \otimes R$ the simplicial spectrum with

$$(X \otimes R)_n = \bigwedge_{x \in X_n} R.$$

By slight abuse of notation we will use the same symbol for the geometric realization of $X \otimes R$.

*Remarks 6.6.2.* – As the smash product is the coproduct in $\mathcal{C}_S$, the simplicial structure maps of $X \otimes R$ are induced from the ones on $X$.

– As $X$ is pointed, $X \otimes R$ comes with maps

$$R \to X \otimes R \to R$$

whose composition is the identity.

– The commutative multiplication on $R$ induces a commutative multiplication on $X \otimes R$; hence $X \otimes R$ is an augmented commutative (simplicial) $R$-algebra spectrum.

– One could also use the fact that the spectra of [94] are tensored over topological spaces or, similarly, that symmetric spectra [133] in topological spaces are enriched over simplicial sets and over topological spaces. This gives an equivalent situation. It is for instance shown in [94, Corollary VII.3.4] that $|X \otimes A| \simeq |X| \otimes A$ for simplicial spaces $X$ and commutative $R$-algebra spectra $A$.

– The above definition can be extended to tensoring with an arbitrary pointed simplicial set by expressing such a simplicial set as the colimit of its finite pointed simplicial subcomplexes.

There are many important special cases of this construction.

**Definition 6.6.3.**

1. For the simplicial 1-sphere $X = \mathbb{S}^1$ the commutative $R$-algebra spectrum $\mathbb{S}^1 \otimes R$ is the *topological Hochschild homology of R* and is denoted by $\mathsf{THH}(R)$.
2. More generally, for an $n$-sphere, we denote by $\mathsf{THH}^{[n]}(R)$ the spectrum $\mathbb{S}^n \otimes R$; this is called *topological Hochschild homology of order n*.
3. If $\mathbb{T}^n$ denotes the torus $(\mathbb{S}^1)^n$, then $\mathbb{T}^n \otimes R$ is the *n-torus homology of R*.

For the small model of the simplicial 1-sphere with just one non-degenerate 0- and 1-simplex we have $(\mathbb{S}^1)_n = \{0, 1, \dots, n\}$ and the simplicial spectrum $\mathbb{S}^1 \otimes R$ is precisely the cyclic bar construction on $R$:

$$R \rightleftarrows R \wedge R \mathrel{\substack{\longrightarrow \\[-0.5ex] \longrightarrow \\[-0.5ex] \longrightarrow}} R \wedge R \wedge R \mathrel{\substack{\longrightarrow \\[-0.5ex] \longrightarrow \\[-0.5ex] \longrightarrow \\[-0.5ex] \longrightarrow}} \cdots,$$

where the degeneracy map $s_i \colon R^{n+1} \to R^{n+2}$ inserts the unit map $\eta \colon S \to R$ after the $i$-th factor of $R$ and the face maps $d_i \colon R^{n+1} \to R^n$ for $0 \leqslant i < n$ are given by the multiplication in $R$ of the $i$-th and $(i+1)$-st smash factor. The last face map $d_n$ cyclically permutes the smash factors to bring the last one to the front and then it multiplies the former factors numbered $n$ and $0$.

As for Hochschild homology you should think about this as a genuine cyclic object:

$$
\begin{array}{ccc}
 & R & \\
R \wedge & \wedge & \wedge \cdot \\
\wedge & R & \vdots \\
R & & \\
 & \wedge & \cdots
\end{array}
$$

The original definition of THH is due to Marcel Bökstedt [52]. McClure, Schwänzl and Vogt [203] show that for an $E_\infty$-ring spectrum $R$, $\mathsf{THH}(R)$ is equivalent to tensoring $R$ with the topological 1-sphere. Kuhn systematically studies constructions like $X \otimes R$ in a reduced setting [150] for pointed spaces $X$. So the above definition is an unreduced variant of this that uses simplicial sets instead of topological spaces.

**Lemma 6.6.4.** *Let $X$ and $Y$ be finite simplicial pointed sets. Then*

$$(X \times Y) \otimes R \simeq X \otimes (Y \otimes R).$$

*Proof.* Observe that

$$((X \times Y) \otimes R)_n = \bigwedge_{(x,y) \in X_n \times Y_n} R \cong \bigwedge_{x \in X_n} \left( \bigwedge_{y \in Y_n} R \right)$$

and this is the diagonal of the bisimplicial spectrum

$$([m], [\ell]) \mapsto (X \otimes ((Y \otimes R)_\ell))_m$$

in degree $n$.                                                                                    □

One of the important features of $\mathsf{THH}(R)$ is that it receives a trace map from algebraic $K$-theory (see (6.6.1)), which we can now write as

$$\mathrm{tr} \colon K(R) \to \mathbb{S}^1 \otimes R.$$

Taking higher-dimensional tori gives targets for iterated trace maps. Algebraic $K$-theory of a commutative ring spectrum is again a commutative ring spectrum and the trace map is a map of commutative ring spectra; hence one can iterate the process of forming $K$-theory and traces. If we denote by $K^n(R)$ the $n$-fold iteration, then, since we have the product formula from Lemma 6.6.4, we get an iterated trace to $\mathbb{T}^n \otimes R$. Explicitly, for $n = 2$ this is

$$K(K(R)) \to \mathbb{S}^1 \otimes (\mathbb{S}^1 \otimes R) \simeq (\mathbb{S}^1 \times \mathbb{S}^1) \otimes R = \mathbb{T}^2 \otimes R.$$

There are variants of Definition 6.6.1: As we work with pointed simplicial sets, we can glue an $R$-module to the base point and use the $R$-module structure for the face maps. A second variant is to work relative to some commutative ring spectrum $R$: in Definition 6.6.1 the smash products were over the sphere spectrum, but if we work with a commutative $R$-algebra spectrum $A$, then we can take smash products over $R$ instead of $S$. Recall that $\wedge_R$ is the coproduct in the category of commutative $R$-algebra spectra, $\mathcal{C}_R$.

**Definition 6.6.5.** Let $R$ be a cofibrant commutative ring spectrum, $A$ a cofibrant commutative $R$-algebra spectrum, $M$ an $A$-module spectrum over $R$ and let $X$ be a finite pointed simplicial set. We denote by $\mathcal{L}_X^R(A; M)$ the simplicial spectrum with

$$\mathcal{L}_X^R(A; M)_n = M \wedge_R \bigwedge_{x \in X_n \setminus *} {}_R A.$$

We call $\mathcal{L}_X^R(A; M)$ the *Loday construction of $A$ over $R$ with coefficients in $M$*.

As $M$ is just an $A$-module spectrum, the resulting simplicial spectrum and also its realization carries an $A$-module structure over $R$, but no multiplicative structure in general. However, if we place a commutative $A$-algebra $C$ at the basepoint, then the resulting spectrum is an augmented commutative $C$-algebra spectrum.

We will see in Section 6.8 that for instance

$$\mathrm{THH}^R(A) := \mathcal{L}^R_{\mathbb{S}^1}(A)$$

measures properties of $A$ as a commutative $R$-algebra spectrum. The case of $X = \mathbb{S}^0$ gives

$$\mathcal{L}^R_{\mathbb{S}^0}(A) = A \wedge_R A,$$

so there is a Künneth spectral sequence [94, IV.4.1] for calculating its homotopy groups.

An important example of a Loday construction is Pirashvili's construction of *higher-order Hochschild homology*. He works with discrete commutative $k$-algebras $A$ and $A$-modules $M$ and defines $\mathrm{HH}^k_X(A;M)$ [223, §5.1]. For $X = \mathbb{S}^n$ this is his notion of higher-order Hochschild homology (in his notation $H^{[n]}(A;M)$). In our setting this corresponds to $\mathcal{L}^{Hk}_X(HA;HM)$ if $A$ is flat over $k$.

## Examples

1. A classical example of a THH-calculation is the one of $H\mathbb{Z}$ and $H\mathbb{F}_p$ by Marcel Bökstedt ([51]; see [161, Chapter 13] and the references for published accounts of these results):

**Proposition 6.6.6.** $\qquad \mathrm{THH}_*(H\mathbb{F}_p) \cong \mathbb{F}_p[\mu], \quad |\mu| = 2.$

$$\mathrm{THH}_i(H\mathbb{Z}) \cong \begin{cases} \mathbb{Z} & \text{if } i = 0, \\ \mathbb{Z}/j\mathbb{Z} & \text{if } i = 2j - 1, \\ 0 & \text{otherwise.} \end{cases}$$

A crucial ingredient for these and many other calculations of THH is Bökstedt's spectral sequence: If $R$ is a commutative ring spectrum and if $E_*$ is a homotopy commutative ring spectrum such that $E_*(R)$ is flat over $E_*$ then there is a multiplicative spectral sequence

$$E^2_{p,q} = \mathrm{HH}^{E_*}_{p,q}(E_*(R)) \Rightarrow E_{p+q}\mathrm{THH}(R).$$

Here $\mathrm{HH}_{p,q}$ denotes Hochschild homology in homological degree $p$ and internal degree $q$ ([51], [94, Theorem IV.1.9]).

2. If we apply THH to Eilenberg–Mac Lane spectra of number rings, Lindenstrauss and Madsen show that THH detects arithmetic properties:

**Proposition 6.6.7** [160, Theorem 1.1]. *Let $K$ be a number field and let $\mathcal{O}_K$ be its ring of integers. Then*

$$\mathrm{THH}_n(H\mathcal{O}_K) = \begin{cases} \mathcal{O}_K & \text{if } n = 0, \\ \mathcal{D}^{-1}_{\mathcal{O}_K}/\ell\mathcal{O}_K & \text{if } n = 2\ell - 1, \\ 0 & \text{otherwise.} \end{cases}$$

Here, $\mathcal{D}_{\mathcal{O}_K}^{-1}$ is the inverse different. This is the set of those $x \in K$ such that the trace $\mathrm{tr}(xy)$ is an integer for all $y \in \mathcal{O}_K$. The inverse different detects ramified primes.

Dundas, Lindenstrauss and I calculate higher-order THH of number rings with reduced coefficients in [82, Theorem 4.3].

3. For a suspension spectrum on a based (Moore) loop space, $\Sigma_+^\infty \Omega_M X$, the cyclic bar construction reduces to the suspension spectrum of the cyclic bar construction on $\Omega_M X$ and Goodwillie [109, Proof of Theorem V.1.1] identifies the latter with the free loop space on $X$, $LX$. Hence one obtains

$$\mathsf{THH}(\Sigma_+^\infty \Omega_M X) \simeq \Sigma_+^\infty LX.$$

4. Let $R$ be a ring spectrum and let $\Pi$ be a pointed monoid. Hesselholt and Madsen show that $\mathsf{THH}(R[\Pi])$ splits as

$$\mathsf{THH}(R[\Pi]) \simeq \mathsf{THH}(R) \wedge |N^{cy}\Pi|,$$

where $|N^{cy}\Pi|$ denotes the cyclic nerve of $\Pi$ [119, Theorem 7.1].

5. As a sample calculation for second order THH Dundas, Lindenstrauss and I get [83, Theorem 2.1]:

$$\mathsf{THH}_*^{[2]}(H\mathbb{Z}_{(p)}) \cong \mathbb{Z}_{(p)}[x_1, x_2, \ldots ]/(p^n x_n, x_n^p - p x_{n+1}, n \geqslant 1) \qquad (6.6.2)$$

with $|x_1| = 2p$.

6. At an odd prime $KU_{(p)}$ splits as

$$KU_{(p)} \simeq \bigvee_{i=0}^{p-2} \Sigma^{2i} L.$$

Here, $L$ is the Adams summand of $KU_{(p)}$ with $\pi_*(L) \cong \mathbb{Z}_{(p)}[v_1^{\pm 1}]$ and $|v_1| = 2p - 2$. For consistency we set $L = KU_{(2)}$ at the prime 2. We denote by $ku$, $\ell$ and $ko$ the connective covers of $KU$, $L$ and $KO$.

McClure and Staffeldt determine the mod $p$-homotopy of $\mathsf{THH}(\ell)$ at odd primes [202] and they show that $\mathsf{THH}(L)_p \simeq L_p \vee (\Sigma L_p)_{\mathbb{Q}}$ [202, Corollary 7.2, Theorem 8.1].

Ausoni [13] determines the mod $p$ and mod $v_1$ homotopy of $\mathsf{THH}(ku)$ as an input for his work on $K(ku)$.

Angeltveit, Hill and Lawson show [9, Theorem 2.6] that for all primes,

$$\mathsf{THH}_*(\ell) \cong \ell_* \oplus \Sigma^{2p-1} F \oplus T$$

as $\ell_*$-modules, where $F$ is a torsionfree summand and $T$ is an infinite direct sum of torsion modules concentrated in even degrees. They describe $F$ explicitly using a rational calculation. Determining the torsion is way more involved [9, Theorem 2.8]. The calculation of $\mathsf{THH}_*(\ell)$ uses the method of *dueling Bockstein spectral sequences* for the Bockstein spectral sequences associated to

$$
\begin{array}{ccc}
\ell & \longrightarrow & \ell/p \\
\downarrow & & \downarrow \\
\ell/v_1 = H\mathbb{Z}_{(p)} & \longrightarrow & H\mathbb{F}_p = \ell/(p, v_1)
\end{array}
$$

They describe the 2-local homotopy groups of $\mathsf{THH}(ko)$ [9, §7] by first determining $\mathsf{THH}_*(ko; ku)$ and then using the Bockstein spectral sequence associated to the cofiber sequence $\Sigma ko \to ko \to ku$.

Again, things are way easier for the periodic versions (see [13, Proposition 7.13] and [9, Corollary 7.9]):

$$\mathsf{THH}(KO) \simeq KO \vee \Sigma KO_\mathbb{Q}, \quad \mathsf{THH}(KU) \simeq KU \vee \Sigma KU_\mathbb{Q}.$$

7. John Greenlees uses a generalization of the concept of Gorenstein maps of commutative rings to the spectral world in order to determine Gorenstein descent properties for cofiber sequences of connective commutative ring spectra [110, Theorem 7.4].

## Topological Hochschild homology of Thom spectra

We start with a general statement about $X \otimes M^I(f)$ if $M^I(f)$ is a Thom spectrum associated to an $E_\infty$-map to $BG_{hI}$ with $BG_{hI}$ as in (6.4.8) with $R = S$; hence $G$ is a cofibrant replacement of $GL_1^I(S)$.

**Theorem 6.6.8 [259, Theorem 1.1].** *For any pointed simplicial set $X$ and any map of grouplike $E_\infty$-spaces $f: A \to BG_{hI}$ there is an equivalence of $E_\infty$-ring spectra,*

$$X \otimes M^I(f) \simeq M^I(f) \wedge \Omega^\infty(a \wedge |X|)_+,$$

*where $a$ is the spectrum associated to $A$ with $\Omega^\infty a = A$.*

This result generalizes [45], where the case of $X = S^1$ is covered. In general, for $X = S^n$ Theorem 6.6.8 determines the higher-order topological Hochschild homology of $M^I(f)$ [259, (1.6)] as

$$\mathsf{THH}^{[n]}(M^I(f)) \simeq M^I(f) \wedge B^n A_+.$$

As an example, for the canonical map $f: BU \to BG_{hI}$ one obtains

$$X \otimes MU \simeq MU \wedge \Omega^\infty(bu \wedge |X|),$$
$$\mathsf{THH}^{[n]}(MU) \simeq MU \wedge \Omega^\infty(bu \wedge S^n) \simeq MU \wedge B^n BU_+.$$

There is also a statement about THH of Thom spectra associated to single loop maps in [45, Theorem 1]. We state the relative version of this, so in the following $G$ is a cofibrant replacement of $GL_1^I(R)$.

**Theorem 6.6.9 [31, Theorem 6.6].** *Assume that $R$ is a commutative symmetric ring* *spectrum that is semistable and S-cofibrant. Let $M^I(f)$ be a Thom spectrum associated to a map $f: M \to BG_{hI}$ of topological monoids, where $M$ is grouplike and well-pointed. Then*

$$\mathsf{THH}^R(M^I(f)) \simeq M^I(L^\eta(B(f))).$$

Here, $M^I(L^\eta(B(f)))$ is the Thom spectrum associated to the map

Note that $BBG_{hI}$ is an $H$-group, so we can split the free loop space $LBBG_{hI}$ into the base space and the based loops

$$LBBG_{hI} \simeq BBG_{hI} \times \Omega BBG_{hI}$$

and the second factor is equivalent to $BG_{hI}$. As usual, $\eta$ denotes the Hopf map $\eta: \mathbb{S}^3 \to \mathbb{S}^2$ and it induces a map $\eta: BBG_{hI} \to BG_{hI}$ as above via

$$BBG_{hI} \simeq \Omega^2 B^4 G_{hI} \to \Omega^3 B^4 G_{hI} \simeq BG_{hI}$$

by reducing the loop coordinates by precomposition.

For quotient spectra, this result gives a new way of calculating $\mathsf{THH}^R(R/I)$. For related results see [8] and in the case where $R/I$ is commutative see [83, §7].

A second example comes from viewing $H\mathbb{Z}_{(p)}$ as a Thom spectrum associated to a 2-fold loop map $\Omega^2(\mathbb{S}^3\langle 3\rangle) \to BG_{hI}$, which allows for a determination of $\mathsf{THH}(H\mathbb{Z}_{(p)})$ as $H\mathbb{Z}_{(p)} \wedge \Omega(\mathbb{S}^3\langle 3\rangle)_+$ [45, Theorem 3.8] and an additive equivalence

$$\mathsf{THH}^{[2]}(H\mathbb{Z}_{(p)}) \simeq H\mathbb{Z}_{(p)} \wedge \mathbb{S}^3\langle 3\rangle_+.$$

This gives a geometric interpretation of (6.6.2), but without an identification of the multiplicative structure. See also [149, §4], where Klang presents related results, using the framework of factorization homology.

## Topological Hochschild cohomology as a derived center

In the discrete case, i.e., for a commutative ring $k$ and a $k$-algebra $A$ one can describe the center of $A$,

$$Z(A) = \{b \in A, ab = ba \text{ for all } a \in A\},$$

as the set of $A$-bimodule maps from $A$ to $A$. If $f$ is such a map, $f: A \to A$ with $f(cad) = cf(a)d$ for all $a, c, d \in A$, then $f$ is determined by $f(1) =: b$ and this $b$ satisfies

$$ab = af(1) = f(a \cdot 1) = f(a) = f(1 \cdot a) = f(1)a = ba,$$

so the set of such morphisms gives rise to an element in the center; conversely, for any $b \in Z(A)$ we get such an $f$ by setting $f(1) = b$.

Hochschild cohomology of $A$ over $k$ can be described as

$$HH^*_k(A) = Ext^*_{A \otimes_k A^o}(A, A)$$

if $A$ is $k$-projective. Hence $HH^0_k(A) = Z(A)$ and the Hochschild cohomology of $A$ is the *derived center of A*. Hochschild cohomology has a graded commutative algebra structure via a cup product, but the solved Deligne conjecture [204] says that the Hochschild cochain complex is in general not a differential graded commutative algebra, but that it has an $E_2$-algebra structure.

For ring spectra there is no homotopically meaningful definition of a center: requiring equality translates to an equalizer diagram and this wouldn't be homotopy invariant. For a commutative ring spectrum $R$ and an $R$-algebra spectrum $A$ this equalizer corresponds precisely to taking not just $R$-module endomorphisms but $A$-bimodule endomorphisms. So a homotopy invariant version is as follows.

**Definition 6.6.10.** For a commutative ring spectrum $R$ and an $R$-algebra spectrum $A$, the *topological Hochschild cohomology groups of A over R* are

$$THH^*_R(A) = \pi_* Ext_{A \wedge_R A^o}(A, A)$$

and the *derived center of A over R* is

$$THH_R(A) = Ext_{A \wedge_R A^o}(A, A).$$

Here, $Ext_{A \wedge_R A^o}(A, A)$ denotes the derived endomorphism spectrum of $A$ as an $A$-bimodule [94, IV §1].

McClure and Smith's proof of the Deligne conjecture also provides a spectrum version for topological Hochschild cohomology, giving the derived center an $E_2$-structure:

**Theorem 6.6.11** [204]. *If $A$ is an associative $R$-algebra spectrum, then $THH_R(A)$ is an $E_2$-ring spectrum.*

An important example of a calculation of such a derived center is Angeltveit's calculation of $THH_{E_n}(K_n)$. Here $E_n$ denotes Morava $E$-theory with

$$\pi_*(E_n) \cong W(\mathbb{F}_q)[[u_1, \ldots, u_{n-1}]][u^{\pm 1}],$$

where the $u_i$ are deformation parameters for the height-$n$ Honda formal group law with $|u_i| = 0$ and $u$ is a periodicity element with $|u| = 2$. The sequence of elements $(p, u_1, \ldots, u_{n-1})$ is a regular sequence and $K_n$ is the 2-periodic version of Morava $K$-theory:

$$K_n = E_n/(p, u_1, \ldots, u_{n-1}), \quad (K_n)_* = \mathbb{F}_q[u^{\pm 1}].$$

Angeltveit shows that the derived center of $K_n$ over $E_n$ depends on the chosen $A_\infty$-algebra structure of $K_n$ over $E_n$:

**Theorem 6.6.12** [8, Theorems 5.21, 5.22]. 1. *For any prime $p$ and any $n \geqslant 1$ there is an $A_\infty$-structure on $K_n$ such that $THH_{E_n}(K_n) \simeq E_n$.*

2. *For $n = 1$ and any $d$ with $1 \leqslant d < p - 1$ and any $a$ with $1 \leqslant a \leqslant p - 1$ there is an $A_\infty$-structure on $K_1$ with*

$$\mathsf{THH}^*_{E_1}(K_1) \cong \pi_*(E_1)[[q]]/(p + a(uq)^d).$$

Here, the structure in statement 1 occurs as the one coming from the *least commutative $A_\infty$-structure* on $K_n$ (see [8, Theorem 5.8] for a precise statement). The case $n = 1, p = 2$ of statement 1 is due to Baker and Lazarev [16, Proof of Theorem 3.1] who show that at the prime 2

$$\mathsf{THH}_{KU_2}(K(1)) \simeq KU_2.$$

## Topological André–Quillen homology

We will first sketch the definition of ordinary André–Quillen homology. See [225] for the original account and [134] for a very readable modern introduction.

**Definition 6.6.13.** Let $k$ be a commutative ring with unit and let $A$ be a commutative $k$-algebra. The *$A$-module of Kähler differentials of $A$ over $k$* is the $A$-module generated by elements $d(a)$ for $a \in A$ subject to the relations that $d$ is $k$-linear and satisfies the Leibniz rule:

$$d(ab) = d(a)b + ad(b).$$

This $A$-module is denoted by $\Omega^1_{A|k}$.

The conditions imply $d(1) = d(1 \cdot 1) = 2d(1)$ and hence $d(1) = 0$. For a polynomial algebra $A = k[x_1, \ldots, x_n]$ the $A$-module $\Omega^1_{k[x_1,\ldots,x_n]|k}$ is freely generated by $dx_1, \ldots, dx_n$. By induction one shows $d(x_i^m) = mx_i^{m-1}d(x_i)$ for all $m \geqslant 2$.

Consider for instance the $\mathbb{F}_p$-algebra $\mathbb{F}_p[x]/(x^p - x)$. Then the module of Kähler differentials is generated by $d(x)$. However, as we are in characteristic $p$ we get

$$d(x) = d(x^p) = px^{p-1}d(x) = 0$$

and hence $\Omega^1_{\mathbb{F}_p[x]/(x^p-x)|\mathbb{F}_p} = 0$.

*Remark 6.6.14.* For a commutative $k$-algebra $A$ there is an isomorphism between $\Omega^1_{A|k}$ and the first Hochschild homology group of $A$ over $k$: Every $a \otimes b$ in Hochschild chain degree one is a cycle and if you send $a \otimes b$ to $ad(b)$ then this gives a well-defined map modulo Hochschild boundaries and it induces an isomorphism $\mathsf{HH}^k_1(A) \cong \Omega^1_{A|k}$ [161, Proposition 1.1.10].

**Definition 6.6.15.** Let $M$ be an $A$-module. A *$k$-linear derivation from $A$ to $M$* is a $k$-linear map $\delta \colon A \to M$ which satisfies the Leibniz rule.

The set of all such derivations, $\mathrm{Der}_k(A, M)$, is an $A$-submodule of the $A$-module of all $k$-linear maps. The symbol $d$ in the definition of $\Omega^1_{A|k}$ satisfies the conditions of a derivation; hence the map

$$d \colon A \to \Omega^1_{A|k}, \quad a \mapsto da$$

is a derivation, in fact, it is the *universal derivation*.

Proposition 6.6.16 [134]. *For all A-modules M the canonical map*

$$\operatorname{Hom}_A(\Omega^1_{A|k}, M) \to \operatorname{Der}_k(A, M), \quad f \mapsto f \circ d,$$

*is an A-linear isomorphism.*

There is another crucial reformulation of the above isomorphism: $\operatorname{Der}_k(A, M)$ can also be identified with the morphisms of commutative $k$-algebras over $A$ from $A$ to the square-zero extension $A \oplus M$. The latter is the commutative augmented $A$-algebra with underlying module $A \oplus M$ with multiplication

$$(a_1, m_1)(a_2, m_2) = (a_1 a_2, a_1 m_2 + a_2 m_1), \quad a_1, a_2 \in A, m_1, m_2 \in M.$$

A derivation $\delta \colon A \to M$ corresponds to the map into the second component of $A \oplus M$.

The idea of André–Quillen homology is to take the derived functor of $A \mapsto M \otimes_A \Omega^1_{A|k}$. But in which sense? As $A$ is a commutative algebra, we need a resolution of $A$ as such an algebra. The category of differential graded commutative $k$-algebras in general doesn't have a (right-induced) model structure, so instead one works with *simplicial resolutions*. The category of simplicial commutative $k$-algebras *does* have a nice model structure. Let $P_\bullet \to A$ be a cofibrant resolution. Each $P_n$ can be chosen to be a polynomial algebra [134, §4].

Definition 6.6.17. The *André–Quillen homology of A over k with coefficients in M* is

$$\operatorname{AQ}_*(A|k : M) = \pi_*(M \otimes_{P_\bullet} \Omega^1_{P_\bullet|k}).$$

A definition of $\Omega^1_{A|k}$ in terms of generators and relations is not suitable for a generalization to commutative ring spectra. Instead we use the following description:

Lemma 6.6.18. *Denote by I the kernel of the multiplication map $\mu \colon A \otimes_k A \to A$. Then $\Omega^1_{A|k}$ is isomorphic to $I/I^2$.*

*Proof.* The ideal $I$ is generated by elements of the form $a \otimes 1 - 1 \otimes a$. Such an element is identified with $d(a)$. Taking the quotient by $I^2$ corresponds to the Leibniz rule for $d$.    □

The ideal $I$ can also be viewed as a non-unital $k$-algebra and $I/I^2$ is the *module of indecomposables of I*. This definition translates to brave new commutative rings. Basterra's work is formulated in the setting of [94]:

Definition 6.6.19. Let $A$ be a commutative $R$-algebra spectrum.

– We define $I(A \wedge_R A)$ as the pullback

$$
\begin{array}{ccc}
I(A \wedge_R A) & \cdots\cdots\!\!\rightarrow & A \wedge_R A \\
\vdots & & \downarrow{\scriptstyle \mu} \\
* & \longrightarrow & A
\end{array}
$$

- If $N$ is a non-unital commutative $R$-algebra spectrum, then its $R$-*module of indecomposables*, $Q(N)$, is defined as the pushout

$$
\begin{array}{ccc}
N \wedge_R N & \longrightarrow & * \\
{\scriptstyle \mu_N}\downarrow & & \vdots \\
N & \cdots\cdots\cdots\rightarrow & Q(N)
\end{array}
$$

- For an $A$-module spectrum $M$ we define the *topological André–Quillen homology of $A$ over $R$ with coefficients in $M$* as

$$\mathrm{TAQ}(A|R;M) = \mathbf{L}Q(\mathbf{R}I(A \wedge_R A)) \tag{6.6.3}$$

and denote its homotopy groups as $\mathrm{TAQ}_*(A|R;M)$. We use the abbreviation $\Omega_{A|R}$ for $\mathbf{L}Q(\mathbf{R}I(A \wedge_R A))$.

Thus for $\Omega_{A|R}$ we take homotopy invariant versions of the kernel of the multiplication map followed by taking indecomposables by applying the right derived functor of $I$ and the left derived functor of $Q$.

**Definition 6.6.20.** Dually, *topological André–Quillen cohomology of $A$ over $R$ with coefficients in $M$* is $F_A(\Omega_{A|R}, M)$ and we set $\mathrm{TAQ}^n(A|R;M) = \pi_{-n}F_A(\Omega_{A|R}, M)$.

Basterra proves [27, Proposition 3.2] that maps from $\Omega_{A|R}$ to $M$ in the homotopy category of $A$-modules correspond to maps in the homotopy category of commutative $R$-algebra maps over $A$ from $A$ to $A \vee M$, where $A \vee M$ carries the square-zero multiplication.

For example, if $f\colon B \to BGL_1(S)$ is an infinite loop map and $M(f)$ is the associated Thom spectrum, then Basterra and Mandell show [28, Theorem 5 and Corollary] that

$$\mathrm{TAQ}(M(f)) \simeq M(f) \wedge b,$$

where $\Omega^\infty b \simeq B$. In the case of an $E_\infty$-space $B$ the spherical group ring $\Sigma_+^\infty B$ has

$$\mathrm{TAQ}(\Sigma_+^\infty B) \simeq \Sigma_+ B \wedge b.$$

## 6.7 How do we recognize ring spectra as being (non) commutative?

If you have a concrete model of a homotopy type, say in symmetric spectra, then you can be lucky and this model possesses a commutative structure and you should be able to check this by hand. Of course you could also try to disprove commutativity by showing that your spectrum doesn't have power operations as in (6.2.2) and this has been done in many cases, but sometimes you might need a different approach.

## Obstructions via filtrations and resolutions

An obstruction theory for $A_\infty$-structures on homotopy ring spectra was developed as early as 1989 [247] by Alan Robinson. Obstruction theories for $E_\infty$-structures came much later: Goerss–Hopkins [107] and Robinson [246] independently developed one with obstruction groups that later turned out to be isomorphic [30]. The idea is to use a filtration or resolution of an operad such that the corresponding filtration quotients or the corresponding spectral sequence give rise to obstruction groups that contain obstructions for lifting a partial structure to a full $E_\infty$-ring structure ([246, Theorem 5.6] and [107, Corollary 5.9]). The Goerss–Hopkins approach also allows one to calculate the homotopy groups of the derived $E_\infty$ mapping space between two such $E_\infty$-ring spectra [107, Theorem 4.5].

The obstruction groups have as input the algebra of cooperations $E_*E$ of a spectrum $E$ and they compute André–Quillen cohomology groups of the graded commutative $E_*$-algebra $E_*E$ in the setting of differential graded (or simplicial) $E_\infty$-algebras. See [185] or [106, §2.4] for background on these cohomology groups and see [30, §2] for the comparison results. In Robinson's setting these groups are called $\Gamma$-cohomology. The obstruction groups vanish if for instance $E_*E$ is étale as an $E_*$-algebra.

If you prefer to work with explicit chain complexes, then there are several equivalent ones computing $\Gamma$-cohomology groups in Robinson's setting (see [246, §2.5], [250, §6], [222, §2]) and therefore, by the comparison result from [30, Theorem 2.6], computing the obstruction groups in the Goerss–Hopkins setting as well.

There is another version of obstruction theory for promoting a homotopy $T$-algebra structure to an actual one, where $T$ is a monad, by Johnson and Noel [139]. This includes obstructions for operadic structures on spectra but also includes for instance group actions. Noel shows that in certain situations the obstruction theory [139] can be compared to the one of [107].

We list some important applications:

1. The development of the Hopkins–Miller and Goerss–Hopkins obstruction theory was motivated by the Morava-$E$-theory spectra $E_n$, also known as Lubin–Tate spectra, and their variants. These are Landweber exact cohomology theories that govern the deformation theory of height $n$ formal group laws. In [234] an obstruction theory was established leading to a proof that the $E_n$ are $A_\infty$-spectra and that the Morava stabilizer group $\mathbb{G}_n$ acts on $E_n$ via maps of $A_\infty$-spectra. In [107] the corresponding obstruction theory for $E_\infty$-structures was developed and [107, Corollaries 7.6, 7.7] shows that the $\mathbb{G}_n$-action is via $E_\infty$-maps.

2. It was known that $KU$ and $KO$ are $E_\infty$-spectra and it was also known that the $p$-completed Adams summand $L_p$ is $E_\infty$. In [18] Andy Baker and I use Robinson's version of the $E_\infty$-obstruction theory to show that these $E_\infty$-structures are unique and that the $p$-local Adams summand also has a unique $E_\infty$-structure. Uniqueness also holds for the connective covers [19]. It is important to have uniqueness results for $E_\infty$-structures because calculations can depend on a choice of such a structure.

3. For an $E_\infty$-ring spectrum $R$ there is a $\theta$-algebra structure on its $p$-adic $K$-theory,

$\pi_* L_{K(1)}(KU_p \wedge R)$ [106, Theorem 2.2.4], and in good cases

$$\pi_* L_{K(1)}(KU_p \wedge R) \cong \lim_k (KU_p)_*(R \wedge M(p^k)),$$

where $M(p^k)$ is the mod-$p^k$ Moore spectrum. The study of such structures was initiated by McClure in [63, Chapter IX]. There is a variant of the Goerss–Hopkins obstruction theory for realizing for instance a $\theta$-algebra (see [106, §2.4.4] and [153, Theorem 5.14]) as a $K(1)$-local $E_\infty$-ring spectrum.

There is one for realizing an $E_\infty$-$Hk$-algebra spectrum with a fixed Dyer–Lashof structure on its homotopy [217, Proposition 2.2] (for $k$ a field of characteristic $p$). Other variants can be found in the literature.

The $\theta$-algebra version was successfully applied by Lawson and Naumann [153] to show that $BP\langle 2 \rangle$ at 2 has an $E_\infty$-structure. By a different method Hill and Lawson [122, Theorem 4.2] find a commutative model for $BP\langle 2 \rangle$ at the prime 3.

4. Mathew, Naumann and Noel use operations in Morava-$E$-theory to prove May's nilpotence conjecture:

**Theorem 6.7.1** [188, Theorem A]. *If $R$ is an $H_\infty$-ring spectrum and if $x \in \pi_*(R)$ is in the kernel of the Hurewicz homomorphism $\pi_*(R) \to H_*(R; \mathbb{Z})$, then $x$ is nilpotent.*

They use this — among many other applications — for the following result about $E_\infty$-ring spectra:

**Theorem 6.7.2** [188, Proposition 4.2]. *If $R$ is an $E_\infty$-ring spectrum and if there is an $m \in \mathbb{Z}$, $m \neq 0$ with $m \cdot 1 = 0 \in \pi_0(R)$, then, for all primes $p$ and all $n \geq 1$,*

$$K(n)_*(R) \cong 0.$$

Lawson observed that using $K(n)$-techniques (see [231] for background) this implies that for finite $E_\infty$-ring spectra $R$ either the rational homology is non-trivial or $R$ is weakly contractible, because if $H_*(R; \mathbb{Q}) \cong 0$, then by the above result all the Morava $K$-theories also vanish on $R$, but then the finiteness of $R$ implies weak contractibility (see [188, Remark 4.3] for the full argument).

The Dyer–Lashof variant is for instance important when one wants to decide whether a given $H_\infty$-map can be upgraded to an $E_\infty$-map: roughly speaking, an $H_\infty$-spectrum is like an $E_\infty$-spectrum in the homotopy category. You can find applications of this approach for instance in Noel's work [217] and in [139].

Other spectra, such as $BP$, come with homology operations just because they sit in the right place: analyzing the maps $MU \to BP \to H\mathbb{F}_p$ gives [63, p. 63] that $(H\mathbb{F}_p)_*(BP)$ embeds into the dual of the Steenrod algebra such that $(H\mathbb{F}_p)_*(BP)$ is closed under the action of the Dyer–Lashof algebra — even without establishing a structured multiplication on $BP$. This led Lawson [152] to look for the right obstructions for an $E_\infty$-structure of $BP$ at 2 via secondary operations (see Theorem 6.7.5).

## Obstructions via Postnikov towers

A different approach to obstruction theory is to consider Postnikov towers in the world of commutative ring spectra [27] or in the setting of $E_n$-algebras [29].

To this end Basterra uses TAQ-cohomology to lift ordinary $k$-invariants of a connective commutative ring spectrum to $k$-invariants in a multiplicative Postnikov tower:

Assume that $R$ is a connective commutative ring spectrum. Then there is a map of commutative ring spectra

$$p_0 \colon R \to H(\pi_0(R))$$

and without loss of generality we can assume that $p_0$ is a cofibration of commutative ring spectra that realizes the identity on $\pi_0$, i.e., $\pi_0(p_0) = \mathrm{id}_{\pi_0(R)}$.

If we abbreviate $\pi_0(R)$ to $B$ and if $M$ is a $B$-module, an element in $\mathrm{TAQ}^n(A|R; HM)$ corresponds to a morphism $\varphi \colon A \to A \vee \Sigma^n HM$ in the homotopy category of $R$-algebra spectra over $A$ and we can form the pullback of

$$
\begin{array}{c}
A \\
\downarrow {\scriptstyle i_A} \\
A \xrightarrow{\ \varphi\ } A \vee \Sigma^n HM
\end{array}
$$

If we postcompose $\varphi$ with the projection map to $\Sigma^n HM$

$$A \xrightarrow{\ \varphi\ } A \vee \Sigma^n HM \longrightarrow \Sigma^n HM \tag{6.7.1}$$

such a TAQ-class forgets to an Ext-class in $\mathrm{Ext}_R^n(A; HM)$, specifically to an ordinary cohomology class if $R$ is the sphere spectrum. Basterra shows that this projection maps $k$-invariants in the world of commutative ring spectra to ordinary $k$-invariants of the underlying spectrum.

**Theorem 6.7.3 [27, Theorem 8.1].** *For any connective commutative ring spectrum $A$ there is a sequence of commutative ring spectra $A_i$, $\pi_0(A)$-modules $M_i$ and elements*

$$\tilde{k}_i \in \mathrm{TAQ}^{i+2}(A_i|S; HM_{i+1})$$

*such that*

- *$A_0 = H\pi_0(A)$ and $A_{i+1}$ is the pullback of $A_i$ with respect to $\tilde{k}_i$,*
- *$\pi_j A_i = 0$ for all $j > i$,*
- *there are maps of commutative ring spectra $\lambda_i \colon A \to A_i$ which induce an isomorphism in homotopy groups up to degree $i$ such that the diagram*

$$
\begin{array}{ccc}
 & & A_{i+1} \\
 & {\scriptstyle \lambda_{i+1}} \nearrow & \downarrow \\
A & \xrightarrow[{\scriptstyle \lambda_i}]{} & A_i
\end{array}
$$

*commutes in the homotopy category of commutative ring spectra.*

You start with $A_0 = H\pi_0(A)$ and then you have to find a suitable map $A_0 \to A_0 \vee \Sigma^2 H(\pi_1(A))$ as a starting point of the multiplicative Postnikov tower.

Basterra's result can be used as an obstruction theory as follows. If $A$ is a connective spectrum then it has an ordinary Postnikov tower with $k$-invariants living in ordinary cohomology groups

$$k_i \in H^{i+2}(A_i; \pi_{i+1}(A)).$$

You can then investigate whether it is possible to find multiplicative $k$-invariants

$$\tilde{k}_i \in \mathsf{TAQ}^{i+2}(A_i|S; H\pi_{i+1}(A))$$

that forget to the $k_i$'s under the map from (6.7.1).

Using Postnikov towers for $E_n$-algebra spectra, Basterra and Mandell show:

**Theorem 6.7.4** [29, Theorem 1.1]. *The Brown Peterson spectrum, BP, has an $E_4$-structure at every prime.*

This ensures by the main result of [184] that the derived category of $BP$-module spectra has a symmetric monoidal smash product. Tyler Lawson, however, showed that there are certain secondary operations in the $\mathbb{F}_2$-homology of every such spectrum with an $E_{12}$-structure and he could show that these are not present in the $\mathbb{F}_2$-homology of $BP$ at 2. Let $BP\langle n \rangle$ denote the spectrum $BP/(v_{n+1}, v_{n+2}, \dots)$.

**Theorem 6.7.5** [152, Theorem 1.1.2]. *The Brown–Peterson spectrum at the prime 2 does not possess an $E_n$-structure for any $n$ with $12 \leqslant n \leqslant \infty$. The truncated Brown–Peterson spectrum $BP\langle n \rangle$ for $n \geqslant 4$ cannot have an $E_n$-structure for any $n$ with $12 \leqslant n \leqslant \infty$.*

See [271] for the corresponding results at odd primes.

### Realization of $E_\infty$-spectra via derived algebraic geometry

There is a completely different important and highly successful approach to realization problems, using *derived algebraic geometry*, for which see Chapter 8 of this volume.

## 6.8    What are étale maps?

We first recall the algebraic notion of an étale $k$-algebra from [161, E.1]: Let $k$ be a commutative ring and let $A$ be a finitely generated commutative $k$-algebra. Then $A$ is *étale* if $A$ is flat over $k$ and if the module of Kähler differentials $\Omega^1_{A|k}$ is trivial. If $\Omega^1_{A|k} = 0$, then $k \to A$ is called *unramified*. A $k$-algebra $B$ (not necessarily commutative) is called *separable* if the multiplication map

$$B \otimes_k B^o \to B$$

has a section as a $B$-bimodule map. In algebra, a commutative separable algebra has Hochschild homology concentrated in homological degree zero, in particular the module of Kähler differentials is trivial.

## Rognes' Galois extensions of commutative ring spectra

**Definition 6.8.1** [251, Definition 4.1.3].   Let $A \to B$ be a map of commutative ring spectra and let $G$ be a finite group acting on $B$ via commutative $A$-algebra maps. Assume that $S \to A \to B$ is a sequence of cofibrations in the model structure on commutative ring spectra of [94, Corollary VII.4.10]. Then $A \to B$ is a $G$-Galois extension if

1. the canonical map $\iota \colon A \to B^{hG}$ is a weak equivalence and

2. $$h \colon B \wedge_A B \to \prod_G B \qquad\qquad (6.8.1)$$

is a weak equivalence.

The first condition is the familiar fixed points condition from classical Galois theory of fields. The map $\iota$ comes from taking the adjoint of the map

$$A \wedge EG_+ \xrightarrow{\mathrm{id} \wedge p} A \wedge S^0 \cong A \longrightarrow B,$$

where $p \colon EG_+ \to S^0$ collapses $EG$ to the non-base point of $S^0$.

The map $h$ is adjoint to the composite

$$B \wedge_A B \wedge G_+ \to B \wedge_A B \to B$$

that comes from the $G$-action on the right factor of $B \wedge_A B$ followed by the multiplication in $B$. (Informally, if smashes were tensors, then $h(b_1 \otimes b_2) = (b_1 \cdot g(b_2))_{g \in G}$.) Note that $\prod_G B$ is isomorphic to $F(G_+, B)$, so we could rewrite the condition in (6.8.1) as the requirement that

$$h \colon B \wedge_A B \to F(G_+, B)$$

is a weak equivalence.

The condition that the map $h$ from (6.8.1) is a weak equivalence is crucial. It is also necessary for Galois extensions of discrete commutative rings in order to ensure that the extension is unramified. For instance, $\mathbb{Z} \subset \mathbb{Z}[i]$ satisfies $\mathbb{Z}[i]^{C_2} = \mathbb{Z}$, but $h \colon \mathbb{Z}[i] \otimes_{\mathbb{Z}} \mathbb{Z}[i] \to \mathbb{Z}[i] \times \mathbb{Z}[i]$ is not surjective: $h$ detects the ramification at the prime 2. Therefore $\mathbb{Z} \to \mathbb{Z}[i]$ is *not* a $C_2$-Galois extension but $\mathbb{Z}[\frac{1}{2}] \to \mathbb{Z}[\frac{1}{2}, i]$ is $C_2$-Galois.

Galois extensions of commutative ring spectra can have rather bad properties as modules. So the following definition is actually an additional assumption (this does not happen in the discrete setting).

**Definition 6.8.2** [251, Definition 4.3.1].   A Galois extension $A \to B$ is *faithful* if it is faithful as an $A$-module, i.e., for every $A$-module $M$ with $M \wedge_A B \simeq *$ we have $M \simeq *$.

Important examples of Galois extensions of commutative ring spectra are the following. By $C_n$ we denote the cyclic group of order $n$.

1. The concept of Galois extensions of commutative ring spectra corresponds to the one for commutative rings via the Eilenberg–Mac Lane spectrum functor [251, Proposition 4.2]: Let $R \to T$ be a homomorphism of discrete commutative rings and

let $G$ be a finite group acting on $T$ via $R$-algebra homomorphisms. Then $R \to T$ is a $G$-Galois extension of commutative rings if and only if $HR \to HT$ is a $G$-Galois extension of commutative ring spectra.

2. The complexification of real vector bundles gives rise to a map of commutative ring spectra $KO \to KU$ from real to complex topological $K$-theory. There is a $C_2$-action on $KU$ corresponding to complex conjugation of complex vector bundles. Rognes shows [251, Proposition 5.3.1] that this turns $KO \to KU$ into a $C_2$-Galois extension.

3. At an odd prime $p$ there is a $p$-adic Adams operation on $KU_p$ that gives rise to a $C_{p-1}$-action on $KU_p$ such that $L_p \to KU_p$ is a $C_{p-1}$-Galois extension [251, §5.5.4].

4. There is a notion of pro-Galois extensions of commutative ring spectra and $L_{K(n)}S \to E_n$ is a $K(n)$-local pro-Galois extension with the extended Morava stabilizer group as the Galois group [251, Theorem 5.4.4].

5. Let $p$ be an arbitrary prime. The projection map $\pi: EC_p \to BC_p$ induces a map on function spectra

$$F(\pi_+, H\mathbb{F}_p): F((BC_p)_+, H\mathbb{F}_p) \to F((EC_p)_+, H\mathbb{F}_p) \sim H\mathbb{F}_p$$

which identifies $H\mathbb{F}_p$ as a $C_p$-Galois extension over $F((BC_p)_+, H\mathbb{F}_p)$ [251, Proposition 5.6.3]. Hence in the world of commutative ring spectra group cohomology sits between $S$ and $H\mathbb{F}_p$ as the base of a Galois extension! Beware, this Galois extension is not faithful. This observation is due to Ben Wieland: the Tate construction $H\mathbb{F}_p^{tC_p}$ isn't trivial and it is actually killed by the Galois extension (in the spectral sequence you augment a Laurent generator to zero).

6. Studying elliptic curves with level structures gives $C_2$-Galois extensions $\mathrm{TMF}_0(3) \to \mathrm{TMF}_1(3)$ and $\mathrm{Tmf}_0(3) \to \mathrm{Tmf}_1(3)$ [187, Theorems 7.6, 7.12]. For $\mathrm{TMF}_1(3)$ and $\mathrm{Tmf}_1(3)$ you consider elliptic curves with one chosen point of exact order 3 and for $\mathrm{TMF}_0(3)$ and $\mathrm{Tmf}_0(3)$ you only remember a subgroup of order 3. As $C_2 \cong \mathbb{Z}/3\mathbb{Z}^\times$ this gives a $C_2$-action. This can be made rigorous; see [121, 122, 187].

## Notions of étale morphisms

Weibel–Geller [298] show that for an étale extension of commutative rings $\varphi: A \to B$ Hochschild homology satisfies *étale descent*: The map $\mathrm{HH}(\varphi)_*$ induces an isomorphism

$$B \otimes_A \mathrm{HH}_*(A) \cong \mathrm{HH}_*(B) \tag{6.8.2}$$

and for finite $G$-Galois extensions $\varphi: A \to B$ one obtains *Galois descent*:

$$\mathrm{HH}_*(A) \cong \mathrm{HH}_*(B)^G. \tag{6.8.3}$$

It is easy to see that for a $G$-Galois extension of discrete commutative rings $\varphi: A \to B$ with finite $G$, the induced extension of graded commutative rings $\mathrm{HH}_*(\varphi): \mathrm{HH}_*(A) \to \mathrm{HH}_*(B)$ is again $G$-Galois. In addition to having the right fixed-point property it satisfies

$$HH_*(B) \otimes_{HH_*(A)} HH_*(B) \cong B \otimes_A HH_*(A) \otimes_{HH_*(A)} B \otimes_A HH_*(A)$$
$$\cong B \otimes_A B \otimes_A HH_*(A)$$
$$\cong \prod_G B \otimes_A HH_*(A)$$
$$\cong \prod_G HH_*(B).$$

If $\varphi \colon A \to B$ is étale, then the module of Kähler differentials $\Omega^1_{B|A}$ is trivial and it can be easily seen that the map $B \to HH^A_*(B)$ is an isomorphism and that André–Quillen homology of $B$ over $A$ is trivial, because étale algebras are smooth.

For commutative ring spectra the situation is different. There are several non-equivalent notions of étale maps:

**Definition 6.8.3.**  Let $\varphi \colon A \to B$ be a morphism of commutative ring spectra.

1. [168, Definition 7.5.1.4] We call $\varphi$ *Lurie-étale* if $\pi_0(\varphi) \colon \pi_0(A) \to \pi_0(B)$ is an étale map of commutative rings and if the canonical map

$$\pi_*(A) \otimes_{\pi_0(A)} \pi_0(B) \to \pi_*(B)$$

   is an isomorphism. In Chapter 8, this will be the only notion of étale map considered, and the adjective "Lurie" will be dropped.

2. [201, Definiton 3.2], [251, Definition 9.2.1] The morphism $\varphi$ is *(formally)* THH-*étale* if $B \to THH^A(B)$ is a weak equivalence.

3. [201, Definiton 3.2], [251, Definition 9.4.1] We define $\varphi$ to be *(formally)* TAQ-*étale* if $TAQ(B|A)$ is weakly equivalent to $*$.

*Remark 6.8.4.*
  - Rognes [251] reserves the labels THH-étale and TAQ-étale for maps that, in addition to the conditions above, identify $B$ as a dualizable $A$-module.
  - The condition of being Lurie-étale is strong and is a very algebraic one. It is for instance not satisfied by the $C_2$-Galois extension $KO \to KU$ because on the level of homotopy groups this extension is rather appalling, compare (6.1.1).
  - McCarthy and Minasian show that THH-étale implies TAQ-étale and they show that for $n > 1$ the map $H\mathbb{F}_p \to F(K(\mathbb{Z}/p\mathbb{Z}, n)_+, H\mathbb{F}_p)$ is a TAQ-étale morphism that is not THH-étale. They attribute this example to Mandell [201, Example 3.5]. Minasian [211, Corollary 2.8] proves that both notions are equivalent for morphisms between connective commutative ring spectra.
  - For connective spectra, the notion of Lurie-étaleness has good features [168, §7.5] and Mathew shows in [186, Corollary 3.1] that one can use [165, Lemma 8.9] to show that under some finiteness condition TAQ-étaleness implies Lurie-étaleness in the connective case.

**Definition 6.8.5** [251, Definition 9.1.1].  Let $C$ be a cofibrant associative $A$-algebra spectrum. Then $C$ is *separable* if the multiplication map $\mu \colon C \wedge_A C^o \to C$ has a section in the homotopy category of $C$-bimodule spectra.

**Proposition 6.8.6** [251, Lemma 9.2.6]. *If C is a commutative separable A-algebra spectrum, then C is* THH*-étale.*

*Proof.* Recall from Remark 6.6.2 that $\text{THH}^A(C)$ is an augmented commutative $C$-algebra spectrum, so the composite of the unit map $C \to \text{THH}^A(C)$ with the augmentation

$$C \to \text{THH}^A(C) \to C$$

is the identity. We also get a splitting in the homotopy category of $C$-bimodule spectra,

$$C \xrightarrow{s} C \wedge_A C \xrightarrow{\mu} C,,$$

i.e., the above composite is the identity on $C$. Taking the derived smash product $C \wedge_{C \wedge_A C}^L (-)$ of the above sequence gives the sequence

$$\text{THH}^A(C) \to C \to \text{THH}^A(C),$$

in which the last map is equivalent to the unit map of $\text{THH}^A(C)$ and whose composite is the identity. So the unit map $C \to \text{THH}^A(C)$ has a right and a left inverse in the homotopy category of $C$-module spectra.  □

**Definition 6.8.7.** Let $A \to B$ be a map of commutative ring spectra and let $G$ be a finite group acting on $B$ via maps of commutative $A$-algebra spectra. Assume that $S \to A \to B$ is a sequence of cofibrations in the model structure on commutative ring spectra of [94, Corollary VII.4.10]. Then $A \to B$ is *unramified* if

$$h \colon B \wedge_A B \to \prod_G B$$

is a weak equivalence.

**Proposition 6.8.8** (compare [251, Lemma 9.1.2]). *If $A \to B$ is unramified, then $B$ is separable over $A$.*

*Proof.* The canonical inclusion map $i \colon B \to F(G_+, B)$ can be modeled by the pointed map from $G_+$ to $S^0$ that sends the neutral element $e \in G$ to the non-basepoint of $S^0$ and sends all other elements to the basepoint. We define a section to the multiplication map of $B$ to be

$$B \xrightarrow{i} F(G_+, B) \xleftarrow{h, \sim} B \wedge_A B.$$

Note that $h$ is not a $B$-bimodule map, but we are only interested in its $e$-component of $F(G_+, B)$.  □

Thus we can conclude that unramified maps of commutative ring spectra are THH-étale and that the failure of the map $B \to \text{THH}^A(B)$ to be a weak equivalence detects ramification. This idea was exploited in [83] in order to show that the inclusion of the Adams summand $\ell \to ku_{(p)}$ is tamely ramified [83, Theorem 4.1]. Sagave also identifies this map as being log-étale [255, Theorem 1.6].

## Versions of étale descent

Transferring the Geller–Weibel result to the setting of commutative ring spectra, it seems natural to define two versions of descent:

**Definition 6.8.9.** In the following $\varphi\colon A \to B$ is a cofibration and $A$ is cofibrant.

– The morphism $\varphi\colon A \to B$ satisfies *étale descent* if the canonical morphism

$$B \wedge_A \mathrm{THH}(A) \to \mathrm{THH}(B) \tag{6.8.4}$$

is a weak equivalence.

– If $\varphi\colon A \to B$ is a map of commutative ring spectra and if $G$ is a finite group acting on $B$ via commutative $A$-algebra maps, then we say that $\varphi$ satisfies *Galois descent* if the map

$$\mathrm{THH}(A) \to \mathrm{THH}(B)^{hG} \tag{6.8.5}$$

is a weak equivalence.

Akhil Mathew clarifies the relationship between the different notions of étale morphisms and the notions of descent. He proves that Lurie-étale morphisms satisfy étale descent [186, Theorem 1.3] and that for a faithful $G$-Galois extension with finite Galois group $G$, both descent properties are equivalent [186, Proposition 4.3] and they are in turn equivalent to the property that $\mathrm{THH}(A) \to \mathrm{THH}(B)$ is again a $G$-Galois extension.

Moreover, he shows that the morphism

$$\varphi\colon F(\mathbb{S}^1_+, H\mathbb{F}_p) \to F(\mathbb{S}^1_+, H\mathbb{F}_p)$$

that is induced by the degree-$p$ map on $\mathbb{S}^1$ is a faithful $C_p$-Galois extension, but that it does *not* satisfy étale descent [186, Theorem 2.1] and hence it doesn't satisfy Galois descent.

The Hopf fibration $\mathbb{S}^1 \to \mathbb{S}^3 \to \mathbb{S}^2$ is a principal $\mathbb{S}^1$-bundle. The corresponding morphism of commutative $H\mathbb{Q}$-algebra spectra of cochains

$$F(\eta, H\mathbb{Q})\colon F(\mathbb{S}^2_+, H\mathbb{Q}) \to F(\mathbb{S}^3_+, H\mathbb{Q})$$

is therefore an $\mathbb{S}^1$-Galois extension [251, Proposition 5.6.3].

In joint work with Christian Ausoni we show that the morphism $F(\eta, H\mathbb{Q})$ does not satisfy Galois descent, i.e.,

$$\mathrm{THH}(F(\mathbb{S}^2_+, H\mathbb{Q})) \not\simeq \mathrm{THH}(F(\mathbb{S}^3_+, H\mathbb{Q}))^{h\mathbb{S}^1}.$$

Indeed, the homotopy groups of $\mathrm{THH}(F(\mathbb{S}^2_+, H\mathbb{Q}))$ contain an element in degree $-1$ that is not present in $\pi_*(\mathrm{THH}(F(\mathbb{S}^3_+, H\mathbb{Q}))^{h\mathbb{S}^1})$.

Mathew identifies the problem with étale descent of finite faithful Galois extensions for $\mathrm{THH}$ as being caused by the non-trivial fundamental group of $\mathbb{S}^1$. He shows the following result.

Theorem 6.8.10 [186, Proposition 5.2]. *Let X be a simply connected pointed space and let A → B be a faithful G-Galois extension of commutative ring spectra with finite G. Then the map*

$$B \wedge_A (X \otimes A) \to X \otimes B$$

*is a weak equivalence.*

In particular, higher-order topological Hochschild homology, $\mathrm{THH}^{[n]}$ for $n \geqslant 2$, *does* satisfy étale descent for faithful finite Galois extensions. However, étale descent remains for instance an issue for torus homology.

Sometimes THH *does* satisfy descent, even for ramified maps of commutative ring spectra. For instance, Ausoni shows in [13, Theorem 10.2] that $\mathrm{THH}(\ell_p)$ is $p$-adically equivalent to $\mathrm{THH}(ku_p)^{hC_{p-1}}$ and even that $K(\ell_p)$ is $p$-adically equivalent to $K(ku_p)^{hC_{p-1}}$.

*Remark 6.8.11.* In [69] Clausen, Mathew, Naumann and Noel prove far-reaching Galois descent results for topological Hochschild homology and algebraic $K$-theory; in particular they confirm a Galois descent conjecture for algebraic $K$-theory by Ausoni and Rognes in many important cases. They identify THH as a *weakly additive invariant* (see [69, Definition 3.10]) and prove descent in the form of [69, Theorems 5.1 and 5.6].

## 6.9    Picard and Brauer groups

### Picard groups in the setting of a symmetric monoidal category

Let $(\mathcal{C}, \otimes, 1, \tau)$ be a symmetric monoidal category. An important class of objects in $\mathcal{C}$ are those objects $C$ that have an inverse with respect to $\otimes$, i.e., such that there is an object $C'$ of $\mathcal{C}$ satisfying

$$C \otimes C' \cong 1.$$

One wants to gather such objects in a category and build a space and spectrum out of them:

Definition 6.9.1. The *Picard groupoid of $\mathcal{C}$*, $\mathrm{Picard}(\mathcal{C})$, is the category whose objects are the invertible objects of $\mathcal{C}$ and whose morphisms are isomorphisms between invertible objects.

If $C_1$ and $C_2$ are objects of $\mathrm{Picard}(\mathcal{C})$, then so is $C_1 \otimes C_2$; in fact, $\mathrm{Picard}(\mathcal{C})$ is itself a symmetric monoidal category. But in general, this category might not be small.

Definition 6.9.2. Let $\mathcal{C}$ be as above and assume that $\mathrm{Picard}(\mathcal{C})$ is small. Then $\mathrm{PIC}(\mathcal{C})$ is the classifying space of the symmetric monoidal category $\mathrm{Picard}(\mathcal{C})$ and let $\mathrm{pic}(\mathcal{C})$ denote the connective spectrum associated to the infinite loop space associated to $\mathrm{PIC}(\mathcal{C})$. The *Picard group of $\mathcal{C}$*, $\mathrm{Pic}(\mathcal{C})$, is $\pi_0 \mathrm{PIC}(\mathcal{C})$.

If the Picard groupoid of $C$ is small, then the Picard group can also be described as the set of isomorphism classes of invertible objects of $C$ with the product

$$[C_1] \otimes [C_2] := [C_1 \otimes C_2].$$

The neutral element is the isomorphism class of the unit, $[1]$.

Definition 6.9.3. Let $R$ be a (discrete) commutative ring; we denote by $\mathrm{Pic}(R)$ the Picard group of the symmetric monoidal category of the category of $R$-modules and by $\mathrm{PIC}(R)$ (and $\mathrm{pic}(R)$) the Picard space (and Picard spectrum) of this category.

For instance the Picard group of a ring of integers in a number ring is its ideal class group.

## Picard group for commutative ring spectra

For commutative ring spectra $R$, the above definition of $\mathrm{PIC}(R)$ and $\mathrm{pic}(R)$ would either be much too rigid (if one chose $C$ to be the category of $R$-module spectra and isomorphisms) or not strict enough (if one took $C$ to be the homotopy category of $R$-module spectra). See [189, §2] for an adequate background for a suitable definition and see [103, §4] for a dictionary how to pass from a commutative ring spectrum $R$ and its category of modules to the $\infty$-categorical setting. Instead of working with symmetric monoidal categories, one uses presentable symmetric monoidal $\infty$-categories $C$. Then the Picard $\infty$-groupoid of $C$ is the maximal subgroupoid of the underlying $\infty$-category of $C$ spanned by the invertible objects. This groupoid is equivalent to a grouplike $E_\infty$-space $\mathrm{PIC}(C)$ and hence there is a connective ring spectrum, $\mathrm{pic}(C)$, associated to $C$ [103, §5].

Let $R$ be a commutative ring spectrum. The operadic nerve of the category of cofibrant-fibrant $R$-modules is a stable presentable symmetric monoidal $\infty$-category [168, Proposition 4.1.3.10] and we will abbreviate this as the $\infty$-category of $R$-modules, $R$mod.

Definition 6.9.4. The *Picard group of a commutative ring spectrum R*, $\mathrm{Pic}(R)$, is the group $\pi_0(\mathrm{PIC}(R\mathrm{mod}))$.

Again, these Picard groups can also be described as the set of isomorphism classes of invertible $R$-modules in the homotopy category of $R$-module spectra.

The Picard space $\mathrm{PIC}(R)$ is a delooping of the units of $R$ ([189, §2.2], [289, §5]): There is an equivalence

$$\mathrm{PIC}(R) \simeq \mathrm{Pic}(R) \times BGL_1(R).$$

*Remark 6.9.5.* There is a map $\mathrm{Pic}(\pi_* R) \to \mathrm{Pic}(R)$ that realizes an element in the algebraic Picard group of invertible graded $\pi_* R$-modules as a module over $R$ and in many cases this map is an isomorphism [17, Theorem 43]. In this case we call $\mathrm{Pic}(R)$ *algebraic*. A notable exception comes from Galois extensions of ring spectra: As in algebra, if $A \to B$ is a $G$-Galois extension of commutative ring spectra with abelian

Galois group $G$, then $[B] \in \text{Pic}(A[G])$ [251, Proposition 6.5.2]. But for instance $[KU_*]$ is certainly *not* an element in the algebraic Picard group $\text{Pic}(KO_*[C_2])$; see (6.1.1).

The equivalence classes of suspensions of $R$ are always in $\text{Pic}(R)$, but if $R$ is periodic, these suspensions don't generate a free abelian group. Let us mention some crucial examples of Picard groups of commutative ring spectra:

- The Picard group of the initial commutative ring spectrum $S$ is $\text{Pic}(S) \cong \mathbb{Z}$, where $n \in \mathbb{Z}$ corresponds to the class of $S^n$ [129].
- For connective commutative ring spectra the Picard group of $R$ is algebraic; see [17, Theorem 21], [189, Theorem 2.4.4].
- For periodic real and complex $K$-theory the Picard groups just notice the suspensions of the ground ring: the Picard group of $KU$ is algebraic, with $\text{Pic}(KU) \cong \mathbb{Z}/2\mathbb{Z}$, and $\text{Pic}(KO) \cong \mathbb{Z}/8\mathbb{Z}$ (Hopkins, [189, Example 7.1.1] and [103, §7]).
- The same applies to the periodic version of the spectrum of topological modular forms: $\text{Pic}(\text{TMF}) \cong \mathbb{Z}/576\mathbb{Z}$ [189, Theorem A]. But for Tmf, the spectrum of topological forms that mediates between TMF and its connective version tmf, one gets [189, Theorem B]

$$\text{Pic}(\text{Tmf}) \cong \mathbb{Z} \oplus \mathbb{Z}/24\mathbb{Z},$$

  where the copy of the integers comes from the suspensions of Tmf and the generator of the $\mathbb{Z}/24\mathbb{Z}$-summand is described in [189, Construction 8.4.2].
- Using Galois descent techniques for pic, Heard, Mathew and Stojanoska prove in [116, Theorem 1.5] that, for any odd prime and any finite subgroup $G$ of the full Morava stabilizer group $\mathbb{G}_{p-1}$, the Picard group of $E_{p-1}^{hG}$ is a cyclic group generated by the suspension of $E_{p-1}^{hG}$.

A Picard group that contains more elements than just the ones coming from suspensions of the commutative ring spectrum says that there are more self-equivalences of the homotopy category of $R$-modules than the standard suspensions. One might view these as twisted suspensions. Gepner and Lawson explore the concept of having a Picard grading on the category of $R$-module spectra and they develop a Pic-resolution model category structure in the sense of Bousfield [103, §3.2].

## Descent method and local versions

A crucial method for calculating Picard groups is Galois descent. If $A \to B$ is a $G$-Galois extension (for $G$ finite), then for the Picard spectra and spaces the following equivalences hold [103, 189]:

$$\text{pic}(A) \simeq \tau_{\geqslant 0}\text{pic}(B)^{hG}, \quad \text{PIC}(A) \simeq \text{PIC}(B)^{hG}. \tag{6.9.1}$$

Here, $\tau_{\geqslant 0}$ denotes the connective cover of a spectrum. In general, the extension $B$ is easier to understand than $A$; for instance, in the case of the $C_2$-Galois extension $KO \to KU$, one obtains information about $\text{pic}(A)$ using the homotopy fixed point spectral sequence

$$H^{-s}(G; \pi_t\text{pic}(B)) \Rightarrow \pi_{t-s}(\text{pic}(B)^{hG}).$$

In [121, §6], for instance, Hill and Meier use Galois descent to determine the Picard groups of $\mathsf{TMF}_0(3)$ and $\mathsf{Tmf}_0(3)$:

**Theorem 6.9.6** [121, Theorems 6.9, 6.12].

$$\mathsf{Pic}(\mathsf{TMF}_0(3)) \cong \mathbb{Z}/48\mathbb{Z}, \quad \mathsf{Pic}(\mathsf{Tmf}_0(3)) \cong \mathbb{Z} \oplus \mathbb{Z}/8\mathbb{Z}.$$

Hopkins–Mahowald–Sadofsky started the investigation of the Picard groups of the $K(n)$-local homotopy categories for varying $n$ [129]. They denote these Picard groups by $\mathsf{Pic}_n$. Note that the relevant symmetric monoidal product for fixed $n$ is

$$X \otimes Y = L_{K(n)}(X \wedge Y)$$

for $K(n)$-local $X$ and $Y$. They determined $\mathsf{Pic}_1$ for all primes $p$:

**Theorem 6.9.7** [129, Theorem 3.3, Proposition 2.7].

- *At the prime* 2, $\mathsf{Pic}_1 \cong \mathbb{Z}_2^\times \times \mathbb{Z}/4\mathbb{Z}$.
- *For all odd primes* $p$, $\mathsf{Pic}_1 \cong \mathbb{Z}_p \times \mathbb{Z}/q\mathbb{Z}$ *with* $q = 2p - 2$.

In the $K(n)$-local setting the notion of algebraic elements in $\mathsf{Pic}_n$ is slightly more involved. Hopkins, Mahowald and Sadofsky show [129] (see also [108, Theorem 2.4]) that a $K(n)$-local spectrum $X$ is $K(n)$-locally invertible if and only if $\pi_*(L_{K(n)}(E_n \wedge X))$ is a free $(E_n)_*$-module of rank one and if and only if $\pi_*(L_{K(n)}(E_n \wedge X))$ is invertible as a continuous module over the completed group ring $(E_n)_*[[\mathbb{G}_n]]$. Here, $\mathbb{G}_n$ is the full Morava-stabilizer group. Hence applying $\pi_*(L_{K(n)}(E_n \wedge -))$ gives a map from $\mathsf{Pic}_n$ to the Picard group of continuous $(E_n)_*[[\mathbb{G}_n]]$-modules and this group is called $\mathsf{Pic}_n^{\mathrm{alg}}$. The kernel of the map, $\kappa_n$, collects the exotic elements in $\mathsf{Pic}_n$:

$$0 \to \kappa_n \to \mathsf{Pic}_n \to \mathsf{Pic}_n^{\mathrm{alg}}.$$

For odd primes, all elements in $\mathsf{Pic}_1$ can be detected algebraically but for $p = 2$ one has a non-trivial element in $\kappa_1$. See [108] for $\mathsf{Pic}_2$ at $p = 3$ and a general overview. There is ongoing work on $\mathsf{Pic}_2$ at $p = 2$ by Agnès Beaudry, Irina Bobkova, Paul Goerss and Hans-Werner Henn.

## Brauer groups of commutative rings

Probably most of you will know the definition of the Brauer group of a field. But as for many features that we want to transfer to the spectral world we need to consider algebraic concepts developed for commutative rings (not fields).

Azumaya started to think about general Brauer groups [14] in the setting of local rings. A general definition of the Brauer group of a commutative ring $R$ was given by Auslander and Goldman [12] as Morita equivalence classes of Azumaya algebras. The Brauer group was then globalized to schemes by Grothendieck [114]. He also shows that the Brauer group of the initial ring $\mathbb{Z}$ is trivial; this is a byproduct of his identification of Brauer groups of number rings in [114, III, Proposition (2.4)].

### Brave new Brauer groups

Baker and Lazarev define in [16] what an Azumaya algebra spectrum is. We use one version of this definition in [20] to develop Brauer groups for commutative ring spectra. Related concepts can be found in [138] and [291].

Fix a cofibrant commutative ring spectrum $R$.

**Definition 6.9.8.** A cofibrant associative $R$-algebra $A$ is called an *Azumaya $R$-algebra spectrum* if $A$ is dualizable and faithful as an $R$-module spectrum and if the canonical map

$$A \wedge_R A^o \to F_R(A, A)$$

is a weak equivalence.

We list some crucial properties of Azumaya algebra spectra. For the first property recall the discussion of derived centers from Definition 6.6.10.

**Proposition 6.9.9.**

1. [16, Proposition 2.3] *If $A$ is an Azumaya $R$-algebra spectrum, then $A$ is homotopically central over $R$, i.e., $R \to \mathsf{THH}_R(A)$ is a weak equivalence.*

2. [20, Proposition 1.5] *If $A$ is Azumaya over $R$ and if $C$ is a cofibrant commutative $R$-algebra then $A \wedge_R C$ is Azumaya over $C$. Conversely, if $C$ is as above and dualizable and faithful as an $R$-module, then $A \wedge_R C$ being Azumaya over $C$ implies that $A$ is Azumaya over $R$.*

   *If $A$ and $B$ are Azumaya over $R$, then $A \wedge_R B$ is also Azumaya over $R$.*

3. [20, 2.2] *If $M$ is a faithful, dualizable, cofibrant $R$-module, then (a cofibrant replacement of) $F_R(M, M)$ is an $R$-Azumaya algebra spectrum.*

Thus the endomorphism Azumaya algebras are the ones that are always there and you want to ignore them.

**Definition 6.9.10.** Let $A$ and $B$ be two Azumaya $R$-algebra spectra. We call them *Brauer equivalent* if there are dualizable, faithful $R$-modules $N$ and $M$ such that there is an $R$-algebra equivalence

$$A \wedge_R F_R(M, M) \simeq B \wedge_R F_R(N, N).$$

We denote by $Br(R)$ the set of Brauer equivalence classes of $R$-Azumaya algebra spectra.

Note that $Br(R)$ is an abelian group with multiplication induced by the smash product over $R$. Johnson shows [138, Lemma 5.7] that one can reduce the above relation to what he calls *Eilenberg–Watts equivalence*. This implies that one can still think about the Brauer group of a commutative ring spectrum as the Morita equivalence classes of Azumaya algebra spectra.

We showed a Galois descent result [20, Proposition 3.3], saying that under a natural condition you can descend an Azumaya algebra $C$ over $B$ to an Azumaya algebra $C^{hG}$ over $A$ if $A \to B$ is a faithful $G$-Galois extension with finite Galois group $G$.

## Examples of Brauer groups

As we know that $Br(\mathbb{Z}) = 0$, we conjectured [20] that the Brauer group of the initial ring spectrum is also trivial. This conjecture was proven in [10, Corollary 7.17]. The authors actually showed a much stronger result:

**Theorem 6.9.11** [10, Theorem 7.16]. *If $R$ is a connective commutative ring spectrum such that $\pi_0(R)$ is either $\mathbb{Z}$ or the Witt vectors $W(\mathbb{F}_q)$, then the Brauer group of $R$ is trivial.*

Different approaches — see [10, Definition 7.1], [103, §5], and [289] — can be used to construct a *Brauer space*, $Br_R$, for a commutative ring spectrum $R$ and to show that this space is a delooping of the Picard space, $\mathsf{PIC}$

$$\Omega Br_R \simeq \mathsf{PIC}(R)$$

with $\pi_0(Br_R) \cong Br(R)$.

An important question in the classical context of Brauer groups of schemes is to which extent these groups can be controlled by the second étale cohomology group. See the introduction of [291] for a nice overview. Toën shows that for quasi-compact and quasi-separated schemes $X$ one can identify the *derived Brauer group of $X$* with $H^1_{\text{ét}}(X; \mathbb{G}_m) \times H^2_{\text{ét}}(X; \mathbb{G}_m)$. The work of Antieau and Gepner [10, §7.4] relates Brauer groups of connective commutative ring spectra to étale cohomology groups by establishing a spectral sequence starting from étale cohomology groups for étale sheaves over a connective commutative ring spectrum converging to the homotopy groups of the Brauer space [10, Theorem 7.12].

The integral version of the quaternions gives a non-trivial element in $Br(S[\frac{1}{2}])$ [20, Proposition 6.3]. Antieau and Gepner show in [10, Corollary 7.18]

$$Br(S[\tfrac{1}{p}]) \cong \mathbb{Z}/2\mathbb{Z} \quad \text{for all primes } p$$

and they prove the existence of a short exact sequence

$$0 \to Br(S_{(p)}) \to \mathbb{Z}/2\mathbb{Z} \oplus \bigoplus_{q \neq p} \mathbb{Q}/\mathbb{Z} \to \mathbb{Q}/\mathbb{Z} \to 0$$

by applying [10, Corollary 7.13], where they calculate the homotopy groups of the Brauer space of any connective commutative ring spectrum $R$ in terms of étale cohomology groups and the homotopy groups of $R$.

They use the classical exact sequence for the Brauer group of the rationals [114, §2] coming from the Albert–Brauer–Hasse–Noether theorem:

$$0 \to Br(\mathbb{Q}) \to \mathbb{Z}/2\mathbb{Z} \oplus \bigoplus_{p \text{ prime}} Br(\mathbb{Q}_p) \to \mathbb{Q}/\mathbb{Z} \to 0,$$

with $Br(\mathbb{Q}_p) = \mathbb{Q}/\mathbb{Z}$. This determines $Br(\mathbb{Z}[\frac{1}{p}])$ and $Br(\mathbb{Z}_{(p)})$ and this in turn gives the above result for the sphere spectra with $p$ inverted or localized at $p$.

In [20, Theorem 10.1] we show that the $K(n)$-local Brauer group of the $K(n)$-local sphere is non-trivial at least for odd primes and $n > 1$.

Gepner and Lawson prove a version of Galois descent for a suitable $\infty$-category of Azumaya algebras:

Theorem 6.9.12 [103, Theorem 6.15]. *There is an equivalence of symmetric monoidal* $\infty$-*categories*

$$Az_A \to (Az_B)^{hG}$$

*for every* $G$-*Galois extension* $A \to B$ *with finite* $G$.

They also construct a map of $\infty$-groupoids $Az_R \to Br_R$ for any commutative ring spectrum $R$ and show that this map is essentially surjective, so that equality in $\pi_0(Br_R)$ corresponds precisely to Morita equivalence. They investigate the algebraic Brauer groups (i.e., the Morita classes of Azumaya algebras over the coefficients) [103, §7.1] of 2-periodic commutative ring spectra with vanishing odd homotopy groups, such as $KU$ or $E_n$, by relating them to the classical Brauer–Wall group of $\pi_0$ of the ring spectrum and they identify a non-trivial Morita class of a quaternion $KO$-algebra that becomes Morita-trivial over $KU$.

There is recent work by Hopkins and Lurie [128] who identify the $K(n)$-local Brauer group of a Lubin–Tate spectrum $E$ at all primes. For odd primes they obtain:

Theorem 6.9.13 [128, Theorem 1.0.11]. *The* $K(n)$-*local Brauer group of* $E$ *is the product of the Brauer–Wall group of the residue field* $\pi_0(E)/m$ *and a group* $Br'(E)$ *which in turn can be expressed as an inverse limit of abelian groups* $Br'_\ell$ *such that the kernel of* $Br'_\ell \to Br'_{\ell-1}$ *is non-canonically isomorphic to* $m^{\ell+2}/m^{\ell+3}$.

One ingredient is their construction of *atomic* $E$-*algebra spectra* [128, Definition 1.0.2] via a Thom spectrum construction relative to $E$ for polarizations of lattices [128, Definition 3.2.1] using the machinery from [6, 5]. Here, the starting point is a lattice $\Lambda$ of finite rank together with a *polarization map*

$$Q\colon K(\Lambda, 1) \to \mathrm{PIC}(E) \simeq \mathrm{Pic}(E) \times BGL_1(E).$$

# 7 An introduction to Bousfield localization

by Tyler Lawson

## 7.1 Introduction

Bousfield localization encodes a wide variety of constructions in homotopy theory, analogous to localization and completion in algebra. Our goal in this chapter is to give an overview of Bousfield localization, sketch how basic results in this area are proved, and illustrate some applications of these techniques. Near the end we will give more details about how localizations are constructed using the small object argument. The underlying methods apply in many contexts, and we provide examples that exhibit a variety of behaviors.

We will begin by discussing categorical localizations. Given a collection of maps in a category, the corresponding localization of that category is formed by making these maps invertible in a universal way; this technique is often applied to discard irrelevant information and focus on a particular type of phenomenon. In certain cases, localization can be carried out internally to the category itself: this happens when there is a sufficiently ample collection of objects that already see these maps as isomorphisms. This leads naturally to the study of reflective localizations.

Bousfield localization generalizes this by taking place in a category where there are *spaces* of functions, rather than sets, with uniqueness only being true up to contractible choice. Bousfield codified these properties, for spaces in [54] and for spectra in [55]. The definitions are straightforward, but proving that localizations exist takes work, some of it of a set-theoretic nature.

Our presentation is close in spirit to Bousfield's work, but the reader should go to the books of Farjoun [96] and Hirschhorn [124] for more advanced information on this material. We will focus, for the most part, on *left* Bousfield localization, since the techniques there are easier and this is where most of our applications lie. In [25] right Bousfield localization is discussed at greater length.

### Historical background

The story of localization techniques in algebraic topology probably begins with Serre classes of abelian groups [272]. After choosing a class $\mathcal{C}$ of abelian groups that is closed under subobjects, quotients, and extensions, Serre showed that one could effectively ignore groups in $\mathcal{C}$ when studying the homology and homotopy of a simply connected space $X$. In particular, he proved mod-$\mathcal{C}$ versions of the Hurewicz and Whitehead theorems, showed the equivalence between finite generation of homology and homotopy groups, determined the rational homotopy groups of spheres, and significantly reduced the technical overhead in computing the torsion in homotopy groups by allowing one to work with only one prime at a time. His techniques for computing rational homotopy groups only require rational homology groups; $p$-local homotopy groups only require $p$-local homology groups; $p$-completed homotopy groups only require mod-$p$ homology groups.

These techniques received a significant technical upgrade in the late 1960s and early 1970s, starting with the work of Quillen on rational homotopy theory [228] and work of Sullivan and Bousfield–Kan on localization and completion of spaces [287, 285, 58]. Rather than using Serre's algebraic techniques to break up the homotopy groups $\pi_*X$ and homology groups $H_*X$ into localizations and completions, their insight was that *space-level* versions of these constructions provided a more robust theory. For example, a simply connected space $X$ has an associated space $X_{\mathbb{Q}}$ whose homotopy groups and (positive-degree) homology groups are, themselves, rational homotopy and homology groups of $X$; similarly for Sullivan's $p$-localization $X_{(p)}$ and $p$-completion $X_p^{\wedge}$. Without this, each topological tool requires a proof that it is compatible with Serre's mod-$\mathcal{C}$-theory, such as Serre's mod-$\mathcal{C}$ Hurewicz and Whitehead theorems or mod-$\mathcal{C}$ cup products. Now these are simply consequences of the Hurewicz and Whitehead theorems applied to $X_{\mathbb{Q}}$, and any subsequent developments will automatically come along. Moreover, Sullivan pioneered arithmetic fracture techniques that allowed $X$ to be recovered from its rationalization $X_{\mathbb{Q}}$ and its $p$-adic completions $X_p^{\wedge}$ via a homotopy pullback diagram:

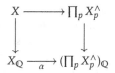

This allows us to reinterpret homotopy theory. We are no longer using rationalization and completion just to understand algebraic invariants of $X$: instead, knowledge of $X$ is equivalent to knowledge of its localizations, completions, and an "arithmetic attaching map" $\alpha$. This entirely changed both the way theorems are proved and the way that we think about the subject. Later, work of Morava, Ravenel, and others made extensive use of localization techniques [213, 230], which today gives an explicit decomposition of the stable homotopy category into layers determined by Quillen's relation to the structure theory of formal group laws [226].

Many of the initial definitions of localization and completion were constructive. One can build $X_{\mathbb{Q}}$ from $X$ by showing that one can replace the basic cells $S^n$ in a CW-decomposition with rationalized spheres $S_{\mathbb{Q}}^n$, or by showing that the Eilenberg–Mac Lane spaces $K(A, n)$ in a Postnikov decomposition can be replaced by rationalized versions $K(A \otimes \mathbb{Q}, n)$. One can instead use Bousfield and Kan's more functorial, but also more technical, construction as the homotopy limit of a cosimplicial space. Quillen's work gives more, in the form of a model structure whose weak equivalences are isomorphisms on rational homology groups. In his work, the map $X \to X_{\mathbb{Q}}$ is a fibrant replacement, and the essential uniqueness of fibrant replacements means that $X_{\mathbb{Q}}$ has a form of universality. It is this universal property that Bousfield localization makes into a definition.

## Notation

We will use $\mathcal{S}$ to denote a convenient category of spaces (one can use simplicial sets, but with appropriate modifications throughout) with internal function objects.

Throughout this article we will often be working in categories enriched in spaces: for any $X$ and $Y$ in $\mathcal{C}$ we will write $\mathrm{Map}_{\mathcal{C}}(X, Y)$ for the mapping space, or just $\mathrm{Map}(X, Y)$ if the ambient category is understood. Letting $[X, Y] = \pi_0 \mathrm{Map}_{\mathcal{C}}(X, Y)$, we obtain an ordinary category called the *homotopy category* h$\mathcal{C}$. Two objects in $\mathcal{C}$ are *homotopy equivalent* if and only if they become isomorphic in h$\mathcal{C}$.

For us, homotopy limits and colimits in the category of spaces are given by the descriptions of Vogt or Bousfield–Kan [295, 58]. A homotopy limit or homotopy colimit in $\mathcal{C}$ is characterized by having a natural weak equivalence of spaces:

$$\mathrm{Map}_{\mathcal{C}}(X, \mathrm{holim}_J Y_j) \simeq \mathrm{holim}_J \mathrm{Map}_{\mathcal{C}}(X, Y_j),$$

$$\mathrm{Map}_{\mathcal{C}}(\mathrm{hocolim}_I X_i, Y) \simeq \mathrm{holim}_I \mathrm{Map}_{\mathcal{C}}(X_i, Y).$$

Since homotopy limit constructions on spaces preserve objectwise weak equivalences of diagrams, homotopy limits and colimits also preserve objectwise homotopy equivalences in $\mathcal{C}$.

Some set theory is unavoidable, but we will not spend a great deal of time with it. For us, a *collection* or *family* may be a proper class, rather than a set. Categories will be what are sometimes called *locally small* categories: the collection of objects may be large, but there is a set of maps between any pair of objects.

## Acknowledgements

The author would like to thank Clark Barwick and Thomas Nikolaus for discussions related to localizations of spaces, as well as David White, Carles Casacuberta, and Javier Gutiérrez for comments and several corrections on an earlier version.

The author was partially supported by NSF grant 1610408 and a grant from the Simons Foundation. The author would like to thank the Isaac Newton Institute for Mathematical Sciences for support and hospitality during the programme HHH when work on this text was undertaken. This work was supported by EPSRC grants EP/K032208/1 and EP/R014604/1.

## 7.2     Motivation from categorical localization

In general, we recall that for an ordinary category $\mathcal{A}$ and a class $\mathcal{W}$ of the maps called *weak equivalences* (or simply *equivalences*), we can attempt to construct a categorical localization $\mathcal{A} \to \mathcal{A}[\mathcal{W}^{-1}]$. See Chapter 2 of this volume for more details. This localization is universal among functors $\mathcal{A} \to \mathcal{D}$ that send the maps in $\mathcal{W}$ to isomorphisms. The category $\mathcal{A}[\mathcal{W}^{-1}]$ is unique up to isomorphism if it exists.[1]

Example 7.2.1.  We will begin by remembering the case of the category $\mathcal{S}$ of spaces, with $\mathcal{W}$ the class of weak homotopy equivalences. The projection $p \colon X \times [0,1] \to X$ is always a weak equivalence with homotopy inverses $i_t$ given by $i_t(x) = (x,t)$. In the localization, we find that homotopic maps are equal: for a homotopy $H$ from $f$ to $g$, we have $f = Hi_0 = Hp^{-1} = Hi_1 = g$. Therefore, localization factors through the homotopy category $h\mathcal{S}$.

However, within the category of spaces we have a collection with special properties: the subcategory $\mathcal{S}^{CW}$ of CW-complexes. For any CW-complex $K$, weak equivalences $X \to Y$ induce bijections $[K,X] \to [K,Y]$ — this can be proved, for example, inductively on the cells of $K$ — and any space $X$ has a CW-complex $K$ with a weak homotopy equivalence $K \to X$. These two properties show, respectively, that the composite

$$h\mathcal{S}^{CW} \to h\mathcal{S} \to \mathcal{S}[\mathcal{W}^{-1}]$$

is fully faithful and essentially surjective. Within the homotopy category $h\mathcal{S}$ we have found a large enough library of special objects, and localization can be done by forcibly moving objects into this subcategory.[2]

Example 7.2.2.  A similar example occurs in the category $\mathcal{K}_R$ of nonnegatively graded cochain complexes of modules over a commutative ring $R$, with $\mathcal{W}$ the class of quasi-isomorphisms. Within $\mathcal{K}_R$ there is a subcategory $\mathcal{K}_R^{Inj}$ of complexes of injective modules. Fundamental results of homological algebra show that for a quasi-isomorphism $A \to B$ and a complex $Q$ of injectives, there exists a bijection $[B,Q] \to [A,Q]$ of chain homotopy classes of maps, and that any complex $A$ has a quasi-isomorphism $A \to Q$ to a complex of injectives. This similarly shows that the composite functor

$$h\mathcal{K}_R^{Inj} \to h\mathcal{K}_R \to \mathcal{K}_R[\mathcal{W}^{-1}]$$

is an equivalence of categories.

---

[1]  For the record, this category also satisfies a 2-categorical universal property: for any $\mathcal{D}$, the map of functor categories

$$\mathrm{Fun}(\mathcal{A}[\mathcal{W}^{-1}], \mathcal{D}) \to \mathrm{Fun}(\mathcal{A}, \mathcal{D})$$

is fully faithful, and the image consists of those functors sending $\mathcal{W}$ to isomorphisms. If we replace "image" with "essential image" in this description, we recover a universal property characterizing $\mathcal{A} \to \mathcal{A}[\mathcal{W}^{-1}]$ up to equivalence of categories rather than up to isomorphism.

[2]  Technically speaking, we often use a result like this to actually show that $\mathcal{S}[\mathcal{W}^{-1}]$ exists.

These examples are at the foundation of Quillen's theory of model categories, and we will return to examples like them when we discuss localization of model categories.

## 7.3 Local objects in categories

In this section we will fix an ordinary category $\mathcal{A}$.

**Definition 7.3.1.** Let $S$ be a class of morphisms in $\mathcal{A}$. An object $Y \in \mathcal{A}$ is *S-local* if, for all $f : A \to B$ in $S$, the map

$$\mathrm{Hom}_{\mathcal{A}}(B, Y) \xrightarrow{f^*} \mathrm{Hom}_{\mathcal{A}}(A, Y)$$

is a bijection. We write $L^S(\mathcal{A})$ for the full subcategory of $S$-local objects.

If $S = \{f : A \to B\}$ consists of just one map, we simply refer to this property as being *f-local* and write $L^f(\mathcal{A})$ for the category of $f$-local objects.

*Remark 7.3.2.* If $S = \{f_\alpha : A_\alpha \to B_\alpha\}$ is a set and $\mathcal{A}$ has coproducts indexed by $S$, then by defining $f = \coprod_\alpha f_\alpha : \coprod A_\alpha \to \coprod B_\alpha$ we find that $S$-local objects are equivalent to $f$-local objects — *so long as* we don't have to worry about cases where $\mathrm{Hom}(A_\alpha, Y)$ could be the empty set. (For example, there is no problem if $\mathcal{A}$ is pointed.)

A special case of localization is when our maps in $S$ are maps to a terminal object.

**Definition 7.3.3.** Suppose $S$ is a class of maps $\{W_\alpha \to *\}$, where $*$ is a terminal object. In this case, we refer to such a localization as a *nullification* of the objects $W_\alpha$.

*Remark 7.3.4.* Nullification often takes place when $\mathcal{A}$ is pointed. If $S$ is a set, $\mathcal{A}$ is pointed, and $\mathcal{A}$ has coproducts, then any coproduct of copies of $*$ is again $*$ and we can again replace nullification of a set of objects with nullification of an individual object.

**Definition 7.3.5.** A map $A \to B$ in $\mathcal{A}$ is an *S-equivalence* if, for all $S$-local objects $Y$, the map

$$\mathrm{Hom}_{\mathcal{A}}(B, Y) \to \mathrm{Hom}_{\mathcal{A}}(A, Y)$$

is a bijection.

The class of $S$-equivalences contains $S$ by definition.

**Definition 7.3.6.** A map $X \to Y$ is an *S-localization* if it is an $S$-equivalence and $Y$ is $S$-local, and under these conditions we say that $X$ *has an S-localization*. If all objects in $\mathcal{A}$ have $S$-localizations, we say that $\mathcal{A}$ *has S-localizations*.

**Proposition 7.3.7.** *Any two S-localizations $f_1 : X \to Y_1$ and $f_2 : X \to Y_2$ are isomorphic under $X$ in $\mathcal{A}$.*

*Proof.* Because $Y_i$ are $S$-local, $\mathrm{Hom}(B, Y_i) \to \mathrm{Hom}(A, Y_i)$ is always an isomorphism for any $S$-equivalence $A \to B$. Applying this to the $S$-equivalences $X \to Y_j$, we get isomorphisms $\mathrm{Hom}(Y_j, Y_i) \to \mathrm{Hom}(X, Y_i)$ in $\mathcal{A}$: any map $X \to Y_i$ has a unique extension to a map $Y_j \to Y_i$. Existence allows us to find maps $Y_1 \to Y_2$ and $Y_2 \to Y_1$ under $X$, and uniqueness allows us to conclude that these two maps are inverse to each other in $\mathcal{A}$.

More concisely, $Y_1$ and $Y_2$ are both initial objects in the comma category of $S$-local objects under $X$ in $\mathcal{A}$, and this universal property forces them to be isomorphic. □

As a result, it is reasonable to call such an object *the* $S$-localization of $X$ and write it as $L^S X$ (or simply $LX$ if $S$ is understood). More generally than this, if $X \to LX$ and $X' \to LX'$ are $S$-localization maps, any map $X \to X'$ in $\mathcal{A}$ extends uniquely to a commutative square. This is encoded by the following result.

**Proposition 7.3.8.** *Let $Loc^S(\mathcal{A})$ be the category of localization morphisms, whose objects are $S$-localization maps $X \to LX$ in $\mathcal{A}$ and whose morphisms are commuting squares. Then the forgetful functor*

$$Loc^S(\mathcal{A}) \to \mathcal{A},$$

*sending $(X \to LX)$ to $X$, is fully faithful. The image consists of those objects $X$ that have $S$-localizations.*

**Proposition 7.3.9.** *The collection of $S$-local objects is closed under limits, and the collection of $S$-equivalences is closed under colimits.*

*Proof.* If $f : A \to B$ is in $S$ and $\{Y_j\}$ is a diagram of $S$-local objects, then

$$\mathrm{Hom}(B, Y_j) \to \mathrm{Hom}(A, Y_j)$$

is a diagram of isomorphisms, and taking limits we find that we have an isomorphism

$$\mathrm{Hom}(B, \lim_J Y_j) \to \mathrm{Hom}(A, \lim_J Y_j).$$

Since $A \to B$ was an arbitrary map in $S$, this shows that $\lim_J Y_j$ is $S$-local.

Similarly, if $\{A_i \to B_i\}$ is a diagram of $S$-equivalences and $Y$ is $S$-local, then

$$\mathrm{Hom}(B_i, Y) \to \mathrm{Hom}(A_i, Y)$$

is a diagram of isomorphisms, and taking limits we find that

$$\mathrm{Hom}(\mathrm{colim}_I B_i, Y) \to \mathrm{Hom}(\mathrm{colim}_I A_i, Y)$$

is also an isomorphism. Since $Y$ was an arbitrary local object, this shows that the map $\mathrm{colim}_I A_i \to \mathrm{colim}_I B_i$ is an $S$-equivalence. □

**Example 7.3.10.** Consider the map $f : \mathbb{N} \to \mathbb{Z}$ in the category of monoids. A monoid $M$ is $f$-local if and only if any monoid homomorphism $\mathbb{N} \to M$ automatically extends to a homomorphism $\mathbb{Z} \to M$, which is the same as asking that every element in $M$ has an inverse. Therefore, $f$-local monoids are precisely *groups*. The natural transformation $M \to M^{gp}$, from a monoid to its group completion, is an $f$-localization.

Example 7.3.11. Consider the map $f \colon F_2 \to \mathbb{Z}^2$, from a free group on two generators $x$ and $y$ to its abelianization. A group $G$ is $f$-local if and only if *every* homomorphism $F_2 \to G$, equivalent to choosing a pair of elements $x$ and $y$ of $G$, can be factored through $\mathbb{Z}^2$, which happens exactly when the commutator $[x, y]$ is sent to the trivial element. Therefore, $f$-local groups are precisely *abelian* groups. The natural transformation $G \to G_{ab}$, from a group to its abelianization, is an $f$-localization.

These two localizations are left adjoints to the inclusion of a subcategory, and this phenomenon is completely general.

**Proposition 7.3.12.** *Let $S$ be a class of morphisms in $\mathcal{A}$, and suppose that $\mathcal{A}$ has $S$-localizations. Then the inclusion $L^S \mathcal{A} \to \mathcal{A}$ is part of an adjoint pair*

$$\mathcal{A} \overset{L}{\underset{}{\rightleftarrows}} L^S \mathcal{A}.$$

*As a result, $L$ is a reflective localization onto the subcategory $L^S \mathcal{A}$.*

*Proof.* In this situation, the functor $Loc^S(\mathcal{A}) \to \mathcal{A}$ is fully faithful and surjective on objects. Therefore, it is an equivalence of categories and we can choose an inverse,[3] functorially sending $X$ to a pair $(X \to LX)$ in $Loc^S(\mathcal{A})$. The composite functor sending $X$ to $LX$ is the desired left adjoint. □

*Remark 7.3.13.* Embedding the category $\mathcal{A}$ as a full subcategory of a larger category can change localization drastically. Consider a set $S$ of maps in $\mathcal{A} \subset \mathcal{B}$. The $S$-local objects of $\mathcal{A}$ are simply the $S$-local objects of $\mathcal{B}$ that happen to be in $\mathcal{A}$, but because there may be more local objects in $\mathcal{B}$ there may be fewer $S$-equivalences in $\mathcal{B}$ than in $\mathcal{A}$. Localization in $\mathcal{B}$ may not preserve objects of $\mathcal{A}$; a localization map in $\mathcal{A}$ might not be an equivalence in $\mathcal{B}$; there might, in general, be no comparison map between the two localizations.

For example, let $S$ be the set of multiplication-by-$p$ maps $\mathbb{Z} \to \mathbb{Z}$ (as $p$ ranges over primes) in the category of finitely generated abelian groups, considered as a full subcategory of all abelian groups. An abelian group is $S$-local if and only if it is a rational vector space, and the only finitely generated group of this form is trivial. A map $A \to B$ of finitely generated abelian groups is an $S$-equivalence in the larger category of all abelian groups if and only if it induces an isomorphism $A \otimes \mathbb{Q} \to B \otimes \mathbb{Q}$, whereas it is *always* an equivalence within the smaller category of finitely generated abelian groups because there is only 0 as a local object to test against. Within all abelian groups, $S$-localization is rationalization, whereas within finitely generated abelian groups, $S$-localization takes all groups to zero.

---

[3] If the category $\mathcal{A}$ is large then we need to be a little bit more honest here, and worry about whether a fully faithful and essentially surjective map between large categories has an inverse equivalence. This depends on our model for set theory: it is asking for us to make a distinguished choice of objects for our inverse functor, which may require an axiom of choice for proper classes. It is an awkward situation, because choosing these inverses isn't categorically interesting unless we can't do it.

## 7.4        Localization using mapping spaces

We now consider the case where $\mathcal{C}$ is a category enriched in spaces. The previous definitions and results apply perfectly well to the homotopy category $h\mathcal{C}$. The following illustrates that the homotopy category may be an inappropriate place to carry out such localizations.

Example 7.4.1. We start with the homotopy category of spaces $h\mathcal{S}$, and fix $n \geq 0$. Suppose that we want to invert the inclusion $S^n \to D^{n+1}$. We fairly readily find that any space $X$ has a map $X \to X'$ such that $[D^{n+1}, X'] \to [S^n, X']$ is an isomorphism: construct $X'$ by attaching $(n+1)$-dimensional cells to $X$ until the $n$-th homotopy group $\pi_n(X', x) = 0$ is trivial at any basepoint.

However, this construction lacks *universality*. If $Y$ is any other space whose $n$-th homotopy groups are trivial, then any map $X \to Y$ can be extended to a map $X' \to Y$ because the attaching maps for the cells of $X'$ are trivial, but this extension is *not unique* up to homotopy: any two extensions $D^{n+1} \to X' \to Y$ of a cell $S^n \to X \to Y$ glue together to an obstruction class in $[S^{n+1}, Y]$. As a result, if we construct two spaces $X'$ and $X''$ as attempted localizations of $X$, we can find maps $X' \to X''$ and $X'' \to X'$ but cannot establish that they are mutually inverse in the homotopy category.

In short, in order for $Y$ to have *uniqueness* for filling maps from $n$-spheres, we have to have *existence* for filling maps from $(n+1)$-spheres. Thus, to make this localization work canonically we would need to enlarge our class $S$ to contain $S^{n+1} \to D^{n+2}$. The same argument then repeats, showing that a canonical localization for $S$ requires that $S$ also contain $S^m \to D^{m+1}$ for $m \geq n$.

The example in the previous section leads to the following principle. In our definitions, we must replace isomorphism on the path components of mapping spaces with homotopy equivalence.

Definition 7.4.2. Let $S$ be a class of morphisms in the category $\mathcal{C}$. An object $Y \in \mathcal{C}$ is *S-local* if, for all $f : A \to B$ in $S$, the map

$$\operatorname{Map}_{\mathcal{C}}(B, Y) \xrightarrow{f^*} \operatorname{Map}_{\mathcal{C}}(A, Y)$$

is a weak equivalence.[4] We write $L^S(\mathcal{C})$ for the full subcategory of $S$-local objects.

If $S = \{f : A \to B\}$ consists of just one map, we simply refer to this property as being *f-local* and write $L^f(\mathcal{C})$ for the category of $f$-local objects.

Definition 7.4.3. A map $A \to B$ in $\mathcal{C}$ is an *S-equivalence* if, for all $S$-local objects $Y$, the map

$$\operatorname{Map}_{\mathcal{C}}(B, Y) \to \operatorname{Map}_{\mathcal{C}}(A, Y)$$

is a weak equivalence.

---

[4] This property of the map $\operatorname{Map}_{\mathcal{C}}(B, Y) \to \operatorname{Map}_{\mathcal{C}}(A, Y)$ only depends on the image of $f : A \to B$ in the homotopy category $h\mathcal{C}$, and so we may simply view $S$ as a collection of representatives for a class of maps $\bar{S}$ in $h\mathcal{C}$.

Definition 7.4.4.  A map $X \to Y$ is an *S-localization* if it is an $S$-equivalence and $Y$ is $S$-local, and under these conditions we say that $X$ *has an S-localization*. If all objects in $C$ have $S$-localizations, we say that $C$ *has S-localizations*.

By applying $\pi_0$ to mapping spaces, we find that some of this passes to the homotopy category.

Proposition 7.4.5.  *Let $\bar{S}$ be the image of $S$ in the homotopy category $hC$. If $Y$ is $S$-local in $C$, then its image in the homotopy category $hC$ is $\bar{S}$-local.*

*Remark 7.4.6.*  An $S$-equivalence in $C$ does not necessarily becomes an $\bar{S}$-equivalence in $hC$ because there is potentially a larger supply of $\bar{S}$-local objects.

Proposition 7.4.7.  *Any two S-localizations $f_1: X \to Y_1$ and $f_2: X \to Y_2$ become isomorphic under $X$ in the homotopy category $hC$.*

*Proof.*  This proceeds exactly as in the proof of Proposition 7.3.7. Applying $\mathrm{Map}_C(-, Y_i)$ to the $S$-equivalence $X \to Y_j$, we find that the maps $X \to Y_i$ extend to maps $Y_j \to Y_i$ which are unique up to homotopy. By first taking $i \neq j$ we construct maps between the $Y_i$ whose restrictions to $X$ are homotopic to the originals, and taking $i = j$ shows that the double composites are homotopic under $X$.  $\square$

*Remark 7.4.8.*  At this point it would be very useful to show that, if they exist, localizations can be made functorial in the spirit of Proposition 7.3.8. There is typically no easy way to produce a functorial localization because many choices are made up to homotopy equivalence, and this leads to coherence issues: for example, if we have a diagram

$$
\begin{array}{ccc}
X & \longrightarrow & X' \\
\downarrow & & \downarrow \\
LX & \dashrightarrow & LX'
\end{array}
$$

where the vertical maps are $S$-localization, then we can construct at best the dotted map together with a *homotopy* between the two double composites. Larger diagrams do get more extensive families of homotopies, but these take work to describe. This is a *rectification problem* and in general it is not solvable without asking for more structure on $C$. The small object argument, which we will discuss in §7.6, can often be done carefully enough to give some form of functorial construction of the localization.

Proposition 7.4.9.  *The following properties hold for a class $S$ of morphisms in $C$.*

1. *The collection of $S$-local objects is closed under equivalence in the homotopy category.*
2. *The collection of $S$-equivalences is closed under equivalence in the homotopy category.*
3. *The collection of $S$-local objects is closed under homotopy limits.*
4. *The collection of $S$-equivalences is closed under homotopy colimits.*
5. *The homotopy pushout of an $S$-equivalence is an $S$-equivalence.*
6. *The $S$-equivalences satisfy the two-out-of-three axiom: given maps $A \xrightarrow{f} B \xrightarrow{g} C$, if any two of $f$, $g$, and $gf$ are $S$-equivalences then so is the third.*

*Proof.* If $X \to Y$ becomes an isomorphism in the homotopy category, then one can choose an inverse map and homotopies between the double composites. Composing with these makes $\mathrm{Map}_\mathcal{C}(-, X) \to \mathrm{Map}_\mathcal{C}(-, Y)$ a homotopy equivalence of functors on $\mathcal{C}$, and so $X$ is $S$-local if and only if $Y$ is.

Similarly, if two maps $f \colon A \to B$ and $f' \colon A' \to B'$ become isomorphic in the homotopy category, there exist homotopy equivalences $A' \to A$ and $B \to B'$ such that the composite $A' \to A \to B \to B'$ is homotopic to $f'$, and applying $\mathrm{Map}_\mathcal{C}(-, Y)$ we obtain the desired result.

If $f \colon A \to B$ is in $S$ and $\{Y_j\}$ is a diagram of $S$-local objects, then

$$\mathrm{Map}_\mathcal{C}(B, Y_j) \to \mathrm{Map}_\mathcal{C}(A, Y_j)$$

is a diagram of weak equivalences of spaces, and taking homotopy limits we find that we have an equivalence

$$\mathrm{Map}_\mathcal{C}(B, \mathrm{holim}_J Y_j) \to \mathrm{Map}_\mathcal{C}(A, \mathrm{holim}_J Y_j).$$

Since $A \to B$ was an arbitrary map in $S$, this shows that $\mathrm{holim}_J Y_j$ is $S$-local.

Similarly, if $\{A_i \to B_i\}$ is a diagram of $S$-equivalences and $Y$ is $S$-local, then

$$\mathrm{Map}_\mathcal{C}(B_i, Y) \to \mathrm{Map}_\mathcal{C}(A_i, Y)$$

is a diagram of weak equivalences of spaces, and so

$$\mathrm{Map}_\mathcal{C}(\mathrm{hocolim}_I B_i, Y) \to \mathrm{Map}_\mathcal{C}(\mathrm{hocolim}_I A_i, Y)$$

is also a weak equivalence. Since $Y$ was an arbitrary $S$-local object, this shows that the map $\mathrm{hocolim}_I A_i \to \mathrm{hocolim}_I B_i$ is an $S$-equivalence.

Suppose that we have a homotopy pushout diagram

$$
\begin{array}{ccc}
A & \xrightarrow{\ f\ } & B \\
\downarrow & & \downarrow \\
A' & \xrightarrow[\ f'\ ]{} & B'
\end{array}
$$

where $f \colon A \to B$ is an $S$-equivalence. Given any $S$-local object $Y$, we get a homotopy pullback diagram

$$
\begin{array}{ccc}
\mathrm{Map}_\mathcal{C}(A, Y) & \longleftarrow & \mathrm{Map}_\mathcal{C}(B, Y) \\
\uparrow & & \uparrow \\
\mathrm{Map}_\mathcal{C}(A', Y) & \longleftarrow & \mathrm{Map}_\mathcal{C}(B', Y)
\end{array}
$$

The top arrow is an equivalence by the assumption that $f$ is an $S$-equivalence, and hence the bottom arrow is an equivalence. Since $Y$ was an arbitrary $S$-local object, we find that $f'$ is an $S$-equivalence.

The 2-out-of-3 property is obtained by first applying $\mathrm{Map}_\mathcal{C}(-, Y)$ to the diagram $A \to B \to C$ and then using the 2-out-of-3 axiom for weak equivalences. $\qquad \square$

If we expand a class $S$ to a larger class $T$ of equivalences, our work so far gives us an automatic relation between $S$-localization and $T$-localization.

Proposition 7.4.10. *Suppose that S and T are classes of morphisms such that every map in S is a T-equivalence. Then the following properties hold.*

1. *Every T-local object is also S-local.*
2. *Every S-equivalence is also a T-equivalence.*
3. *Suppose $X \to L_S X$ is an S-localization and $X \to L_T X$ is a T-localization. Then there exists an essentially unique factorization $X \to L_S X \to L_T X$, and the map $L_S X \to L_T X$ is a T-localization.*

*Proof.* 1. By assumption, every map $f : A \to B$ in $S$ is a $T$-equivalence, and so for any $T$-local object $Y$ we get an equivalence $\mathrm{Map}_{\mathcal{C}}(B, Y) \to \mathrm{Map}_{\mathcal{C}}(A, Y)$. Thus by definition $Y$ is $S$-local.

2. If $f : A \to B$ is an $S$-equivalence, and $Y$ is any $T$-local object, then by the previous point $Y$ is also $S$-local, and so we get an equivalence $\mathrm{Map}_{\mathcal{C}}(B, Y) \to \mathrm{Map}_{\mathcal{C}}(A, Y)$. Since $Y$ was an arbitrary $T$-local object, $f$ is therefore a $T$-equivalence.

3. Since $X \to L_S X$ is an $S$-equivalence, part 2 shows that it is a $T$-equivalence and so we have an equivalence

$$\mathrm{Map}_{\mathcal{C}}(L_S X, L_T X) \to \mathrm{Map}_{\mathcal{C}}(X, L_T X).$$

As a result, the chosen map $X \to L_T X$ has a contractible space of homotopy commuting factorizations $X \to L_S X \to L_T X$. As the maps $X \to L_S X$ and $X \to L_T X$ are both $T$-equivalences, the 2-out-of-3 property implies that $L_S X \to L_T X$ is also a $T$-equivalence whose target is $T$-local. By definition, this makes $L_T X$ into a $T$-localization of $L_S X$. □

## 7.5     Lifting criteria for localizations

In this section we will observe that, if $\mathcal{C}$ has homotopy pushouts, we can characterize local objects in terms of a lifting criterion. To do so, we will need to establish a few preliminaries. Fix a collection $S$ of maps in $\mathcal{C}$.

Proposition 7.5.1. *Suppose that $f : A \to B$ is an S-equivalence, and that $\mathcal{C}$ has homotopy pushouts. Then the map*

$$\mathrm{hocolim}(B \leftarrow A \to B) \to B$$

*is an S-equivalence.*

*Proof.* The map in question is equivalent to the map of homotopy pushouts induced by the diagram

$$
\begin{array}{ccc}
B \xleftarrow{\;f\;} A \xrightarrow{\;f\;} B \\
\Big\| \quad\quad f\Big\downarrow \quad\quad \Big\| \\
B \longleftarrow B \longrightarrow B
\end{array}
$$

However, the vertical maps are $S$-equivalences, and so by Proposition 7.4.9 the map $\mathrm{hocolim}(B \leftarrow A \to B) \to B$ is an $S$-equivalence. □

The lifting criterion we are about to describe rests on the following useful characterization of connectivity of a map.

**Lemma 7.5.2.** *Suppose that* $f: X \to Y$ *is a map of spaces and* $N \geq 0$. *Then* $f$ *is* $N$-*connected if and only if the following two criteria are satisfied:*

1. *The map* $\pi_0(X) \to \pi_0(Y)$ *is surjective.*
2. *The diagonal map* $X \to \text{holim}(X \to Y \leftarrow X)$ *is* $(N-1)$-*connected.*

*Proof.* The map $f$ is $N$-connected if and only if it is surjective on $\pi_0$ and, for all basepoints $x \in X$, the homotopy fiber $Ff$ over $f(x)$ is $(N-1)$-connective.

However, $Ff$ is equivalent to the homotopy fiber of $\text{holim}(X \to Y \leftarrow X) \to X$ over $x$, and so this second condition is equivalent to $\text{holim}(X \to Y \leftarrow X) \to X$ being $N$-connected. The composite $X \to \text{holim}(X \to Y \leftarrow X) \to X$ is the identity, and the map $\text{holim}(X \to Y \leftarrow X) \to X$ is $N$-connected if and only if the map $X \to \text{holim}(X \to Y \leftarrow X)$ is $(N-1)$-connected.   □

**Corollary 7.5.3.** *Suppose that* $C$ *has homotopy pushouts and that we have a map* $f_0: A_0 \to B$ *in* $C$. *Inductively define the* $n$-*fold double mapping cylinder* $f_n$ *as the map*

$$A_n = \text{hocolim}(B \leftarrow A_{n-1} \to B) \to B.$$

*Then an object* $Y$ *is* $f_0$-*local if and only if the maps*

$$\text{Hom}_{hC}(B, Y) \to \text{Hom}_{hC}(A_n, Y)$$

*are surjective; equivalently, for any map* $A_n \to Y$ *there is a map* $B \to Y$ *such that the diagram*

*is homotopy commutative.*

*Proof.* The definition of $A_n$ gives an identification

$$\text{Map}_C(A_n, Y) \simeq \text{holim}[\text{Map}_C(B, Y) \to \text{Map}_C(A_{n-1}, Y) \leftarrow \text{Map}_C(B, Y)].$$

Inductive application of Lemma 7.5.2 shows that the map

$$\text{Map}_C(B, Y) \to \text{Map}_C(A_0, Y)$$

is $N$-connected if and only if the maps

$$\text{Hom}_{hC}(B, Y) \to \text{Hom}_{hC}(A_n, Y)$$

are surjective for $0 \leq n \leq N$. Letting $N$ grow arbitrarily large, we find that $Y$ is $f_0$-local if and only of the maps

$$\text{Hom}_{hC}(B, Y) \to \text{Hom}_{hC}(A_n, Y)$$

are surjective for all $n \geq 0$.   □

Example 7.5.4. Suppose that $C$ has homotopy pushouts and that $f: W \to *$ is a map to a homotopy terminal object of $C$. Then the iterated double mapping cylinders are the maps $\Sigma^t W \to *$, and an object of $C$ is $f$-local if and only if every map $\Sigma^t W \to Y$ factors, up to homotopy, through $*$.

Example 7.5.5. In the category of spaces $S$, the iterated double mapping cylinders $f_n$ of a cofibration $f_0: A \to B$ have a more familiar description as the *pushout-product* maps

$$f_n: (S^{n-1} \times B) \underset{S^{n-1} \times A}{\cup} (D^n \times A) \to D^n \times B \to B.$$

## 7.6    The small object argument

We now sketch how, when we have some form of colimits in our category, Bousfield localizations can often be constructed using the small object argument.

From the previous section we know that we can replace the mapping space criterion for local objects with a lifting criterion when $C$ has homotopy colimits, as follows. Given a map $f_0: A_0 \to B$, we construct iterated double mapping cylinders $f_n: A_n \to B$, and we find that an object is $Y$ is $f_0$-local if and only if we can *solve extension problems*: every map $g: A_n \to Y$ can be extended to a map $\tilde{g}: B \to Y$ up to homotopy. More generally we can enlarge a collection of maps $S$ to a collection $T$ closed under double mapping cylinders, and ask whether $Y$ satisfies an extension property with respect to $T$.

This leads to an inductive method.

1. Start with $Y_0 = Y$.

2. Given $Y_\alpha$, either $Y_\alpha$ is local (in which case we are done) or there exists some set of maps $A_i \to B_i$ in $T$ and maps $g_i: A_i \to Y_\alpha$ which do not extend to $B_i$ up to homotopy. Form the homotopy pushout of the diagram

$$\coprod_i B_i \leftarrow \coprod_i A_i \to Y_\alpha$$

and call it $Y_{\alpha+1}$. The map $Y_\alpha \to Y_{\alpha+1}$ is an $S$-equivalence because it is a homotopy pushout along an $S$-equivalence, and all the extension problems that $Y_\alpha$ had now have solutions in $Y_{\alpha+1}$.

3. Once we have constructed $Y_0, Y_1, Y_2, \ldots$, define $Y_\omega = \operatorname{hocolim} Y_n$. More generally, once we have constructed $Y_\alpha$ for all ordinals $\alpha$ less than some limit ordinal $\beta$, we define $Y_\beta = \operatorname{hocolim} Y_\alpha$. The map $Y \to Y_\beta$ is a homotopy colimit of $S$-equivalences and hence an $S$-equivalence.

The critical thing that we need is that *this procedure can be stopped at some point*, and for this we typically need to know that there will be some ordinal $\beta$ which is so big that any map $A_i \to Y_\beta$ automatically factors, up to homotopy, through some object $Y_\alpha$ with $\alpha < \beta$. This is a categorical *compactness* property of the objects $A_i$, and this argument is called the *small object argument*. If we work on the point-set level this can be addressed using Smith's theory of combinatorial model categories; if we work on the homotopical level this can be addressed using Lurie's theory of presentable $\infty$-categories. We will discuss these approaches in §7.10 and §7.11.

Another important aspect of the small object argument is that it can prove additional properties about localization maps. If $S$ is a collection of maps all satisfying some property $P$ of maps in the homotopy category, and property $P$ is preserved under homotopy pushouts and transfinite homotopy colimits, then this process constructs a localization $Y \to LY$ that also has property $P$. Since localizations are essentially unique, any localization automatically has property $P$ as well.

*Remark 7.6.1.* If our category $\mathcal{C}$ does not have enough colimits, the small object argument may not apply. However, Bousfield localizations may still exist even if this particular construction cannot be applied.

## 7.7    Unstable settings

The classical examples of Bousfield localization are localizations of spaces. It is worthwhile first relating the localization condition to based mapping spaces.

*Proposition 7.7.1.* *Suppose $f\colon A \to B$ is a map of well-pointed spaces with basepoint. Then a space $Y$ is $f$-local in the category of* unbased *spaces if and only if, for all basepoints $y \in Y$, the restriction*
$$f^*\colon \operatorname{Map}_*(B, Y) \to \operatorname{Map}_*(A, Y)$$
*of based mapping spaces is a weak equivalence.*

*Proof.* Evaluation at the basepoint gives a map of fibration sequences
$$
\begin{array}{ccccc}
\operatorname{Map}_*(B,Y) & \longrightarrow & \operatorname{Map}(B,Y) & \longrightarrow & Y \\
\downarrow & & \downarrow & & \| \\
\operatorname{Map}_*(A,Y) & \longrightarrow & \operatorname{Map}(A,Y) & \longrightarrow & Y.
\end{array}
$$
The center vertical map is an isomorphism on $\pi_*$ at any basepoint if and only if the left-hand map is.    $\square$

*Remark 7.7.2.* As $S$-equivalences are preserved under homotopy pushouts and the 2-out-of-3 axiom, we find that any space $Y$ local with respect to $f\colon A \to B$ is also local with respect to the map $B/A \to *$ from the homotopy cofiber to a point, and thus that every path component of $Y$ has a contractible space of based maps $B/A \to Y$. However, we will see shortly that the converse does not hold in general.

*Example 7.7.3.* Let $S$ be the category of spaces, and take $f$ to be the map $S^n \to *$. Then a space $X$ is $f$-local if and only if, for any basepoint $x \in X$, the iterated loop space $\Omega^n X$ at $x$ is weakly contractible. Equivalently, for $n \geq 1$ the space $X$ is $f$-local if and only if it is $(n-1)$-*truncated*: $\pi_k(X, x)$ is trivial for all $k \geq n$ and all $x \in X$.

A map $A \to B$ of CW-complexes, by obstruction theory, is an $f$-equivalence if and only if it is $(n-1)$-connected. Therefore, for $n > 0$ a map $A \to B$ of CW-complexes is an $f$-localization if and only if $\pi_k(A) \to \pi_k(B)$ is an isomorphism for $0 \leq k < n$ and all basepoints, but $\pi_k B$ vanishes for all $k \geq n$ and all basepoints.[5] This characterizes a stage $P_{n-1}(X)$ in the Postnikov tower of $X$.

**Example 7.7.4.** Let $f$ be the inclusion $S^n \vee S^m \to S^n \times S^m$ of spaces. The Cartesian product is formed by attaching an $(n+m)$-cell to $S^n \vee S^m$ along an attaching map given by a Whitehead product $[\iota_n, \iota_m] \in \pi_{n+m-1}(S^n \vee S^m)$. Any map $S^n \vee S^m \to X$, classifying a pair of elements $\alpha \in \pi_n(X)$ and $\beta \in \pi_m(X)$ at some basepoint $x$, sends this attaching map to $[\alpha, \beta]$. The fiber of $\mathrm{Map}(S^n \times S^m, X) \to \mathrm{Map}(S^n \vee S^m, X)$ over the corresponding point is either empty (if $[\alpha, \beta]$ is nontrivial) or equivalent to the iterated loop space $\Omega^{n+m} X$ at $x$ (if $[\alpha, \beta]$ is trivial). A space $X$ is therefore local with respect to $f$ if and only if, at any basepoint, the homotopy groups $\pi_k(X)$ are zero for all $k \geq n+m$ and the Whitehead products

$$\pi_n(X, x) \times \pi_m(X, x) \to \pi_{n+m-1}(X, x)$$

vanish at any basepoint $x$.

Consider the case $n = m = 1$. For a path-connected CW-complex $X$ with fundamental group $G$, the map $X \to K(G_{ab}, 1)$ is an $f$-localization.

**Example 7.7.5.** If $A$ is nonempty, then a space $Y$ is local with respect to $f: \emptyset \to A$ if and only if $Y$ is weakly contractible. All maps are $f$-equivalences, and $X \to *$ is always an $f$-localization.

**Example 7.7.6.** Consider a degree-$p$ map $f: S^1 \to S^1$. A space $Y$ is $f$-local if and only if, at any basepoint of $Y$, the $p$-th power map $\Omega Y \to \Omega Y$ is an equivalence. For this to occur, the $p$-th power map $\pi_1(Y) \to \pi_1(Y)$ must be an isomorphism.[6]

By contrast, let $M(\mathbb{Z}/p, 1)$ be the Moore space constructed as the cofiber of $f$, and consider the map $g: M(\mathbb{Z}/p, 1) \to *$. A space $Y$ is $g$-local if and only if, for any basepoint of $Y$, the homotopy fiber (over the trivial loop) of the $p$-th power map $\Omega Y \to \Omega Y$ is contractible. In particular, the $p$-th power map need only be injective on $\pi_1(Y)$.[7]

---

[5] We should be careful about edge cases. When $n = 0$, $X$ is $(-1)$-truncated if and only if it is either empty or weakly contractible. By convention, $S^{-1} = \emptyset$, and $X$ is $(-2)$-truncated if and only if it is weakly contractible.

When $n = 0$ a map $A \to B$ is an $f$-equivalence if and only if either both $A$ and $B$ are empty or neither of them is, and a map $A \to X$ is an $f$-localization if and only if either $A$ is nonempty and $X$ is weakly contractible, or $A$ and $X$ are both empty. When $n = -1$ any map is an $f$-equivalence, and a map $A \to X$ is an $f$-localization if and only if $X$ is weakly contractible.

[6] We would especially like to thank Carles Casacuberta for a correction. We erroneously stated that a space was $f$-local if, in addition, the $p$-th power maps were isomorphisms on all higher $\pi_n$. This is the result of forgetting that, to show a map $\Omega Y \to \Omega Y$ is an equivalence, we have to vary over choices of basepoint of $\Omega Y$. The correct criterion is this: for any basepoint $y$ and any $\gamma \in \pi_1(Y, y)$, the operator $1 + \gamma + \cdots + \gamma^{p-1}$ must act as an isomorphism on $\pi_n(Y, y)$ [67]. Our only consolation is that we are, apparently, in good company.

[7] This localization is called Anderson localization [65].

**Example 7.7.7** ([215]). Let $S$ be the set of maps $\{K(\mathbb{Z}/p, 1) \to *\}$ as $p$ ranges over the prime numbers. Then the Sullivan conjecture, as proven by Miller [210], is equivalent to the statement that any finite CW-complex $X$ is $S$-local. Since $S$-equivalences are closed under products and homotopy colimits, the expression of $K(\mathbb{Z}/p, n+1)$ as the geometric realization of the bar construction $\{K(\mathbb{Z}/p, n)^q\}$ shows inductively that the maps $K(\mathbb{Z}/p, n) \to *$ are all $S$-equivalences. However, if $Y$ is any nontrivial 1-connected space with finitely generated homotopy groups and a finite Postnikov tower, then $Y$ accepts a nontrivial map from some $K(\mathbb{Z}/p, n)$ and hence cannot be $S$-local. This argument shows that a simply connected finite CW-complex with nonzero mod-$p$ homology has $p$-torsion in infinitely many nonzero homotopy groups, which was conjectured by Serre in the early 1950s and proven by McGibbon and Neisendorfer [209].

Localization still applies to other categories closely related to topological spaces.

**Example 7.7.8.** Let $\mathcal{C}$ be the category of based spaces. A based space $Y$ is local with respect to the based map $* \to S^1$ if and only if the loop space $\Omega Y$ is weakly contractible, or equivalently if and only if the path component $Y_0$ of the basepoint is weakly contractible. A model for the Bousfield localization is given by the mapping cone of the map $Y_0 \to Y$.

**Example 7.7.9.** Fix a discrete group $G$, and consider the category of $G$-spaces: spaces with a continuous action of a group $G$, with maps being $G$-equivariant continuous maps. For example, the empty space has a unique $G$-action, while the orbit spaces $G/H$ have continuous actions under the discrete topology. Every $G$-space has fixed-point subspaces $X^H \cong \mathrm{Map}_G(G/H, X)$ for subgroups $H$ of $G$. In this context, there is an abundance of examples of localizations.

A $G$-space $Y$ is local with respect to $\emptyset \to *$ if and only if the fixed-point subspace $Y^G$ is contractible. A model for Bousfield localization is given by the mapping cone of the map $Y^G \to Y$.

Fix a model for the universal contractible $G$-space $EG$. A $G$-space $Y$ is local with respect to $EG \to *$ if and only if the map from the fixed point space $Y^G$ to the homotopy fixed point space $\mathrm{Map}_G(EG, Y) = Y^{hG}$ is a weak equivalence. Since the projections $EG \times EG \to EG$ are $G$-equivariant homotopy equivalences, a model for the Bousfield localization is the space of nonequivariant maps $\mathrm{Map}(EG, Y)$, with $G$ acting by conjugation.

A $G$-space $Y$ is local with respect to $\emptyset \to G$ if and only if the underlying space $Y$ is contractible. A model for the Bousfield localization is given by the mapping cone of the map $EG \times Y \to Y$, sometimes called $\widetilde{EG} \wedge Y_+$.

**Example 7.7.10.** Fix a collection $S$ of maps and a space $Z$, letting $\mathcal{C}$ be the category of spaces over $Z$. We say that a map $X \to Y$ of spaces over $Z$ is a *fiberwise $S$-equivalence* if the map of homotopy fibers over any point $z \in Z$ is an $S$-equivalence, and refer to the corresponding localizations as *fiberwise $S$-localizations*.

A map $X \to Y$ over $Z$ which is a weak equivalence on underlying spaces is in particular a fiberwise $S$-equivalence. Applying this to the lifting characterization of fibrations, we can find that for an object $Y \to Z$ of $C$ to be fiberwise $S$-local the map $Y \to Z$ must be a fibration. Moreover, for fibrations $Y \to Z$ we can recharacterize being local. Given any map $f \colon A \to B$ in $S$ and any point $z \in Z$, there is a map in $C$ of the form $f_z \colon A \to B \to \{z\} \subset Z$ concentrated entirely over the point $z$; let $S_Z$ be the set of all such maps. A fibration $Y \to Z$ in $C$ is fiberwise $S$-local if and only if it is $S_Z$-local in $C$.

Fiberwise localizations were constructed by Farjoun in [96, 1.F.3]; they are also constructed in [124, §7] and characterized from several perspectives.

Example 7.7.11. The category of topological monoids and continuous homomorphisms has its own homotopy theory. Consider the inclusion $f \colon \mathbb{N} \to \mathbb{Z}$ of discrete monoids. Then $\mathrm{Map}^{mon}(\mathbb{Z}, M) \to \mathrm{Map}^{mon}(\mathbb{N}, M)$ is isomorphic to the map $M^{\times} \to M$ from the space of invertible elements of $M$ to the space $M$.[8] An $f$-local object is a topological group, and localization is a topologized version of group-completion.

However, the map $\mathbb{N} \to \mathbb{Z}$ does not participate well with *weak* equivalences of topological monoids: weakly equivalent topological monoids do not have weakly equivalent spaces of invertible elements because homomorphisms out of $\mathbb{Z}$ are not homotopical. We can get a version that respects weak equivalences in two ways. With model categories, we can factor the map $\mathbb{N} \to \mathbb{Z}$ as $\mathbb{N} \hookrightarrow \mathbb{Z}_c \xrightarrow{\simeq} \mathbb{Z}$ in the category of topological monoids, where $\mathbb{Z}_c$ is a cofibrant topological monoid, and there are explicit models for such. We could instead use coherent multiplications, where a map $\mathbb{Z} \to M$ is no longer required to strictly be a homomorphism but instead be a coherently multiplicative map.

Using either correction, the space $M^{\times}$ of strict units becomes replaced, up to equivalence, by the pullback

$$
\begin{array}{ccc}
M^{inv} & \longrightarrow & M \\
\downarrow & & \downarrow \\
\pi_0(M)^{\times} & \longrightarrow & \pi_0 M
\end{array}
$$

the union of the components of $M$ whose image in $\pi_0(M)$ has an inverse. A local object is then a *grouplike* topological monoid, and localization is homotopy-theoretic group completion. These play a key role the study of iterated loop spaces and algebraic $K$-theory [194, 268, 207].

## 7.8    Stable settings

One of the great benefits of the stable homotopy category, and stable settings in general, is that a map $f \colon X \to Y$ becoming an equivalence is roughly the same as the cofiber $Y/X$ becoming trivial.

---

[8]  As a point-set digression the reader should, as usual, be warned that the source may not have the subspace topology. The space of invertible elements is, instead, homeomorphic to the subspace of $M \times M$ of pairs of elements $(x, y)$ such that $xy = yx = 1$.

We recall the definition of stability from [168, §1.1.1].

**Definition 7.8.1.** The category $C$ is *stable* if it satisfies the following properties:

1. $C$ is (homotopically) *pointed*: there is an object $*$ such that, for all $X \in C$, the spaces $\mathrm{Map}_C(X, *)$ and $\mathrm{Map}_C(*, X)$ are contractible.

2. $C$ has homotopy pushouts of diagrams $* \leftarrow X \to Y$ and homotopy pullbacks of diagrams $* \to Y \leftarrow X$.

   As a special case, we have suspension and loop objects:
   $$\Sigma X = \mathrm{hocolim}(* \leftarrow X \to *), \qquad \Omega X = \mathrm{holim}(* \to X \leftarrow *).$$

3. Suppose that we have a homotopy coherent diagram

$$
\begin{array}{ccc}
X & \longrightarrow & Y \\
\downarrow & & \downarrow \\
* & \longrightarrow & Z
\end{array}
$$

   meaning maps as given and a homotopy between the double composites. Then the induced map
   $$\mathrm{hocolim}(* \leftarrow X \to Y) \to Z$$
   is a homotopy equivalence if and only if the map
   $$X \to \mathrm{holim}(* \to Z \leftarrow Y)$$
   is a homotopy equivalence.

   Taking $Y = *$, we find that a map $X \to \Omega Z$ is an equivalence if and only if the homotopical adjoint $\Sigma X \to Z$ is an equivalence.

**Example 7.8.2.** The category of (cofibrant–fibrant) spectra is the canonical example of a stable category.

**Example 7.8.3.** For any ring $R$, there is a category $\mathcal{K}_R$ of chain complexes of $R$-modules. Any two complexes $C$ and $D$ have a Hom-complex $\mathrm{Hom}_R(C, D)$, and the Dold–Kan correspondence produces a simplicial set $\mathrm{Map}_{\mathcal{K}_R}(C, D)$ whose homotopy groups satisfy
$$\pi_n \mathrm{Map}_{\mathcal{K}_R}(C, D) \cong H_n \mathrm{Hom}_R(C, D)$$
for $n \geq 0$.[9] This gives the category $\mathcal{K}_R$ of complexes an enrichment in simplicial sets, and these mapping spaces make the category $\mathcal{K}_R$ stable. Within this category there are many stable subcategories: categories of complexes which are bounded above or below or both, or with homology groups bounded above or below or both, or which are made up of projectives or injectives, and so on.

We will sometimes write $C_R \subset \mathcal{K}_R$ for the subcategory of cofibrant objects in the projective model structure on $R$, whose homotopy category is the derived category $h(R)$.

---

[9] More generally, if $R[m]$ is the complex equal to $R$ in degree $m$ and zero elsewhere, then for all complexes $C$ we have $[R[m], C]_{h\mathcal{K}_R} \cong H_m(C)$.

Theorem 7.8.4 (see [168, Theorem 1.1.2.14]). *If $C$ is stable, then the homotopy category $hC$ has the structure of a triangulated category.*

In a stable category, every object $Y$ has an equivalence $Y \to \Omega\Sigma Y$. However, there is a natural weak equivalence

$$\begin{aligned} \mathrm{Map}_C(X, \Omega Z) &\simeq \mathrm{holim}[\mathrm{Map}_C(X,*) \to \mathrm{Map}_C(X,Z) \leftarrow \mathrm{Map}_C(X,*)] \\ &\simeq \mathrm{holim}(* \to \mathrm{Map}_C(X,Z) \leftarrow *) \\ &\simeq \Omega \mathrm{Map}_C(X,Z), \end{aligned}$$

and hence the mapping spaces

$$\mathrm{Map}_C(X,Y) \simeq \Omega^n \mathrm{Map}_C(X, \Sigma^n Y)$$

can be extended to be valued in $\Omega$-spectra. This makes it much easier to detect equivalences: we only need to check the homotopy groups of $\Omega^t \mathrm{Map}_C(X,Y)$ at the basepoint.

Definition 7.8.5. Suppose that $C$ is stable and $S$ is a class of maps in $C$. We say that $S$ is *shift-stable* if the image $\bar{S}$ in $hC$ is closed under suspension and desuspension, up to isomorphism.

Proposition 7.8.6. *Suppose that $C$ is stable and $S$ is a shift-stable class of maps $\{f_\alpha \colon A_\alpha \to B_\alpha\}$. Then an object $Y$ in $C$ is $S$-local if and only if the homotopy classes of maps $[B_\alpha/A_\alpha, X]_{hC}$ are trivial.*

*Proof.* The individual fiber sequences

$$\Omega^t \mathrm{Map}_C(B_\alpha/A_\alpha, Y) \to \Omega^t \mathrm{Map}_C(B_\alpha, Y) \to \Omega^t \mathrm{Map}_C(A_\alpha, Y),$$

on homotopy classes of maps, are part of a long exact sequence

$$\cdots \to [\Sigma^t B_\alpha/A_\alpha, Y]_{hC} \to \pi_t \mathrm{Map}_C(B_\alpha, Y) \to \pi_t \mathrm{Map}_C(A_\alpha, Y) \to [\Sigma^{t-1} B_\alpha/A_\alpha, Y]_{hC} \to \cdots$$

from the triangulated structure. We get an isomorphism on homotopy groups if and only if the terms $[\Sigma^t B_\alpha/A_\alpha, Y]_{hC}$ vanish for all values of $t$. $\square$

By contrast with the unstable case where basepoints are a continual issue, these shift-stable localizations in a stable category are always nullifications, and they are *equivalent* to nullifications of the triangulated homotopy category by a class $S$ that is closed under shift operations.

Definition 7.8.7. Suppose that $\mathcal{D}$ is a triangulated category. A full subcategory $\mathcal{T}$ is called a *thick subcategory* if its objects are closed under closed under isomorphism, shifts, cofibers, and retracts. If $\mathcal{D}$ has coproducts, a thick subcategory $\mathcal{T}$ is *localizing* if it is also closed under coproducts.

Proposition 7.8.8. *Suppose that $\mathcal{D}$ is a triangulated category and that $\mathcal{T} \subset \mathcal{D}$ is a thick subcategory. Then there exists a triangulated category $\mathcal{D}/\mathcal{T}$ called the* Verdier quotient *of $\mathcal{D}$ by $\mathcal{T}$, with a functor $\mathcal{D} \to \mathcal{D}/\mathcal{T}$. The Verdier quotient is universal among triangulated categories under $\mathcal{D}$ such that the objects of $\mathcal{T}$ map to trivial objects.*

This universal characterization allows us to strongly relate Bousfield localization of stable categories to localization of the homotopy category.

**Proposition 7.8.9.** *Suppose that $C$ is stable, and that $S$ is a shift-stable collection of maps in $C$.*

1. *An object in $C$ is $S$-local if and only if its image in the homotopy category $hC$ is $S$-local.*
2. *A map in $C$ is an $S$-equivalence if and only if its image in the homotopy category is an $S$-equivalence.*
3. *The subcategories $L^S C$ of $S$-local objects and $T$ of $S$-trivial objects are thick subcategories of $C$.*
4. *The subcategory $T$ of $S$-trivial objects is closed under all coproducts that exist in $C$. If $C$ has small coproducts then $T$ is a localizing subcategory.*
5. *If all objects in $C$ have $S$-localizations, then the left adjoint to the inclusion $hL^S C \to hC$ has a factorization*

$$hC \to hC/hT \to hL^S C.$$

*The latter functor is an equivalence of categories.*

*Remark 7.8.10.* The fact that Bousfield localization of $C$ is determined by a construction purely in terms of $hC$ is special to the stable setting.

*Remark 7.8.11.* This relates Bousfield localization to Verdier quotients in a stable category, but only quotients by a *localizing* subcategory. For a homotopical interpretation of more general Verdier quotients, see [216, §I.3].

**Example 7.8.12.** Let $S$ be the collection of multiplication-by-$m$ maps $S^n \to S^n$ for $n \in \mathbb{Z}$, $m > 0$. A spectrum $Y$ is $S$-local if and only if multiplication by $m$ is an isomorphism on the homotopy groups $\pi_* Y$ for all positive $m$, or equivalently if the maps $\pi_* Y \to \mathbb{Q} \otimes \pi_* Y$ are isomorphisms. Such spectra are called *rational*.

If $Y$ is such a spectrum, we can calculate that the natural map

$$[X, Y] \to \prod_n \operatorname{Hom}(\pi_n X, \pi_n Y)$$

is an isomorphism for any spectrum $X$: because $\pi_n Y$ is a graded vector space, $\operatorname{Hom}(-, \pi_n Y)$ is exact and so both sides are cohomology theories in $X$ that satisfy the wedge axiom and agree on spheres. Therefore, $A \to B$ is an $S$-equivalence if and only if $\mathbb{Q} \otimes \pi_n(A) \to \mathbb{Q} \otimes \pi_n(B)$ is an isomorphism for all $n$, and such maps are called *rational equivalences*. In this case, this is the same as the map $H_*(A; \mathbb{Q}) \to H_*(B; \mathbb{Q})$ being an isomorphism.

This analysis allows us to conclude that $X \to H\mathbb{Q} \wedge X = X_{\mathbb{Q}}$ is a rationalization for all $X$.

**Example 7.8.13.** In the above, we can make $S$ smaller. If $S$ is the set of multiplication-by-$p$ maps $S^n \to S^n$, we similarly find that $S$-local spectra are those whose homotopy groups are $\mathbb{Z}[1/p]$-modules, and that equivalences are those maps which induce isomorphisms on homotopy groups after inverting $p$. The localization of $\mathbb{S}$ is the homotopy colimit

$$\mathbb{S}[1/p] = \text{hocolim}(\mathbb{S} \xrightarrow{p} \mathbb{S} \xrightarrow{p} \mathbb{S} \xrightarrow{p} \cdots),$$

which is also a Moore spectrum for $\mathbb{Z}[1/p]$. We similarly find that $X \to \mathbb{S}[1/p] \wedge X$ is an $\mathbb{S}$-localization for all $X$.

We could also let $S$ be the set of multiplication-by-$m$ maps for $m$ relatively prime to $p$, which replaces the ring $\mathbb{Z}[1/p]$ with the local ring $\mathbb{Z}_{(p)}$ in the above.

**Example 7.8.14.** Fix a commutative ring $R$ and a multiplicatively closed subset $W \subset R$, recalling that localization with respect to $W$ is exact. If we define $S$ to be the set of maps of the form $R[n] \xrightarrow{w} R[n]$ for $w \in W$, a complex $C$ of $R$-modules is $S$-local if and only if the multiplication-by-$w$ maps $H_*(C) \to H_*(C)$ are isomorphisms, or equivalently if and only if $H_*(C) \to W^{-1}H_*(C) \cong H_*(W^{-1}C)$ is an isomorphism. A map $A \to B$ of complexes is an $S$-equivalence if and only if the map $W^{-1}A \to W^{-1}B$ is an equivalence.

The natural map $C \to W^{-1}C \cong W^{-1}R \otimes_R C$ is an $S$-localization.

These examples have such nice properties that it is convenient to axiomatize them.

**Definition 7.8.15.** A stable Bousfield localization on spectra[10] is a *smashing localization* if either of the following equivalent conditions hold.

1. There is a map of spectra $\mathbb{S} \to L\mathbb{S}$ such that, for any $X$, the map $X \to L\mathbb{S} \wedge X$ is a localization.

2. Local objects are closed under arbitrary homotopy colimits.

The equivalence between these two characterizations is not immediately obvious. The first implies the second, because

$$L\mathbb{S} \wedge \text{hocolim}\, X_i \to \text{hocolim}(L\mathbb{S} \wedge X_i)$$

is always an equivalence and the former is always local. The converse follows because the only homotopy-colimit preserving functors on spectra are all equivalent to functors of the form $X \mapsto A \wedge X$ for some $A$, and the resulting localization map $\mathbb{S} \to A$ is of the desired form.

**Example 7.8.16.** A spectrum $Y$ is local for the maps $\mathbb{S}[1/p] \wedge S^n \to *$ if and only if the homotopy limit

$$\text{holim}(\cdots \to Y \xrightarrow{p} Y \xrightarrow{p} Y) \simeq F(\mathbb{S}[1/p], Y)$$

of function spectra is weakly contractible. However, taking homotopy limits of the

---

[10] This definition extends if we have a stable category $\mathcal{C}$ with a symmetric monoidal structure appropriately compatible with the stable structure.

natural fiber sequences

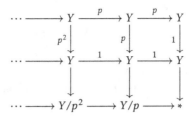

shows that $Y$ is local if and only if the map $Y \to Y_p^\wedge = \text{holim } Y/p^k$ is an equivalence. Therefore, we refer to a spectrum local for these maps as $p$-*complete*; a Bousfield localization of $Y$ will be called the $p$-*completion*; a trivial object is called $p$-*adically trivial*; an equivalence is called a $p$-*adic equivalence*. The above presents $Y_p^\wedge$ as a candidate for the $p$-completion of $Y$.

If we construct the fiber sequence

$$\Sigma^{-1}\mathbb{S}/p^\infty \to \mathbb{S} \to \mathbb{S}[1/p],$$

we find that we can identify $Y_p^\wedge$ with the function spectrum $F(\Sigma^{-1}\mathbb{S}/p^\infty, Y)$. Moreover, the map $Y_p^\wedge \to (Y_p^\wedge)_p^\wedge$ is always an equivalence. Therefore, $Y_p^\wedge$ is always $p$-complete.

If multiplication-by-$p$ is an equivalence on $Z$, then $Z \simeq Z \wedge \mathbb{S}[1/p]$, and so maps $Z \to Y$ are equivalent to maps $Z \to F(\mathbb{S}[1/p], Y)$. For any $Y$ which is $p$-adically complete, this is trivial, so such objects $Z$ are $p$-adically trivial. In particular, the fiber of $Y \to Y_p^\wedge$ is always trivial and so $Y \to Y_p^\wedge$ is a $p$-adic equivalence. Therefore, this is a $p$-adic completion.

If each homotopy group of $Y$ has a bound on the order of $p$-power torsion, we can further identify the homotopy groups of $Y_p^\wedge$ as the ordinary $p$-adic completions of the homotopy groups of $Y$; if the homotopy groups of $Y$ are finitely generated, then $\pi_*(Y_p^\wedge) \to \pi_*(Y) \otimes \mathbb{Z}_p$.[11]

*Remark 7.8.17.* The previous example is not a smashing localization. For any connective spectrum $X$, the map $\mathbb{S}_p^\wedge \wedge X \to X_p^\wedge$ induces the map $\pi_*(X) \otimes \mathbb{Z}_p \to \pi_*(X)_p^\wedge$ on homotopy groups; this is typically only an isomorphism if the homotopy groups $\pi_*(X)$ are finitely generated.

*Example 7.8.18.* For an element $x$ in a commutative ring $R$, let $K_x$ be the complex

$$\cdots \to 0 \to R \to x^{-1}R \to 0 \to \cdots$$

concentrated in degrees 0 and $-1$, with a map $K_X \to R$. For a sequence of elements $(x_1, \ldots, x_n)$, let $K_{(x_1, \ldots, x_n)} = \bigotimes_R K_{x_i}$ be the *stable Koszul complex*. If $y$ is in the ideal generated by $(x_1, \ldots, x_n)$, then the inclusion $K_{(x_1, \ldots, x_n)} \to K_{(x_1, \ldots, x_n, y)}$ is a quasi-isomorphism, and so up to quasi-isomorphism the Koszul complex only depends on the ideal. Let $K_I$ be a cofibrant replacement.

---

[11] In general, the homotopy groups of the $p$-adic completion are somewhat sensitive and one needs to be careful about derived functors of completion.

We say that a complex $C$ is $I$-*complete* if and only if it is local with respect to the shifts of the map $K_I \to R$. This is true if and only if the homology groups of $C$ are $I$-complete in the derived sense. If $R$ is Noetherian and the homology groups of $C$ are finitely generated, this is true if and only if the homology groups of $C$ are $I$-adically complete in the ordinary sense.

These frameworks for the study of localization and completion, and many generalizations of it, were developed by Greenlees and May [112].

Example 7.8.19.  Fix a Noetherian ring $R$, and let $\mathcal{C}$ be the category of unbounded complexes of finitely generated projective left $R$-modules that only have nonzero homology groups in finitely many degrees. Consider the set $S$ of maps $R[n] \to 0$. An object $C$ is $S$-local if and only if its homology groups are trivial.

We can inductively take mapping cones of maps $R[n] \to C$ to construct a localization $C \to LC$, embedding $C$ into an unbounded complex of finitely generated projective modules with trivial homology groups. Therefore, localizations exist in this category.

For two such complexes $C$ and $D$ with trivial homology, we have

$$\mathrm{Hom}_{h\mathcal{C}}(C,D) \cong \lim_n \mathrm{Hom}_R(Z_nC, Z_nD)/\mathrm{Hom}_R(Z_nC, D_{n+1}),$$

where $D_{n+1} \to Z_n(D)$ is the boundary map — a surjective map from a projective module.

This can be interpreted in terms of the stable module category of $R$. Defining $W_n(C) = Z_{-n}(C)$, the short exact sequences $0 \to Z_{-n}(C) \to C_{-n} \to Z_{-n-1}(C) \to 0$ determine isomorphisms $W_n(C) \cong \Omega W_{n+1}(C)$ in the stable module category, assembling the $W_n$ into an "$\Omega$-spectrum". Maps $C \to D$ are then equivalent to maps of $\Omega$-spectra in the stable module category.[12]

## 7.9    Homology localizations

### Homology localization of spaces

Definition 7.9.1.  Suppose $E_*$ is a homology theory on spaces. We say that a map $f : A \to B$ of spaces is an $E_*$-*equivalence* if it induces an isomorphism $f_* : E_* A \to E_* B$. A space is $E_*$-*local* if it is local with respect to the class of $E_*$-equivalences.

Example 7.9.2.  Suppose that $E_*$ is integral homology $H_*$. Any Eilenberg–Mac Lane space $K(A, n)$ is $H_*$-local by the universal coefficient theorem for cohomology. Moreover, any simply connected space $X$ is the homotopy limit of a Postnikov tower built from fibration sequences $P_n X \to P_{n-1} X \to K(\pi_n X, n+1)$. Since local objects are closed under homotopy limits, we find that simply connected spaces are $H_*$-local.[13]

---

[12]  In certain cases, such as for Frobenius algebras, $\Omega$ is an autoequivalence. This definition then simply recovers the stable module category of $R$ by itself. If $R$ has finite projective dimension, $\Omega$-spectrum objects are necessarily trivial.

[13]  This argument can be refined to show that *nilpotent spaces* (where $\pi_1(X)$ is nilpotent, and acts nilpotently on the higher homotopy groups) are $H_*$-local.

*Remark 7.9.3.* This example illustrates a very different approach to the construction of localizations. Because homology isomorphisms are detected by the $K(A,n)$, these spaces are automatically local; therefore, any object built from these using homotopy limits is automatically local. Such objects are often called *nilpotent*. Thus gives us a dual approach to building the Bousfield localization of $X$: construct a natural diagram of nilpotent objects that receive maps from $X$, and try to verify that the homotopy limit is a localization of $X$.

**Example 7.9.4.** Serre's rational Hurewicz theorem implies that a map of simply connected spaces is an isomorphism on rational homology groups if and only if it is an isomorphism on rational homotopy groups. A simply connected space is local for rational homology if and only if it its homotopy groups are rational vector spaces.

The same is not true for general spaces. The map $\mathbb{RP}^2 \to *$ is a rational homology isomorphism, and the covering map $S^2 \to \mathbb{RP}^2$ is an isomorphism on rational homotopy groups, but the composite $S^2 \to *$ is neither. The problem here is the failure of a simple Postnikov tower for $\mathbb{RP}^2$ due to the action of $\pi_1$ on the higher homotopy groups.

**Example 7.9.5.** If $X$ is a connected space with perfect fundamental group, then Quillen's plus-construction gives a map $X \to X^+$ that induces an $H_*$-isomorphism such that $X^+$ is simply connected. This makes $X^+$ into an $H_*$-localization of $X$.

Classically, Quillen's plus-construction can be applied to groups with a perfect subgroup. In order to properly identify the universal property, we need to work in a relative situation.

**Example 7.9.6.** Fix a group $G$, and let $\mathcal{C}$ be the category of spaces over $BG$. Given an abelian group $A$ with $G$-action, there is an associated local coefficient system $\underline{A}$ on $BG$, and so given any object $X \to BG$ of $\mathcal{C}$ we can define the homology groups $H_*(X;\underline{A})$. We say that a map $X \to Y$ over $BG$ is a relative homology equivalence if it induces isomorphisms on homology with coefficients in any $\underline{A}$. Taking $A$ to be the group algebra $\mathbb{Z}[G]$, we find that this is equivalent to the map of homotopy fibers $F_X \to F_Y$ being a homology isomorphism, so this is the same as a *fiberwise $H_*$-equivalence*. If an object $Y$ over $BG$ has simply connected homotopy fiber it is automatically local.

Suppose that $X$ is any connected space such that $\pi_1(X)$ contains a perfect normal subgroup $P$ with quotient group $G$. The homomorphism $\pi_1(X) \to G$ lifts to a map $X \to BG$. The plus-construction with respect to $P$ is a fiber homology equivalence $X \to X^+$ where $X^+ \to BG$ has simply connected homotopy fiber, and thus is a localization in $\mathcal{C}$.

Localization with respect to homology is very difficult to analyze in the case when a space is not simply connected, especially if the space is not *simple* (either the fundamental group is not nilpotent or it does not act nilpotently on the higher homotopy groups). Many natural spaces are not local. Here are some basic tools to prove this.

**Lemma 7.9.7.** *Suppose that $F_n$ is a free group on $n$ generators and $\alpha\colon F_n \to F_n$ is a homomorphism, with induced map $\alpha_{ab}\colon \mathbb{Z}^n \to \mathbb{Z}^n$. Under the identification $\mathrm{Hom}(F_n, G) \cong G^n$ for any group $G$, write $\alpha^*$ for the natural map of sets $G^n \to G^n$.*

*Suppose the map $\alpha_{ab}$ becomes an isomorphism after tensoring with a ring $R$. For any space $X$, a necessary condition for $X$ to be $H_*(-;R)$-local is that $\alpha^*\colon \pi_1(X,x)^n \to \pi_1(X,x)^n$ must be a bijection at any basepoint.*

*Proof.* The map $\alpha_{ab}$, after tensoring with $R$, can be identified with the map on first homology, $H_1(F_n; R) \to H_1(F_n; R)$, induced by $\alpha$. If $\alpha_{ab}$ becomes an isomorphism after tensoring with $R$, then $\alpha\colon K(F_n, 1) \to K(F_n, 1)$ is an $H_*(-;R)$-equivalence. For a space $X$ to be $H_*(-;R)$-local, the induced map

$$\mathrm{Map}_*(K(F_n, 1), X) \to \mathrm{Map}_*(K(F_n, 1), X)$$

must be a weak equivalence. Taking a wedge of circles as our model, we find that the induced map

$$(\Omega X)^n \to (\Omega X)^n$$

must be a weak equivalence. On $\pi_0$, this is the map $\alpha^*$ on $\pi_1(X)^n$. $\qquad\square$

**Example 7.9.8.** For $n \neq 0$, the multiplication-by-$n$ map $\mathbb{Z} \to \mathbb{Z}$ is a rational isomorphism. Therefore, for $X$ to be rationally local, the $n$-th power map $\pi_1(X) \to \pi_1(X)$ should be a bijection: every element $g \in \pi_1(X)$ has a unique $n$-th root $g^{1/n}$. Such groups are called uniquely divisible, or sometimes $\mathbb{Q}$-groups. The structure of free $\mathbb{Q}$-groups was studied in [32].

**Example 7.9.9.** Let $F_2$ be free on generators $x$ and $y$, and define $\alpha\colon F_2 \to F_2$ by

$$\alpha(x) = x^{-9}y^{-20}(y^2x)^{10}, \quad \alpha(y) = x^{-9}y^{10}(yx^{-1})^{-9}.$$

The map $\alpha_{ab}$ is the identity map. Therefore, for a space with fundamental group $G$ to be local with respect to integral homology, any pair of elements $(z, w) \in G$ has to be uniquely of the form $(z, w) = (x^{-9}y^{-20}(y^2x)^{10}, x^{-9}y^{-10}(yx^{-1})^{-9})$ for some $x$ and $y$ in $G$. Most groups do not satisfy this property.

We can use this to show that any space whose fundamental group $G$ has a surjective homomorphism $\phi\colon G \to A_5$ cannot be local with respect to integral homology — in particular, this applies to a free group $F_2$. Choose elements $x$ and $y$ in $G$ with $\phi(x) = (123)$ and $\phi(y) = (12345)$. Then $\phi(y^2x) = (14)(25)$ and $\phi(yx^{-1}) = (145)$, and $\phi \circ \alpha$ is the trivial homomorphism while $\phi$ is surjective.[14]

Several other, more easily defined, maps $\alpha$ can be shown to not be bijective. For example, the map $(x, y) \mapsto (x[x, y], y[x, y])$ can be shown not to be a bijection by using Fox's free differential calculus [99].

**Lemma 7.9.10.** *Let $G$ be a group, $R$ a ring, and $\beta \in \mathbb{Z}[G]$ an element such that the composite ring homomorphism $\mathbb{Z}[G] \xrightarrow{\epsilon} \mathbb{Z} \to R$ sends $\beta$ to zero.*

---

[14] In order to use this *particular* technique to show that $\phi$ was not a bijection, we needed to have a homomorphism $\phi$ whose image was a perfect group — the image of $\alpha_{ab}$ is contained in the kernel of $\phi_{ab}$. This particular map $\alpha$ is complicated because it was reverse-engineered from $\phi$.

*Then, for any based space X with fundamental group G, a necessary condition for X to be $H_*(-;R)$-local is that $\pi_k(X)$ must be complete in the topology defined by $\beta$.[15]*

*Proof.* Fix the space $X$ and basepoint and consider the space $Y = X \vee S^k$. The group $\pi_k(Y)$ is isomorphic to $\pi_k(X) \oplus \mathbb{Z}[G]$, and so the element $\beta \in \mathbb{Z}[G]$ lifts to a map $\beta \colon Y \to Y$ given by the identity on $X$ together with the map $S^k \to Y$ corresponding to the element $(0,\beta) \in \pi_k(X) \oplus \mathbb{Z}[G]$. The induced self-map of

$$H_*(Y;R) \cong H_*(X;R) \oplus \widetilde{H}_*(S^k;R)$$

is given by the identity on $H_*(X;R)$ together with the map $\epsilon(\beta)$ tensored with $R$ on the second factor. If $\epsilon(\beta)$ becomes zero after tensoring with $R$, then this map is zero on the second factor.

Define

$$X' = \operatorname{hocolim}(Y \xrightarrow{\beta} Y \xrightarrow{\beta} \cdots).$$

By construction, the map

$$H_*(X;R) \to H_*(X';R) = \operatorname{colim} H_*(Y;R)$$

is an isomorphism. Therefore, $X \to X'$ is an $H_*(-;R)$-equivalence.

For $X$ to be $H_*(-;R)$-local, the induced map

$$\operatorname{Map}(X',X) \to \operatorname{Map}(X,X)$$

must be a weak equivalence. Taking the fiber over the identity map of $X$, we find that there is an induced equivalence

$$\operatorname{holim}(\cdots \xrightarrow{\beta} \Omega^k X \xrightarrow{\beta} \Omega^k X) \xrightarrow{\sim} *.$$

Using the Milnor $\lim^1$-sequence, we find that all of the homotopy groups of $X$ must be derived-complete with respect to $\beta$. □

*Remark 7.9.11.* If $R = \mathbb{Z}$, then this implies that any element $s \in \mathbb{Z}[G]$ with $\epsilon(s) = \pm 1$ must act invertibly on the higher homotopy groups of $X$, and so the action must factor through a large localization $S^{-1}\mathbb{Z}[G]$.

Example 7.9.12. Consider $X = S^1 \vee S^2$, whose fundamental group is isomorphic to $\mathbb{Z}$ with generator $t$. The second homotopy group satisfies

$$\pi_2(S^1 \vee S^2) \cong \mathbb{Z}[t^{\pm 1}]$$

as a module over $\mathbb{Z}[t^{\pm 1}]$. This is not complete with respect to the ideal generated by $\beta = (t-1)$ even though $\epsilon(\beta) = 0$. Therefore, $S^1 \vee S^2$ is not local with respect to integral homology.

Example 7.9.13. The space $\mathbb{RP}^2$ has fundamental group $\mathbb{Z}/2$ generated by an element $\sigma$, and the second homotopy group $\mathbb{Z}$ satisfies $\sigma(y) = -y$. The element

---

[15] This refers to being *derived* complete in the sense of Example 7.8.18.

$(1 - \sigma)$ has $\epsilon(1 - \sigma) = 0$ and acts as multiplication by 2. Since $\mathbb{Z}$ is not complete in the 2-adic topology we find that $\mathbb{RP}^2$ is not local with respect to integral homology.[16]

**Example 7.9.14.** If $R = \mathbb{Q}$, then any element $S \in \mathbb{Z}[G]$ with $\epsilon(s) \neq 0$ must act invertibly on the higher homotopy groups of $X$ for $X$ to be local with respect to rational homology. The homotopy groups of $K(\mathbb{Q}, 1) \vee (S^3)_\mathbb{Q}$ are $\mathbb{Q}$ in degree 1 and the rational group algebra $\mathbb{Q}[\mathbb{Q}]$ in degree 3. If $t$ is the generator of $\mathbb{Z} \subset \mathbb{Q}$, the element $2t - 1$ has $\epsilon(2t - 1) = 1$ and does not act invertibly on this group algebra. Therefore, this space is not local with respect to rational homology even though its homotopy groups are rational.

*Remark 7.9.15.* Bousfield localization with respect to $E_*$-equivalences leads us to some uncomfortable pressure with our previous notation. At first glance, it is not clear whether being an equivalence on $E_*$-homology is the same as having the same mapping spaces into any $E_*$-local object.[17] To prove this, one needs to prove that there is a sufficient supply of $E_*$-local objects: for any $X$, we need to be able to construct an $E_*$-homology isomorphism $X \to L_E X$ such that $L_E X$ is $E_*$-local. Here is how Bousfield addressed this in [54, Theorem 11.1]. It is essentially a cardinality argument, whose general form is called the Bousfield–Smith cardinality argument in [124, §2.3].

Let $E_*$ be a homology theory on spaces. We then have a class $S$ of $E_*$-equivalences, which are those maps which induce equivalences on $E_*$-homology. Unfortunately, this is a proper class of morphisms, and so we cannot immediately apply the small object argument to construct localizations. Moreover, because we do not know anything about local objects we cannot assert that an $S$-equivalence $X \to Y$ is the same as a map inducing an isomorphism $E_* X \to E_* Y$.

Bousfield addresses this by showing the following. Suppose $K \to L$ is an inclusion of simplicial sets such that $E_* K \to E_* L$ is an isomorphism, and that we choose any simplex $\sigma$ of $L$. Then there exists a subcomplex $L' \subset L$ with the following properties:

1. The simplex $\sigma$ is contained in $L'$.
2. The map $E_*(K \cap L') \to E_*(L')$ is an isomorphism on $E_*$.
3. The complex $L'$ has size bounded by a cardinal $\kappa$, which *depends only on $E$*.

Because of the cardinality bound on $L'$, we can find a *set $T$ of $E_*$-equivalences $A \to B$* so that any such map $K \cap L' \to L'$ must be isomorphic to one of them; an arbitrary $E_*$-equivalence $K \to L$ can then be factored as a (possibly transfinite) sequence of pushouts along the maps in the set $T$ followed by an equivalence. The maps in $T$ are $E_*$-isomorphisms, and an object is $S$-local if and only if it is $T$-local. The small object argument then applies to $T$, allowing us to construct $T$-localizations $Y \to LY$ which are also $E_*$-isomorphisms.

We will see in § 7.10 and § 7.11, in general constructions of Bousfield localization, that this verification is the key step.

---

[16] The homology localization of $\mathbb{RP}^2$ has, in fact, a fiber sequence $(S^2)_2^\wedge \to L\mathbb{RP}^2 \to K(\mathbb{Z}/2, 1)$.

[17] One could, but should not, say it this way: it is not clear that an $(E_*$-equivalence)-equivalence is automatically an $E_*$-equivalence.

## Homology localization of spectra

**Definition 7.9.16.** For a spectrum $E$, a map $f : X \to Y$ is an $E$-*homology equivalence* (or simply an $E$-equivalence) if the corresponding map $E_*X \to E_*Y$ is an isomorphism, and we say that $Z$ is $E$-*trivial* if $E_*Z = 0$. A map $f$ is an $E$-equivalence if and only if the cofiber of $f$ is $E$-trivial.[18]

This is most often employed when $E$ is a ring spectrum.

**Proposition 7.9.17.** *If $E$ has a multiplication $m$: $E \wedge E \to E$ with a left unit $\eta$ : $\mathbb{S} \to E$ in the homotopy category, then any spectrum $Y$ with a unital map $E \wedge Y \to Y$ is $E$-local.*

*Remark 7.9.18.* Such spectra $Y$ are sometimes called *homotopy $E$-modules*. Any spectrum of the form $E \wedge W$ is a homotopy $E$-module.

*Proof.* Any map $f : Z \to Y$ has the following factorization in the homotopy category:

$$ Z \xrightarrow{\eta \wedge 1} E \wedge Z \xrightarrow{1 \wedge f} E \wedge Y \xrightarrow{m} Y $$

If $Z$ has trivial $E$-homology, then $E \wedge Z$ is trivial and so the composite $Z \to Y$ is nullhomotopic. Therefore, $[Z, Y] = 0$ for all $E$-trivial $Z$, as desired. $\square$

**Corollary 7.9.19.** *If $E$ has a multiplication $m$: $E \wedge E \to E$ with a left unit $\eta$: $\mathbb{S} \to E$ in the homotopy category, then any homotopy limit of spectra that admit homotopy $E$-module structures is $E$-local.*

**Example 7.9.20.** A particular case of interest is when $E = H\mathbb{Z}$. Any Eilenberg–Mac Lane spectrum $HA$ is $H\mathbb{Z}$-local, being of the form $H\mathbb{Z} \wedge MA$ for a Moore spectrum for $A$.

Then any connective spectrum $Y$ is $H\mathbb{Z}$-local, as follows. As $H\mathbb{Z}$-local objects form a thick subcategory, any spectrum with finitely many nonzero homotopy groups is therefore $H\mathbb{Z}$-local. If $Y$ is connective then $P_n Y$ is $H\mathbb{Z}$-local due to having a finite Postnikov tower. Therefore, $Y = \operatorname{holim} P_n Y$ is the homotopy limit of $H\mathbb{Z}$-local spectra, and is thus $H\mathbb{Z}$-local.

Similarly, any product of Eilenberg–Mac Lane spectra $\prod \Sigma^n HA_n$ is also $H\mathbb{Z}$-local. Any rational spectrum is of this form.

However, not all spectra are $H\mathbb{Z}$-local. For any prime $p$ and any integer $n > 0$, there are $p$-primary Morava $K$-theories $K(n)$ such that $H\mathbb{Z} \wedge K(n)$ is trivial; these are $H\mathbb{Z}$-acyclic. The complex $K$-theory spectrum $KU$ satisfies the property that $H_*(KU; \mathbb{Z}) \to H_*(KU; \mathbb{Q})$ is an isomorphism: from this we can find that $KU \to KU_{\mathbb{Q}}$ is an $H\mathbb{Z}$-equivalence. The target is also $H\mathbb{Z}$-local because it is rational, and so $KU_{\mathbb{Q}}$ is the $H\mathbb{Z}$-localization of $KU$.

**Example 7.9.21.** We can consider the case where $E = H\mathbb{Z}/p$. By a similar argument, we find that any connective spectrum which is $p$-adically complete in the sense of

---

[18] Again, the definitions of this section can be applied to a stable category $\mathcal{C}$ with a compatible symmetric monoidal structure.

Example 7.8.16 is also $H\mathbb{Z}/p$-complete. Again, in connective cases there is not a difference between being $p$-adically complete and being $H\mathbb{Z}/p$-local.

For nonconnective spectra, these are quite different. The Morava $K$-theories $K(n)$ are $p$-adically complete but $H\mathbb{Z}/p$-trivial. The periodic complex $K$-theory spectrum $KU$ has $\pi_*(KU_p^\wedge) \cong (\pi_* KU)_p^\wedge$, but $KU$ is also $H\mathbb{Z}/p$-trivial.

These localizations have the flavor of completion with respect to an ideal. In some cases we can express them as such.

**Definition 7.9.22.** Suppose that $E$ has a binary multiplication $m$ with a left unit $\eta\colon \mathbb{S} \to E$, and let $j\colon I \to \mathbb{S}$ be the fiber of $\eta\colon \mathbb{S} \to E$. Assemble these into the inverse system

$$\cdots \to I^{\wedge 3} \xrightarrow{\; j \wedge 1 \wedge 1 \;} I \wedge I \xrightarrow{\; j \wedge 1 \;} I \xrightarrow{\; j \;} \mathbb{S}$$

The *E-nilpotent completion* $X_E^\wedge$ is the homotopy limit

$$\mathrm{holim}_n(\mathbb{S}/I^{\wedge n}) \wedge X,$$

with map $X \to X_E^\wedge$ induced by the maps $\mathbb{S} \to \mathbb{S}/I^{\wedge n}$.

**Proposition 7.9.23.** *The $E$-nilpotent completion is always $E$-local.*

*If $E$ is a finite complex, or $X$ and $I$ are connective and $E$ is of finite type, then the map $X \to X_E^\wedge$ is an $E$-localization.*

*Proof.* The cofiber sequence $I \to \mathbb{S} \to E$, after smashing with $I^{\wedge(n-1)}$, becomes a cofiber sequence $I^{\wedge n} \to I^{\wedge(n-1)} \to E \wedge I^{\wedge(n-1)}$, and so there are cofiber sequences

$$\mathbb{S}/I^{\wedge n} \wedge X \to \mathbb{S}/I^{\wedge(n-1)} \wedge X \to E \wedge I^{\wedge(n-1)} \wedge X.$$

By induction on $n$ we find that $\mathbb{S}/I^{\wedge n} \wedge X$ is $E$-local, and so the homotopy limit $X_E^\wedge$ is $E$-local.

After smashing with $E$, the cofiber sequence

$$E \wedge I^{\wedge n} \wedge X \to E \wedge I^{\wedge(n-1)} \wedge X \to E \wedge E \wedge I^{\wedge(n-1)} \wedge X$$

has a retraction of the second map via the (opposite) multiplication of $E$, and so the first map is nullhomotopic. Therefore, the homotopy limit $\mathrm{holim}\, E \wedge (I^{\wedge n} \wedge X)$ is trivial, and from the cofiber sequences

$$E \wedge (I^{\wedge n} \wedge X) \to E \wedge X \to E \wedge (\mathbb{S}/I^{\wedge n} \wedge X)$$

we find that $E \wedge X \to \mathrm{holim}(E \wedge (\mathbb{S}/I^{\wedge n} \wedge X))$ is an equivalence.

This reduces us to proving that the map

$$E \wedge \mathrm{holim}(\mathbb{S}/I^{\wedge n} \wedge X) \to \mathrm{holim}(E \wedge \mathbb{S}/I^{\wedge n} \wedge X)$$

is an equivalence: we can move the smash product with $E$ inside the homotopy limit. This is always true if $E$ is finite or if $E$ is of finite type and the homotopy limit is of connective objects. $\qquad\square$

*Remark 7.9.24.* The spectral sequence arising from the inverse system defining $X_E^\wedge$ is the *generalized Adams–Novikov spectral sequence based on E-homology.* It often abuts to the homotopy groups of the Bousfield localization with respect to $E$.

We can generalize our construction by allowing more general towers with a nilpotence property, after Bousfield in [55], or by extending these methods to the category of modules over a ring spectrum, as Baker–Lazarev did in [15] or Carlsson did in [64].

Example 7.9.25.    For any prime $p$ and any $n > 0$, we have the Johnson–Wilson homology theories $E(n)_*$ and the Morava $K$-theories $K(n)_*$. Associated to these we have $E(n)$-localization functors and $K(n)$-localization functors, as well as categories of $E(n)$-local and $K(n)$-local spectra, which play an essential role in chromatic homotopy theory. Ravenel conjectured, and Devinatz–Hopkins–Smith proved, that the localization $L_{E(n)}$ is a smashing localization [230, 73, 231]. These localizations also have *chromatic fractures* which are built using the following result.

Proposition 7.9.26.    *Suppose that $E$ and $K$ are spectra such that $L_K L_E X$ is always trivial. Then, for all $X$, there is a homotopy pullback diagram*

$$\begin{array}{ccc} L_{E\vee K}X & \longrightarrow & L_E X \\ \downarrow & & \downarrow \\ L_K X & \longrightarrow & L_E L_K X. \end{array}$$

*Proof.* The objects in the diagram

$$L_E X \to L_E L_K X \leftarrow L_K X$$

are either $E$-local or $K$-local, and hence automatically $E \vee K$-local; therefore, the homotopy pullback $P$ is $E \vee K$-local. It then suffices to show that the fiber of the map $X \to P$ is $E \vee K$-trivial, which is equivalent to showing that

$$\begin{array}{ccc} X & \longrightarrow & L_E X \\ \downarrow & & \downarrow \\ L_K X & \longrightarrow & L_E L_K X. \end{array}$$

becomes a homotopy pullback after smashing with $E \vee K$. After smashing with $E$, the horizontal maps become equivalences, and so the diagram is a pullback. After smashing with $K$, the left-hand vertical map is an equivalence and the right-hand vertical map is between trivial objects, so the diagram is also a pullback. Therefore, the diagram becomes a pullback after smashing with $E \vee K$.    □

## 7.10    Model categories

The lifting characterization of local objects from §7.5 falls very naturally into the framework of Quillen's model categories. The groundwork for this is in [54, §10]. See also Chapter 3 of this volume for more on localization of model categories.

Definition 7.10.1. Suppose that $\mathcal{M}$ is a category with a model structure. We say that a second model structure $\mathcal{M}'$ with the same underlying category is a *left Bousfield localization* of $\mathcal{M}$ if $\mathcal{M}'$ has the same family of cofibrations but a larger family of weak equivalences than $\mathcal{M}$.

As a first consequence, note that the identity functor (which is its own right and left adjoint) preserves cofibrations and takes the weak equivalences in $\mathcal{M}$ to weak equivalences in $\mathcal{M}'$. This makes it part of a Quillen adjunction

$$\mathcal{M} \rightleftarrows \mathcal{M}'.$$

This has the immediate consequence that the induced adjunction on homotopy categories is a reflective localization.

Proposition 7.10.2. *Suppose that $L\colon \mathcal{M} \rightleftarrows \mathcal{M}'\colon R$ is the adjunction associated to a left Bousfield localization. Then the right adjoint $R$ identifies the homotopy category $h\mathcal{M}'$ with a full subcategory of $h\mathcal{M}$.*

*Proof.* It is necessary and sufficient to show that the counit $\epsilon\colon LRx \to x$ of the adjunction on homotopy categories is always an isomorphism, for this is the same as asking that, in the factorization

$$\operatorname{Hom}_{h\mathcal{M}}(Rx, Ry) \cong \operatorname{Hom}_{h\mathcal{M}'}(LRx, y) \to \operatorname{Hom}_{h\mathcal{M}'}(x, y),$$

the second map is an isomorphism.

For an object of $y$, the composite functor $LR$ on homotopy categories is calculated as follows: find a fibrant replacement $y \xrightarrow{\sim} y_{f'}$ in $\mathcal{M}'$, apply the identity functor to get to $\mathcal{M}$, find a cofibrant replacement $(y_{f'})_c \xrightarrow{\sim} y_{f'}$ in $\mathcal{M}$, and apply the identity functor to get to $\mathcal{M}'$. The counit of the adjunction is represented in the homotopy category of $\mathcal{M}'$ by the composite

$$(y_{f'})_c \xrightarrow{\sim} y_{f'} \xleftarrow{\sim} y.$$

However, equivalences in $\mathcal{M}$ are automatically equivalences in $\mathcal{M}'$, and so the counit is an isomorphism in the homotopy category of $\mathcal{M}'$. $\square$

Because fibrations and acyclic fibrations are determined by having the right lifting property against acyclic cofibrations and fibrations, the new model structure has the same acyclic fibrations but fewer fibrations. For example, a fibrant object in the left Bousfield localization has to have a lifting property against the cofibrations which are weak equivalences in $\mathcal{M}'$.

The next proposition establishes the connection between left Bousfield localization and ordinary Bousfield localization when both are defined and compatible: the case of a simplicial model category.

Proposition 7.10.3. *Suppose that $\mathcal{M}$ is a simplicially enriched category with two model structures, making $\mathcal{M} \to \mathcal{M}'$ into a left Bousfield localization of simplicial model categories. Let $S$ be the collection of weak equivalences between cofibrant objects in $\mathcal{M}'$. Then, in the category of cofibrant-fibrant objects of $\mathcal{M}$, the objects which are fibrant in $\mathcal{M}'$ are precisely the $S$-local fibrant objects.*

*Proof.* Fix an object $Y$ of $\mathcal{M}'$. For it to be fibrant in $\mathcal{M}'$, it must also be fibrant in $\mathcal{M}$. Suppose $Y$ is a fibrant object in $\mathcal{M}'$. Given any acyclic cofibration $A \to B$ in $\mathcal{M}'$, the map of simplicial sets $\mathrm{Map}_{\mathcal{M}'}(A, Y) \to \mathrm{Map}_{\mathcal{M}'}(B, Y)$ is an acyclic fibration by the SM7 axiom of simplicial model categories. Thus, the functor $\mathrm{Map}_{\mathcal{M}'}(-, Y)$ from $\mathcal{M}'$ to the homotopy category of spaces takes acyclic cofibrations to isomorphisms. Ken Brown's lemma then implies that it also takes weak equivalences between cofibrant objects in $\mathcal{M}'$ to isomorphisms in the homotopy category of spaces.

Suppose that we have a map $f : A \to B$ in $S$ between cofibrant objects of $\mathcal{M}$ that is also a weak equivalence in $\mathcal{M}'$. Then $f$ is also a weak equivalence between cofibrant objects of $\mathcal{M}'$. The induced map $\mathrm{Map}_{\mathcal{M}}(B, Y) \to \mathrm{Map}_{\mathcal{M}}(A, Y)$ is a weak equivalence because the mapping spaces in $\mathcal{M}$ and $\mathcal{M}'$ are the same. Thus, $Y$ is $S$-local. □

We would now like to establish results in the other direction. Namely, given a model category $\mathcal{M}$ and a collection $S$ of maps $A_i \to B_i$ in $\mathcal{M}$, we would like to establish the existence of a Bousfield localization $\mathcal{M}'$ of $\mathcal{M}$. Because we want to work within the already-established homotopy theory of $\mathcal{M}$, we want to use derived mapping spaces out of $A$ and $B$ and replace homotopy lifting properties with strict lifting properties. We assume without loss of generality that our set $S$ is made up of cofibrations $A_i \to B_i$ between cofibrant objects.

**Definition 7.10.4.** Suppose that $\mathcal{M}$ is a simplicial model category, and that $f : A \to B$ is a map. Then the *iterated double mapping cylinders* are the maps

$$(B \otimes \partial \Delta^n) \coprod_{A \otimes \partial \Delta^n} (A \otimes \Delta^n) \to B \otimes \Delta^n.$$

This definition is rigged in such a way that an object $Y$ has the right lifting property with respect to the iterated double mapping cylinders if and only if the map $\mathrm{Map}_{\mathcal{M}}(B, Y) \to \mathrm{Map}_{\mathcal{M}}(A, Y)$ is an acyclic fibration of simplicial sets. One of the equivalent formulations of the SM7 axioms for a simplicial model category is that double mapping cylinders are always cofibrations, as follows.

**Proposition 7.10.5.** *Suppose that $f : A \to B$ is a map. If $f$ is a cofibration, then the iterated double mapping cylinders are cofibrations. If $A$ is also cofibrant, then the iterated double mapping cylinders have cofibrant source.*

*Remark 7.10.6.* If $\mathcal{M}$ does not have a simplicial model structure, we can obtain replacements for these objects by iteratively replacing the maps $B \coprod_A B \to B$ with equivalent cofibrations (see Corollary 7.5.3).

**Definition 7.10.7.** Suppose that $\mathcal{M}$ is a simplicial model category, that $S$ is a collection of maps, and that $T$ is the collection of iterated double mapping cylinders of maps in $S$. We say that a map in $\mathcal{M}$ is an $S$-*cofibration* if it is a cofibration in $\mathcal{M}$, and that it is an $S$-*fibration* if it has the right lifting property with respect to the maps in $T$. If these determine a new model structure $\mathcal{M}'$, we call this the *left Bousfield localization with respect to $S$*.

This gives us two fundamentally different approaches to the process of constructing a left Bousfield localization. In the first, we may try to expand our family of weak equivalences to some new family $\mathcal{W}$; we must then prove that we can construct enough fibrations and fibrant objects to make the model structure work. In the second, we may try to start with some collection of maps $S$ which serve as new "cells" to build acyclic cofibrations, and use them to contract our family of fibrations; we then lose control over the weak equivalences, and typically must work to prove that cofibrations which are weak equivalences can be built out of our new cells.

The most advanced technology available for Bousfield localization is Jeff Smith's theory of combinatorial model categories.

Definition 7.10.8. A model category $\mathcal{M}$ is *cofibrantly generated* if there are sets $I$ and $J$ of maps satisfying the following properties:

1.  the fibrations in $\mathcal{M}$ are the maps that have the right lifting property with respect to $J$;
2.  the acyclic fibrations in $\mathcal{M}$ are the maps that have the right lifting property with respect to $I$;
3.  $I$ permits the small object argument, so that from any object $X$ we can construct a map $X \to X'$, as a transfinite composition of pushouts along coproducts of maps in $I$, that has the right lifting property with respect to $I$;
4.  $J$ also permits the small object argument.

We refer to $I$ as the set of *generating cofibrations* and to $J$ as the set of *generating acyclic cofibrations*.

Further, the cofibrantly generated model category is *combinatorial* if it is also locally presentable, meaning there exists a regular cardinal $\kappa$ and a set $\mathcal{M}_0$ of objects satisfying the following properties:

1.  Any small diagram in $\mathcal{M}$ has a colimit.
2.  For any object $x$ in $\mathcal{M}_0$, the functor $\mathrm{Hom}_{\mathcal{M}}(x,-)$ commutes with $\kappa$-filtered colimits.
3.  Every object in $\mathcal{M}$ is a $\kappa$-filtered colimit of objects in $\mathcal{M}_0$.

Theorem 7.10.9 (Dugger's theorem [79]). *Any combinatorial model category is Quillen equivalent to a left proper simplicial model category.*

*Remark 7.10.10.* The axioms of a cofibrantly generated model category and a locally presentable category have nontrivial overlap. In one direction, the model category axioms already ask that $\mathcal{M}$ has all colimits. In the other direction, being locally presentable means that *every* set of maps admits the small object argument.

Example 7.10.11. Simplicial sets are the motivating example of a combinatorial model category. Fibrations and acyclic fibrations are defined as having the right lifting property with respect to the generating acyclic cofibrations $\Lambda_i^n \to \Delta^n$ and the generating cofibrations $\partial\Delta^n \to \Delta^n$. The category is also locally presentable because it is generated by finite simplicial sets. Every simplicial set is the filtered colimit of its finite subobjects; there are only countably many isomorphism classes of finite simplicial sets; for any finite simplicial set $X$, $\text{Hom}(X, -)$ commutes with filtered colimits.

Theorem 7.10.12 (Smith's theorem [35, 25, 169]). *Suppose that $\mathcal{M}$ is a locally presentable category with a family $\mathcal{W}$ of* weak equivalences *and a set $I$ of* generating cofibrations. *Call those maps which have the right lifting property with respect to $I$ the* acyclic fibrations, *and those maps which have the left lifting property with respect to acyclic fibrations the* cofibrations. *Suppose further that*

1. *$\mathcal{W}$ satisfies the 2-out-of-3 axiom;*
2. *acyclic fibrations are in $\mathcal{W}$;*
3. *the class of cofibrations which are in $\mathcal{W}$ is closed under pushout and transfinite composition; and*
4. *maps in $\mathcal{W}$ are closed under $\kappa$-filtered colimits for some regular cardinal $\kappa$, and generated under $\kappa$-filtered colimits by some set of maps in $\mathcal{W}$.*

*Then there exists a combinatorial model structure on $\mathcal{M}$ with set $I$ of generating cofibrations and set $\mathcal{W}$ of weak equivalences. This model structure on $\mathcal{M}$ has cofibrant and fibrant replacement functors. Moreover, any combinatorial model structure arises in this fashion.*

Corollary 7.10.13. *Suppose that $\mathcal{M}$ is a combinatorial model category with set $I$ of generating cofibrations and class $\mathcal{W}$ of weak equivalences. Given a functor $E\colon \mathcal{M} \to \mathcal{D}$ factoring through the homotopy category $h\mathcal{M}$, define a map to be an $E$-equivalence if its image under $E$ is an isomorphism. Then there exists a left Bousfield localization $\mathcal{M}_E$, whose equivalences are the $E$-equivalences, if the following conditions hold:*

1. *$E$-equivalence is preserved by transfinite composition along cofibrations.*
2. *Pushouts of $E$-acyclic cofibrations are $E$-equivalences.*
3. *There exists a set of $E$-acyclic cofibrations that generate all $E$-acyclic cofibrations under $\kappa$-filtered colimits.*

*Proof.* The 2-out-of-3 axiom is automatic: if two of $E(g)$, $E(f)$ and $E(gf) = E(g)E(f)$ are isomorphisms, then so is the third. The fact that $E$ factors through the homotopy category automatically implies that acyclic fibrations are taken by $E$ to isomorphisms. $\square$

Example 7.10.14. Let $E_*$ be a homology theory on the category of simplicial sets. The excision and direct limit axioms for homology imply that $E$-equivalences are preserved by homotopy pushouts and transfinite compositions. Therefore, the verification that we have a model structure is immediately reduced to the core of the Bousfield–Smith cardinality argument of Example 7.9.15: that there is a set of $E$-acyclic cofibrations generating all others under filtered colimits.

The great utility of combinatorial model structures is that they allow us to *build* new model categories: categories of diagrams and Bousfield localizations.

Theorem 7.10.15 ([169, A.2.8.2, A.3.3.2]). *Suppose that* $\mathcal{M}$ *is a combinatorial model category and that* $I$ *is a small category. Then there exists a* projective *(resp.* injective*) model structure on the functor category* $\mathcal{M}^I$, *where a natural transformation of diagrams is an equivalence or fibration (resp. cofibration) if and only if it is an objectwise equivalence or fibration (resp. cofibration).*

*If* $\mathcal{M}$ *is a simplicial model category, then the natural simplicial enrichment on* $\mathcal{M}^I$ *makes the injective and projective model structures into simplicial model categories.*

Theorem 7.10.16 ([169, A.3.7.3]). *Suppose that* $\mathcal{M}$ *is a left proper combinatorial simplicial model category and that* $S$ *is a set of cofibrations in* $\mathcal{M}$. *Let* $S^{-1}\mathcal{M}$ *have the same underlying category as* $\mathcal{M}$ *and the same cofibrations, but with weak equivalences the S-equivalences.*

*Then* $S^{-1}\mathcal{M}$ *has the structure of a left proper combinatorial model category, whose fibrant objects are precisely the S-local fibrant objects of* $\mathcal{M}$.

## 7.11    Presentable ∞-categories

Bousfield localization for model categories has the useful property that it *keeps the category in place* and merely changes the equivalences. One cost is that making localization canonical or extending monoidal structures to localized objects takes hard work. By contrast, localization for ∞-categories has the useful property that it is genuinely *defined by a universal property*, automatically making localization canonical and making it much easier to extend a monoidal structure to local objects without rectifying structure. Of course, this comes at the cost of coming to grips with coherent category theory itself.

The homotopy theory of presentable ∞-categories is equivalent, in a precise sense, to the homotopy theory of combinatorial model categories [169, A.3.7.6]. However, by contrast with our techniques for Bousfield localization using model categories and fibrant replacement functors, it allows us to rephrase some of our localization techniques in a way that connects more directly with the homotopical techniques that we originally used in §7.5.

In this section, we will let $\mathcal{C}$ be an ∞-category in the sense of [169]. (It is outside the scope of this chapter to give a technically correct discussion of ∞-categories; please see Chapter 2 of this volume for a more detailed introduction to them). The study of ∞-categories is equivalent to the study of categories with morphism spaces, and where possible we will attempt to make connection with classical techniques. With this in mind, if $\mathcal{C}$ is an enriched category we will say that a *coherent diagram* $I \to \mathcal{C}$ is a coherent functor in the sense of Vogt [295]. This is equivalent to either the notion of a functor $\mathfrak{C}[I] \to \mathcal{C}$ from a certain simplicially enriched category or to the notion of a functor $I \to N\mathcal{C}$ of simplicial sets to the coherent nerve in the sense of [169]. As before a *homotopy colimit* for such a diagram is based on classical homotopy limits and colimits in spaces, and is characterized by having natural weak equivalences

$$\mathrm{Map}_{\mathcal{C}}(\mathrm{hocolim}_I F(i), Y) \simeq \mathrm{holim}_I \mathrm{Map}_{\mathcal{C}}(F(i), Y).$$

**Definition 7.11.1** ([169, 5.5.1.1]). An $\infty$-category $C$ is *presentable* if there exists a regular cardinal $\kappa$ and a set $C_0$ of objects satisfying the following properties:

1. Any small diagram in $C$ has a homotopy colimit.
2. For any object $x$ in $C_0$, the functor $\mathrm{Hom}_C(x, -)$ commutes with $\kappa$-filtered homotopy colimits.
3. Every object in $C$ is a $\kappa$-filtered homotopy colimit of objects in $C_0$.

This definition is precisely parallel to the definition of local presentability in an ordinary category (see Definition 7.10.8). In essence, $C$ is a large category that is formally generated under colimits by a small category.

Given such an $\infty$-category $C$ and a collection $S$ of morphisms in $C$, it makes sense to define the $S$-local objects and $S$-equivalences just as in §7.4: an object $Y$ is $S$-local if and only if the mapping spaces $\mathrm{Map}_C(-, Y)$ take maps in $S$ to equivalences of spaces.

**Definition 7.11.2** ([169, 5.5.4.5]). Suppose that $C$ is an $\infty$-category with small colimits and that $\mathcal{W}$ is a collection of maps in $C$. We say that $\mathcal{W}$ is *strongly saturated* if it satisfies the following conditions:

1. Given a homotopy pushout diagram

$$
\begin{array}{ccc}
C & \xrightarrow{\ f\ } & D \\
\downarrow & & \downarrow \\
C' & \xrightarrow[\ f'\ ]{} & D'
\end{array}
$$

   if $f$ is in $\mathcal{W}$ then so is $f'$.
2. The class $\mathcal{W}$ is closed under homotopy colimits.
3. The class $\mathcal{W}$ is closed under equivalence, and its image in the homotopy category satisfies the 2-out-of-3 axiom.

**Proposition 7.11.3** ([169, 5.5.4.7]). *Given a set $S$ of morphisms in $C$, there is a smallest saturated class of morphisms containing $S$. We denote it by $\bar{S}$. If $\mathcal{W} = \bar{S}$ for some set $S$, then we say that $\mathcal{W}$ is of small generation.*

**Example 7.11.4.** Suppose that $E\colon C \to C'$ is a functor of $\infty$-categories that preserves homotopy colimits. Then the set $\mathcal{W}^E$ of maps in $C$ that map to equivalences is strongly saturated.

The presentability axioms for an $\infty$-category provide a homotopical version of what we needed to construct localizations by ensuring that the small object argument goes through. As a result, we obtain a result on the existence of Bousfield localizations for presentable $\infty$-categories.

**Theorem 7.11.5** ([169, 5.5.4.15]). *Let $C$ be a presentable $\infty$-category and $S$ a set of morphisms in $C$, generating the saturated class $\bar{S}$. Let $L^S C$ be the full subcategory of $S$-local objects. Then*

1. *for every object $C \in C$, there is a map $C \to C'$ in $\bar{S}$ such that $C'$ is $S$-local;*

2. *the ∞-category $L^S C$ is presentable;*
3. *the inclusion $L^S C \to C$ has a (homotopical) left adjoint L;*
4. *the class of S-equivalences coincides with both the saturated class $\overline{S}$ and the set of maps taken to equivalences by L.*

*Remark 7.11.6.* The homotopical left adjoint can be rephrased as follows. If we write $Loc^S(C)$ for the category of $S$-localizations $C \to C'$, then the forgetful functor

$$Loc^S(C) \to C,$$

sending $(C \to C')$ to $C$, is an equivalence of categories (in fact, a trivial fibration of quasicategories). By choosing a section, given by $C \mapsto (C \to LC)$, we obtain a localization functor $L$.

As in the case of Bousfield localization of combinatorial model categories, this connects the two approaches to Bousfield localization. We can start with a set $S$ of generating equivalences and construct localizations from those, so for a given class $\mathcal{W}$ of weak equivalences we are reduced to showing that $\mathcal{W}$ is generated by a set $S$ of maps. Moreover, if the maps in $S$ all happen to be in a particular saturated class, then so are the maps in $\mathcal{W}$.

## 7.12    Multiplicative properties

Many of the categories where we carry out Bousfield localization have monoidal structures, and under good circumstances localization is compatible with them. In this section we will briefly discuss the circumstances under which this is true. The interested reader should consult [301, 66] for further information.

### Enriched monoidal structures

In order to begin work, we need a monoidal or symmetric monoidal structure on $C$ that respects morphism spaces.

**Definition 7.12.1.** Suppose $C$ is a category enriched in spaces. The structure of an *enriched monoidal category* on $C$ consists of a functor $\otimes\colon C \times C \to C$ of enriched categories, a unit object $\mathbb{I}$ of $C$, and natural associativity and commutativity isomorphisms that satisfy the axioms for a monoidal category.

A compatible symmetric monoidal structure on $C$ is defined similarly.

Throughout this section we will fix such an enriched monoidal category $C$.

**Definition 7.12.2.** Suppose that $S$ is a class of morphisms in $C$. We say that $S$-equivalences are *compatible with the monoidal structure* (or simply that $S$ is compatible) if, for any $S$-equivalence $f\colon Y \to Y'$ and any object $X \in C$, the maps $id_X \otimes f$ and $f \otimes id_X$ are $S$-equivalences.

**Proposition 7.12.3.** *Suppose that S is compatible with the monoidal structure. Then localization respects the monoidal structure: any choices of localization give an equivalence*

$$L(X_1 \otimes \cdots \otimes X_n) \to L(LX_1 \otimes \cdots \otimes LX_n).$$

*Proof.* By induction, the map $X_1 \otimes \cdots \otimes X_n \to LX_1 \otimes \cdots \otimes LX_n$ is an $S$-equivalence, and therefore any $S$-localization of the latter is equivalent to any $S$-localization of the former. □

**Corollary 7.12.4.** *The monoidal structure on the homotopy category of $C$ induces a monoidal structure on the homotopy category of the localization $L^S C$, making any localization functor into a monoidal functor. If $C$ was symmetric monoidal, then so is the localization.*

*Remark 7.12.5.* The inclusion $L^S C \to C$ is almost never monoidal. For example, it usually does not preserve the unit.

**Example 7.12.6.** Let $C$ be the category of spaces with cartesian product, and let $E_*$ be a homology theory. Then any map $X \to X'$ which induces an isomorphism on $E_*$-homology also induces isomorphisms $E_*(X \times Y) \to E_*(X' \times Y)$ for any CW-complex $Y$: one can prove this inductively on the cells of $Y$. Therefore, $E$-homology equivalences are compatible with the Cartesian product monoidal structure.

Similarly, $E$-homology equivalences are compatible with the smash product on based spaces (using that based spaces are built from $S^0$) or the smash product on spectra (using that all spectra are built from spheres $S^n$).

**Example 7.12.7.** Let $C$ be the category of spectra, and $f$ be the map $S^n \to *$. Then $f$-equivalences are maps inducing isomorphisms in degree strictly less than $n$. This is not compatible with the smash product on spectra: for example, smashing with $\Sigma^{-1}\mathbb{S}$ does not preserve $f$-equivalences. If one restricts to the subcategory of *connective* spectra, however, one finds that $f$-equivalences are compatible with the smash product.

**Example 7.12.8.** Consider the map $f : S^n \to *$ of spaces, so that $S$-equivalences are maps inducing an isomorphism on all homotopy groups in degrees less than $n$. This map is compatible with several symmetric monoidal structures, such as

1. spaces with Cartesian product,
2. spaces with disjoint union,
3. based spaces with wedge product. and
4. based spaces with smash product.

Despite the usefulness of these results, the existence of a (symmetric) monoidal localization functor on the homotopy category does not, by itself, allow us to extend very structured multiplication from an object $X$ to its localization $LX$. To counter this we typically require the theory of operads. See Chapter 5 of this volume for an introduction to operads.

**Definition 7.12.9.** Suppose that $C$ is (symmetric) monoidal, and that $X$ is an object of $C$. The endomorphism operad $\text{End}_C(X)$ is the (symmetric) sequence of spaces $\text{Map}_C(X \otimes \cdots \otimes X, X)$, with (symmetric) operad structure given by composition.

Given a map $f \colon X \to Y$, the endomorphism operad $\text{End}_C(f)$ is the (symmetric) sequence which in degree $n$ is the pullback diagram

$$
\begin{array}{ccc}
\text{End}_C(f)_n & \longrightarrow & \text{Map}_C(X \otimes \cdots \otimes X, X) \\
\downarrow & & \downarrow \\
\text{Map}_C(Y \otimes \cdots \otimes Y, Y) & \longrightarrow & \text{Map}_C(X \otimes \cdots \otimes X, Y)
\end{array}
$$

The space $\text{End}_C(f)_n$ is the space of strictly commutative diagrams

$$
\begin{array}{ccc}
X^{\otimes n} & \longrightarrow & X \\
f^{\otimes n} \downarrow & & \downarrow f \\
Y^{\otimes n} & \longrightarrow & Y
\end{array}
$$

and as such the operad structure is given by composition.

The operad $\text{End}_C(f)$ has forgetful maps to $\text{End}_C(X)$ and $\text{End}_C(Y)$. The following results are a special case of [66, 6.1].

**Proposition 7.12.10.** *Suppose that the (symmetric) monoidal structure on $C$ is compatible with $S$ and that $f \colon X \to LX$ is an $S$-localization. If the maps $\text{Map}_C(LX^{\otimes n}, LX) \to \text{Map}_C(X^{\otimes n}, LX)$ are fibrations for all $n \geq 0$, then in the diagram of operads*

$$
\text{End}_C(X) \leftarrow \text{End}_C(f) \to \text{End}_C(LX),
$$

*the left-hand arrow is an equivalence on the level of underlying spaces.*

*Proof.* This is merely the observation that $\text{End}_C(f) \to \text{End}_C(X)$ is, level by level, a homotopy pullback of the equivalences $\text{Map}_C(LX^{\otimes n}, LX) \to \text{Map}_C(X^{\otimes n}, LX)$. $\square$

This condition then allows us to lift structured multiplication.

**Corollary 7.12.11.** *Suppose a (symmetric) operad $\mathcal{O}$ acts on $X$ via a map $C \to \text{End}_C(X)$. Then there exists a weak equivalence $\mathcal{O}' \to \mathcal{O}$ of operads and an action of $\mathcal{O}'$ on $LX$ such that $f$ is a map of $\mathcal{O}'$-algebras.*

*Proof.* We define $\mathcal{O}'$ to be the fiber product of the diagram $\mathcal{O} \to \text{End}_C(X) \leftarrow \text{End}_C(f)$. The map $\mathcal{O}' \to \mathcal{O}$ is an equivalence by the fibration condition, and the map of operads $\mathcal{O}' \to \text{End}_C(f)$ precisely states that $f$ is a map of $\mathcal{O}'$-algebras.[19] $\square$

This means that $A_\infty$ and $E_\infty$ multiplications on $X$ extend automatically to $A_\infty$ and $E_\infty$ multiplications on $LX$. However, this is the best we can do in general: lifting more refined multiplicative structures requires stronger assumptions.

In cases where the category $C$ has more structure, it is typically easier to verify that $S$ is compatible with the monoidal structure.

---

[19] If $\mathcal{O}$ happens to be a cofibrant (symmetric) operad $\mathcal{O}$ in Berger–Moerdijk's model structure [39] we can do better. Any map $\mathcal{O} \to \text{End}_C(X)$ lifts, up to homotopy, to a map $\mathcal{O} \to \text{End}_C(f) \to \text{End}_C(LX)$.

**Proposition 7.12.12.** *Suppose that the monoidal structure on $C$ has internal function objects $F^L(X, Y)$ and $F^R(X, Y)$ that are adjoint to the monoidal structure: there are isomorphisms*

$$\mathrm{Map}_C(X, F^L(Y, Z)) \cong \mathrm{Map}(X \otimes Y, Z) \cong \mathrm{Map}_C(Y, F^R(X, Z))$$

*that are natural in $X$, $Y$, and $Z$. Then $S$ is compatible with the monoidal structure on $C$ if and only if, for any $f : A \to B$ in $S$ and any object $X \in C$, the maps $id_X \otimes f$ and $f \otimes id_X$ are $S$-equivalences.*

*Proof.* Suppose that for any $f : A \to B$ in $S$ and any object $X \in C$, the maps $id_X \otimes f$ are $S$-equivalences. Using the unit isomorphisms, we find that if $Z$ is $S$-local the maps in the diagram

$$
\begin{array}{ccc}
\mathrm{Map}_C(X \otimes B, Z) & \longrightarrow & \mathrm{Map}_C(X \otimes A, Z) \\
\downarrow & & \downarrow \\
\mathrm{Map}_C(B, F^R(X, Z)) & \longrightarrow & Map_C(A, F^R(X, Z))
\end{array}
$$

are equivalences. Thus, $F^R(X, Z)$ is $S$-local, and so for any $S$-equivalence $f : Y \to Y'$ the maps in the diagram

$$
\begin{array}{ccc}
\mathrm{Map}_C(X \otimes Y', Z) & \longrightarrow & \mathrm{Map}_C(X \otimes Y, Z) \\
\downarrow & & \downarrow \\
\mathrm{Map}_C(Y', F^R(X, Z)) & \longrightarrow & Map_C(Y, F^R(X, Z))
\end{array}
$$

are all equivalences. Similar considerations apply to $F^L$.     □

## Monoidal model categories

The necessary conditions for compatibility between model structures and monoidal structures were determined by Schwede–Shipley [267] and Hovey [130, §4.2], in the symmetric and nonsymmetric cases respectively. This structure allows us, after [267], to construct model structures on categories of algebras and modules in $M'$ such that the localization functor $M \to M'$ preserves this structure. See, also, Chapter 3 of this volume for a detailed discussion of monoidal model categories.

**Definition 7.12.13.** A *(symmetric) monoidal model category* $M$ is a model category with a (symmetric) monoidal closed structure[20] satisfying the following axioms.

1. (Pushout-product) Given cofibrations $i : A \to A$ and $j : B \to B'$ in $M$, the induced pushout-product map

$$i \boxtimes j : (A \otimes B') \coprod_{A \otimes B} (A' \otimes B) \to A' \otimes B'$$

is a cofibration, which is acyclic if either $i$ or $j$ is.

[20] Analogously to the previous section, this means that the symmetric monoidal structure must have left and right function objects which are adjoints in each variable.

2. (Unit) Let $Q\mathbb{I} \to \mathbb{I}$ be a cofibrant replacement of the unit. Then the natural maps $Q\mathbb{I} \otimes X \to X \leftarrow X \otimes Q\mathbb{I}$ are isomorphisms for all cofibrant $X$.

**Proposition 7.12.14.** *Suppose that $\mathcal{M}$ is a monoidal model category. Then, for cofibrant objects $X$, the functors $X \otimes (-)$ and $(-) \otimes X$ preserve cofibrations, acyclic cofibrations, and weak equivalences between cofibrant objects.*

*Proof.* Since $\otimes$ has adjoints, it preserves colimits in each variable. In particular, any object tensored with an initial object of $\mathcal{M}$ is an initial object of $\mathcal{M}$. Applying the pushout-product axiom to the map $\emptyset \to X$ in either variable, we find that the two functors in question preserve cofibrations and acyclic cofibrations. By Ken Brown's lemma, they also automatically take weak equivalences between cofibrant objects to weak equivalences. □

This connects with the previous section, which only asked that the tensor product preserved equivalences in each variable. The pushout-product axiom for monoidal model categories looks stronger, in principle, but Proposition 7.12.14 has a partial converse.

**Proposition 7.12.15.** *Suppose that $j \colon B \to B'$ is a map such that $(-) \otimes B$ preserves acyclic cofibrations and that $(-) \otimes B'$ preserves weak equivalences between cofibrant objects. If $i$ is an acyclic cofibration with cofibrant source, then the pushout-product map $i \boxtimes j$ is an equivalence.*

*Proof.* Without loss of generality, let $i \colon A \to A'$ be an acyclic cofibration and $j \colon B \to B'$ a cofibration, with all four objects cofibrant. Then the pushout-product $i \boxtimes j$ is part of the following diagram:

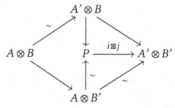

The upper-left and lower-right maps are equivalences because they are obtained by tensoring an acyclic cofibration with the cofibrant objects $B$ and $B'$. The map $A \otimes B' \to P$ is the pushout of an acyclic cofibration, and so it is an acyclic cofibration. Therefore, by the 2-out-of-3 axiom the map $i \boxtimes j$ is an equivalence. □

The adjunction isomorphism $\mathrm{Hom}_{\mathcal{M}}(X \otimes Y, Z) \cong \mathrm{Hom}_{\mathcal{M}}(X, F^R(Y, Z))$, and similarly for the left, allows us to rephrase the pushout-product axiom in multiple ways.

**Proposition 7.12.16** ([130, 4.2.2]). *The following are equivalent for a model category $\mathcal{M}$ with a closed monoidal structure.*

1. *The model category $\mathcal{M}$ satisfies the pushout-product axiom.*

2. *For a cofibration* $i\colon A \to B$ *and a fibration* $p\colon X \to Y$ *in* $\mathcal{M}$, *the induced map*

$$F^R(B,X) \to F^R(B,Y) \times_{F^R(A,Y)} F^R(A,X)$$

*is a fibration, which is acyclic if either* $i$ *or* $p$ *are.*

3. *For a cofibration* $i\colon A \to B$ *and a fibration* $p\colon X \to Y$ *in* $\mathcal{M}$, *the induced map*

$$F^L(B,X) \to F^L(B,Y) \times_{F^L(A,Y)} F^L(A,X)$$

*is a fibration, which is acyclic if either* $i$ *or* $p$ *are.*

**Corollary 7.12.17** ([130, 4.2.5]). *Suppose that* $\mathcal{M}$ *is a cofibrantly generated model category with a closed monoidal structure, a set* $I$ *of generating cofibrations and* $J$ *of generating acyclic cofibrations. Then the pushout-product axiom for* $\mathcal{M}$ *holds if and only if the pushout-product takes* $I \times I$ *to cofibrations in* $\mathcal{M}$ *and takes both* $I \times J$ *and* $J \times I$ *to acyclic cofibrations.*

Because left Bousfield localization doesn't change the cofibrations in a model structure, to verify compatibility of a monoidal model structure with a localization we are reduced to a few key verifications.

**Proposition 7.12.18** ([300, 4.5, 4.6], [301]). *Suppose that* $\mathcal{M}$ *is a (symmetric) monoidal closed model category with left Bousfield localization* $\mathcal{M}'$. *Then* $\mathcal{M}'$ *is compatibly a (symmetric) monoidal model category if and only if, for cofibrations* $i$ *and* $j$ *such that one is acyclic, the pushout-product map* $i \boxtimes j$ *is acyclic.*

*If* $\mathcal{M}'$ *is cofibrantly generated, then it suffices to check that the pushout-product of a generating acyclic cofibration with a generating cofibration, in either order, is a weak equivalence.*

*Remark 7.12.19.* If the generating cofibrations and generating acyclic cofibrations of $\mathcal{M}'$ have cofibrant source, then by Proposition 7.12.15 we only need to show that tensoring with the sources or target of any map in $I$ or $J$ takes generating cofibrations in $\mathcal{M}'$ to weak equivalences.

*Remark 7.12.20.* Bousfield localization of stable model categories has been more extensively studied by Barnes and Roitzheim [22, 21]. To have homotopical control over *commutative* algebra objects in a symmetric monoidal model category, one needs to obtain control over the extended power constructions; see [299].

## Monoidal ∞-categories

We will begin by giving a brief background on monoidal structures on ∞-categories which is light on technical details.

Recall that a *multicategory* $\mathcal{O}$ is equivalent to the following data:

1. a collection of objects of $\mathcal{O}$;
2. for any object $Y$ and indexed set of objects $\{X_s\}_{s\in S}$ of $\mathcal{O}$, a space $\mathrm{Map}_{\mathcal{O}}(\{X_s\}_{s\in S}; Y)$ of multimaps; and

3. for a surjection $p \colon S \to T$ of finite sets, natural composition maps

$$\mathrm{Map}_{\mathcal{O}}(\{Y_t\}_{t \in T}; Z) \times \prod_{t \in T} \mathrm{Map}_{\mathcal{O}}(\{X_s\}_{s \in p^{-1}(t)}; Y_t) \to \mathrm{Map}_{\mathcal{O}}(\{X_s\}_{s \in S}; Z)$$

that are compatible with composing surjections $S \to T \to U$.

*Remark 7.12.21.* As a special case, for $\sigma$ a permutation of $S$ there is an isomorphism $\mathrm{Map}_{\mathcal{O}}(\{X_s\}_{s \in S}; Y) \to \mathrm{Map}_{\mathcal{O}}(\{X_{\sigma(s)}\}_{s \in S}; Y)$, and the composition operations are appropriately equivariant with respect to these isomorphisms.

For such a multicategory, we could give a prototype definition of an $\mathcal{O}$-monoidal $\infty$-category $\mathcal{C}$ as an enriched functor from $\mathcal{O}$ to $\infty$-categories. This data specifies, for each object $X$ of $\mathcal{O}$, a category $\mathcal{C}_X$. For each object $Y$ and indexed set $\{X_s\}_{s \in S}$ of objects, there is a specified continuous map from $\mathrm{Map}_{\mathcal{O}}(\{X_s\}_{s \in S}; Y)$ to the space of functors $\prod_{s \in S} \mathcal{C}_{X_s} \to \mathcal{C}_Y$. Moreover, these maps must be compatible with composition on both sides.

The definition of an $\infty$-operad $\mathcal{O}$ and an $\mathcal{O}$-monoidal $\infty$-category $\mathcal{C}$ is slightly different from this [168, §2.1]. Roughly, it is an *unstraightened* definition where the spaces of multimaps in $\mathcal{O}$ and the product functors on $\mathcal{C}$ are only specified up to a contractible space of choices; the technical details are related in spirit to Segal's work [268]. Even though the functors induced from $\mathcal{O}$ are specified only up to contractible indeterminacy, it still makes sense to ask about compatibility of the monoidal structure with localization.

The following very general result encodes the situations under which homotopical localization is compatible with monoidal structures.

**Theorem 7.12.22** ([168, 2.2.1.9]). *Let $\mathcal{O}^{\otimes}$ be an $\infty$-operad and $\mathcal{C}$ an $\mathcal{O}$-monoidal $\infty$-category. Suppose that for all objects $X$ of $\mathcal{O}$ we have a localization functor $L_X \colon \mathcal{C}_X \to \mathcal{C}_X$, and that for any map $\alpha \colon \{X_s\}_{s \in S} \to Y$ in $\mathcal{O}^{\otimes}$ the induced functor $\prod_{s \in S} \mathcal{C}_{X_s} \to \mathcal{C}_Y$ preserves $L$-equivalences in each variable. Then there exists an $\mathcal{O}$-monoidal structure on the category $L\mathcal{C}$ of local objects making the localization $L \colon \mathcal{C} \to L\mathcal{C}$ into an $\mathcal{O}$-monoidal functor.*

**Corollary 7.12.23.** *Suppose that $\mathcal{C}$ is a (symmetric) monoidal $\infty$-category and that $L$ is a localization functor on $\mathcal{C}$ such that $L(X \otimes Y) \to L(LX \otimes LY)$ is always an equivalence. Then the subcategory $L\mathcal{C}$ of local objects has the structure of a (symmetric) monoidal $\infty$-category and any localization functor $L$ has the structure of a (symmetric) monoidal functor.*

**Example 7.12.24.** In the category of spaces, we can use the mapping space adjunctions and find that for any $S$-local object $Z$, we have

$$\begin{aligned}
\mathrm{Map}(X \times Y, Z) &\simeq \mathrm{Map}(X, \mathrm{Map}(Y, Z)) \\
&\simeq \mathrm{Map}(X, \mathrm{Map}(LY, Z)) \\
&\simeq \mathrm{Map}(X \times LY, Z)
\end{aligned}$$

and similarly on the other side, showing that $LX \times LY$ is a localization of $X \times Y$. This gives the cartesian product on spaces the special property that it is compatible with *all* localization functors.

Example 7.12.25. Fix an $E_n$-operad $\mathcal{O}$ and an $\mathcal{O}$-algebra $B$ in spaces representing an $n$-fold loop space. Consider the category $\mathcal{C}$ of functors $B \to \mathcal{S}$, viewed as local systems of spaces over $B$. Then the category $\mathcal{C}$ has a *Day convolution*, developed by Glasman [105] in the $E_\infty$-case and by Lurie [168, §2.2.6] in general, making $\mathcal{C}$ into an $\mathcal{O}$-monoidal category. The category $\mathcal{C}$ is equivalent (via unstraightening) to the category of spaces over $B$. In these terms the $\mathcal{O}$-monoidal structure is given by maps

$$\mathcal{O}(n) \to \mathrm{Map}(B^n, B)$$
$$\to \mathrm{Fun}((\mathcal{S}_{/B})^n, \mathcal{S}_{/B})$$

that respect composition. Here $f \in \mathcal{O}(n)$ first goes to $f \colon B^n \to B$, then to the functor sending $\{X_i \to B\}$ to the map

$$\prod X_i \to B^n \xrightarrow{f} B.$$

An $\mathcal{O}$-algebra in $\mathcal{C}$ is equivalent to an $E_n$-space $X$ with a map $X \to B$ of $E_n$-spaces.

Suppose $L$ is a Bousfield localization on spaces and that $B$ is connected. Consider the associated pointwise localization on the functor category $\mathcal{C}$ (which corresponds to the fiberwise localization on spaces over $B$). All operations in $\mathcal{O}$ are, up to homotopy, composites of the binary multiplication operation, and so it suffices to show that this preserves localization. However, if the maps $X_i \to B$ have homotopy fibers $F_i$, then the homotopy fiber of the map $X_1 \times X_2 \to B \times B \to B$ is, up to equivalence, the geometric realization of the bar construction

$$B(F_1, \Omega B, F_2).$$

Since any localization preserves homotopy colimits and products of spaces, this bar construction preserves it also. Therefore, fiberwise localization is an $E_n$-monoidal functor on the category of spaces over $B$.[21]

---

[21] For *grouplike* $E_n$-spaces over $B$, this is roughly the statement that we can take $n$-fold classifying spaces, apply the fiberwise localization, and then take $n$-fold loop spaces.

# 8  Spectral algebraic geometry

by Charles Rezk

## 8.1  Introduction

This chapter is a very modest introduction to some of the ideas of spectral algebraic geometry, following the approach due to Lurie. The goal is to introduce a few of the basic ideas and definitions, with the goal of understanding Lurie's characterization of highly structured elliptic cohomology theories.

### A motivating example: elliptic cohomology theories

*Generalized cohomology theories* are functors which take values in some abelian category. Traditionally, we consider ones which take values in *abelian groups,* but we can work more generally. For instance, take cohomology theories which take values in *sheaves of graded abelian groups (or rings)* on some given topological space, or in *sheaves of graded $\mathcal{O}_S$-modules (or rings)* on $S$, where $S$ is a scheme, or possibly a more general kind of geometric object, such as a *Deligne–Mumford stack,* and $\mathcal{O}_S$ is its *structure sheaf.*

Given a scheme (or Deligne–Mumford stack) $S$, it is easy to construct an example of a cohomology theory taking values in graded $\mathcal{O}_S$-algebras; for instance, using ordinary cohomology, we can form

$$\mathcal{F}^*(X) := \left( U \mapsto H^*(X, \mathcal{O}_S(U)) \right),$$

which is a presheaf of graded $\mathcal{O}_S$-algebras on $S$, which in turn can be sheafified into a sheaf of graded $\mathcal{O}_S$-modules on $S$.

A more interesting example is given by *elliptic cohomology theories.* These consist of

1. an elliptic curve $\pi \colon C \to S$ (which is in particular an algebraic group with an identity section $e \colon S \to C$),

2. a multiplicative generalized cohomology theory $\mathcal{F}^*$ taking values in sheaves graded commutative $\mathcal{O}_S$-algebras, which is *even and weakly 2-periodic* in the sense that $\mathcal{F}^{\mathrm{odd}}(\mathrm{point}) \approx 0$ while $\mathcal{F}^0(\mathrm{point}) \approx \mathcal{O}_S$ and $\mathcal{F}^2(\mathrm{point})$ is an invertible $\mathcal{O}_S$-module, together with

3. a choice of isomorphism

$$\alpha \colon \mathrm{Spf}\,\mathcal{F}^0(\mathbb{CP}^\infty) \xrightarrow{\sim} C_e^\wedge$$

of *formal groups*, where the right-hand side denotes the formal completion of the elliptic curve $\pi \colon C \to S$ at the identity section.

This is easiest to think about when $S$ is affine, i.e., $S = \mathrm{Spec}\,A$ for some ring $A$. Then the above data corresponds exactly to what is known as an *elliptic spectrum* [4]: a weakly 2-periodic spectrum $E$ with $\pi_0 E = A$, together an isomorphism of formal groups $\mathrm{Spf}\,E^0\mathbb{CP}^\infty \approx C_e^\wedge$, where $C$ is an elliptic curve defined over the ring $A$. Many such elliptic spectra exist, including some which are structured commutative ring spectra.

For a more general elliptic cohomology theory defined over some base scheme (or stack) $S$, one may ask that it be "represented" by a *sheaf of (commutative ring) spectra* on $S$, which I'll call $\mathcal{O}_S^{\mathrm{top}}$. E.g., for an open subset $U$ of the scheme $S$, and a finite CW-complex $X$, we would have

$$\mathcal{F}^q(X)(U) \approx \pi_0\,\mathrm{Map}_{\mathrm{Spectra}}(\Sigma^{-q}\Sigma^\infty X, \mathcal{O}_S^{\mathrm{top}}(U))$$

where $\mathcal{O}_S^{\mathrm{top}}(U) \in \mathrm{Spectra}$ are the sections of $\mathcal{O}_S^{\mathrm{top}}(U)$ over $U$.

Goerss, Hopkins, and Miller showed that such an object exists, where $S = \mathcal{M}_{\mathrm{Ell}}$ is the moduli stack of (smooth) elliptic curves, and $C \to S$ is the universal elliptic curve. This can be viewed as giving a "universal" example of an elliptic cohomology theory. As a consequence you can take global sections of $\mathcal{O}_S^{\mathrm{top}}$ over the entire moduli stack $S$, obtaining a ring spectrum called $TMF$, the **topological modular forms**. (There is also an extension of this theory to the "compactification" of $\mathcal{M}_{\mathrm{Ell}}$, the moduli stack of *generalized* elliptic curves; I will not discuss this version of the theory here.)

From the point of view of spectral algebraic geometry, the pair $(\mathcal{M}_{\mathrm{Ell}}, \mathcal{O}^{\mathrm{top}})$ is an example of a *nonconnective spectral Deligne–Mumford stack*, i.e., an object in *spectral algebraic geometry*.

Lurie proves a further result, which precisely characterizes the nonconnective spectral Deligne–Mumford stack $S = (\mathcal{M}_{\mathrm{Ell}}, \mathcal{O}^{\mathrm{top}})$. Namely, it is the classifying object for a suitable type of "derived elliptic curve", called an *oriented elliptic curve*. More precisely, for each nonconnective spectral Deligne–Mumford stack $X$ there is an equivalence of $\infty$-groupoids

$$\mathrm{Map}_{\mathrm{SpDM^{nc}}}(X, S) \approx \{\text{oriented elliptic curves over } X\},$$

natural in $X$; here $\mathrm{SpDM}^{\mathrm{nc}}$ denotes the $\infty$-category of nonconnective spectral Deligne–Mumford stacks. In particular, there is a "universal" oriented elliptic curve $C \to S$.

## Organization of this chapter

We describe some of the basic concepts of spectral algebraic geometry. This chapter is written for algebraic topologists, with the example of elliptic cohomology as a prime motivation. This chapter will only give an overview of some of the ideas. I'll give precise definitions and complete proofs when I can (rarely); more often, I will try to give an idea of a definition and/or proof, sometimes by appealing to an explicit example, or to a "classical" analogue.

I will not try to describe applications to geometry or representation theory. The reader should look at Lurie's introduction to [170], as well as Toën's survey [290], to get a better idea of motivations from classical geometry.

We will follow Lurie's approach. This was originally presented in the book *Higher Topos Theory* [169], together with the sequence of "DAG" preprints [163]. Some of the DAG preprints have been incorporated in/superseded by the book *Higher Algebra* [168], while others have been absorbed by the book-in-progress *Spectral Algebraic Geometry* [170]. I try to use notation consistent with [170], and give references to it when possible (references are to the February 2018 version). Note that [170] is still under construction and its numbering and organization is likely to change. Lurie's approach to elliptic cohomology is sketched in [162], and described in detail in [166] and [167].

Derived algebraic geometry had its origins in problems in algebraic geometry, and was first pursued by geometers. We note in particular the work of Toën and Vezzosi, which develops a theory broadly similar to Lurie's; the aforementioned survey [290] is a good introduction.

## Notation and terminology

I'll use the "naive" language of $\infty$-categories pretty freely. When I say "category" I really mean "$\infty$-category", unless "1-category" or "ordinary category" is explicitly indicated. An "isomorphism" in an $\infty$-category is the same thing as an "equivalence"; I use the two terms interchangeably. Sometimes I will say that a construction is "essentially unique", which means it is defined up to contractible choice.

I write $\mathrm{Cat}_\infty$ and $\widehat{\mathrm{Cat}}_\infty$ for the $\infty$-categories of small and locally small $\infty$-categories respectively. I write $\mathcal{S}$ for the $\infty$-category of small $\infty$-groupoids. "Sets" are implicitly identified with the full subcategory of "0-truncated $\infty$-groupoids": thus, Set $\approx \tau_{\leq 0}\mathcal{S} \subseteq \mathcal{S}$. I write $\mathrm{Map}_{\mathcal{C}}(X, Y)$ for the space (= $\infty$-groupoid) of maps between two objects in an $\infty$-category $\mathcal{C}$. I use the notations $\mathcal{C}_{X/}$ and $\mathcal{C}_{/X}$ for the slice categories under and over an object $X$ of $\mathcal{C}$.

I will consistently notate adjoint pairs of functors in the following way. In

$$L: \mathcal{C} \rightleftarrows \mathcal{D} : R \qquad \text{or} \qquad R: \mathcal{D} \leftrightarrows \mathcal{C} : L,$$

the arrow corresponding to the left adjoint is always *above* that for the right adjoint.

I use the notation $C \rightarrowtail D$ for a *fully faithful* functor, and $C \twoheadrightarrow D$ for a *localization* functor, i.e., the universal example of formally inverting a class of arrows in $C$. Note that any adjoint (left or right) of a fully faithful functor is a localization, and any adjoint (left or right) of a localization functor is fully faithful.

I'd like to thank those who suffered through some talks I gave based on an early version of this at University of Illinois, and for the corrections which have been provided by various people, including a careful and detailed list of errata from Ko Aoki.

## 8.2     The notion of an ∞-topos

A scheme is a particular kind of *ringed space*, i.e., a topological space equipped with a sheaf of rings. Spectral algebraic geometry replaces "rings" with an ∞-categorical generalization, namely *commutative ring spectra*, which (following Lurie) we will here call $\mathbb{E}_\infty$-*rings*. Similarly, spectral algebraic geometry replaces "topological space" with its ∞-categorical generalization, which is called an ∞-*topos*.

The key observation motivating ∞-topoi is that a topological space $X$ is determined[1] by the ∞-category of *sheaves* of ∞-groupoids on $X$. I will try to justify this in the next few sections.

The notion of ∞-topos is itself a generalization of a more classical notion, that of a *1-topos* (or *Grothendieck topos*), which can be thought of as the 1-categorical generalization of topological space. I will not have much to say about these, instead passing directly to the ∞-case (but see (8.2) below). However, the theory of ∞-topoi does parallel the classical case in many respects; a good introduction to 1-topoi is [173].

There is a great deal to say about ∞-topoi, so I'll try to say as little as possible. Note that to merely understand the basic definitions of spectral algebraic geometry, only a small part of the theory is necessary: much as, to understand the definition of a scheme, you need enough topology to understand the "Zariski spectrum" of a ring, without any need to inhale large quantities of esoteric results in point-set topology.

We refer to a functor $F\colon \mathcal{C}^{\mathrm{op}} \to \mathcal{S}$ as a **presheaf** of ∞-groupoids on $\mathcal{C}$, and write

$$\mathrm{PSh}(\mathcal{C}) = \mathrm{Fun}(\mathcal{C}^{\mathrm{op}}, \mathcal{S})$$

for the ∞-category of presheaves.

We first describe two examples of ∞-topoi arising from "classical" constructions.

### The ∞-topos of a topological space

Let $X$ be a topological space, with $\mathrm{Open}_X$ = its poset of open subsets. A **sheaf** of ∞-groupoids on $X$ is a functor $F\colon \mathrm{Open}_X^{\mathrm{op}} \to \mathcal{S}$ such that, for every open cover $\{U_i \to U\}_{i\in I}$ of an element $U$ of $\mathrm{Open}_X$, the evident map

$$F(U) \xrightarrow{\sim} \lim_\Delta \left[ [n] \mapsto \prod_{i_0,\dots,i_n \in I} F(U_{i_0} \cap \cdots \cap U_{i_n}) \right] \tag{8.2.1}$$

is an equivalence; the target is the limit of functor $\Delta \to \mathcal{S}$, i.e., of a cosimplicial space. We let $\mathrm{Shv}(X) \subseteq \mathrm{PSh}(\mathrm{Open}_X)$ denote the full subcategory of sheaves. It turns out that this embedding admits a left adjoint $a\colon \mathrm{PSh}(\mathrm{Open}_X) \to \mathrm{Shv}(X)$ which is **left exact**, i.e., $a$ preserves finite limits.

### The ∞-topos of sheaves on the étale site of a scheme

Let $X$ be a scheme, and let $\mathrm{\acute{E}t}_X$ = a full subcategory of the category of schemes over $X$ spanned by a suitable collection of étale morphisms $U \to X$, (e.g., morphisms which

---

[1]  This is not exactly true; see (8.5) below.

factor as $U \xrightarrow{f} V \rightarrowtail X$ where $f$ is a finitely presented étale map to an open affine subset of $X$). An **étale cover** is a collection of étale maps $\{U_i \to U\}_{i \in I}$ in $\text{Ét}_X$ which are jointly surjective on Zariski spectra. We get full subcategory $\text{Shv}(X^{\text{ét}}) \subseteq \text{PSh}(\text{Ét}_X)$ of **étale sheaves** on $X$, whose objects are functors $F \colon \text{Ét}_X^{\text{op}} \to \mathcal{S}$ such that the evident map

$$F(U) \xrightarrow{\sim} \lim_{\Delta}\left[[n] \mapsto \prod_{i_0,\dots,i_n \in I} F(U_{i_0} \times_X \cdots \times_X U_{i_n})\right]$$

is an equivalence for every étale cover. (This makes sense because $\text{Ét}_X$ is an essentially small category which is closed under finite limits.) As in (8.2), the embedding $\text{Shv}(X^{\text{ét}}) \subseteq \text{PSh}(\text{Ét}_X)$ admits a left exact left adjoint.

## Definition of ∞-topos

An ∞-**topos** is an ∞-category $\mathcal{X}$ such that

1. there exists a small ∞-category $\mathcal{C}$, and
2. an adjoint pair

$$i \colon \mathcal{X} \overset{\longleftarrow}{\underset{\longrightarrow}{}} \text{PSh}(\mathcal{C}) \colon a$$

where the right adjoint $i$ is fully faithful (whence $a$ is a localization), and such that
3. $i$ is **accessible**, i.e., there exists a regular cardinal $\lambda$ such that $i$ preserves all $\lambda$-filtered colimits, and
4. $a$ is left exact.

*Remark 8.2.1* (Presentable ∞-categories). An $\mathcal{X}$ for which there exists data (1)–(3) is called a **presentable** ∞-**category** [169, 5.5]. This class includes many familiar examples such as: small ∞-groupoids, chain complexes of modules, spectra, $\mathbb{E}_\infty$-ring spectra, functors from a small ∞-category to a presentable ∞-category, etc. (Note: [169, 5.5.0.1] defines this a little differently, but it is equivalent to what I just said by [169, 5.5.1.1].)

All presentable ∞-categories are complete and cocomplete. The "presentation" $(\mathcal{C}, i, a)$ of $\mathcal{X}$ leads to an explicit recipe for computing limits and colimits in $\mathcal{X}$: apply $i$ to your diagram in $\mathcal{X}$ to get a diagram in $\text{PSh}(\mathcal{C})$, take limits or colimits there, and apply $a$ to get the desired answer. (Since $i$ is a fully faithful right adjoint, the last step of applying $a$ is not even needed when computing limits.)

*Remark 8.2.2* (Adjoint functors between presentable ∞-categories). It turns out that a very strong form of an "adjoint functor theorem" applies to presentable ∞-categories [169, 5.5.2.9].

1. If $\mathcal{A}$ is presentable, then a functor $F \colon \mathcal{A} \to \mathcal{B}$ admits a right adjoint if and only if it preserves small colimits.
2. If $\mathcal{A}$ and $\mathcal{B}$ are presentable, then a functor $F \colon \mathcal{A} \to \mathcal{B}$ admits a left adjoint if and only if it preserves small limits and is accessible.

In particular, if $\mathcal{A}$ is presentable, then a functor $\mathcal{A}^{\mathrm{op}} \to \mathcal{S}$ to $\infty$-groupoids is *representable* if and only if it preserves limits, and $\mathcal{A} \to \mathcal{S}$ is *corepresentable* if and only if it preserves limits and is accessible.

*Remark 8.2.3.* The presentation $(\mathcal{C}, i, a)$ is not part of the structure of an $\infty$-topos (or presentable $\infty$-category): it merely needs to exist, and it is not in any sense unique.

Any presheaf category $\mathrm{PSh}(\mathcal{C})$ is an $\infty$-topos, and in particular $\mathcal{S}$ is one.

Both the examples (8.2) and (8.2) given above are $\infty$-topoi. They are special cases of sheaves on a *Grothendieck topology* on an $\infty$-category $\mathcal{C}$; see (8.5) below and [169, 6.1, 6.2].

## Relation to the classical notion of topos

Recall that an object $U$ of any $\infty$-category $\mathcal{X}$ is 0-**truncated** if $\mathrm{Map}_{\mathcal{X}}(-, U)$ takes values in $\tau_{\leq 0}\mathcal{S} \subseteq \mathcal{S}$, i.e., in "sets". For an $\infty$-topos $\mathcal{X}$, its full subcategory $\mathcal{X}^{\heartsuit} \subseteq \mathcal{X}$ of 0-truncated objects is called the **underlying 1-topos** of $\mathcal{X}$. This $\mathcal{X}^{\heartsuit}$ is equivalent to a 1-category, and is a "classical" topos in the sense of Grothendieck; in fact all Grothendieck topoi arise from $\infty$-topoi in this way.

For instance, if $X$ is a topological space then $\mathrm{Shv}(X)^{\heartsuit}$ is the 1-category of sheaves of *sets* on $X$.

Example 8.2.4. As we'll see (8.4), the slice category $\mathcal{S}_{/X}$ is an $\infty$-topos for any $X \in \mathcal{S}$, and it is easy to verify that $(\mathcal{S}_{/X})^{\heartsuit} \approx \mathrm{Fun}(\Pi_1 X, \mathrm{Set})$. Thus $(\mathcal{S}_{/X})^{\heartsuit}$ only depends on the fundamental groupoid of $X$, while $\mathcal{S}_{/X}$ itself depends on the homotopy type of $X$. Thus, non-equivalent $\infty$-topoi can share the same underlying 1-topos.

## 8.3     Sheaves on an $\infty$-topos

There is an obvious notion of sheaves on a topological space which take values in an arbitrary complete $\infty$-category $\mathcal{A}$. These are functors $F \colon \mathrm{Open}_X^{\mathrm{op}} \to \mathcal{A}$ which satisfy the "sheaf condition", i.e., that the map in (8.2.1) is an equivalence for every open cover. We can reformulate this definition so that it depends only on the $\infty$-category $\mathcal{X} = \mathrm{Shv}(X)$, rather than on the category of open sets in $X$. This leads to a definition of $\mathcal{A}$-valued sheaf which makes sense in an arbitrary $\infty$-topos.

### Sheaves valued in an $\infty$-category

For a general $\infty$-topos, an $\mathcal{A}$-**valued sheaf** on $\mathcal{X}$ is a limit preserving functor $F \colon \mathcal{X}^{\mathrm{op}} \to \mathcal{A}$. These objects form a full subcategory $\mathrm{Shv}_{\mathcal{A}}(\mathcal{X}) \subseteq \mathrm{Fun}(\mathcal{X}^{\mathrm{op}}, \mathcal{A})$.

Example 8.3.1 ($\mathcal{A}$-valued sheaves on a presheaf $\infty$-topos). If $\mathcal{X} = \mathrm{PSh}(\mathcal{C})$, then $\mathrm{Shv}_{\mathcal{A}}(\mathcal{X})$ is equivalent to the category $\mathrm{Fun}(\mathcal{C}^{\mathrm{op}}, \mathcal{A})$ of "$\mathcal{A}$-valued presheaves" on $\mathcal{C}$. This is because the Yoneda embedding $\rho \colon \mathcal{C} \to \mathrm{PSh}(\mathcal{C})$ is the "free colimit completion" of $\mathcal{C}$ [169, 5.1.5]: for any cocomplete $\mathcal{B}$, restriction along $\rho$ gives an equivalence

$$\mathrm{Fun}(\mathrm{PSh}(\mathcal{C}), \mathcal{B}) \supseteq \mathrm{Fun}^{\mathrm{colim\ pres.}}(\mathrm{PSh}(\mathcal{C}), \mathcal{B}) \xrightarrow{\sim} \mathrm{Fun}(\mathcal{C}, \mathcal{B})$$

between the full subcategory of *colimit preserving* functors $\mathrm{PSh}(\mathcal{C}) \to \mathcal{B}$ and all functors $\mathcal{C} \to \mathcal{B}$; the inverse of this equivalence is defined by left Kan extension along $\rho$. Taking $\mathcal{B} = \mathcal{A}^{\mathrm{op}}$ we obtain the equivalence

$$\mathrm{Fun}(\mathrm{PSh}(\mathcal{C})^{\mathrm{op}}, \mathcal{A}) \supseteq \mathrm{Fun}^{\mathrm{lim\ pres.}}(\mathrm{PSh}(\mathcal{C})^{\mathrm{op}}, \mathcal{A}) \xrightarrow{\sim} \mathrm{Fun}(\mathcal{C}^{\mathrm{op}}, \mathcal{A}).$$

**Example 8.3.2** ($\mathcal{A}$-valued sheaves on a space, revisited). For $\mathcal{X} = \mathrm{Shv}(X)$ the two definitions coincide: limit preserving functors $F' \colon \mathcal{X}^{\mathrm{op}} \to \mathcal{A}$ correspond to functors $F \colon \mathrm{Open}_X^{\mathrm{op}} \to \mathcal{A}$ satisfying the sheaf condition.

To see this, recall the adjoint pair $i \colon \mathrm{Shv}(X) \xleftrightarrow{\hspace{1.2em}} \mathrm{PSh}(\mathrm{Open}_X) \colon a$. For each open cover $\mathcal{U} = \{U_i \to U\}_{i \in I}$ in $X$, the functor $a$ carries the evident map

$$s_{\mathcal{U}} \colon \mathrm{colim}_{\Delta^{\mathrm{op}}}\left[ [n] \mapsto \coprod_{i_0, \dots, i_n \in I} \rho_{U_{i_0} \cap \cdots \cap U_{i_n}} \right] \to \rho_U$$

in $\mathrm{PSh}(\mathrm{Open}_X)$ to an isomorphism in $\mathrm{Shv}(X)$, where $\rho_U := \mathrm{Map}_{\mathrm{Open}_X}(-, U)$ denotes the representable functor. (Proof: applying $\mathrm{Map}_{\mathrm{PSh}(\mathrm{Open}_X)}(-, F)$ to this exactly recovers the map (8.2.1) exhibiting the sheaf condition for a presheaf $F$, and if $F'$ is a sheaf we have $\mathrm{Map}_{\mathrm{PSh}(\mathrm{Open}_X)}(-, iF') = \mathrm{Map}_{\mathrm{Shv}(X)}(a(-), F')$.)

More is true: the functor $a$ is the *initial example* of a colimit preserving functor which takes all such maps $s_{\mathcal{U}}$ to isomorphisms. (In the terminology of [169, 5.5.4] $\mathrm{Shv}(X)$ is the *localization* of $\mathrm{PSh}(\mathrm{Open}_X)$ with respect to the *strongly saturated class generated* by $\{s_{\mathcal{U}}\}$, and universality is [169, 5.5.4.20].)

Thus, objects $F \in \mathrm{Shv}_{\mathcal{A}}(\mathcal{X})$ coincide with limit preserving $F' \colon \mathrm{PSh}(\mathrm{Open}_X)^{\mathrm{op}} \to \mathcal{A}$ such that $F'(s_{\mathcal{U}})$ is an equivalence for every open cover $\mathcal{U}$, which coincide with functors $F \colon \mathrm{Open}_X^{\mathrm{op}} \to \mathcal{A}$ satisfying the sheaf condition.

**Example 8.3.3** (Sheaves of $\infty$-groupoids). Every limit preserving functor $\mathcal{X}^{\mathrm{op}} \to \mathcal{S}$ is representable by an object of $\mathcal{X}$ (8.2.2). Therefore, the Yoneda embedding restricts to an equivalence $\mathcal{X} \xrightarrow{\sim} \mathrm{Shv}_{\mathcal{S}}(\mathcal{X}) \subseteq \mathrm{Fun}(\mathcal{X}^{\mathrm{op}}, \mathcal{S})$: the underlying $\infty$-category of the $\infty$-topos $\mathcal{X}$ is also the category of sheaves of $\infty$-groupoids on $\mathcal{X}$.

**Example 8.3.4** (Sheaves of sets). We have that $\mathrm{Shv}_{\mathrm{Set}}(\mathcal{X}) \approx \mathcal{X}^{\heartsuit}$.

*Remark 8.3.5* (Sheaves of $\infty$-groupoids as "generalized open sets"). The above displays the first instance of a philosophy you encounter a lot of in this theory. For an $\infty$-topos $\mathcal{X}$, objects $U \in \mathcal{X}$ can be thought of *either* as "sheaves of $\infty$-groupoids" on $\mathcal{X}$ via $\mathcal{X} \approx \mathrm{Shv}_{\mathcal{S}}(\mathcal{X})$, *or* as "generalized open sets of $\mathcal{X}$", in the sense that it makes sense to evaluate any sheaf $F \in \mathrm{Shv}_{\mathcal{A}}(\mathcal{X})$ at any object $U$.

Given an $\mathcal{A}$-valued sheaf $F \colon \mathcal{X}^{\mathrm{op}} \to \mathcal{A}$ on $\mathcal{X}$, its **global sections** are defined to be

$$\Gamma(\mathcal{X}, F) := F(1_{\mathcal{X}}).$$

## 8.4        Slices of $\infty$-topoi

We give a quick tour through some basic general constructions and properties involving $\infty$-topoi. First, we look at slices of $\infty$-topoi, which give more examples of $\infty$-topoi.

## Slices of ∞-topoi are ∞-topos

Given an object $U$ in an ∞-category $\mathcal{X}$, we get a slice ∞-category $\mathcal{X}_{/U}$.

**Proposition 8.4.1** ([169, 6.3.5.1]). *Every slice $\mathcal{X}_{/U}$ of an ∞-topos $\mathcal{X}$ is an ∞-topos.*

*Proof.* Choose a presentation $(\mathcal{C}, i, a)$ of $\mathcal{X}$ with fully faithful $i \colon \mathcal{X} \rightarrowtail \mathrm{PSh}(\mathcal{C})$, which induces a fully faithful $i' \colon \mathcal{X}_{/U} \rightarrowtail \mathrm{PSh}(\mathcal{C})_{/iU}$, which furthermore admits a left adjoint $a'$ induced by $a$ (since $U \to aiU$ is an equivalence). The functor $a'$ is seen to be accessible and left exact since $a$ is.

Note that $\mathrm{PSh}(\mathcal{C})_{/iU}$ is itself equivalent to presheaves on $\mathcal{C}/iU := \mathcal{C} \times_{\mathrm{PSh}(\mathcal{C})} \mathrm{PSh}(\mathcal{C})_{/iU}$, which is itself a equivalent to small ∞-category. We therefore obtain a presentation for $\mathcal{X}_{/U}$ as a full subcategory of $\mathrm{PSh}(\mathcal{C}/iU)$. $\quad\square$

**Example 8.4.2.** Let $X$ be a topological space. The Yoneda functor $\mathrm{Open}_X \to \mathrm{Shv}(X)$ factors through the full subcategory $\mathrm{Shv}(X)$. Thus for any open set $U$ of $X$, we have the representable sheaf $\rho_U \in \mathrm{Shv}(X)$, which we simply denote $U$ by abuse of notation. It is straightforward to show that $\mathrm{Shv}(X)_{/U} \approx \mathrm{Shv}(U)$: the slice category over the sheaf $U$ is exactly sheaves on the topological space $U$.

*Remark 8.4.3* (Relativized notions). Any morphism $f \colon V \to U$ in an ∞-topos $\mathcal{X}$ is also an object in an ∞-topos (namely $\mathcal{X}_{/U}$). Thus any general concept defined for objects in an ∞-topos can be "relativized" to a concept defined on morphisms (assuming the definition is preserved by equivalence of ∞-topoi). Conversely, any concept defined for morphisms in an arbitrary ∞-topos can be specialized to objects, by applying it to projection maps $U \to 1$.

## Colimits are universal in ∞-topoi

Given a morphism $f \colon U \to V$ in an ∞-topos $\mathcal{X}$, we get an induced **pullback functor** $f^* \colon \mathcal{X}_{/V} \to \mathcal{X}_{/U}$, which on objects sends $V' \to V$ to $V' \times_V U \to U$.

**Proposition 8.4.4.** *Colimits are "universal" in ∞-topoi; i.e., $f^* \colon \mathcal{X}_{/V} \to \mathcal{X}_{/U}$ preserves small colimits.*

*Proof.* The statement of the proposition only involves colimits and *finite* limits in $\mathcal{X}$. Thus via a choice of presentation $(\mathcal{C}, i, a)$ for $\mathcal{X}$ we can reduce to the case of $\mathcal{X} = \mathrm{PSh}(\mathcal{C})$. As colimits and limits of presheaves are computed "objectwise", we can reduce to the case of infinity groupoids $\mathcal{X} = \mathcal{S}$. In this case the statement is "well-known" [169, 6.1.3.14]. $\quad\square$

## ∞-topoi have internal homs

A consequence of universality of colimits is that $U \times (-) \colon \mathcal{X} \to \mathcal{X}$ is colimit preserving, and therefore (8.2.2) has a right adjoint which we may denote $[U, -] \colon \mathcal{X} \to \mathcal{X}$. This is an internal function object, so any ∞-topos is *cartesian closed*, and so is *locally cartesian closed* (i.e., every slice is cartesian closed).

## ∞-topoi have descent

Given any $\infty$-category $\mathcal{X}$, let $\mathrm{Cart}(\mathcal{X}) \subseteq \mathrm{Fun}(\{0 \to 1\}, \mathcal{X})$ denote the (non-full) subcategory of the arrow category of $\mathcal{X}$, consisting of all the objects, and morphisms $f \to g$ which are pullback squares in $\mathcal{X}$. This is a subcategory because pullback squares paste together.

We say that $\mathcal{X}$ has **descent** if $\mathrm{Cart}(\mathcal{X})$ has small colimits, and if the inclusion functor $\mathrm{Cart}(\mathcal{X}) \to \mathrm{Fun}(\{0 \to 1\}, \mathcal{X})$ preserves small colimits.

**Proposition 8.4.5 (Descent [169, 6.1.3]).** *Every $\infty$-topos has descent.*

Let's spell out the consequences of this. Suppose given a functor $F: \mathcal{C} \to \mathcal{X}$ from a small $\infty$-category to an $\infty$-topos. We obtain a family of slice categories $\mathcal{X}_{/F(c)}$, which is a contravariant functor of $\mathcal{C}$ via the functors $F(\alpha)^*: \mathcal{X}_{/F(c')} \to \mathcal{X}_{/F(c)}$ for $\alpha: c \to c'$ in $\mathcal{C}$. This functor $\mathcal{C}^{\mathrm{op}} \to \widehat{\mathrm{Cat}}_\infty$ extends to a cone $(\mathcal{C}^{\triangleright})^{\mathrm{op}} \to \widehat{\mathrm{Cat}}_\infty$, where the value at the cone point is the slice category $\mathcal{X}_{/\overline{F}}$ over the colimit $\overline{F} = \mathrm{colim}_{c \in \mathcal{C}} F(c)$ of $F$.[2]

We can also form the limit $\lim_{c \in \mathcal{C}^{\mathrm{op}}} \mathcal{X}_{/F(c)}$ in $\widehat{\mathrm{Cat}}_\infty$. An object of this limit amounts to: a functor $A: \mathcal{C} \to \mathcal{X}$ and a natural transformation $f: A \to F$ such that for each $\alpha: c \to c'$ in $\mathcal{C}$ the square

$$
\begin{array}{ccc}
A(c') & \xrightarrow{\ A(\alpha)\ } & A(c) \\
\downarrow & & \downarrow \\
F(c') & \xrightarrow[\ F(\alpha)\ ]{} & F(c)
\end{array}
$$

is a pullback in $\mathcal{X}$. Descent implies the following.

**Proposition 8.4.6.** *The functor*

$$
\mathcal{X}_{/\overline{F}} \to \lim_{c \in \mathcal{C}^{\mathrm{op}}} \mathcal{X}_{/F(c)}
$$

*sending $\overline{A} \to \overline{F}$ to $\left( c \mapsto (\overline{A} \times_{\overline{F}} F(c) \to F(c)) \right)$ is an equivalence. The inverse equivalence is a functor which sends $(A \to F) \in \lim_{\mathcal{C}^{\mathrm{op}}} \mathcal{X}_{/F(c)}$ to the object of $\mathcal{X}_{/\overline{F}}$ represented by the evident map*

$$
\mathrm{colim}_{\mathcal{C}} A \to \mathrm{colim}_{\mathcal{C}} F.
$$

Thus, descent in an $\infty$-topos has a very beautiful interpretation in terms of the definition of "sheaves on $\mathcal{X}$" as functors: the functor $\mathcal{X}^{\mathrm{op}} \to \widehat{\mathrm{Cat}}_\infty$ which sends $U \mapsto \mathcal{X}_{/U}$ is limit preserving, and so is a *sheaf* on $\mathcal{X}$ valued in locally small $\infty$-categories.

**Example 8.4.7.** Let $X$ be a topological space. Recall that (after identifying an open set $U$ with its representable sheaf on $X$), we have that $\mathrm{Shv}(X)_{/U} \approx \mathrm{Shv}(U)$. If $U$ and

---

[2] This is not a complete description of a functor $(\mathcal{C}^{\triangleright})^{\mathrm{op}} \to \widehat{\mathrm{Cat}}_\infty$, as there is also "higher coherence" data to keep track of. A correct description is implemented using the theory of *Cartesian fibrations* [169, 2.4]. I am not going to try to be precise about such matters here.

$V$ are open sets of $X$, then $U \cup V$ is the pushout of $U \leftarrow U \cap V \rightarrow V$ as sheaves. Given this, descent says that there is an equivalence

$$\operatorname{Shv}(U \cup V) \xrightarrow{\sim} \lim [\operatorname{Shv}(U) \rightarrow \operatorname{Shv}(U \cap V) \leftarrow \operatorname{Shv}(V)].$$

That is, the category of sheaves of $\infty$-groupoids on $U \cup V$ is equivalent to a category of "descent data" involving sheaves on $U$, $V$, and $U \cap V$.

This particular example works "the same way" in the classical topos $\operatorname{Shv}(X)^\heartsuit$ of sheaves of sets on $X$: the category of sheaves of sets on $U \cup V$ can be reconstructed from appropriate descent data, i.e., as an $\infty$-categorical pullback of a diagram of categories of sheaves of sets on $U$, $V$, and $U \cap V$. However, 1-categorical descent in this form fails for general pushout diagrams in $\operatorname{Shv}(X)^\heartsuit$. This is one way in which the theory of $\infty$-topoi shows advantages over the classical theory.

## 8.5    Truncation and connectivity in $\infty$-topoi

### $n$-Truncation and $n$-connectivity in $\infty$-categories

An $\infty$-groupoid $X$ is $n$-**truncated** if

$$\pi_k(X, x_0) \approx \{*\} \qquad \text{for all } k > n \text{ and all } x_0 \in X.$$

In particular, 0-truncated $\infty$-groupoids are equivalent to sets (discrete spaces), while $(-1)$-truncated $\infty$-groupoids are equivalent to either the empty set $\varnothing$ or the terminal object. By fiat, $(-2)$-truncated $\infty$-groupoids are those which are equivalent to the terminal object.

An object $X \in \mathcal{A}$ in a general $\infty$-category is $n$-**truncated** if $\operatorname{Map}_{\mathcal{A}}(A, X)$ is an $n$-truncated $\infty$-groupoid for all objects $A$ in $\mathcal{A}$. We relativize to the notion of $n$-**truncated morphism**: i.e., an $f : X \rightarrow Y$ which is $n$-truncated as an object of the slice $\mathcal{A}_{/Y}$. I write $\tau_{\leq n} \mathcal{A} \subseteq \mathcal{A}$ for the full subcategory of $n$-truncated objects.

In many $\infty$-categories (including all presentable $\infty$-categories and thus all $\infty$-topoi), there is an $n$-**truncation functor** which associates to each object $X$ the initial example $X \rightarrow \tau_{\leq n} X$ of a map to an $n$-truncated object. When this exists, the essential image of the $n$-truncation functor $\tau_{\leq n} : \mathcal{A} \rightarrow \mathcal{A}$ is $\tau_{\leq n} \mathcal{A}$, and we have an adjoint pair $\tau_{\leq n} \mathcal{A} \xhookleftarrow{\rightleftarrows} \mathcal{A}$.

Relativized, we obtain for a morphism $f : X \rightarrow Y$ in $\mathcal{A}$ an $n$-truncation factorization

$$X \xrightarrow{g} \tau_{\leq n}(f) \xrightarrow{h} Y,$$

so that $h$ is the initial example of an $n$-truncated map over $Y$ which factors $f$.

Following Lurie, we say that an object $U$ in an $\infty$-category is $n$-**connective** if $\tau_{\leq n-1} U \approx 1$. Likewise an $n$-**connective morphism** $f : X \rightarrow Y$ in $\mathcal{A}$ is one which is an $n$-connective object of $\mathcal{A}_{/Y}$.

*Remark 8.5.1.* In $\mathcal{S}$, an $n$-connective object is the same as what is usually called an $(n-1)$-connected space (so 1-connective means connected). However, an $n$-connective map is the same as what is classically called an $n$-connected map of spaces.

The $n$-truncation factorization is in fact a factorization into "$(n+1)$-connective followed by $n$-truncated".

**Proposition 8.5.2.** *If $X \xrightarrow{g} \tau_{\leq n}(f) \xrightarrow{h} Y$ is the $n$-truncation factorization of $f : X \to Y$ in $\mathcal{A}$, then $g$ is an $(n+1)$-connective map in $\mathcal{A}$. (Assuming all the relevant truncations exist in $\mathcal{A}$.)*

*Proof.* By replacing $\mathcal{A}$ with $\mathcal{A}_{/Y}$, we can assume $Y \approx 1$. Thus we need to show that $g : X \to \tau_{\leq n} X$, the "absolute" $n$-truncation of the object $X$, is also the "relative" $n$-truncation of the map $g$, i.e., that in the $n$-truncation factorization

$$X \xrightarrow{g'} \tau_{\leq n}(g) \xrightarrow{g''} \tau_{\leq n} X$$

of the object $g$ of $\mathcal{A}_{/\tau_{\leq n} X}$, the map $g''$ is an equivalence.

Both $\tau_{\leq n} X \to 1$ and $g''$ are $n$-truncated maps of $\mathcal{A}$, from which it is straightforward to show that $\tau_{\leq n}(g)$ is an $n$-truncated object of $\mathcal{A}$. Thus, the universal property for $g : X \to \tau_{\leq n} X$ gives $s : \tau_{\leq n} X \to \tau_{\leq n}(g)$ such that $sg = g'$ and $g''s = \mathrm{id}_{\tau_{\leq n} X}$. The universal property for $g' : X \to \tau_{\leq n}(g)$ then implies that $sg'' = \mathrm{id}_{\tau_{\leq n}(g)}$. $\qquad\square$

*Remark 8.5.3.* $n$-truncation of objects in an $\infty$-topos preserves finite products, as can be seen by choosing a presentation and reducing to the case of $\mathcal{S}$ [169, 6.5.1.2].

## Čech nerves and effective epimorphisms

For $\infty$-topoi, the case of truncation when $n = -1$ is especially important. An $(-1)$-truncated map in an $\infty$-category is the same thing as a **monomorphism**, i.e., a map $i : A \to B$ such that all the fibers of all induced maps $\mathrm{Map}(C,A) \to \mathrm{Map}(C,B)$ are either empty or contractible. Equivalently, $i$ is a monomorphism if and only if the diagonal map $A \to A \times_B A$ is an equivalence (if the pullback exists), if and only if either projection $A \times_B A \to A$ is an equivalence.

In an $\infty$-topos, an **effective epimorphism** is defined to be a 0-connective morphism. The $(-1)$-truncation factorization in an $\infty$-topos (also called **epi/mono factorization**) can be computed using Čech nerves.

Given a morphism $f : U \to V$ in an $\infty$-topos $\mathcal{X}$, its **Čech nerve** is an augmented simplicial object $\check{C}(f) : \Delta_+^{\mathrm{op}} \to \mathcal{X}$ of the form

$$\cdots \Longrightarrow U \times_V U \times_V U \Longrightarrow U \times_V U \rightrightarrows U \xrightarrow{f} V$$

**Proposition 8.5.4.** *Given a map $f : U \to V$ in an $\infty$-topos, the factorization*

$$U \xrightarrow{p} \mathrm{colim}_{\Delta^{\mathrm{op}}} \check{C}(f) \xrightarrow{i} V$$

*defined by taking the colimit of the underlying simplicial object of the Čech nerve is equivalent to the factorization of $f$ into an effective epimorphism $p$ followed by a monomorphism $i$.*

*Proof.* Without loss of generality assume $V \approx 1$ (since the slice $\mathcal{X}_{/V}$ is an $\infty$-topos). Write $\overline{U} = \mathrm{colim}_{\Delta^{\mathrm{op}}} \check{C}(F) = \mathrm{colim}\,[[n] \mapsto U^{n+1}]$. Because colimits are universal in an $\infty$-topos (8.4.4), we have that $\overline{U} \times U^{k+1} \approx \mathrm{colim}\,[[n] \mapsto U^{n+1} \times U^{k+1}]$. For any $k \geq 0$ the augmented simplicial object $[n] \mapsto U^{n+1} \times U^{k+1}$ admits a contracting homotopy, so $\overline{U} \times U^{k+1} \xrightarrow{\sim} U^{k+1}$. Universality of colimits again gives $\overline{U} \times \overline{U} \xrightarrow{\sim} \overline{U}$, whence $\overline{U} \to 1$ is monomorphism, i.e., $\overline{U}$ is a $(-1)$-truncated object

To show that $p \colon U \to \overline{U}$ is the universal $(-1)$-truncation is easy: for any $f \colon U \to Z$ to a $(-1)$-truncated object, we have

$$\mathrm{Map}_{\mathcal{X}_{U/}}(p, f) \approx \lim_{\Delta}[[n] \mapsto \mathrm{Map}_{\mathcal{X}_{U/}}(U \to U^{n+1}, f)],$$

which is easy to evaluate since all the mapping spaces must be contractible if non-empty, since $Z$ is $(-1)$-truncated. $\qquad\square$

**Warning 8.5.5.** In an $\infty$-topos the class of *effective epimorphisms* contains, but is *not equal* to the class of *epimorphisms*. This is very unlike the classical case: in a 1-topos the two classes coincide.

*Remark 8.5.6* (Covers). A set $\{U_i\}$ of objects in an $\infty$-topos $\mathcal{X}$ is called a **cover** of $\mathcal{X}$ if $\bigsqcup U_i \to 1$ is an effective epimorphism in $\mathcal{X}$. We also speak of a cover of an object $V$ in $\mathcal{X}$, which is a set $\{U_i \to V\}$ of maps in $\mathcal{X}$ such that $\bigsqcup U_i \to V$ is an effective epi.

If $X$ is a topological space, then a set $\{U_i\} \subseteq \mathrm{Open}_X$ of open sets of $X$ is a open cover of $X$ if and only if the corresponding set $\{U_i\} \subseteq \mathrm{Shv}(X)$ of sheaves on $X$ is a cover in the above sense.

Sometimes we see the following condition on a collection $\{U_i\}$ of objects in $\mathcal{X}$: that it generates $\mathcal{X}$ under small colimits. This condition implies that there exists a subset of $\{U_i\}$ which covers $\mathcal{X}$.

*Example 8.5.7* (Effective epis in $\infty$-groupoids). A map in $\mathcal{S}$ is an effective epimorphism if and only if it induces a surjection on sets of path components. The epi/mono factorization of a map $f \colon U \to V$ in $\mathcal{S}$ is through $\overline{U} \subseteq V$, the disjoint union of path components of $V$ which are in the image of $f$.

## Homotopy groups

Given a *pointed* object $(U, u_0 \colon 1 \to U)$ in an $\infty$-topos $\mathcal{X}$, there is an object $(U, u_0)^K$ in $\mathcal{X}$ for every pointed space $K \in \mathcal{S}_*$, which represents the functor

$$\mathrm{Map}_{\mathcal{S}_*}(K, \mathrm{Map}_{\mathcal{X}}(-, U)) \colon \mathcal{X}^{\mathrm{op}} \to \mathcal{S}$$

(which is clearly limit preserving, so by (8.2.2) defines a $\mathcal{S}$-valued sheaf on $\mathcal{X}$). We let

$$\pi_n(U, u_0) := \tau_{\leq 0}((U, u_0)^{S^n}) \in \mathcal{X}^{\heartsuit},$$

the $n$th **homotopy sheaf** of $(U, u_0)$. This is in general a sheaf of based sets on $\mathcal{X}$, a sheaf of groups for $n \geq 1$, and a sheaf of abelian groups for $n \geq 2$.

*Remark 8.5.8.* An object $U$ in an ∞-topos can easily fail to have "enough" global sections, or even any global sections. Thus it is often necessary to use a more sophisticated formulation of homotopy sheaves of $U$ allowing for arbitrary "local" choices of basepoint. These are objects $\pi_n U \in (\mathcal{X}_{/U})^\heartsuit$, defined as the homotopy sheaves (as defined above) of $(\text{proj}_2\colon U \times U \to U, \Delta\colon U \to U \times U)$ in $\mathcal{X}_{/U}$, the projection map "pointed" by the diagonal map. See [169, 6.5.1].

For instance, with this more sophisticated definition, an object $U$ is $n$-connective if and only if $\pi_k U \approx 1$ for all $k < n$ [169, 6.5.1.12].

Example 8.5.9 (Eilenberg–Mac Lane objects and sheaf cohomology). An **Eilenberg–Mac Lane object** of dimension $n$ is a *pointed* object $(K, k_0)$ in $\mathcal{X}$ such that $K$ is both $n$-truncated and $n$-connective. One can show [169, 7.2.2.12] that taking $(K, k_0) \mapsto \pi_n(K, k_0)$ gives a correspondence between Eilenberg–Mac Lane objects of dimension $n$ and: abelian group objects in $\mathcal{X}^\heartsuit$ (if $n \geq 2$), group objects in $\mathcal{X}^\heartsuit$ (if $n = 1$), and pointed objects in $\mathcal{X}^\heartsuit$ (if $n = 0$).

Thus, given a sheaf $A$ of (classical) abelian groups on $\mathcal{X}$, we can define the cohomology group

$$H^n(\mathcal{X}; A) := \pi_0 \operatorname{Map}_{\mathcal{X}}(1, K(A, n))$$

of the ∞-topos $\mathcal{X}$.

## ∞-connectedness and hypercompletion

An object or morphism is ∞-**connected** if it is $n$-connective for all $n$. It turns out that the obvious analogue of the "Whitehead theorem" can fail in an ∞-topos: ∞-connected maps need not be equivalences.

We say that an object $U$ in $\mathcal{X}$ is **hypercomplete** if $\operatorname{Map}(V', U) \to \operatorname{Map}(V, U)$ is an equivalence for any ∞-connected map $V \to V'$.

Example 8.5.10. All $n$-truncated objects are hypercomplete, for any $n$. Any limit of hypercomplete objects is hypercomplete.

We write $\mathcal{X}^{\text{hyp}} \subseteq \mathcal{X}$ for the full subcategory of hypercomplete objects of $\mathcal{X}$. It turns out that the inclusion is accessible, and admits a left adjoint which is itself left exact. So $\mathcal{X}^{\text{hyp}}$ is an ∞-topos in its own right [169, 6.5.2].

We say that $\mathcal{X}$ is itself **hypercomplete** if all ∞-connected maps are equivalences, i.e., if $\mathcal{X}^{\text{hyp}} = \mathcal{X}$.

Example 8.5.11. Any presheaf ∞-category is hypercomplete, including $\mathcal{S}$ itself.

## Truncation towers

Given an object $U$ in $\mathcal{X}$, we may consider the tower

$$U \to \cdots \to \tau_{\leq n} U \to \tau_{\leq n-1} U \to \cdots \to \tau_{\leq -1} U \to *$$

of truncations of $U$. There is a limit $U_\infty := \lim \tau_{\leq n} U$, together with a tautological map $U \to U_\infty$. It is generally not the case that $U \to U_\infty$ is an equivalence. For instance, $U_\infty$ is necessarily hypercomplete, whereas $U$ may not be. Furthermore, even if $U$ is hypercomplete, $U \to U_\infty$ can fail to be an equivalence.

There are various general conditions which ensure that $U \xrightarrow{\sim} U_\infty$ for all objects $U$ in $\mathcal{X}$ (and in fact ensure a stronger fact, called convergence of Postnikov towers). For instance, this is the case when $\mathcal{X}$ is *locally of homotopy dimension* $\leq n$ for some $n$ [169, 7.2.1.12]. (Say $\mathcal{X}$ is of **homotopy dimension** $\leq n$ if every $n$-connective object $U \in \mathcal{X}$ admits a global section $1 \to U$. We say $\mathcal{X}$ is **locally of homotopy dimension** $\leq n$ if there exists a set $\{U_i\}$ of objects which generate $\mathcal{X}$ under colimits and such that each $\mathcal{X}_{/U_i}$ is of homotopy dimension $\leq n$.)

## Constructing ∞-topoi

We defined an ∞-topos $\mathcal{X}$ to be an ∞-category which admits a *presentation* $(\mathcal{C}, i, a)$. It is natural to ask: given a small ∞-category $\mathcal{C}$, can we classify the presentations of ∞-topoi which use it?

Given any left exact accessible localization $\mathcal{X} \subseteq \mathrm{PSh}(\mathcal{C})$, let $\mathcal{T}$ denote the collection of morphisms $j$ in $\mathrm{PSh}(\mathcal{C})$ which

1. are monomorphisms of the form $S \rightarrowtail \rho_C$ for some object $C$ of $\mathcal{C}$, and
2. are such that $a(j)$ is an isomorphism in $\mathcal{X}$.

The class of maps $\mathcal{T}$ is an example of a *Grothendieck topology* on $\mathcal{C}$. When $\mathcal{C}$ is a 1-category this precisely recovers the classical notion of a Grothendieck topology on a 1-category.

It can be shown [169, 6.4.1.5] that if $F \in \mathrm{PSh}(\mathcal{C})$ is $n$-truncated for some $n < \infty$, then $F \in \mathcal{X}$ if and only if $F(j)$ is an isomorphism for all $j \in \mathcal{T}$. That is, the $n$-*truncated objects* in left exact accessible localizations of $\mathrm{PSh}(\mathcal{C})$ are entirely determined by $\mathcal{T}$.

Conversely, given a Grothendieck topology $\mathcal{T}$ on $\mathcal{C}$, the full subcategory $\mathrm{Shv}(\mathcal{C}, \mathcal{T}) := \{F \mid F(j) \text{ iso for all } j \in \mathcal{T}\} \subseteq \mathrm{PSh}(\mathcal{C})$ is an example of an ∞-topos. This includes the examples (8.2) and (8.2).

A general left exact localization of $\mathrm{PSh}(\mathcal{C})$ can be obtained by (i) choosing a Grothendieck topology $\mathcal{T}$ on $\mathcal{C}$, and then (ii) possibly localizing $\mathrm{Shv}(\mathcal{C}, \mathcal{T})$ further with respect to a suitable class of ∞-connected maps [169, 6.5.2.20].

*Remark 8.5.12* (1-localic reflection). Given any classical topos, i.e., a 1-topos $\mathcal{X}_1$, we can upgrade it to an ∞-topos denoted $\mathrm{Shv}_S(\mathcal{X}_1)$; this is called its **1-localic reflection**. In general this can be difficult to describe. In the case that $\mathcal{X}_1 \approx \mathrm{Shv}_{\mathrm{Set}}(\mathcal{C}, \mathcal{T})$ is an identification of $\mathcal{X}_1$ as a category of sheaves of sets on a 1-category $\mathcal{C}$ equipped with a Grothendieck topology $\mathcal{T}$, and if $\mathcal{C}$ *has finite limits*, then $\mathrm{Shv}_S(\mathcal{X}_1) := \mathrm{Shv}(\mathcal{C}, \mathcal{T})$ is the 1-localic reflection of $\mathcal{X}_1$ [169, 6.4.5, esp. 6.4.5.6].

For instance, we constructed $\mathrm{Shv}(X)$ and $\mathrm{Shv}(X^{\text{ét}})$, sheaves on a topological space or on the étale site of a scheme, in exactly this way, so they are 1-localic.

As can be seen from (8.2.4), an ∞-topos $\mathcal{X}$ is not generally equivalent to the 1-localic reflection of $\mathrm{Shv}_S(\mathcal{X}^\heartsuit)$ of its underlying 1-topos $\mathcal{X}^\heartsuit$.

*Warning:* $\mathrm{Shv}_S(\mathcal{X}_1)$ is not the same as the construction of (8.3): it is *not* equivalent to limit preserving functors $\mathcal{X}_1^{\mathrm{op}} \to S$.

*Remark 8.5.13* (Simplicial presheaves). Given a small 1-category $\mathcal{C}$ with a Grothendieck topology $\mathcal{T}$, Jardine [135] produced a model category structure on the category $\mathrm{Fun}(\mathcal{C}^{\mathrm{op}}, \mathrm{sSet})$ of presheaves of simplicial sets. The ∞-category associated to that model category is equivalent to what we have called $\mathrm{Shv}(\mathcal{C}, \mathcal{T})^{\mathrm{hyp}}$ [169, 6.5.2].

## 8.6    Morphisms of ∞-topoi

To justify the claim that ∞-topoi are the ∞-categorical generalization of topological spaces, we need an appropriate notion of morphism between ∞-topoi that generalizes the notion of continuous map. This is called a *geometric morphism*. In fact, I won't consider any other kind of morphism between ∞-topoi here.

### Geometric morphisms

A **geometric morphism** (or just **morphism**) of ∞-topoi $f : \mathcal{X} \to \mathcal{Y}$ is an adjoint pair of functors

$$f_* : \mathcal{X} \leftrightarrows \mathcal{Y} : f^*$$

such that the left adjoint $f^*$ is left exact (i.e., preserves finite limits). The functor $f_*$ is **direct image**, and $f_*$ is **pullback** or **preimage**.

The collection of geometric morphisms from $\mathcal{X}$ to $\mathcal{Y}$, together with natural transformations between the *left* adjoints of the geometric morphisms, forms an ∞-category, sometimes denoted $\mathrm{Fun}^*(\mathcal{Y}, \mathcal{X})$. We note that this ∞-category is *not in general* equivalent to a small ∞-category, although it is in some cases; it is always an *accessible* ∞-category [169, 6.3.1.13]. We will mostly be concerned with the maximal ∞-groupoid inside this ∞-category, which we denote $\mathrm{Map}_{\infty\mathcal{T}\mathrm{op}}(\mathcal{X}, \mathcal{Y})$, and regard as mapping spaces of $\infty\mathcal{T}\mathrm{op}$, the ∞-category of ∞-topoi.

*Remark 8.6.1.* Since ∞-topoi are presentable ∞-categories, to construct a geometric morphism $f : \mathcal{X} \to \mathcal{Y}$ it suffices to produce a functor $f^* : \mathcal{Y} \to \mathcal{X}$ which preserves colimits and finite limits; presentability then implies (8.2.2) that a right adjoint $f_*$ exists. Typically, having a "presentation" for $\mathcal{Y}$ gives an explicit recipe for describing colimit preserving $f^*$, so constructing morphisms amounts to finding such functors which also preserve finite limits.

Example 8.6.2 (The terminal ∞-topos). The ∞-category $S$ of infinity groupoids is the *terminal* ∞-topos, i.e., there is an essentially unique geometric morphism $\mathcal{X} \to S$ from any ∞-topos. To see this, note that a colimit preserving $\pi^* : S \to \mathcal{X}$ is precisely determined by its value on the terminal object $1_S$ of $S$, while to preserve finite limits it is necessary that $\pi^*$ take $1_S$ to the terminal object of $\mathcal{X}$. This is also sufficient, by the fact the colimits are universal in $\mathcal{X}$ (8.4.4). Thus $\mathrm{Map}_{\infty\mathcal{T}\mathrm{op}}(\mathcal{X}, S) \approx *$ for any $\mathcal{X}$.

Example 8.6.3. Every presentation of an $\infty$-topos $\mathcal{X} \subseteq \mathrm{PSh}(\mathcal{C})$ as in (8.2) corresponds to a geometric morphism $\mathcal{X} \to \mathrm{PSh}(\mathcal{C})$.

Example 8.6.4. Hypercompletion (8.5) gives a geometric morphism $\mathcal{X}^{\mathrm{hyp}} \to \mathcal{X}$.

## Continuous maps vs. geometric morphisms

Let $\mathcal{X} = \mathrm{Shv}(X)$ for some topological space $X$, and let $\mathcal{Y}$ be any $\infty$-topos. We can describe $\mathrm{Fun}^*(\mathrm{Shv}(X), \mathcal{Y})$ as follows. It is equivalent to the full subcategory of

$$\mathrm{Fun}^{\mathrm{colim\ pres.}}(\mathrm{PSh}(\mathrm{Open}_X), \mathcal{Y}) \xrightarrow{\sim} \mathrm{Fun}(\mathrm{Open}_X, \mathcal{Y}),$$

spanned by those $\phi \colon \mathrm{Open}_X \to \mathcal{Y}$ such that

1. for each open cover $\{U_i \to U\}$, the map $\coprod_i \phi(U_i) \to \phi(U)$ is an effective epi in $\mathcal{Y}$,
2. $\phi(X) \approx *$, and
3. $\phi(U \cap V) \approx \phi(U) \times_{\phi(X)} \phi(V)$.

Condition (1) ensures that $\mathrm{PSh}(\mathrm{Open}_X) \to \mathcal{Y}$ factors through the localization

$$a \colon \mathrm{PSh}(\mathrm{Open}_X) \twoheadrightarrow \mathrm{Shv}(X),$$

while conditions (2) and (3) ensure that the resulting functor $f^* \colon \mathrm{Shv}(X) \to \mathcal{Y}$ preserves finite limits. (This is a special case of [169, 6.1.5.2].)

Note that since $U \cap U \approx U$, (2) and (3) imply that each $\phi(U) \to \phi(X) \approx *$, is a monomorphism, i.e., that each $\phi(U)$ is a $(-1)$-truncated object of $\mathcal{Y}$.

For instance, if $\mathcal{Y} = \mathrm{Shv}(Y)$ for some topological space $Y$, then $\tau_{\leq -1}\mathcal{Y} \approx \mathrm{Open}_Y$. Under this identification, morphisms of topoi $\mathcal{Y} \to \mathcal{X}$ correspond to functors $\mathrm{Open}_X \to \mathrm{Open}_Y$ which (1) take covers to covers, (2) take $X$ to $Y$, and (3) preserve finite intersections.

Example 8.6.5. If $X$ is a scheme, we have both $\mathrm{Shv}(X^{\mathrm{Zar}})$ (sheaves on the underlying Zariski space of $X$) and $\mathrm{Shv}(X^{\mathrm{ét}})$ (sheaves in the étale topology (8.2)). There is an evident geometric morphism $\mathrm{Shv}(X^{\mathrm{ét}}) \to \mathrm{Shv}(X^{\mathrm{Zar}})$ induced by $\mathrm{Open}_{X^{\mathrm{Zar}}} \to \mathrm{Shv}(X^{\mathrm{ét}})$ sending an open set to the étale sheaf it represents.

A space $X$ is **sober** if every irreducible closed subset is the closure of a unique point (e.g., Hausdorff spaces, or the Zariski space of a scheme). One can show that if $X$ is sober, then

$$\mathrm{Map}_{\infty\mathcal{T}\mathrm{op}}(\mathrm{Shv}(Y), \mathrm{Shv}(X)) \approx (\text{set of continuous maps } Y \to X).$$

This justifies the assertion that "$\infty$-topos" is a generalization of the notion of a topological space.

*Remark 8.6.6.* The sobriety condition is necessary. For instance, if $Y = \{*\}$, then the $\phi \colon \mathrm{Open}_X \to \mathrm{Open}_Y \approx \{0 \to 1\}$ satisfying (1)–(3) are in bijective correspondence with irreducible closed $C \subseteq X$: we have

$$\left(\phi \leftrightarrow C\right) \iff \left(C = \bigcap_{\phi(U)=0}(X \smallsetminus U)\right) \iff \left(\phi(U) = 0 \text{ iff } U \cap C = \varnothing\right).$$

That is, the underlying point set of $X$ can be recovered from $\mathrm{Open}_X$ only if $X$ is sober.

## Locales

We see that it is not quite correct to say that ∞-topoi generalize topological spaces; rather, they generalize *locales*.

A **locale** is a poset $\mathcal{O}$ equipped with all the formal algebraic properties of the poset of open sets of a space: i.e., it is a complete lattice such that finite meets distribute over infinite joins. A map $f : \mathcal{O}' \to \mathcal{O}$ of locales is a function $f^* : \mathcal{O} \to \mathcal{O}'$ which preserves all joins and all finite meets. Any locale $\mathcal{O}$ has an ∞-category of sheaves $\mathrm{Shv}(\mathcal{O})$ (defined exactly as sheaves on a space), and $\mathrm{Map}_{\infty\mathcal{T}\mathrm{op}}(\mathrm{Shv}(\mathcal{O}), \mathrm{Shv}(\mathcal{O}')) \approx$ {locale maps $\mathcal{O} \to \mathcal{O}'$}.

Every topological space determines a locale, though not every locale comes from a space. From the point of view of sheaf theory, a space is indistinguishable from its locale. For spaces we care about (i.e., sober spaces), we can recover their point sets from their locale, and this is good enough for us.

*Remark 8.6.7.* From the point of view that "objects in an ∞-topos are generalized open sets" (8.3.5), the preimage functor $f^* : \mathcal{Y} \to \mathcal{X}$ of a geometric morphism is the operation of "preimage of generalized open sets".

*Remark 8.6.8.* Every ∞-topos $\mathcal{X}$ has an associated locale, whose lattice of "open sets" $\mathrm{Open}_{\mathcal{X}}$ consists precisely of the $(-1)$-truncated objects of $\mathcal{X}$.

## Limits and colimits of ∞-topoi

The ∞-category of ∞-topoi itself (remarkably) has all small limits and colimits.

Colimits are easy to describe (modulo the technical issues involved in making precise statements; see [169, 6.3.2]): given $F : \mathcal{C} \to \infty\mathcal{T}\mathrm{op}$, consider the functor $F^* : \mathcal{C}^{\mathrm{op}} \to \widehat{\mathrm{Cat}}_{\infty}$ which sends an arrow $\alpha : C \to C'$ to the left adjoint $F(\alpha)^* : F(C') \to F(C)$ of the geometric morphism. Then the underlying ∞-category of the colimit of $F$ in ∞-topoi is just the limit of the diagram $F^*$ of ∞-categories.

Limits are more difficult. As we have seen, the terminal object in $\infty\mathcal{T}\mathrm{op}$ is $\mathcal{S}$. *Filtered* limits are computed by a pointwise construction much like colimits [169, 6.3.3]. To get general limits we also need pullbacks; see [169, 6.3.4] for details.

*Remark 8.6.9.* The product of two ∞-topoi $\mathcal{X}$ and $\mathcal{Y}$ has a nice description. It is equivalent to

$$\mathrm{Fun}^{\text{lim pres./lim pres.}}(\mathcal{X}^{\mathrm{op}} \times \mathcal{Y}^{\mathrm{op}}, \mathcal{S}) \subseteq \mathrm{Fun}(\mathcal{X}^{\mathrm{op}} \times \mathcal{Y}^{\mathrm{op}}, \mathcal{S}),$$

the full subcategory consisting of functors $F$ which preserve limits separately in each variable, i.e., such that $F(\mathrm{colim}_i U_i, V) \xrightarrow{\sim} \lim_i F(U_i, V)$ and $F(U, \mathrm{colim}_j V_j) \xrightarrow{\sim} \lim_j F(U, V_j)$. This ∞-category is also equivalent to both of

$$\mathrm{Fun}^{\text{lim pres.}}(\mathcal{X}^{\mathrm{op}}, \mathcal{Y}) \approx \mathrm{Fun}^{\text{lim pres.}}(\mathcal{Y}^{\mathrm{op}}, \mathcal{X}),$$

by the adjoint functor theorem for presentable $\infty$-categories (8.2.2). That is,

$$\mathcal{X} \times^{\infty Top} \mathcal{Y} \approx \mathrm{Shv}_{\mathcal{Y}}(\mathcal{X}) \approx \mathrm{Shv}_{\mathcal{X}}(\mathcal{Y})$$

[168, 4.8.1.18]. This construction is a special case of the "tensor product" of presentable $\infty$-categories; see [168, 4.8].

*Remark 8.6.10.* Recall that in scheme theory, the underlying topological space of the pullback of schemes is *not* usually equivalent to the pullback of the underlying spaces of the schemes, as is already easily seen in the case of affine schemes. The analogous fact applies in the setting of derived geometry. Thus, we won't actually need to worry about general limits of $\infty$-topoi.

## Sheaves and geometric morphisms

We are going to be interested in sheaves on $\infty$-topoi with values in things like *spectra* or $\mathbb{E}_\infty$-*ring spectra*. Thus we need to know how these behave under geometric morphisms.

For any complete $\infty$-category $\mathcal{A}$, any geometric morphism $f : \mathcal{X} \to \mathcal{Y}$ induces a **direct image** functor $f_* : \mathrm{Shv}_{\mathcal{A}}(\mathcal{X}) \to \mathrm{Shv}_{\mathcal{A}}(\mathcal{Y})$, which is defined by precomposition with $f^*$. That is, it sends a limit preserving $F : \mathcal{X}^{\mathrm{op}} \to \mathcal{A}$ to the composite functor

$$\mathcal{Y}^{\mathrm{op}} \xrightarrow{(f^*)^{\mathrm{op}}} \mathcal{X}^{\mathrm{op}} \xrightarrow{F} \mathcal{A},$$

which is limit preserving because $f^*$ is colimit preserving. The construction $F \mapsto f_* F$ is itself limit preserving, and thus, if $\mathcal{A}$ is presentable, admits a left adjoint $f^*$.

The left adjoint $f^*$ is in general difficult to describe explicitly. However, in many of the cases we are interested in (e.g., spectra, $\mathbb{E}_\infty$-rings, topological abelian groups) $\mathcal{A}$ is a **compactly generated $\infty$-category** (see [169, 5.5.7]). This means[3] that there exists a small and finite cocomplete $\mathcal{A}_0$, and a left exact functor $\mathcal{A}_0 \to \mathcal{A}$ inducing an equivalence

$$A \mapsto \mathrm{Map}_{\mathcal{A}}(-, A) : \mathcal{A} \xrightarrow{\sim} \mathrm{Fun}^{\mathrm{lex}}((\mathcal{A}_0)^{\mathrm{op}}, \mathcal{S}) \subseteq \mathrm{Fun}((\mathcal{A}_0)^{\mathrm{op}}, \mathcal{S}),$$

where "lex" indicates the full subcategory of left exact (= finite limit preserving) functors.

Example 8.6.11. For instance, if $\mathcal{A} = \mathrm{Sp}$ is the $\infty$-category of spectra, we can take $\mathcal{A}_0$ to be the full subcategory of "finite" spectra, i.e., those built from finitely many cells.

For such $\mathcal{A}$, we then have equivalences

$$\mathrm{Shv}_{\mathcal{A}}(\mathcal{X}) = \mathrm{Fun}^{\mathrm{lim.\ pres}}(\mathcal{X}^{\mathrm{op}}, \mathcal{A}) \approx \mathrm{Fun}^{\mathrm{lim.\ pres}}(\mathcal{A}^{\mathrm{op}}, \mathcal{X}) \approx \mathrm{Fun}^{\mathrm{lex}}((\mathcal{A}_0)^{\mathrm{op}}, \mathcal{X}),$$

(where the middle equivalence sends a limit preserving functor $\mathcal{X}^{\mathrm{op}} \to \mathcal{A}$ to the right adjoint of its opposite, using (8.2.2)). It turns out that in this case a geometric morphism $f : \mathcal{X} \to \mathcal{Y}$ induces direct image and pullback functors $\mathrm{Shv}_{\mathcal{A}}(\mathcal{X}) \leftrightarrows \mathrm{Shv}_{\mathcal{A}}(\mathcal{Y})$ by

---

[3] To see this combine [169, 5.3.5.10] and [169, 5.5.1.9].

*postcomposition* with $f_*: \mathcal{X} \to \mathcal{Y}$ and $f^*: \mathcal{Y} \to \mathcal{X}$ respectively (defined because both of these are left exact). (See [163, V 1.1.8].)

*Remark 8.6.12* (Descent for sheaves).  An immediate consequence of this is *descent* for sheaves with values in compactly generated $\infty$-categories $\mathcal{A}$: if $\mathcal{X} \approx \operatorname{colim}_i \mathcal{X}_i$ in $\infty\mathcal{T}op$, then $\operatorname{Shv}_{\mathcal{A}}(\mathcal{X}) \approx \lim_i \operatorname{Shv}_{\mathcal{A}}(\mathcal{X}_i)$, where the limit is taken over pullback functors. In particular, if $U \approx \operatorname{colim}_i U_i$ in $\mathcal{X}$, then $\operatorname{Shv}_{\mathcal{A}}(\mathcal{X}_{/U}) \approx \lim_i \operatorname{Shv}_{\mathcal{A}}(\mathcal{X}_{/U_i})$.

## 8.7    Étale morphisms

Any morphism $f: U \to V$ in $\mathcal{X}$ gives rise to a geometric morphism, denoted $f: \mathcal{X}_{/U} \to \mathcal{X}_{/V}$, where the left exact left adjoint $f^*$ is defined by pullback along $f$. (We already met this functor in (8.4).) In particular, for any $U \in \mathcal{X}$ there is a geometric morphism $\pi: \mathcal{X}_{/U} \to \mathcal{X}$.

### Maps to slices of ∞-topoi

**Proposition 8.7.1.**  *Given $U \in \mathcal{X}$ and a geometric morphism $f: \mathcal{Y} \to \mathcal{X}$, there is an equivalence*

$$\left\{ \begin{array}{c} \mathcal{X}_{/U} \\ {}^{s}{\nearrow} \quad {\downarrow}{\pi} \\ \mathcal{Y} \xrightarrow{\ f\ } \mathcal{X} \end{array} \right\} \xrightarrow{\sim} \left\{ 1 \dashrightarrow f^*U \right\}$$

*between the $\infty$-category of "sections" of $\pi$ over $\mathcal{Y}$, and the $\infty$-groupoid of global sections of $f^*U$ on $\mathcal{Y}$. It is defined by sending $s$ to $s^*(t)$, where $t: 1 \to \pi^*U$ is the map in $\mathcal{X}_{/U}$ represented by the diagonal map $\Delta: U \to U \times U$. (See [169, 6.3.5.5] for a more precise statement and proof.)*

As a consequence, we see that $U \mapsto \mathcal{X}_{/U}$ describes a fully faithful functor $\mathcal{X} \rightarrowtail \infty\mathcal{T}op_{/\mathcal{X}}$. Thus, objects of $\mathcal{X}$, which as we have seen (8.3.5) can be thought of as "generalized open sets" of $\mathcal{X}$, can also be identified with particular kinds of geometric morphisms to $\mathcal{X}$, and we lose no information by doing so.

Example 8.7.2 (Espace étalé).  Given a sheaf of *sets* $F$ on a topological space $X$, the **espace étalé** of $F$ is a topological space $X_F$ equipped with a map $\pi: X_F \to X$, defined so that $\operatorname{Open}_{X_F} = \coprod_{U \in \operatorname{Open}_X} F(U)$. It is not hard to show that $\operatorname{Shv}(X_F) \approx \operatorname{Shv}(X)_{/F}$, and that there is a bijection between maps $F \to F'$ in $\operatorname{Shv}_{\operatorname{Set}}(X)$, and maps $X_F \to X_{F'}$ of topological spaces which are compatible with the projection to $X$.

Any local homeomorphism $f: Y \to X$ of spaces is equivalent to the espace étalé of a sheaf of sets. Local homeomorphisms are also called *étale* maps of spaces, which motivates the terminology of the next section.

## Étale morphisms of ∞-topoi

A geometric morphism is **étale** if it is equivalent to a morphism of the form $\pi\colon \mathcal{X}_{/U} \to \mathcal{X}$ for some ∞-topos $\mathcal{X}$ and object $U \in \mathcal{X}$. This class includes the geometric morphism $\mathcal{X}_{/U} \to \mathcal{X}_{/V}$ induced by a map $f\colon U \to V$ in $\mathcal{X}$, as $f$ also represents object of the ∞-topos $\mathcal{X}_{/V}$.

*Remark 8.7.3* (Pullbacks of étale morphisms). Pullbacks of étale morphisms of ∞-topoi are étale: (8.7.1) implies a pullback diagram

$$
\begin{array}{ccc}
\mathcal{Y}_{/f^*U} & \longrightarrow & \mathcal{X}_{/U} \\
\downarrow & & \downarrow \\
\mathcal{Y} & \xrightarrow{\;f\;} & \mathcal{X}
\end{array}
$$

in ∞$\mathcal{T}$op.

*Remark 8.7.4* (Characterization of étale morphisms). For any étale morphism $f\colon \mathcal{Y} \to \mathcal{X}$, the pullback functor $f^*$ admits a left adjoint $f_!\colon \mathcal{Y} \to \mathcal{X}$. In the case of the projection $\pi\colon \mathcal{X}_{/U} \to \mathcal{X}$, this is the evident functor which on objects sends $V \to U$ to $V$.

The left adjoint $f_!$ associated to an étale morphism $f\colon \mathcal{Y} \to \mathcal{X}$ is conservative, and has the property that the evident map $f_!(f^*U \times_{f^*V} Z) \xrightarrow{\sim} U \times_V f_!Z$ is an equivalence for all $Z \in \mathcal{Y}$ and all $U \to V$ and $f_!Z \to V$ in $\mathcal{X}$. Furthermore, étale morphisms $f$ are characterized by the existence of an $f_!$ with these properties [169, 6.3.5.11].

*Remark 8.7.5* ("Restriction" of sheaves along étale maps). For an étale morphism $f\colon \mathcal{Y} \to \mathcal{X}$ and any ∞-category $\mathcal{A}$, the induced functor $f^*\colon \mathrm{Shv}_{\mathcal{A}}(\mathcal{X}) \to \mathrm{Shv}_{\mathcal{A}}(\mathcal{Y})$ on $\mathcal{A}$-valued sheaves admits a very simple description using $f_!$: it sends $F\colon \mathcal{X}^{\mathrm{op}} \to \mathcal{A}$ to $F(f_!)^{\mathrm{op}}\colon \mathcal{Y}^{\mathrm{op}} \to \mathcal{A}$. When $f$ is the projection $\mathcal{X}_{/U} \to \mathcal{X}$ this amounts to saying that $(f^*F)(V \to U) \approx F(V)$. It is easy to think of this as a "restriction" functor, so sometimes we will use the notation "$F|_U$" for $f^*F$ in this case.

## Colimits along étale maps of ∞-topoi

Let ∞$\mathcal{T}$op$_{\text{ét}} \subseteq$ ∞$\mathcal{T}$op denote the (non-full) subcategory consisting of étale morphisms between arbitrary ∞-topoi.

**Proposition 8.7.6** ([169, 6.3.5.13]). *The ∞-category* ∞$\mathcal{T}$op$_{\text{ét}}$ *has all small colimits, and the inclusion* ∞$\mathcal{T}$op$_{\text{ét}} \to$ ∞$\mathcal{T}$op *preserves small colimits.*

For instance, given an ∞-topos $\mathcal{X}$, the descent property (8.4.5), plus the fact that colimits in ∞$\mathcal{T}$op are computed as limits in $\widehat{\mathrm{Cat}}_\infty$ (8.5), implies that the functor

$$
U \mapsto \mathcal{X}_{/U}\colon \mathcal{X} \to \infty\mathcal{T}\mathrm{op}
$$

is itself colimit preserving. This functor clearly factors through the subcategory ∞$\mathcal{T}$op$_{\text{ét}}$. In fact, every colimit in ∞$\mathcal{T}$op$_{\text{ét}}$ is equivalent to one of this form.

**Example 8.7.7.** Any equivalence of ∞-topoi is étale. Thus, if $\mathcal{X}\colon \mathcal{G} \to \infty\mathcal{T}\mathrm{op}$ is a functor from a small ∞-groupoid $\mathcal{G}$, it factors through ∞$\mathcal{T}$op$_{\text{ét}} \to$ ∞$\mathcal{T}$op, so its colimit is a "quotient $\infty$-topos" $\mathcal{X}/\!/\mathcal{G}$, with the property that $\mathcal{X}(c) \to \mathcal{X}/\!/\mathcal{G}$ is étale for all objects $c \in \mathcal{G}$.

For instance, let $\mathcal{X} = \mathrm{Shv}(X)$ be the $\infty$-topos of sheaves on a topological space $X$, and let $G$ be a discrete group acting on $X$. Then $\mathcal{X}/\!/\mathcal{G}$ is equivalent to an $\infty$-category of "$G$-equivariant sheaves on $X$", and the projection map $\pi \colon \mathcal{X} \to \mathcal{X}/\!/\mathcal{G}$ is étale.

*Remark 8.7.8.* The proof of (8.7.6) is pretty technical, but ultimately it is a generalization of the following observation: given open immersions $U \leftarrow W \to V$ of topological spaces, the pushout $X$ in spaces can be constructed so that a basis of open sets is described by the category $\mathrm{colim}[\mathrm{Open}_U \leftarrow \mathrm{Open}_W \to \mathrm{Open}_V]$.

## 8.8    Spectra and commutative ring spectra

Now that we have $\infty$-categorical versions of spaces, we can put sheaves of spectra or commutative ring spectra on them. In this section I collect some notation and observations about these; some familiarity with spectra and structured ring spectra on the part of the reader is assumed.

### Spectra

We write $\mathrm{Sp}$ for the $\infty$-category of spectra. It is an example of a *stable $\infty$-category* [168, 1.1.1.9], and so is pointed, has suspension and loop functors which are inverse to each other, has fiber sequences and cofiber sequences which coincide, and so forth.

The $\infty$-category $\mathrm{Sp}$ has a symmetric monoidal structure with respect to "smash product", here denoted "$\otimes$", with unit object being the sphere spectrum $\mathbb{S}$. The monoidal structure is closed, so there are internal hom objects.

We write $\Omega^{\infty-n} \colon \mathrm{Sp} \to \mathcal{S}$ for the usual "forgetful" functors, and define homotopy groups of spectra by $\pi_n X = \pi_{n+k}\Omega^{\infty-k}X$ for $n \in \mathbb{Z}$, and any $k \geq -n$. We say that a spectrum $X$ is $n$-**truncated** if $\Omega^{\infty-k}X \approx 1$, or equivalently if $\pi_k X \approx 0$ for $k < n$. We say a spectrum is $n$-**connective** if $\pi_k X \approx 0$ for $k > n$, and **connective** if 0-connective.

We write $\mathrm{Sp}_{\leq n}$ and $\mathrm{Sp}_{\geq n}$ respectively for the full subcategories in $\mathrm{Sp}$ of $n$-truncated and $n$-connective objects. The intersection

$$\mathrm{Sp}^\heartsuit = \mathrm{Sp}_{\geq 0} \cap \mathrm{Sp}_{\leq 0}$$

is equivalent to the ordinary category of abelian groups: every abelian group $A$ corresponds to an Eilenberg–MacLane spectrum in $\mathrm{Sp}^\heartsuit$, which we also denote $A$ by abuse of notation.

**Warning 8.8.1.** The notion of $n$-truncated spectrum described above is *not* the same as the general notion of $n$-truncation in an $\infty$-category that we described earlier (8.5): since every spectrum is a suspension of one, every $n$-truncated object in $\mathrm{Sp}$ (in the earlier sense) is equivalent to 0. The pair $(\mathrm{Sp}_{\leq 0}, \mathrm{Sp}_{\geq 0})$ is instead an example of a *t-structure* on $\mathrm{Sp}$ [168, 1.2.1].

## Commutative ring spectra

By an $\mathbb{E}_\infty$-**ring**, we mean a commutative ring object with respect to the symmetric monoidal structure on the $\infty$-category of spectra. The $\infty$-category of commutative rings is denoted CAlg. (We are following the notation and terminology of [170] here. This notion of $\mathbb{E}_\infty$-ring is an $\infty$-categorical manifestation of the notion of structured commutative ring spectrum/commutative S-algebra as defined in, e.g., [94].)

Given $A \in$ CAlg we write $\text{CAlg}_A = \text{CAlg}_{A/}$ for the category of $\mathbb{E}_\infty$-rings under $A$, also called **commutative $A$-algebras**. The initial $\mathbb{E}_\infty$-algebra is the sphere spectrum $\mathbb{S}$, so $\text{CAlg} = \text{CAlg}_\mathbb{S}$.

There is a forgetful functor CAlg $\to$ Sp which is conservative. The homotopy groups of an $\mathbb{E}_\infty$-algebra are those of its underlying spectrum, and likewise we may speak of an $\mathbb{E}_\infty$-ring being $n$-truncated or $n$-connective by reference to its underlying spectrum. In particular we distinguish the full subcategory $\text{CAlg}^{\text{cn}}$ of **connective** $\mathbb{E}_\infty$-rings, i.e., those $A \in$ CAlg such that $\pi_k A \approx 0$ for $k < 0$.

We further consider the full subcategory $\text{CAlg}^\heartsuit$ of $\mathbb{E}_\infty$-algebras which are both 0-connective and 0-truncated. This is equivalent to the ordinary category of commutative rings, so we will identify an ordinary commutative ring with its corresponding Eilenberg–Mac Lane spectrum in $\text{CAlg}^\heartsuit$.

We have adjoint pairs

$$\text{CAlg}^\heartsuit \xhookleftarrow{\quad\longrightarrow\quad} \text{CAlg}^{\text{cn}} \xleftarrow{\quad\longrightarrow\quad} \text{CAlg}$$

of fully faithful and localization functors relating these subcategories; the localization functors of these pairs are denoted $\tau_{\geq 0} : \text{CAlg} \to \text{CAlg}^{\text{cn}}$ and $\tau_{\leq 0} : \text{CAlg}^{\text{cn}} \to \text{CAlg}^\heartsuit$. Note that $\mathbb{S} \in \text{CAlg}^{\text{cn}}$ and that $\mathbb{S} \to \tau_{\leq 0} \mathbb{S} \approx \mathbb{Z}$.

*Remark 8.8.2* (General truncation of $\mathbb{E}_\infty$-rings). The $\infty$-category CAlg of $\mathbb{E}_\infty$-rings, being a presentable $\infty$-category, has $n$-truncation functors $\tau_{\leq n} : \text{CAlg} \to \text{CAlg}$ for $n \geq -1$ (8.5). However, these are not generally compatible with the $n$-truncation functors on spectra defined in (8.8). For example, the periodic complex $K$-theory spectrum $KU$ admits the structure of an $\mathbb{E}_\infty$-ring, but its $n$th truncation as an $\mathbb{E}_\infty$-ring is equivalent to 0 for all $n \geq -1$.

However, the $n$-truncation functors on CAlg restrict to functors on connective $\mathbb{E}_\infty$-rings $\tau_{\leq n} : \text{CAlg}^{\text{cn}} \to \text{CAlg}^{\text{cn}}$, which are in fact the $n$-truncation functors for $\text{CAlg}^{\text{cn}}$, and which are in fact compatible with $n$-truncation of the underlying spectra.

## Modules

To each $\mathbb{E}_\infty$-ring $A$ there is an associated $\infty$-category of (left) modules $\text{Mod}_A$, which is itself closed symmetric monoidal: we write $M \otimes_A N$ for the monoidal product and $\underline{\text{Hom}}_A(M, N)$ for the internal hom. We have that $\text{Mod}_\mathbb{S} \approx \text{Sp}$, an equivalence of symmetric monoidal $\infty$-categories.

*Example 8.8.3.* If $A \in \text{CAlg}^\heartsuit$ is an ordinary ring, then $\text{Mod}_A$ is equivalent to the $\infty$-category obtained from chain complexes of $A$-modules and quasi-isomorphisms

[266]. Thus, the homotopy category of $\mathrm{Mod}_A$ is the *derived category* of the ring $A$. The tensor product on $\mathrm{Mod}_A$ corresponds to the *derived* tensor product of complexes.

*Remark 8.8.4* ($\mathbb{Z}$-modules are abelian groups). We will write $\mathrm{Mod}_{\mathbb{Z}}^{cn} \subseteq \mathrm{Mod}_{\mathbb{Z}}$ for the full subcategory of $(-1)$-connected $\mathbb{Z}$-modules. The $\infty$-category $\mathrm{Mod}_{\mathbb{Z}}^{cn}$ is equivalent to those obtained from each of the following examples by inverting the evident weak equivalences: $(-1)$-connected chain complexes of abelian groups, simplicial abelian groups, topological abelian groups.

An object $X$ in an $\infty$-category $\mathcal{A}$ is called an **abelian group object** if it represents a functor $\mathcal{A}^{\mathrm{op}} \to \mathrm{Mod}_{\mathbb{Z}}^{cn}$.

Every commutative $A$-algebra has an underlying $A$-module. The coproduct of $A$-algebras coincides with tensor product of $A$-modules. For this reason, we typically denote coproduct in $\mathrm{CAlg}_A$ by $B \otimes_A C$.

The homotopy groups $\pi_* M$ of an $A$-module are automatically a graded $\pi_* A$-module. To get a feel for how these things behave, it is useful to be aware of two spectral sequences:

$$E_2 = \mathrm{Tor}_*^{\pi_* A}(\pi_* M, \pi_* N) \Longrightarrow \pi_*(M \otimes_A N),$$
$$E_2 = \mathrm{Ext}_{\pi_* A}^*(\pi_* M, \pi_* N) \Longrightarrow \pi_* \underline{\mathrm{Hom}}_A(M, N).$$

The Tor spectral sequence satisfies complete convergence, while the Ext spectral sequence satisfies conditional convergence [94, Ch. IV].

## Flat modules and $\mathbb{E}_\infty$-rings

An $A$-module $M$ is said to be **flat** if

1. $\pi_0 M$ is flat as a $\pi_0 A$-module, and
2. the evident maps $\pi_0 M \otimes_{\pi_0 A} \pi_n A \to \pi_n M$ are isomorphisms for all $n$.

Likewise, a map $A \to B$ of $\mathbb{E}_\infty$-rings is **flat** if $B$ is flat as an $A$-module, In view of the tor spectral sequence, we see that if $A \to B$ is flat then $\pi_*(B \otimes_A N) \approx \pi_0 B \otimes_{\pi_0 A} \pi_* N$ for $N \in \mathrm{Mod}_A$.

*Remark 8.8.5* (Flatness and connective covers). Let's pause to note the following. Consider the map $\tau_{\geq 0} A \to A$ from the connective cover to an $\mathbb{E}_\infty$-ring $A$. The base change functor $A \otimes_{\tau_{\geq 0} A} - : \mathrm{Mod}_{\tau_{\geq 0} A} \to \mathrm{Mod}_A$ restricts to an *equivalence*

$$\mathrm{Mod}_{\tau_{\geq 0} A}^{\flat} \xrightarrow{\sim} \mathrm{Mod}_A^{\flat}$$

of full subcategories of *flat* modules; the inverse equivalence sends an $A$-module $N$ to its connective cover $\tau_{\geq 0} N$ viewed as a $\tau_{\geq 0} A$-module. Similarly, we obtain an equivalence

$$\mathrm{CAlg}_{\tau_{\geq 0} A}^{\flat} \xrightarrow{\sim} \mathrm{CAlg}_A^{\flat}$$

of full subcategories of algebras which are flat over the ground ring. Thus, any flat morphism of $\mathbb{E}_\infty$-rings is a base change of one between connective $\mathbb{E}_\infty$-rings. This phenomenon turns out to extend to nonconnective spectral Deligne–Mumford stacks (8.13.6).

## Examples of $\mathbb{E}_\infty$-rings

Example 8.8.6 (Polynomial rings). Given any space $K$, we obtain a spectrum $\mathbb{S}[K] =$ the suspension spectrum of $K_+$. If $K$ is equipped with the structure of an $\mathbb{E}_\infty$-space (i.e., space with an action by an $\mathbb{E}_\infty$-operad), then $\mathbb{S}[K]$ is equipped with a corresponding structure of $\mathbb{E}_\infty$-ring. A particular example of this is when $K$ is a discrete commutative monoid.

For instance, we can form **polynomial rings**: $\mathbb{S}[x] := \mathbb{S}[\mathbb{Z}_{\geq 0}]$, and more generally $A[x_1, \ldots, x_n] := A \otimes \mathbb{S}[(\mathbb{Z}_{\geq 0})^n] \approx A \otimes \mathbb{S}[\mathbb{Z}_{\geq 0}]^{\otimes n}$. We have

$$\pi_*\big(A[x_1, \ldots, x_n]\big) \approx (\pi_* A)[x_1, \ldots, x_n].$$

Thus, $A[x_1, \ldots, x_n]$ is a flat $A$-algebra. In particular, if $A$ is an ordinary ring, then $A[x_1, \ldots, x_n]$ is also an ordinary ring.

Example 8.8.7 (Free rings). Let $\mathbb{S}\{x\}$ denote the **free $\mathbb{E}_\infty$-ring on one generator**, which is characterized by the existence of isomorphisms

$$\mathrm{Map}_{\mathrm{CAlg}}(\mathbb{S}\{x\}, R) \xrightarrow{\sim} \Omega^\infty(R)$$

natural in $R \in \mathrm{CAlg}$. We have that $\mathbb{S}\{x\} \approx \mathbb{S}[\coprod_k B\Sigma_k]$.

We may similarly define $A\{x_1, \ldots, x_n\} := A \otimes \mathbb{S}\{x\}^{\otimes n}$, the free commutative $A$-algebra on $n$ generators.

There is a canonical map $A\{x_1, \ldots, x_n\} \to A[x_1, \ldots, x_n]$ from the free ring to the polynomial ring. It is generally not an equivalence, but is an equivalence if $\mathbb{Q} \subseteq \pi_0 A$. When $A$ is connective so is $A\{x_1, \ldots, x_n\}$, and then $\pi_0\big(A\{x_1, \ldots, x_n\}\big) \approx \pi_0 A[x_1, \ldots, x_n]$; however, no such isomorphism on $\pi_0$ holds for general non-connective $\mathbb{E}_\infty$-rings.

## $\mathbb{E}_\infty$-rings of finite characteristic

We note the following curious fact, conjectured by May and proved by Hopkins; see [188]. It is a generalization of the Nishida nilpotence theorem, which is the special case $R = \mathbb{S}$.

Theorem 8.8.8.  *For any $R \in \mathrm{CAlg}$, all elements in the kernel of the evident map $\pi_* R \to \pi_*(R \otimes \mathbb{Z})$ are nilpotent. In particular, $R \otimes \mathbb{Z} \approx 0$ implies $R \approx 0$.*

Many spectra which arise in chromatic homotopy theory have the property that $R \otimes \mathbb{Z} \xrightarrow{\sim} R \otimes \mathbb{Q}$; e.g., if $R \approx L_n^f R$ for some $n$ at some prime $p$. Therefore, if $R \in \mathrm{CAlg}$ is such that $R_{(p)} \approx L_n^f R_{(p)} \not\approx 0$ for some prime $p$ and some $n < \infty$, then $1 \in \pi_0 R$ has infinite order. So there are no non-trivial $\mathbb{E}_\infty$-rings of finite characteristic in chromatic homotopy.

A related result of Hopkins–Mahowald is: any $R \in \mathrm{CAlg}$ such that $p = 0 \in \pi_0 R$ admits the structure of a $\mathbb{Z}/p$-module [188, Theorem 4.18]. In particular, the underlying spectrum of an $\mathbb{E}_\infty$-ring of positive characteristic $p$ is always a product of Eilenberg–MacLane spectra.

## Other kinds of commutative rings

We note several other flavors of commutative ring which can be used in derived versions of algebraic geometry.

1. Given an ordinary ring $R$, there is a notion of chain-level $\mathbb{E}_\infty$-$R$-algebra, consisting of an unbounded chain complex of abelian groups equipped with the action of a chain-level $\mathbb{E}_\infty$-operad. The resulting $\infty$-category of chain level $\mathbb{E}_\infty$-$R$-algebras is equivalent to $\mathrm{CAlg}_R$ [236].

2. Over any ordinary ring $R$ we may consider the category of differential graded commutative $R$-algebras. In general it is not possible to extract a useful $\infty$-category from this notion. However, it is possible when $R \supseteq \mathbb{Q}$, in which case the resulting $\infty$-category is equivalent to $\mathrm{CAlg}_R$.

3. The category of *simplicial commutative rings* gives rise to an $\infty$-category $\mathrm{CAlg}^\Delta$. This $\infty$-category is related to $\mathrm{CAlg}_\mathbb{Z}$ but is quite distinct from it. In fact, there is a conservative "forgetful" functor

$$\mathrm{CAlg}^\Delta \to \mathrm{CAlg}^{cn}_\mathbb{Z}$$

which is both limit and colimit preserving. This implies that simplicial commutative rings are intrinsically connective objects, and that pushouts in $\mathrm{CAlg}^\Delta$ are computed as tensor products on underlying $\mathbb{Z}$-modules.

   However, the above functor is far from being an equivalence. For instance, the "free simplicial commutative ring on one generator" maps to $\mathbb{Z}[x] \in \mathrm{CAlg}^{cn}_\mathbb{Z}$, rather than to $\mathbb{Z}\{x\}$. See [170, 25.1].

## Spectrally ringed $\infty$-topoi

The categories Sp and CAlg are presentable $\infty$-categories (and in fact are compactly generated), so it is straightforward to consider sheaves on an $\infty$-topos valued in each of these. For any such sheaf $\mathcal{O}$ on $\mathcal{X}$ we have homotopy sheaves $\pi_k\mathcal{O}$ on $\mathcal{X}^\heartsuit$.

A **spectrally ringed $\infty$-topos** is a pair $X = (\mathcal{X}, \mathcal{O}_X)$ consisting of an $\infty$-topos $\mathcal{X}$ and a sheaf $\mathcal{O}_X \in \mathrm{Shv}_{\mathrm{CAlg}}(\mathcal{X})$ of $\mathbb{E}_\infty$-rings. These are objects of an $\infty$-category $\infty\mathcal{T}\mathrm{op}_{\mathrm{CAlg}}$, in which morphisms $X \to Y$ are pairs consisting of a geometric morphism $f : \mathcal{X} \to \mathcal{Y}$ together with a map $\phi : \mathcal{O}_Y \to f_*\mathcal{O}_X$ of sheaves of $\mathbb{E}_\infty$-rings on $\mathcal{Y}$ (see [170, 1.4.1.3]).

## 8.9    The étale site of a commutative ring

Our objects of study will be spectrally ringed $\infty$-topoi which are "locally affine". There are two such notions of affine we can use here, corresponding in the classical case to the Zariski and étale topologies of a ring. We are going to focus on the étale case (which is in some sense strictly more general). Thus, in this section we describe the spectrally ringed $\infty$-topos $\mathrm{Sp\acute{e}t}\, A$ associated to an $\mathbb{E}_\infty$-ring $A$. It is an "étale topology version" of an analogous construction of a spectrally ringed $\infty$-topos $\mathrm{Spec}\, A$, which generalizes the classical construction of affine schemes.

*Warning:* this notion of "étale" map of rings is not to be confused with that of étale maps of $\infty$-topoi (8.7), though the notions will be linked later on (8.13).

## Étale maps of $\mathbb{E}_\infty$-rings

A map $R \to S$ of ordinary commutative rings is **étale** if:

1.  $S$ is finitely presented over $R$,
2.  $R \to S$ is flat, and
3.  the fold map $S \otimes_R S \to S$ is projection onto a factor (or equivalently, there exists idempotent $e \in S \otimes_R S$ inducing $(S \otimes_R S)[e^{-1}] \xrightarrow{\sim} S$).

Example 8.9.1. If $K$ is a field, then $K \to R$ is étale if and only if $R \approx \prod_{i=1}^d F_i$, where each $K \to F_i$ is a finite separable field extension.

We say that a map $A \to B$ of $\mathbb{E}_\infty$-rings is **étale** if

1.  the underlying map $\pi_0 A \to \pi_0 B$ of ordinary commutative rings is étale, and
2.  $\pi_n A \otimes_{\pi_0 A} \pi_0 B \to \pi_n B$ is an isomorphism for all $n$ (so that $A \to B$ is flat in the sense of (8.8)).

*Remark 8.9.2.* If $A \in \mathrm{CAlg}^\heartsuit$ is an ordinary commutative ring, then the two notions of étale coincide.

Theorem 8.9.3 (Goerss–Hopkins–Miller). *Let $A \in \mathrm{CAlg}$.*

1.  *For every étale map $\psi \colon \pi_0 A \to B_0$ of ordinary rings, there exists an étale map $\phi \colon A \to B$ of $\mathbb{E}_\infty$-rings and an isomorphism $\pi_0 B \approx B_0$ with respect to which $\pi_0 \phi \colon \pi_0 A \to \pi_0 B$ is identified with $\psi$.*
2.  *Let $\phi \colon A \to B$ be an étale map of $\mathbb{E}_\infty$-rings. Then for every $C \in \mathrm{CAlg}_A$, the evident map*

$$\mathrm{Map}_{\mathrm{CAlg}_A}(B, C) \to \mathrm{Map}_{\mathrm{CAlg}_{\pi_0 A}^\heartsuit}(\pi_0 B, \pi_0 C)$$

*is an equivalence.*

See [168, 7.5.4] for a proof of a generalized formulation of this.

*Remark 8.9.4.* A consequence of this theorem is that $\mathrm{Map}_{\mathrm{CAlg}_A}(B, C)$ is a *set* (i.e., 0-truncated) whenever $\phi \colon A \to B$ is étale. This consequence can be proved directly from the definition of étale morphism. In fact, when $\phi$ is étale, then the evident map $B \otimes_{(B \otimes_A B)} B \to B$ must be an equivalence (using that both $A \to B$ and $B \otimes_A B \to B$ are flat). Writing $X = \mathrm{Map}_{\mathrm{CAlg}_A}(B, C)$, this equivalence implies that $X \to X \times_{(X \times X)} X$ is an equivalence, which says exactly that $X$ is 0-truncated.

*Remark 8.9.5.* Given an étale morphism $\pi_0 A \to B_0$ of ordinary rings, it is not hard to show that the functor $CAlg_A \to Set \subseteq S$ defined by $Map_{CAlg_{\pi_0 A}^{\heartsuit}}(B_0, \pi_0(-))$ preserves limits[4] and is accessible, so is corepresentable by a $B \in CAlg_A$. The hard part of (8.9.3) is to show that $B_0 \to \pi_0 B$ is an isomorphism.

*Remark 8.9.6.* Statement (2) of the theorem is equivalent to: for every étale map $A \to B$ and $R \in CAlg$, the square

$$
\begin{array}{ccc}
Map_{CAlg}(B, R) & \longrightarrow & Map_{CAlg}(\pi_0 B, \pi_0 R) \\
\downarrow & & \downarrow \\
Map_{CAlg}(A, R) & \longrightarrow & Map_{CAlg}(\pi_0 A, \pi_0 R)
\end{array}
$$

is a pullback of $\infty$-groupoids.

Let $CAlg_A^{\text{ét}} \subseteq CAlg_A$ be the full subcategory of $A$-algebras whose objects are maps $A \to B$ which are étale. As we have seen, it is equivalent to a 1-category.

*Remark 8.9.7.* If $A \xrightarrow{f} B \xrightarrow{g} C$ are maps of $\mathbb{E}_\infty$-rings such that $f$ and $gf$ are étale, then $g$ is also étale [168, 7.5.1.7]. Thus every morphism in $CAlg_A^{\text{ét}}$ is itself étale.

**Corollary 8.9.8.** *For any $A \in CAlg$, the functor $CAlg_A^{\text{ét}} \to CAlg_{\pi_0 A}^{\text{ét}}$ defined by taking $\pi_0$ is an equivalence of $\infty$-categories.*

**Example 8.9.9** (Localization of $\mathbb{E}_\infty$-rings). Let $A \in CAlg$, and suppose $f \in \pi_0 A$. Then $\pi_0 A \to (\pi_0 A)[f^{-1}]$ is an étale morphism of commutative rings. By (8.9.3), (i) there exists a map $A \to A[f^{-1}]$ of $\mathbb{E}_\infty$-rings such that (i) $\pi_*(A[f^{-1}]) \approx (\pi_* A)[f^{-1}]$, and (ii) for any $C \in CAlg$, $Map_{CAlg}(A[f^{-1}], C) \to Map_{CAlg}(A, C)$ is the inclusion of those path components consisting of $\phi: A \to C$ which take $f$ to a unit in $\pi_0 C$.

This special case predates the proof of the Goerss–Hopkins–Miller theorem for $\mathbb{E}_\infty$-rings. In fact, one can in a similar way invert any multiplicative subset $S \subseteq \pi_* A$ of the *graded* homotopy ring to obtain $A_S$ with $\pi_*(A_S) \approx (\pi_* A)_S$.

**Example 8.9.10** (Adjoining primitive roots of unity). Here is a hands-on construction of an étale morphism, due to [260]. Given any $\mathbb{E}_\infty$-ring $A$, prime $p$, and $k \geq 1$, consider the group ring $B' := A[\mathbb{Z}/p^k]$ (8.8.6), with $\pi_0 B' \approx (\pi_0 A)[t]/(t^{p^k} - 1)$. Let $f = \sum_{j=0}^{p-1}(1 - t^{jp^{k-1}})$ in $\pi_0 B'$, and note that $f^2 = pf$. Formally inverting $f$ we obtain

$$B := B'[f^{-1}], \quad \text{with} \quad \pi_* B \approx (\pi_* A)[\tfrac{1}{p}, t]/(1 + t^{p^{k-1}} + \cdots + t^{(p-1)p^{k-1}}).$$

It turns out that $A \to B$ is an étale morphism, and $\pi_0 B$ is obtained from $\pi_0 A$ by (i) inverting $p$ and (ii) adjoining a primitive $p^k$th root of unity.

*Remark 8.9.11.* In general, you can always construct étale maps of $\mathbb{E}_\infty$-rings using "generators and relations" (using free rings (8.8.7)), which in fact leads to an alternate proof of (8.9.3); see [170, B.1]. In particular, this shows that every étale map in $CAlg$ is a base change of one between compact objects in $CAlg$ ([170, B.1.3.3] with $R = \mathbb{S}$).

---

[4] Using the fact that étale maps of rings are also "formally étale".

(An object $A$ in an $\infty$-category is **compact** if $\mathrm{Map}_A(A,-) \colon \mathcal{A} \to \mathcal{S}$ preserves filtered colimits.)

### The étale site of an $\mathbb{E}_\infty$-ring

Given $A \in \mathrm{CAlg}$, consider the category $\mathrm{CAlg}_A^{\text{ét}}$ of étale morphisms under $A$. A finite set $\{A \to A_i\}_{i=1}^d$ of maps in $\mathrm{CAlg}_A^{\text{ét}}$ is an **étale cover** if $\pi_0 A \to \prod_{i=1}^d \pi_0 A_i$ is *faithfully flat*.

We define $\mathrm{Shv}_A^{\text{ét}} \subseteq \mathrm{Fun}(\mathrm{CAlg}_A^{\text{ét}}, \mathcal{S})$ to be the full subcategory of functors $F$ such that

$$F(A) \to \lim_\Delta \Big[[n] \mapsto \prod_{i_0,\dots,i_n} F(A_{i_0} \otimes_A \cdots \otimes_A A_{i_n})\Big]$$

is an equivalence for every étale cover $\{A \to A_i\}_i$ in $\mathrm{CAlg}_R^{\text{ét}}$. This $\mathrm{Shv}_A^{\text{ét}}$ is an $\infty$-topos; in fact, it is equivalent to the $\infty$-topos $\mathrm{Shv}_{\pi_0 A}^{\text{ét}}$ of étale sheaves on the ordinary commutative ring $\pi_0 A$. I'll call its objects of **sheaves on the étale site** of $A$.

### The étale spectrum of an $\mathbb{E}_\infty$-ring

Let $\mathcal{O} \colon \mathrm{CAlg}_A^{\text{ét}} \to \mathrm{CAlg}$ denote the forgetful functor.

**Proposition 8.9.12.** *The functor $\mathcal{O}$ is a sheaf of $\mathbb{E}_\infty$-rings on the étale site of $A$.*

We thus define the **étale spectrum** of $A \in \mathrm{CAlg}$ to be the spectrally ringed $\infty$-topos $\mathrm{Sp\acute{e}t}\, A = (\mathrm{Shv}_A^{\text{ét}}, \mathcal{O})$.

*Proof of* (8.9.12). We must show that for every finite étale cover $\{A \to A_i\}_{i=1}^d$ the evident map

$$A \to \lim_\Delta \Big[[n] \mapsto \prod_{i_0,\dots,i_n} A_{i_0} \otimes_A \cdots \otimes_A A_{i_n}\Big]$$

is an equivalence of $\mathbb{E}_\infty$-rings. This is a special case of a much more general statement, called *flat descent* for $\mathbb{E}_\infty$-rings; see [170, D.5] for the general theory.

In this case, the proof amounts to computing the spectral sequence computing the homotopy groups of the inverse limit, whose $E_1$-term takes the form

$$E_1^{s,t} = \pi_t(A_{i_0} \otimes_A \cdots \otimes_A A_{i_s}) \approx \pi_t A \otimes_{\pi_0 A} (\pi_0 A_{i_0} \otimes_{\pi_0 A} \cdots \otimes_{\pi_0 A} \pi_0 A_{i_s})$$

because étale morphisms are flat. The classical version of flat descent for ordinary rings implies that

$$E_2^{s,t} \approx H^s[\pi_t A \otimes_{\pi_0 A} (\pi_0 A_{i_0} \otimes_{\pi_0 A} \cdots \otimes_{\pi_0 A} \pi_0 A_{i_s})] \approx \begin{cases} \pi_t A & \text{if } s = 0, \\ 0 & \text{if } s > 0, \end{cases}$$

so the spectral sequence collapses to a single line at $E_2$. The claim follows because the inverse limit spectral sequence has conditional convergence. $\qquad\square$

*Remark 8.9.13.* We actually have that $\mathcal{O}$ is a *hypercomplete* sheaf of spectra $\mathrm{Shv}_A^{\text{ét}}$. In fact, the argument of the proof of (8.9.12) shows that for each $n \geq 0$ the presheaf $\tau_{\leq n}\mathcal{O}\colon A \mapsto \tau_{\leq n}A$ of spectra obtained by truncation is a sheaf on the étale site, whence $\mathcal{O} \approx \lim_n \tau_{\leq n}\mathcal{O}$; this relies on the fact that $\mathrm{CAlg}_{\tau_{\leq n}A}^{\text{ét}} \approx \mathrm{CAlg}_{\pi_0 A}^{\text{ét}}$ for all $n \geq 0$, so all these rings have the same étale site.

## The Zariski site and spectrum of an $\mathbb{E}_\infty$-ring

In the above we can replace $\mathrm{CAlg}_A^{\text{ét}}$ with the full subcategory $\mathrm{CAlg}_A^{\mathrm{Zar}}$ spanned by objects equivalent to localizations $A \to A[f^{-1}]$. Then $\{A \to A[f_i^{-1}]\}_{i=1}^d$ is a **Zariski cover** if $\pi_0 A \to \prod_{i=1}^d \pi_0 A[f_i^{-1}]$ is faithfully flat; equivalently, if $(f_1, \ldots, f_d)\pi_0 A = \pi_0 A$. We obtain an $\infty$-topos $\mathrm{Shv}_A^{\mathrm{Zar}} \subseteq \mathrm{Fun}(\mathrm{CAlg}_A^{\mathrm{Zar}}, \mathcal{S})$ of Zariski sheaves. We have $\mathrm{Shv}_A^{\mathrm{Zar}} \approx \mathrm{Shv}_{\pi_0 A}^{\mathrm{Zar}}$, and these are equivalent to the $\infty$-categories of sheaves on a topological space, namely the prime ideal spectrum of $\pi_0 A$ equipped with the Zariski topology.

We can likewise define the **Zariski spectrum** to be the spectrally ringed $\infty$-topos $\mathrm{Spec}\, A = (\mathrm{Shv}_A^{\mathrm{Zar}}, \mathcal{O})$, as the forgetful functor $\mathcal{O}\colon \mathrm{CAlg}_A^{\mathrm{Zar}} \to \mathrm{CAlg}$ is sheaf of $\mathbb{E}_\infty$-rings on the Zariski site.

Example 8.9.14 (Points in étale site vs. the Zariski site). To get a sense of the difference between the Zariski and étale sites, let's compare $\mathrm{Map}_{\infty Top}(\mathcal{S}, \mathrm{Shv}_A^{\mathrm{Zar}})$ with $\mathrm{Map}_{\infty Top}(\mathcal{S}, \mathrm{Shv}_A^{\text{ét}})$. (A map of $\infty$-topoi of the form $\mathcal{S} \to \mathcal{X}$ is called a **point** of $\mathcal{X}$.)

First, suppose $K \in \mathrm{CAlg}^\heartsuit$ is an ordinary field. Then $\mathrm{CAlg}_K^{\mathrm{Zar}} \approx 1$, so $\mathrm{Shv}_K^{\mathrm{Zar}} \approx \mathcal{S}$, so there is a unique map $\mathcal{S} \to \mathrm{Shv}_K^{\mathrm{Zar}}$ of $\infty$-topoi. On the other hand, any separable closure $K \to K^{\mathrm{sep}}$ induces a geometric morphism $f\colon \mathcal{S} \to \mathrm{Shv}_K^{\text{ét}}$, characterized by the property that $f^*U \approx \mathrm{Map}_{\mathrm{CAlg}_K}(R, K^{\mathrm{sep}})$ when $U \in \mathrm{Shv}_K^{\text{ét}}$ is the sheaf represented by a map $K \to R \in \mathrm{CAlg}_K^{\text{ét}}$. Therefore,

$$\mathrm{Map}_{\infty Top}(\mathcal{S}, \mathrm{Shv}_K^{\text{ét}}) \approx B\mathrm{Gal}(K),$$

the classifying space of the absolute Galois group of $K$ viewed as an $\infty$-groupoid.

For general $A \in \mathrm{CAlg}$, the $\infty$-groupoid $\mathrm{Map}_{\infty Top}(\mathcal{S}, \mathrm{Shv}_A^{\mathrm{Zar}})$ is equivalent to the *set* $|\mathrm{Spec}\, A|$ of prime ideals in $\pi_0 A$ (i.e., the prime ideal spectrum as a discrete set), while $\mathrm{Map}_{\infty Top}(\mathcal{S}, \mathrm{Shv}_A^{\text{ét}})$ is equivalent to a *1-groupoid* whose objects are pairs $(\mathfrak{p}, \pi_0 A/\mathfrak{p} \to F)$ consisting of a prime ideal $\mathfrak{p} \subset \pi_0 A$ and a separable closure $F$ of the residue field $\pi_0 A/\mathfrak{p}$.

## 8.10 Spectral Deligne–Mumford stacks

We can now define the main notion, that of a *spectral Deligne–Mumford stack*.

First note that given a spectrally ringed $\infty$-topos $X = (\mathcal{X}, \mathcal{O}_X)$ and an object $U \in \mathcal{X}$, we obtain a new spectrally ringed $\infty$-topos

$$X_U := (\mathcal{X}_{/U}, \mathcal{O}_X|_U)$$

where $\mathcal{O}_X|_U := \pi^*\mathcal{O}_X$ is the preimage of $\mathcal{O}_X$ along the projection $\pi\colon \mathcal{X}_{/U} \to \mathcal{X}$. Furthermore, this comes with an evident map $X_U \to X$ of spectrally ringed $\infty$-topoi.

**Example 8.10.1.** If $X = \operatorname{Sp\acute{e}t} A = (\operatorname{Shv}_A^{\mathrm{\acute{e}t}}, \mathcal{O})$ and $U \in \operatorname{Shv}_A^{\mathrm{\acute{e}t}} \subseteq \operatorname{PSh}((\operatorname{CAlg}_A^{\mathrm{\acute{e}t}})^{\mathrm{op}})$ is the sheaf represented by an étale map $(A \to B) \in \operatorname{CAlg}_A^{\mathrm{\acute{e}t}}$, then $X_U \approx ((\operatorname{Shv}_A^{\mathrm{\acute{e}t}})_{/U}, \mathcal{O}|_U) \approx (\operatorname{Shv}_B^{\mathrm{\acute{e}t}}, \mathcal{O}) = \operatorname{Sp\acute{e}t} B$.

## The definition of spectral Deligne–Mumford stacks

We say that a spectrally ringed $\infty$-topos $X = (\mathcal{X}, \mathcal{O}_X)$ is **affine** if it is isomorphic to $\operatorname{Sp\acute{e}t} A$ for some $A \in \operatorname{CAlg}$. Likewise, we say that an object $U \in \mathcal{X}$ is affine if $X_U$ (as defined above) is affine.

A **nonconnective spectral Deligne–Mumford (DM) stack** is a spectrally ringed $\infty$-topos $X = (\mathcal{X}, \mathcal{O}_X)$ for which there exists a set of objects $\{U_i\}$ in $\mathcal{X}$ such that

1. the set $\{U_i\}$ covers $\mathcal{X}$ (i.e., $\coprod U_i \to 1$ is effective epi in $\mathcal{X}$), and
2. each $U_i$ is affine.

*Remark 8.10.2.* The structure sheaf of a nonconnective spectral DM stack is always hypercomplete, as a consequence of the fact that this is so in the affine case (8.9.13).

A **spectral Deligne–Mumford (DM) stack** is a nonconnective DM stack $(\mathcal{X}, \mathcal{O}_X)$ such that the sheaf $\mathcal{O}_X$ is connective; i.e., such that the homotopy sheaves $\pi_k\mathcal{O}_X \in \mathcal{X}^\heartsuit$ satisfy $\pi_k\mathcal{O}_X \approx 0$ for $k < 0$.

*Remark 8.10.3.* $\operatorname{Sp\acute{e}t} A$ is always a nonconnective spectral DM stack, and is a spectral DM stack if and only if $A$ is connective.

*Remark 8.10.4.* If $X = (\mathcal{X}, \mathcal{O}_X)$ is a nonconnective spectral DM stack and $U \in \mathcal{X}$, then $X_U$ is also a nonconnective spectral DM stack. Furthermore, if $X$ is a spectral DM stack, so is $X_U$.

This is a consequence of the following claim: for a nonconnective spectral DM stack $X$, the collection $\mathcal{A} = \{V_j\}$ of all affine objects in $\mathcal{X}$ generates $\mathcal{X}$ under colimits [170, 1.4.7.9]. In particular, this implies that for any $U$ we can find a set of maps of the form $V_j \to U$ with all $V_j \in \mathcal{A}$ which is a cover of $\mathcal{X}_{/U}$ (8.5.6).

Here's a proof that affines generate $\mathcal{X}$ under colimits. First note that if $X \approx \operatorname{Sp\acute{e}t} A$ is itself affine, then $\mathcal{X} \approx \operatorname{Shv}_A^{\mathrm{\acute{e}t}}$ which is manifestly generated by affines (i.e., by the image of $(\operatorname{CAlg}_A^{\mathrm{\acute{e}t}})^{\mathrm{op}} \rightarrowtail \operatorname{Shv}_A^{\mathrm{\acute{e}t}}$ (8.10.1)). In the general case, if $\{U_i\}$ is an affine cover of $\mathcal{X}$, choose for each $i$ a set $\{V_{i,j} \to U_i\}$ of affine objects of $\mathcal{X}_{/U_i}$ which generate $\mathcal{X}_{/U_i}$ under colimits. Then the collection $\{V_{i,j}\}$ in $\mathcal{X}$ is a collection of affines which generate $\mathcal{X}$ under colimits (since $(X_{U_i})_{V_{i,j}} \approx X_{V_{i,j}}$).

## Spectral schemes

We can carry out an analogous definition using the Zariski topology. A special case of this is a **nonconnective spectral scheme**, which is a spectrally ringed $\infty$-topos $X = (\mathcal{X}, \mathcal{O}_X)$ such that

1. $\mathcal{X} \approx \mathrm{Shv}(X_{\mathrm{top}})$ for some topological space $X_{\mathrm{top}}$, and
2. there exists an open cover $\{U_i\}$ of $X_{\mathrm{top}}$ such that $X_{U_i} \approx \mathrm{Spec}\, A_i$ for some $A_i \in$ CAlg.

It is a **spectral scheme** if also $\pi_k \mathcal{O}_X \approx 0$ for $k < 0$. (This is not the definition given as [170, 1.1.2.8], but is equivalent to it by [170, 1.1.6.3, 1.1.6.4].)

## 8.11     Morphisms of spectral DM stacks

We need to work rather harder to get the correct notion of morphism of spectral DM stacks. Our goal is produce a category $\mathrm{SpDM}^{\mathrm{nc}}$ of nonconnective spectral DM stacks which includes $\mathrm{Sp\acute{e}t}\, R$ for any $R \in$ CAlg, with the property that

$$\mathrm{Map}_{\mathrm{SpDM}^{\mathrm{nc}}}(\mathrm{Sp\acute{e}t}\, S, \mathrm{Sp\acute{e}t}\, R) \approx \mathrm{Map}_{\mathrm{CAlg}}(R, S).$$

More generally, we would like to have

$$\mathrm{Map}_{\mathrm{SpDM}^{\mathrm{nc}}}(X, \mathrm{Sp\acute{e}t}\, R) \approx \mathrm{Map}_{\mathrm{CAlg}}(R, \Gamma(\mathcal{X}, \mathcal{O}_X)),$$

for any object $X \in \mathrm{SpDM}^{\mathrm{nc}}$, where $\Gamma(\mathcal{X}, \mathcal{O}_X) \in$ CAlg is the global sections of the structure sheaf $\mathcal{O}_X$.

Let's make this more precise. Given a map $(f, \psi)\colon X \to \mathrm{Sp\acute{e}t}\, R$ of spectrally ringed $\infty$-topoi, we obtain a map $R \to \Gamma(\mathcal{X}, \mathcal{O}_X)$ of $\mathbb{E}_\infty$-rings, by evaluating the composite of

$$\mathcal{O} \to f_* f^* \mathcal{O} \xrightarrow{f_*(\psi)} f_* \mathcal{O}_X$$

at global sections over $\mathrm{Shv}_R^{\mathrm{\acute{e}t}}$. Thus we get a map of $\infty$-groupoids

$$\mathrm{Map}_{\infty \mathcal{T}\mathrm{op}_{\mathrm{CAlg}}}(X, \mathrm{Sp\acute{e}t}\, R) \to \mathrm{Map}_{\mathrm{CAlg}}(R, \Gamma(\mathcal{X}, \mathcal{O}_X)). \qquad (8.11.1)$$

This map is rarely an equivalence, even when $X$ is affine. It turns out that we obtain an equivalence when we require $X$ to be *strictly Henselian*, and restrict to a full subgroupoid $\mathrm{Map}_{\infty \mathcal{T}\mathrm{op}_{\mathrm{CAlg}}^{\mathrm{sHen}}}(X, \mathrm{Sp\acute{e}t}\, R) \subseteq \mathrm{Map}_{\infty \mathcal{T}\mathrm{op}_{\mathrm{CAlg}}}(X, \mathrm{Sp\acute{e}t}\, R)$, consisting of *local* maps.

### Solution sheaves

To carry out this definition, we need to think locally. Given a spectrally ringed $\infty$-topos $X = (\mathcal{X}, \mathcal{O}_X)$ and an $\mathbb{E}_\infty$-ring $R$, define the **solution sheaf** $\mathrm{Sol}_R(\mathcal{O}_X) \in \mathrm{Fun}^{\mathrm{lim.\ pres.}}(\mathcal{X}^{\mathrm{op}}, \mathcal{S}) \approx \mathcal{X}$ by

$$\mathrm{Sol}_R(\mathcal{O}_X)(U) := \mathrm{Map}_{\mathrm{CAlg}}(R, \mathcal{O}_X(U)).$$

Note that $R \mapsto \mathrm{Sol}_R(\mathcal{O}_X)$ is itself a functor $\mathrm{CAlg}^{\mathrm{op}} \to \mathcal{X}$, and is limit preserving.

We obtain a map of sheaves of $\infty$-groupoids on $\mathcal{X}$,

$$\left[ U \mapsto \mathrm{Map}_{\infty \mathcal{T}\mathrm{op}_{\mathrm{CAlg}}}(X_U, \mathrm{Sp\acute{e}t}\, R) \right] \to \mathrm{Sol}_R(\mathcal{O}_X),$$

which can be thought of as a "local" version of the map (8.11.1), since evaluating the above map at the terminal object $U = 1_\mathcal{X}$ of $\mathcal{X}$ recovers the map (8.11.1).

### Strictly Henselian sheaves

A sheaf $\mathcal{O}$ of $\mathbb{E}_\infty$-rings is **strictly Henselian** if for every étale cover $\{R \to R_i\}$ in $\mathrm{CAlg}^{\text{ét}}$, the induced map

$$\coprod \mathrm{Sol}_{R_i}(\mathcal{O}) \to \mathrm{Sol}_R(\mathcal{O}) \tag{8.11.2}$$

is an effective epi in $\mathcal{X}$. (This is not the definition of [170, 1.4.2.1], but is equivalent to it by [170, 1.4.3.9].)

*Remark 8.11.1.* The strictly Henselian condition on $\mathcal{O}$ gives rise to a map

$$\mathrm{Map}_{\mathrm{CAlg}}(R, \Gamma(\mathcal{X}, \mathcal{O})) \to \mathrm{Map}_{\infty \mathrm{Top}}(\mathcal{X}, \mathrm{Shv}_R^{\text{ét}}),$$

i.e., from an $\mathbb{E}_\infty$-ring map $\overline{\alpha} \colon R \to \Gamma(\mathcal{X}, \mathcal{O})$ we can get a map $\mathcal{X} \to \mathrm{Shv}_R^{\text{ét}}$ of $\infty$-topoi. To see how this works, note that in the diagram

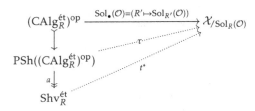

there is an essentially unique colimit preserving functor $\tau$ extending $\mathrm{Sol}_\bullet(\mathcal{O})$. The strictly Henselian condition on $\mathcal{O}$ implies that $\tau$ factors through an essentially unique colimit preserving functor $t^*$. Because $\mathrm{Sol}_\bullet(\mathcal{O})$ preserves limits, $t^*$ preserves finite limits. That is, $t^*$ is the preimage of a geometric morphism $t \colon \mathrm{Shv}_R^{\text{ét}} \to \mathcal{X}_{\mathrm{Sol}_R(\mathcal{O})}$.

An $\mathbb{E}_\infty$-ring map $\alpha \colon R \to \Gamma(\mathcal{X}, \mathcal{O})$ corresponds to a section $1_\mathcal{X} \to \mathrm{Sol}_R(\mathcal{O})$, which induces an étale geometric morphism $\overline{\alpha} \colon \mathcal{X} \to \mathcal{X}_{/\mathrm{Sol}_R(\mathcal{O})}$. The composite $t \circ \overline{\alpha}$ is the desired map of $\infty$-topoi.

*Remark 8.11.2.* It can be shown [170, 1.4.3.8] that the map in (8.11.2) is a pullback of $\coprod \mathrm{Sol}_{\pi_0 R_i}(\pi_0 \mathcal{O}) \to \mathrm{Sol}_{\pi_0 R}(\pi_0 \mathcal{O})$, so $\mathcal{O}$ is strictly Henselian (or local) if and only if $\pi_0 \mathcal{O}$ is so; this recovers the definition in [170, 1.4.2.1]. (The proof rather subtle: you need to use the fact that every étale map is a base change of an étale map between compact objects in $\mathrm{CAlg}$ (8.9.11), in order to reduce to the case of the pullback square (8.9.6) of mapping spaces. The issue here is that it is not the case that $f^* \mathrm{Sol}_R(\mathcal{O}) \to \mathrm{Sol}_R(f^*\mathcal{O})$ is an isomorphism in general, unless $R$ is a compact object of $\mathrm{CAlg}$.)

There is an analogous definition of **local** sheaf, in which étale covers are replaced with Zariski covers in the definition given above.

Example 8.11.3 (Local and strictly Henselian sheaves on a point). Let $\mathcal{X} = \mathcal{S}$, so $\mathrm{Shv}_{\mathrm{CAlg}}(\mathcal{S}) \approx \mathrm{CAlg}$, and for $\mathcal{O} \in \mathrm{CAlg}$ we have $\mathrm{Sol}_R(\mathcal{O}) \approx \mathrm{Map}(R, \mathcal{O}) \in \mathcal{S}$.

From the definitions and the universal property of localization maps $R \to R[f^{-1}]$ in $\mathrm{CAlg}$, we see that $\mathcal{O}$ is local if and only if, for every pair $(R, \{f_1, \ldots, f_d\} \subseteq \pi_0 R)$ consisting of $R \in \mathrm{CAlg}$ such that $(f_1, \ldots, f_d)\pi_0 R = \pi_0 R$, every map $\alpha \colon R \to \mathcal{O}$ in $\mathrm{CAlg}$ is such that $\alpha(f_k)$ is an invertible element of $\pi_0 \mathcal{O}$ for some $k \in \{1, \ldots, d\}$.

It follows that $\mathcal{O}$ must be a local sheaf whenever $\pi_0\mathcal{O}$ is a local ring in the usual sense. The converse also holds: if $\mathcal{O}$ is a local sheaf, apply the condition with $(R = 0, \varnothing \subseteq \pi_0 R)$ to see that $\pi_0\mathcal{O} \not\approx 0$, and with $(R = \mathbb{S}\{x,y\}[(x+y)^{-1}], \{x,y\} \subseteq \pi_0 R)$ to see that $\mathfrak{m} := \pi_0\mathcal{O} \smallsetminus (\pi_0\mathcal{O})^\times$ is an ideal.

A similar argument shows that $\mathcal{O} \in \mathrm{Shv}_{\mathrm{CAlg}}(\mathcal{S})$ is strictly Henselian if and only if $\pi_0\mathcal{O}$ is a strictly Henselian ring in the classical sense, i.e., as defined in [281, Tag 04GE].

## Spectral DM stacks are strictly Henselian

For an affine object $X = \mathrm{Sp\acute{e}t}\, A = (\mathrm{Shv}_A^{\mathrm{\acute{e}t}}, \mathcal{O})$, we see that $\mathrm{Sol}_R(\mathcal{O})(U) \approx \mathrm{Map}_{\mathrm{CAlg}}(R, B)$ when $U \in \mathrm{Shv}_A^{\mathrm{\acute{e}t}}$ is the object represented by the étale $A$-algebra $A \to B$. Using this it is straightforward to show that $\mathcal{O}$ is strictly Henselian.

*Remark 8.11.4* (Spectral DM stacks are strictly Henselian). Observe that $\pi^* \mathrm{Sol}_R(\mathcal{O}) \approx \mathrm{Sol}_R(\pi^*\mathcal{O})$ when $\pi \colon \mathcal{X}_{/U} \to \mathcal{X}$ is the etale map of $\infty$-topoi associated to an object $U \in \mathcal{X}$. (Use (8.7.5).) Given this it is straightforward to prove that any nonconnective DM stack is strictly Henselian.

## The category of strictly Henselian spectrally ringed ∞-topoi

We let $\infty\mathcal{T}\mathrm{op}_{\mathrm{CAlg}}^{\mathrm{sHen}}$ denote the (non-full) subcategory of $\infty\mathcal{T}\mathrm{op}_{\mathrm{CAlg}}$ whose *objects* are $X = (\mathcal{X}, \mathcal{O}_X)$ such that $\mathcal{O}_X$ is strictly Henselian, and whose *morphisms* $f \colon (\mathcal{X}, \mathcal{O}_X) \to (\mathcal{Y}, \mathcal{O}_Y)$ are such that

$$\begin{array}{ccc} f^* \mathrm{Sol}_{R'}(\mathcal{O}_Y) & \longrightarrow & \mathrm{Sol}_{R'}(\mathcal{O}_X) \\ \downarrow & & \downarrow \\ f^* \mathrm{Sol}_R(\mathcal{O}_Y) & \longrightarrow & \mathrm{Sol}_R(\mathcal{O}_X) \end{array}$$

is a pullback in $\mathcal{X}$ for every étale map $R \to R'$ in CAlg. Such morphisms are called **local**.

*Remark 8.11.5.* This is different than the definition given as [170, 1.4.2.1], but is equivalent by [170, 1.4.3.9].

*Remark 8.11.6.* If $X = (\mathcal{X}, \mathcal{O}_X)$ is a nonconnective spectral DM stack and $U \in \mathcal{X}$, then the evident map $X_U \to X$ of spectrally ringed $\infty$-topoi is local.

We can now state our goal.

**Theorem 8.11.7.** *For any strictly Henselian spectrally ringed $\infty$-topos $X = (\mathcal{X}, \mathcal{O}_X)$ and $\mathbb{E}_\infty$-ring $R$, the evident map*

$$\mathrm{Map}_{\infty\mathcal{T}\mathrm{op}_{\mathrm{CAlg}}^{\mathrm{sHen}}}(X, \mathrm{Sp\acute{e}t}\, R) \xrightarrow{\sim} \mathrm{Map}_{\mathrm{CAlg}}(R, \Gamma(\mathcal{X}, \mathcal{O}_X))$$

*is an equivalence.*

*Sketch proof.* This is [170, 1.4.2.4]. Here is a brief sketch.

Geometric morphisms $f \colon \mathrm{Shv}_R^{\text{ét}} \to \mathcal{X}$ correspond (by restriction to representable sheaves) exactly to left-exact functors $\chi \colon (\mathrm{CAlg}_R^{\text{ét}})^{\mathrm{op}} \to \mathcal{X}$ which send étale covers to effective epis. Given such an $f$, maps $\phi \colon \mathcal{O} \to f_*\mathcal{O}_\mathcal{X}$ of sheaves of $\mathbb{E}_\infty$-rings on $\mathrm{Shv}_R^{\text{ét}}$ correspond to natural transformations $\phi' \colon \chi \to \mathrm{Sol}_\bullet(\mathcal{O})$ of functors; to see this, use the evident equivalence $\mathrm{Map}_\chi(\chi(R'), \mathrm{Sol}_{R'}(\mathcal{O})) \approx \mathrm{Map}_{\mathrm{CAlg}}(R', \mathcal{O}_X(\chi(R')))$ for $R' \in \mathrm{CAlg}_R^{\text{ét}}$, and that $f_*\mathcal{O}_X|(\mathrm{CAlg}_R^{\text{ét}})^{\mathrm{op}} \approx \mathcal{O}_X \circ \chi$ as functors $(\mathrm{CAlg}_R^{\text{ét}})^{\mathrm{op}} \to \mathrm{CAlg}$.

One shows that if $\phi$ is local, then $\phi'$ is Cartesian, i.e., $\phi'$ takes morphisms in $(\mathrm{CAlg}_R^{\text{ét}})^{\mathrm{op}}$ to pullback squares of sheaves. But since $(\mathrm{CAlg}_R^{\text{ét}})$ has $R$ as a terminal object, we discover that pairs $(\chi, \phi')$ with $\phi'$ Cartesian correspond exactly to maps $1_\chi = \chi(R) \to \mathrm{Sol}_R(\mathcal{O}_\mathcal{X})$, i.e., to maps $R \to \Gamma(\mathcal{X}, \mathcal{O}_X)$ of $\mathbb{E}_\infty$-rings. In particular, we learn that $\mathrm{Map}_{\infty\mathcal{T}\!\mathrm{op}_{\mathrm{CAlg}}^{\mathrm{sHen}}}(X, \mathrm{Spét}\,R) \to \mathrm{Map}_{\mathrm{CAlg}}(R, \Gamma(\mathcal{X}, \mathcal{O}_X))$ is a monomorphism.

Finally, given a map $\alpha \colon R \to \Gamma(\mathcal{X}, \mathcal{O}_X)$, there is an explicit procedure to construct a morphism $X \to \mathrm{Spét}\,R$ in $\infty\mathcal{T}\!\mathrm{op}_{\mathrm{CAlg}}^{\mathrm{sHen}}$ which projects to $\alpha$; the underlying map $\mathcal{X} \to \mathrm{Shv}_R^{\text{ét}}$ of $\infty$-topoi is produced by the procedure of (8.11.1). □

## The category of locally spectrally ringed ∞-topoi

We can play the same game with "local" replacing "strictly Henselian" as the condition on objects, resulting in a full subcategory $\infty\mathcal{T}\!\mathrm{op}_{\mathrm{CAlg}}^{\mathrm{loc}}$ of $\infty\mathcal{T}\!\mathrm{op}_{\mathrm{CAlg}}^{\mathrm{sHen}}$ and a version of (8.11.7) with Spét replaced with Spec [170, 1.1.5].

## 8.12    The category of spectral DM stacks

We have achieved our goal. We have full subcategories

$$\mathrm{SpDM} \subseteq \mathrm{SpDM}^{\mathrm{nc}} \subseteq \infty\mathcal{T}\!\mathrm{op}_{\mathrm{CAlg}}^{\mathrm{sHen}}$$

of spectral DM stacks and nonconnective DM stacks respectively, inside the $\infty$-category of strictly Henselian spectrally ringed $\infty$-topoi and local maps, which is itself a non-full subcategory of the category $\infty\mathcal{T}\!\mathrm{op}_{\mathrm{CAlg}}$ of spectrally ringed $\infty$-topoi. By (8.11.7) we see that there are adjoint pairs

$$\mathrm{Spét} \colon \mathrm{CAlg}^{\mathrm{op}} \rightleftarrows \mathrm{SpDM}^{\mathrm{nc}} \colon \Gamma \quad \text{and} \quad \mathrm{Spét} \colon (\mathrm{CAlg}^{\mathrm{cn}})^{\mathrm{op}} \rightleftarrows \mathrm{SpDM} \colon \Gamma.$$

*Remark 8.12.1.* There are analogous full subcategories of *spectral schemes* and *nonconnective spectral schemes* in $\infty\mathcal{T}\!\mathrm{op}_{\mathrm{CAlg}}^{\mathrm{loc}}$.

## Finite limits of DM stacks

The categories SpDM and SpDM$^{\mathrm{nc}}$ have finite limits, and finite limits are preserved by the functors $\mathrm{Spét} \colon (\mathrm{CAlg}^{\mathrm{cn}})^{\mathrm{op}} \to \mathrm{SpDM}$ and $\mathrm{Spét} \colon \mathrm{CAlg}^{\mathrm{op}} \to \mathrm{SpDM}^{\mathrm{nc}}$. In particular, for a diagram $B \leftarrow A \to B'$ of rings, we have

$$\mathrm{Spét}(B \otimes_A B') \approx \mathrm{Spét}\,B \times_{\mathrm{Spét}\,A} \mathrm{Spét}\,B',$$

as an immediate consequence of (8.11.7). (See [170, 1.4.11.1], [163, V 2.3.21].)

## Connective covers and truncation of DM stacks

The adjoint pairs

$$\mathrm{CAlg}^{\heartsuit} \rightleftarrows \mathrm{CAlg}^{\mathrm{cn}} \rightleftarrows \mathrm{CAlg}$$

relating classical, connective, and arbitrary $\mathbb{E}_{\infty}$-rings are paralleled by adjoint pairs

$$\mathrm{SpDM}^{\leq 0} \rightleftarrows \mathrm{SpDM} \rightleftarrows \mathrm{SpDM}^{\mathrm{nc}}$$

where $\mathrm{SpDM}^{\leq 0}$ is the $\infty$-category of 0-**truncated spectral DM stacks**, consisting of $X = (\mathcal{X}, \mathcal{O}_X)$ such that $\pi_q \mathcal{O}_X \approx 0$ for $q \neq 0$. The localization functors are obtained respectively by 0-truncating or taking connective cover of the structure sheaf [170, 1.4.5–6].

## Classical objects as spectral DM stacks

We would like to connect this spectral geometry to some more "classical" (i.e., 1-categorical) kind of algebraic geometry.

Note that objects of $\mathrm{SpDM}^{\leq 0}$ are $\infty$-topoi $\mathcal{X}$ equipped with structure sheaves $\mathcal{O}_X$ of *classical* rings. However, the $\infty$-topos $\mathcal{X}$ is not necessarily a "classical" one, i.e., is not necessarily equivalent to the 1-localic $\infty$-topos $\mathrm{Shv}_{\mathcal{S}}(\mathcal{X}^{\heartsuit})$ (8.5.12). So 0-truncated spectral DM stacks are not necessarily classical objects.

The classical analogue of spectral Deligne–Mumford stack is a **Deligne–Mumford stack**, which is a pair $X_0 = (\mathcal{X}, \mathcal{O}_{X_0})$ consisting of a 1-topos $\mathcal{X}$ with a sheaf $\mathcal{O}_{X_0}$ of ordinary commutative rings on it, which is "locally" affine, i.e., there exists a set $\{U_i\}$ of objects in $\mathcal{X}$ such that (i) $\bigsqcup U_i \to 1$ is effective epi in $\mathcal{X}$ and (ii) $(\mathcal{X}_{/U_i}, \mathcal{O}|_{U_i}) \approx ((\mathrm{Shv}^{\mathrm{\acute{e}t}}_{A_i})^{\heartsuit}, \mathcal{O})$ for some ordinary ring $A_i$.

Given a nonconnective spectral DM stack $X = (\mathcal{X}, \mathcal{O}_X)$, we can form $X_{\mathrm{DM}} := (\mathcal{X}^{\heartsuit}, \pi_0 \mathcal{O}_X)$, which is in fact a classical Deligne–Mumford stack, called the **underlying DM stack** of $X$.

Conversely, given a classical DM stack $X_0 = (\mathcal{X}, \mathcal{O})$, we can upgrade it to a 0-truncated spectral DM stack

$$X_{\mathrm{SpDM}} = (\mathrm{Shv}_{\mathcal{S}}(\mathcal{X}), \mathcal{O}')$$

where $\mathrm{Shv}_{\mathcal{S}}(\mathcal{X})$ is the 1-localic reflection of $\mathcal{X}$ (8.5.12), and $\mathcal{O}'$ is the sheaf of connective $\mathbb{E}_{\infty}$-rings represented by the composite functor

$$\mathrm{Shv}_{\mathcal{S}}(\mathcal{X})^{\mathrm{op}} \xrightarrow{(\tau_{\leq 0})^{\mathrm{op}}} \mathcal{X}^{\mathrm{op}} \xrightarrow{\mathcal{O}} \mathrm{CAlg}^{\heartsuit} \rightarrowtail \mathrm{CAlg}^{\mathrm{cn}}.$$

It turns out that this construction describes a fully faithful embedding

$$(\text{classical DM stacks}) \rightarrowtail \mathrm{SpDM}^{\leq 0}.$$

See [170, 1.4.8] for more on the relation between DM stacks and spectral DM stacks.

**Example 8.12.2.** Here is a simple example which exhibits some of these phenomena. Let $K \in \mathrm{CAlg}^{\heartsuit}$ be an ordinary separably closed field, so that $\mathrm{Shv}^{\mathrm{\acute{e}t}}_K \approx \mathcal{S}$. Then

Spét $K \approx (\mathcal{S}, K)$, where $K \in \text{CAlg} \approx \text{Shv}_{\text{CAlg}}(\mathcal{S})$, is an example of a 0-truncated spectral DM stack, whose $\infty$-topos is equivalent to sheaves on the 1-point space. It corresponds to the classical DM stack associated to $K$.

For any $\infty$-groupoid $U \in \mathcal{S}$ we can form $(\text{Spét } K)_U = (\mathcal{S}_{/U}, \pi^*K)$, i.e., $\infty$-groupoids over $U$ equipped with the constant sheaf associated to $K$. Then $(\text{Spét } K)_U$ is also a 0-truncated spectral DM stack. If $U$ is not a 1-truncated space, then $\mathcal{S}_{/U}$ is not 1-localic, and $(\text{Spét } K)_U$ does not arise as a classical DM stack in this case.

In short, spectral DM stacks expand DM stacks in *two* ways: spectral DM stacks are allowed to have underlying $\infty$-topoi which are not classical, i.e., not 1-localic, and spectral DM stacks are also allowed to have structure sheaves which are not classical, i.e., not merely sheaves of ordinary rings.

### Relation to schemes and spectral schemes

There are analogous statements for spectral schemes [170, 1.1]. Thus, a morphism of nonconnective spectral schemes is just a morphism $(X, \mathcal{O}_X) \to (Y, \mathcal{O}_Y)$ of spectrally ringed $\infty$-topoi which is local; we get full subcategories $\text{SpSch} \subseteq \text{SpSch}^{\text{nc}} \subseteq \infty\text{Top}_{\text{CAlg}}^{\text{loc}}$; we have fully faithful $\text{Spec}: (\text{CAlg}^{\text{cn}})^{\text{op}} \to \text{SpSch}$ and $\text{Spec}: \text{CAlg}^{\text{op}} \to \text{SpSch}^{\text{nc}}$; and we have fully faithful embeddings

$$\text{Sch} \xrightarrow{\sim} \text{SpSch}^{\leq 0} \rightarrowtail \text{SpSch},$$

where $\text{Sch}^{\leq 0}$ is the full subcategory of 0-truncated spectral schemes. In this case we have an equivalence $\text{Sch} \approx \text{SpSch}^{\leq 0}$, since underlying topos of a spectral scheme is already assumed to be a space.

What is the relation between spectral schemes and spectral DM stacks? Note that although both spectral schemes and spectral DM stacks are both types of spectrally ringed $\infty$-topoi, there is very little overlap between the two classes. What is true [170, 1.6.6] is that there exist fully faithful functors

$$\text{SpSch} \rightarrowtail \text{SpDM} \quad \text{and} \quad \text{SpSch}^{\text{nc}} \rightarrowtail \text{SpDM}^{\text{nc}}$$

which promote spectral schemes to spectral DM stacks. Objects in the essentially image of these functors are called **schematic**, and this property is easy to characterize: $X = (\mathcal{X}, \mathcal{O}_X)$ is schematic if and only if there exists a set $\{U_i\}$ of $(-1)$-*truncated* objects of $\mathcal{X}$ which are affine and which cover $\mathcal{X}$ [170, 1.6.7.3].

## 8.13    Étale and flat morphisms of spectral DM stacks

### Étale morphisms in spectral geometry

A map $(\mathcal{X}, \mathcal{O}_X) \to (\mathcal{Y}, \mathcal{O}_Y)$ of spectrally ringed $\infty$-topoi is called **étale** if

1.  the underlying map $f: \mathcal{X} \to \mathcal{Y}$ of $\infty$-topoi is étale (8.7), and
2.  the map $f^*\mathcal{O}_Y \to \mathcal{O}_X$ is an isomorphism in $\text{Shv}_{\text{CAlg}}(\mathcal{X})$.

For instance, for any $X = (\mathcal{X}, \mathcal{O}_X)$ and $U \in \mathcal{X}$, the projection map $X_U \to X$ is étale in this sense, where $X_U = (\mathcal{X}_{/U}, \mathcal{O}_X|_{\mathcal{U}})$. In fact, any etale morphism of spectrally ringed $\infty$-topoi is equivalent to one of this form.

If $f : X \to Y$ is an étale map of spectrally ringed $\infty$-topoi and $Y \in \mathrm{SpDM}^{\mathrm{nc}}$, then also $X \in \mathrm{SpDM}^{\mathrm{nc}}$ (8.10.4), and in fact $f$ is a morphism of $\mathrm{SpDM}^{\mathrm{nc}}$ (8.11.6).

This terminology turns out to be compatible with that of "étale map of $\mathbb{E}_\infty$-rings".

**Proposition 8.13.1** ([170, 1.4.10.2]). *A map $A \to B$ of $\mathbb{E}_\infty$-rings is étale if and only if the corresponding map $\mathrm{Sp\acute{e}t}\, B \to \mathrm{Sp\acute{e}t}\, A$ is étale.*

We have the following for "lifting" maps over étale morphisms.

**Proposition 8.13.2.** *Given nonconnective spectral DM stacks $X = (\mathcal{X}, \mathcal{O}_X)$ and $Y = (\mathcal{Y}, \mathcal{O}_Y)$, a map $f : Y \to X$ of nonconnective spectral DM stacks, and an object $U \in \mathcal{X}$, there is an equivalence*

$$\left\{ \begin{array}{c} \phantom{Y} \xrightarrow{\phantom{xx}} X_U \\ s \nearrow \quad \downarrow \pi \\ Y \xrightarrow{\phantom{xx}f\phantom{xx}} X \end{array} \right\} \xrightarrow{\sim} \left\{ 1 \dashrightarrow f^*U \right\}$$

*between the $\infty$-groupoid of "sections" of $\pi$ over $Y$ in $\mathrm{SpDM}^{\mathrm{nc}}$, and the $\infty$-groupoid of global sections of $f^*U$ on $\mathcal{Y}$.*

*Proof sketch.* A map $s : Y \to X_U$ consists of a geometric morphism $s : \mathcal{Y} \to \mathcal{X}_{/U}$ together with a local map $\widetilde{s} : s^* \mathcal{O}_{X_U} \to \mathcal{O}_Y$ of sheaves of $\mathbb{E}_\infty$-rings. We already know (8.7.1) that geometric morphisms $s$ which lift $f$ correspond exactly to global sections of $f^*U$. We then have that $s^* \mathcal{O}_{X_U} = s^* \pi^* \mathcal{O}_X \approx f^* \mathcal{O}_X$, so there is an evident map $s^* \mathcal{O}_{X_U} \to \mathcal{O}_Y$, namely the one equivalent to the map $\widetilde{f} : f^* \mathcal{O}_X \to \mathcal{O}_Y$ which is part of the description of $f : Y \to X$. This is in fact the unique map making the diagram commute in $\mathrm{SpDM}^{\mathrm{nc}}$. (See [170, 21.4.6].) □

**Corollary 8.13.3.** *For any $f : Y \to X$ in $\mathrm{SpDM}^{\mathrm{nc}}$ and $U \in \mathcal{X}$ the square*

$$\begin{array}{ccc} Y_{f^*U} & \longrightarrow & X_U \\ \downarrow & & \downarrow \\ Y & \xrightarrow{\phantom{xx}f\phantom{xx}} & X \end{array}$$

*is a pullback in $\mathrm{SpDM}^{\mathrm{nc}}$. It is a pullback in $\mathrm{SpDM}$ if $X, Y \in \mathrm{SpDM}$.*

## Colimits along étale maps of spectral DM stacks

It turns out that we can "glue" spectral DM stacks along étale maps, much as one can construct new schemes by gluing together ones along open immersions.

Let $\mathrm{SpDM}_{\mathrm{\acute{e}t}} \subseteq \mathrm{SpDM}$ and $\mathrm{SpDM}^{\mathrm{nc}}_{\mathrm{\acute{e}t}} \subseteq \mathrm{SpDM}^{\mathrm{nc}}$ be the (non-full) subcategories containing just the étale maps.

**Proposition 8.13.4.** *The categories $\mathrm{SpDM}_{\mathrm{\acute{e}t}}$ and $\mathrm{SpDM}^{\mathrm{nc}}_{\mathrm{\acute{e}t}}$ have all small colimits, and the inclusions $\mathrm{SpDM}_{\mathrm{\acute{e}t}} \to \mathrm{SpDM}$ and $\mathrm{SpDM}^{\mathrm{nc}}_{\mathrm{\acute{e}t}} \to \mathrm{SpDM}^{\mathrm{nc}}$ preserves colimits.*

*Proof.* Here is a brief sketch; I'll describe the nonconnective case. (See [170, 21.4.4] or [163, V 2.3.5] for more details.)

Suppose $\left(c \mapsto X_c = (\mathcal{X}_c, \mathcal{O}_{X_c})\right) \colon \mathcal{C} \to \mathrm{SpDM}^{\mathrm{nc}}_{\text{ét}}$ is a functor from a small $\infty$-category. We know (8.7.6) that we can form the colimit $\mathcal{X} := \mathrm{colim}^{\infty\mathcal{T}op}_{c \in \mathcal{C}} \mathcal{X}_c$ of $\infty$-topoi, and that each $\mathcal{X}_c \to \mathcal{X}$ is étale. In fact, there exists a functor $U \colon \mathcal{C} \to \mathcal{X}$ so that $(c \mapsto \mathcal{X}_c)$ is equivalent to $(c \mapsto \mathcal{X}_{/U_c})$ as functors $\mathcal{C} \to \infty\mathcal{T}op_{/\mathcal{X}}$.

We also know (8.6.12) that we have descent for sheaves of $\mathbb{E}_\infty$-rings. That is, $\mathrm{Shv}_{\mathrm{CAlg}}(\mathcal{X}) \approx \lim_{c \in \mathcal{C}} \mathrm{Shv}_{\mathrm{CAlg}}(\mathcal{X}_c)$, so there exists $\mathcal{O}_X \in \mathrm{Shv}_{\mathrm{CAlg}}(\mathcal{X})$ together with a compatible family of equivalences $\pi_c^* \mathcal{O}_X \xrightarrow{\sim} \mathcal{O}_{X_c}$. In particular, we obtain a cone $\mathcal{C}^{\triangleright} \to \infty\mathcal{T}op_{\mathrm{CAlg}}$, which in fact lands in the non-full subcategory consisting of étale maps. This cone is a colimit cone, presenting $X = (\mathcal{X}, \mathcal{O}_X)$ as the colimit of the diagram in spectrally ringed $\infty$-topoi.

To show that $X$ is a nonconnective spectral DM stack, we need a set $\{V_j\}$ of objects in $X$ such that each $X_{V_j}$ is affine, and $\bigsqcup V_j \to 1$ is an effective epi in $\mathcal{X}$. This is straightforward: there are sets $\{V_{c,i} \to U_c\}$ of maps for each object $c \in \mathcal{C}$ such that each $X_{V_{c,i}}$ is affine and $\bigsqcup_i V_{c,i} \to U_c$ is effective epi in $\mathcal{X}_{/U_c}$, so just take the union $\bigcup_c \{V_{c,i}\}$.

Finally, show that the maps $X_{U_i} \to X$ of the cone are *local*, so that the cone factors through $\mathcal{C}^{\triangleright} \to \mathrm{SpDM}^{\mathrm{nc}}$; this amounts to the fact that being "local" is itself a local condition in the domain. □

## Spectral DM stacks are colimits of affine objects

We obtain the following interesting consequence: every nonconnective spectral DM stack $X = (\mathcal{X}, \mathcal{O}_X)$ is a colimit of a small diagram of affines. That is,

$$X \approx \mathrm{colim}^{\mathrm{SpDM}^{\mathrm{nc}}}_{c \in \mathcal{C}} X_{U_c}$$

where $c \mapsto U_c \colon \mathcal{C} \to \mathcal{X}$ is a functor such that $\mathrm{colim}_{c \in \mathcal{C}} U_c \approx 1$ and each $U_c$ is affine (which exists by (8.10.4)), and so each $X_{U_c} \approx \mathrm{Sp\acute{e}t}\, A_c$ for some $\mathbb{E}_\infty$-ring $A_c$. Analogous remarks apply to spectral DM stacks, which have the form

$$X \approx \mathrm{colim}^{\mathrm{SpDM}}_{c \in \mathcal{C}} X_{U_c}$$

with each $X_{U_c} \approx \mathrm{Sp\acute{e}t}\, A_c$ for some connective $\mathbb{E}_\infty$-ring $A_c$.

## Flat morphisms in spectral geometry

A map $f \colon Y \to X$ of nonconnective spectral DM stacks is **flat** if for every commutative square

$$\begin{array}{ccc} \mathrm{Sp\acute{e}t}\, B & \longrightarrow & Y \\ {\scriptstyle g}\downarrow & & \downarrow{\scriptstyle f} \\ \mathrm{Sp\acute{e}t}\, A & \longrightarrow & X \end{array}$$

in $\mathrm{SpDM}^{\mathrm{nc}}$ such that the horizontal maps are étale, the map $g$ is induced by a flat morphism $A \to B$ of $\mathbb{E}_\infty$-rings [170, 2.8.2].

It is immediate that the base change of any flat morphism is flat. Also, if $Y \to X$ is flat and $X$ is a spectral DM stack, then $Y$ is a spectral DM stack.

*Remark 8.13.5.* A map $\mathrm{Sp\acute{e}t}\, B \to \mathrm{Sp\acute{e}t}\, A$ of nonconnective spectral DM stacks is flat in the above sense if and only if $A \to B$ is a flat morphism of $\mathbb{E}_\infty$-rings.

Given $A \in \mathrm{CAlg}$, let $\mathrm{SpDM}^{\mathrm{nc}}_A = (\mathrm{SpDM}^{\mathrm{nc}})_{/\mathrm{Sp\acute{e}t}\, A}$, and let $\mathrm{SpDM}^{\flat}_A \subseteq \mathrm{SpDM}^{\mathrm{nc}}_A$ denote the full subcategory spanned by objects which are flat morphisms $X \to \mathrm{Sp\acute{e}t}\, A$. It turns out that although the functor $\mathrm{SpDM}^{\mathrm{nc}}_{\tau_{\geq 0}A} \to \mathrm{SpDM}^{\mathrm{nc}}_A$ induced by base change is not an equivalence, it induces an equivalence on full subcategories of flat objects flat objects.

**Proposition 8.13.6.** *Base change induces an equivalence of $\infty$-categories*

$$\mathrm{SpDM}^{\flat}_{\tau_{\geq 0}A} \xrightarrow{\sim} \mathrm{SpDM}^{\flat}_A.$$

*Proof.* See [170, 2.8.2]. The inverse equivalence sends $X \to \mathrm{Sp\acute{e}t}\, A$ to $\tau_{\geq 0}X \to \mathrm{Sp\acute{e}t}(\tau_{\geq 0}A)$; compare (8.8.5). $\qquad\square$

## 8.14   Affine space and projective space

Let's think about two basic examples: affine $n$-space and projective $n$-space. It turns out that these come in two distinct versions, depending on whether we use polynomial rings (8.8.6) or free rings (8.8.7).

### Affine spaces

Given a connective $\mathbb{E}_\infty$-ring $R \in \mathrm{CAlg}^{\mathrm{cn}}$, define affine $n$-space over $R$ to be the affine spectral DM stack

$$\mathbf{A}^n_R := \mathrm{Sp\acute{e}t}\, R[x_1,\ldots,x_n]$$

on a polynomial ring (8.8.6) over $R$. When $R \in \mathrm{CAlg}^{\heartsuit}$ is an ordinary ring, this is the "usual" affine $n$-space. In general, $\mathbf{A}^n_R \approx \mathbf{A}^n_{\mathbb{S}} \times_{\mathrm{Sp\acute{e}t}\,\mathbb{S}} \mathrm{Sp\acute{e}t}\, R$.

What are the "points" of $\mathbf{A}^n_{\mathbb{S}}$? If $B$ is an *ordinary* ring, then

$$\mathbf{A}^n_{\mathbb{S}}(B) = \mathrm{Map}_{\mathrm{SpDM}_{/\mathrm{Sp\acute{e}t}\, R}}(\mathrm{Sp\acute{e}t}\, B, \mathbf{A}^n_{\mathbb{S}}) \approx \mathrm{Map}_{\mathrm{CAlg}}(\mathbb{S}[x_1,\ldots,x_n], B) \approx B^n.$$

However, if $B$ is not an ordinary ring, then things can be very different. For instance, the image of the evident map

$$\mathbf{A}^n_{\mathbb{S}}(\mathbb{S}) \to \mathbf{A}^n_{\mathbb{S}}(\mathbb{Z}) \approx \mathbb{Z}^n$$

consists exactly of the ordered $n$-tuples $(a_1,\ldots,a_n) \in \mathbb{Z}^n$ such that each $a_i \in \{0,1\}$.[5]

---

[5] Here's a quick proof. We need to understand the image of
$\mathrm{Map}_{\mathrm{CAlg}}(\mathbb{S}[x],\mathbb{S}) \to \mathrm{Map}_{\mathrm{CAlg}^{\heartsuit}}(\mathbb{S}[x],\mathbb{Z}) \approx \mathbb{Z}$ induced by evaluation at $x \in \pi_0\mathbb{S}[x]$. It is straightforward to construct maps realizing $0$ or $1$. To show these are the only possibilities, argue as

From this, we see that $\mathbf{A}_\mathbb{S}^n$ is *not* a group object with respect to addition; i.e., there is no map $\mathbb{S}[x] \to \mathbb{S}[x] \otimes_\mathbb{S} \mathbb{S}[x]$ of $\mathbb{E}_\infty$-rings which on $\pi_0$ sends $x \mapsto x \otimes 1 + 1 \otimes x$.

It is however true that $\mathbf{A}_\mathbb{S}^1$ is a monoid object under multiplication (the coproduct on $\mathbb{S}[x]$ is obtained by applying suspension spectrum to the diagonal map on $\mathbb{Z}_{\geq 0}$). Likewise,

$$\mathbf{G}_m := \mathrm{Sp\acute{e}t}\, \mathbb{S}[x, x^{-1}]$$

is an abelian group object in spectral DM stacks.

There is another affine $n$-space, which I'll call the *smooth affine space*, namely

$$\mathbf{A}_{\mathrm{sm}}^n := \mathrm{Sp\acute{e}t}\, \mathbb{S}\{x_1, \ldots, x_n\},$$

defined using a free ring (8.8.7) instead of a polynomial ring. The points of this are easier to explain:

$$\mathbf{A}_{\mathrm{sm}}^n(B) = \mathrm{Map}_{\mathrm{SpDM}}(\mathrm{Sp\acute{e}t}\, B, \mathbf{A}_{\mathrm{sm}}^n) \approx \mathrm{Map}_{\mathrm{CAlg}}(\mathbb{S}\{x_1, \ldots, x_n\}, B) \approx (\Omega^\infty B)^n.$$

The evident map $\mathbf{A}_\mathbb{S}^n \to \mathbf{A}_{\mathrm{sm}}^n$, though not an equivalence, becomes an equivalence after base-change to any $R \in \mathrm{CAlg}_\mathbb{Q}$.

## Projective spaces

Given $R \in \mathrm{CAlg}_R^{\mathrm{cn}}$ we define projective $n$-space as follows [170, 5.4.1]. Let $[n] = \{0, 1, \ldots, n\}$, and let $P^\circ([n])$ denote the poset of *non-empty* subsets. For each $I \in P^\circ([n])$ let

$$M_I := \{(m_0, \ldots, m_n) \in \mathbb{Z}^{n+1} \mid m_0 + \cdots + m_n = 0,\ m_i \geq 0 \text{ if } i \in I\}.$$

We obtain a functor $P^\circ([n])^{\mathrm{op}} \to \mathrm{SpDM}_{\mathrm{\acute{e}t}}$ by

$$I \mapsto \mathrm{Sp\acute{e}t}(R[M_I]).$$

Define $\mathbf{P}_R^n := \mathrm{colim}_{I \in P^0([n])^{\mathrm{op}}} \mathrm{Sp\acute{e}t}(R[M_I])$, which exists by (8.13.4).

Example 8.14.1. $\mathbf{P}_R^1$ is the colimit of

$$\mathrm{Sp\acute{e}t}(R[x]) \leftarrow \mathrm{Sp\acute{e}t}(R[x, x^{-1}]) \to \mathrm{Sp\acute{e}t}(R[x^{-1}]).$$

This construction is compatible with base change, and for ordinary rings $R$ recovers the "usual" projective $n$-space. You can use the same idea to construct spectral versions of toric varieties.

As for affine $n$-space, it is difficult to understand the functor that $\mathbf{P}_R^n$ represents when $R$ is not an ordinary ring. On the other hand, one can import some of the classical apparatus associated to projective spaces. For instance, there are quasicoherent sheaves

---

follows. Given $f: \mathbb{S}[x] \to \mathbb{S}$, tensor with complex $K$-theory $KU$ and take $p$-completions. The $\pi_0$ of $p$-complete commutative $KU$-algebras carries a natural "Adams operation" $\psi^p$, which is a ring endomorphism such that $\psi^p(a) \equiv a^p \mod p$, and on $\pi_0(KU[x])_p^\wedge$ acts via $\psi^p(f(x)) = f(x^p)$. Using this we can show that $a \in \mathbb{Z}$ is in the image if and only if $a^p = a$ for all primes $p$.

The same kind of argument shows that if $R = \mathbb{S}[\frac{1}{n}, \zeta_n]$ where $\zeta_n$ is a primitive $n$th root of unity as in (8.9.10), then the image of $\mathrm{Map}_{\mathrm{CAlg}}(\mathbb{S}[x], R) \to \pi_0 R = \mathbb{Z}[\frac{1}{n}, \zeta_n]$ is $\{0\} \cup \{\zeta_n^k \mid 0 \leq k < n\}$.

$\mathcal{O}(m)$ over $\mathbf{P}_R^n$ for any $R \in \mathrm{CAlg}^{\mathrm{cn}}$, constructed exactly as their classical counterparts, and $\Gamma(\mathbf{P}_R^n, \mathcal{O}(m))$ has the expected value [170, 5.4.2.6].

There is another projective $n$-space, the **smooth projective space** $\mathbf{P}_{\mathrm{sm}}^n$, defined to be the spectral DM stack representing a functor $R \mapsto \{\text{"lines in } R^{n+1}\text{"}\}$; see [170, 19.2.6].

## 8.15        Functor of points

We have defined an $\infty$-category $\mathrm{SpDM}^{\mathrm{nc}}$ of nonconnective spectral DM stacks. However, we have not yet shown that it is a *locally small* $\infty$-category: the definition of morphism involves morphisms of underlying $\infty$-topoi, and $\infty\mathcal{T}\mathrm{op}$ is not locally small. However, it is true that $\mathrm{SpDM}^{\mathrm{nc}}$ is locally small.

**Proposition 8.15.1.** *For any $X, Y \in \mathrm{SpDM}^{\mathrm{nc}}$, the space $\mathrm{Map}_{\mathrm{SpDM}^{\mathrm{nc}}}(Y, X)$ is essentially small, i.e., equivalent to a small $\infty$-groupoid.*

Note that when $X$ is affine, (8.11.7) already implies that $\mathrm{Map}_{\mathrm{SpDM}^{\mathrm{nc}}}(Y, X)$ is essentially small: $\mathrm{Map}_{\mathrm{SpDM}^{\mathrm{nc}}}(Y, \mathrm{Sp\acute{e}t}\, B) \approx \mathrm{Map}_{\mathrm{CAlg}}(B, \Gamma(\mathcal{Y}, \mathcal{O}_Y))$.

Given this proposition, we can define the **functor of points** of a nonconnective spectral DM stack:

$$h_X^{\mathrm{nc}} : \mathrm{CAlg} \to \mathcal{S} \quad \text{by} \quad h_X^{\mathrm{nc}}(A) := \mathrm{Map}_{\mathrm{SpDM}^{\mathrm{nc}}}(\mathrm{Sp\acute{e}t}\, A, X).$$

For a spectral DM stack, we consider the restriction of $h_X^{\mathrm{nc}}$ to connective $\mathbb{E}_\infty$-rings:

$$h_X : \mathrm{CAlg}^{\mathrm{cn}} \to \mathcal{S} \quad \text{by} \quad h_X(A) := \mathrm{Map}_{\mathrm{SpDM}}(\mathrm{Sp\acute{e}t}\, A, X).$$

Note that if $B \in \mathrm{CAlg}$, then $h_{\mathrm{Sp\acute{e}t}\, B}^{\mathrm{nc}} \approx \mathrm{Map}_{\mathrm{CAlg}}(B, -)$ by the Yoneda lemma, and similarly in the connective case.

**Proposition 8.15.2** ([170, 1.6.4.3]). *The functors*

$$X \mapsto h_X^{\mathrm{nc}} : \mathrm{SpDM}^{\mathrm{nc}} \rightarrowtail \mathrm{Fun}(\mathrm{CAlg}, \mathcal{S}) \quad and \quad X \mapsto h_X : \mathrm{SpDM} \rightarrowtail \mathrm{Fun}(\mathrm{CAlg}^{\mathrm{cn}}, \mathcal{S})$$

*are fully faithful.*

I'll sketch proofs of these below (giving arguments only in the nonconnective case).

### Sheaves of maps into a spectral DM stack

To prove (8.15.1) that $\mathrm{Map}_{\mathrm{SpDM}^{\mathrm{nc}}}(Y, X)$ is essentially small, we can immediately reduce to the case that $Y$ is affine, since every nonconnective spectral DM stack is a colimit of a small diagram of affines (8.8). So assume $Y = \mathrm{Sp\acute{e}t}\, A$ for some $A \in \mathrm{CAlg}$.

Given a nonconnective spectral DM stack $X$, consider the functor

$$H_X^A : \mathrm{CAlg}_A^{\mathrm{\acute{e}t}} \to \widehat{\mathcal{S}} \quad \text{defined by} \quad H_X^A(A') := \mathrm{Map}_{\mathrm{SpDM}^{\mathrm{nc}}}(\mathrm{Sp\acute{e}t}\, A', X).$$

This functor is in fact an object of the full subcategory

$$\widehat{\mathrm{Shv}_A^{\mathrm{\acute{e}t}}} \subseteq \mathrm{Fun}(\mathrm{CAlg}_A^{\mathrm{\acute{e}t}}, \widehat{\mathcal{S}})$$

of sheaves on the étale site of $A$ taking values in the category $\widehat{S}$ of "large" $\infty$-groupoids; this is because for an étale cover $\{R \to R_i\}$ in $\mathrm{CAlg}_A^{\text{ét}}$, the evident map

$$\mathrm{colim}_{\Delta^{\text{op}}}^{\mathrm{SpDM}^{\text{nc}}}\left([n] \mapsto \coprod \mathrm{Spét}\, R_{i_0} \times_{\mathrm{Spét}\, R} \cdots \times_{\mathrm{Spét}\, R} \mathrm{Spét}\, R_{i_n}\right) \xrightarrow{\sim} \mathrm{Spét}\, R$$

is an equivalence by (8.13.4), which exactly provides the sheaf condition for $H_X^A$.

Note: the $\infty$-category $\widehat{\mathrm{Shv}}_A^{\text{ét}}$, although not locally small, behaves in many respects like an $\infty$-topos. For instance, it has descent for small diagrams, and in particular small colimits are universal in $\widehat{\mathrm{Shv}}_A^{\text{ét}}$. Furthermore, the inclusion $\mathrm{Shv}_A^{\text{ét}} \subseteq \widehat{\mathrm{Shv}}_A^{\text{ét}}$ preserves small colimits.

The key fact we need is the following.

**Proposition 8.15.3.** *The functor*

$$X \mapsto H_X^A : \mathrm{SpDM}_{\text{ét}}^{\text{nc}} \to \widehat{\mathrm{Shv}}_A^{\text{ét}}$$

*preserves small colimits.*

Recall (8.8) that $X \approx \mathrm{colim}_{c \in \mathcal{C}}\, V_c$ for some functor $V : \mathcal{C} \to \mathrm{SpDM}_{\text{ét}}^{\text{nc}}$ from a small $\infty$-category. Writing $V_c = \mathrm{Spét}\, B_c$, the proposition gives us the "formula"

$$\mathrm{Map}_{\mathrm{SpDM}^{\text{nc}}}(\mathrm{Spét}\, A, X) \approx (aF)(A),$$

where $aF$ is the sheafification of the presheaf $F : \mathrm{CAlg}_A^{\text{ét}} \to \widehat{S}$ defined by

$$F(A') = \mathrm{colim}_c\, H_{V_c}^A(A') \approx \mathrm{colim}_c\, \mathrm{Map}_{\mathrm{SpDM}^{\text{nc}}}(\mathrm{Spét}\, A', V_c) \approx \mathrm{colim}_c\, \mathrm{Map}_{\mathrm{CAlg}}(B_c, A'),$$

where the colimit is taken in $\widehat{S}$. Since $\mathcal{C}$ and each $\mathrm{Map}_{\mathrm{CAlg}}(B_c, A')$ are small, we see that the value $F(A)$ is a small $\infty$-groupoid, as desired.

*Sketch proof of* (8.15.3). Let $X = \mathrm{colim}_{c \in \mathcal{C}}^{\mathrm{SpDM}_{\text{ét}}^{\text{nc}}}\, V_c$ with $V : \mathcal{C} \to \mathrm{SpDM}_{\text{ét}}^{\text{nc}}$. If $\mathcal{X}$ is the underlying $\infty$-topos of $X$, then we can factor this functor through a functor $U : \mathcal{C} \to \mathcal{X}$, so that $V_c = X_{U_c}$ and $\mathrm{colim}_{c \in \mathcal{C}}^{\mathcal{X}}\, U_c \approx 1$.

To show that

$$\mathrm{colim}_{c \in \mathcal{C}}^{\widehat{\mathrm{Shv}}_A^{\text{ét}}}\, H_{X_{U_c}}^A \xrightarrow{\sim} H_X^A,$$

it suffices to show that for any small sheaf $V \in \mathrm{Shv}_A^{\text{ét}}$ and any map $f : V \to H_X^A$ in $\widehat{\mathrm{Shv}}_A^{\text{ét}}$, the map

$$\mathrm{colim}_{c \in \mathcal{C}}^{\widehat{\mathrm{Shv}}_A^{\text{ét}}}\left(H_{X_{U_c}}^A \times_{H_X^A} V\right) \to V$$

induced by base change along $f$ is an equivalence. (This is using descent in $\widehat{\mathrm{Shv}}_A^{\text{ét}}$, and the fact that any small sheaf is a small colimit of representables $\mathrm{Map}_{\mathrm{CAlg}_A^{\text{ét}}}(B, -)$, which are themselves small sheaves.)

Note that for small sheaves $V \in \mathrm{Shv}_A^{\text{ét}}$ there is a natural equivalence

$$\mathrm{Hom}_{\mathrm{SpDM}^{\text{nc}}}((\mathrm{Spét}\, A)_V, X) \xrightarrow{\sim} \mathrm{Hom}_{\widehat{\mathrm{Shv}}_A^{\text{ét}}}(V, H_X^A),$$

This is because $V \mapsto (\mathrm{Sp\acute{e}t}\,A)_v$ is colimit preserving (8.13.4) and the map is certainly an equivalence when $V$ is representable. So let $g \colon (\mathrm{Sp\acute{e}t}\,A)_V \to X$ be the map corresponding to $f \colon V \to H_X^A$, and use (8.13.3) to obtain for any $U \in \mathcal{X}$ a pullback square

$$
\begin{array}{ccc}
(\mathrm{Sp\acute{e}t}\,A)_{g^*U} & \longrightarrow & X_U \\
\downarrow & & \downarrow \\
(\mathrm{Sp\acute{e}t}\,A)_V & \xrightarrow{\ g\ } & X
\end{array}
$$

in $\mathrm{SpDM}^{nc}$, which on applying the functor $Y \mapsto H_Y^A$ gives a pullback square

$$
\begin{array}{ccc}
g^*U & \longrightarrow & H_{X_U}^A \\
\downarrow & & \downarrow \\
V & \xrightarrow{\ f\ } & H_X^A
\end{array}
$$

in $\widehat{\mathrm{Shv}_A^{\text{ét}}}$. Because $g^* \colon \mathcal{X} \to (\mathrm{Shv}_A^{\text{ét}})_{/V}$ is colimit preserving, we see that we get an equivalence $\mathrm{colim}_{c \in C}^{(\mathrm{Shv}_A^{\text{ét}})_{/V}} g^*(U_c) \xrightarrow{\sim} 1_{(\mathrm{Shv}_A^{\text{ét}})_{/V}}$, and the claim follows. $\qquad\square$

**Example 8.15.4** (Geometric points). Let $K$ be a (classical) separable field, so that $\mathrm{Shv}_K^{\text{ét}} \approx \mathcal{S}$. If $X = \mathrm{colim}_{c \in C}\,\mathrm{Sp\acute{e}t}\,B_c$ is a colimit of affines along étale morphisms, then our "formula" reduces to

$$
\mathrm{Map}_{\mathrm{SpDM}^{nc}}(\mathrm{Sp\acute{e}t}\,K, X) \approx \mathrm{colim}_{c \in C}\,\mathrm{Map}_{\mathrm{CAlg}}(B_c, K).
$$

## Functor of points

Here is an idea of a proof of (8.15.2) (in the nonconnective case; the connective case is similar); see [163, V 2.4] which proves a more general statement in the framework of "geometries", or [170, 8.1.5] which proves a generalization to formal geometry. We want to show that

$$
\mathrm{Map}_{\mathrm{SpDM}^{nc}}(Y, X) \to \mathrm{Map}_{\mathrm{Fun}(\mathrm{CAlg}, \mathcal{S})}(h_Y^{nc}, h_X^{nc})
$$

is an equivalence for all $X, Y \in \mathrm{SpDM}^{nc}$. Since nonconnective spectral DM stacks are colimits of small diagrams of affines along étale maps (8.13.4), we reduce to the case of affine $Y = \mathrm{Sp\acute{e}t}\,B$. Furthermore, if $X = \mathrm{Sp\acute{e}t}\,A$ is also affine, then $\mathrm{Map}_{\mathrm{SpDM}^{nc}}(Y, X) \approx \mathrm{Map}_{\mathrm{CAlg}}(A, B)$ by (8.11.7), and since $h_Y^{nc} \approx \mathrm{Map}_{\mathrm{CAlg}}(B, -)$ we see that the map is an equivalence by Yoneda.

Note that the composite functor $\mathrm{CAlg}_A^{\text{ét}} \to \mathrm{CAlg} \xrightarrow{h_X^{nc}} \mathcal{S}$ is precisely the functor $H_X^A$ of the previous section. Thus $h_X^{nc}$ lives in the full subcategory

$$
\mathrm{Shv}^{\text{ét}} \subseteq \mathrm{Fun}(\mathrm{CAlg}, \mathcal{S})
$$

spanned by $F$ such that $F|_{\mathrm{CAlg}_A^{\text{ét}}}$ is an étale sheaf for all $A \in \mathrm{CAlg}$.

It turns out that $\mathrm{Shv}^{\text{ét}}$ is equivalent to the $\infty$-category of sections of a Cartesian fibration $\mathcal{D} \to \mathrm{CAlg}$, whose fiber over $A \in \mathrm{CAlg}$ is equivalent to $\mathrm{Shv}_A^{\text{ét}}$. Thus, by a standard argument, we see that (8.15.3) implies that

$$X \mapsto h_X^{\text{nc}}: \mathrm{SpDM}_{\text{ét}}^{\text{nc}} \to \mathrm{Shv}^{\text{ét}}$$

preserves colimits. The result then follows using $X$ is also a colimit of a small diagram of affines along étale maps.

## 8.16     Formal spectral geometry

Let's briefly describe the generalization of these ideas to the spectral analogue of *formal* geometry.

### Adic $\mathbb{E}_\infty$-rings

An **adic $\mathbb{E}_\infty$-ring** is a connective $\mathbb{E}_\infty$-ring $A$ equipped with a topology on $\pi_0 A$ which is equal to the $I$-adic topology for some finitely generated ideal $I \subseteq \pi_0 A$. A map of adic $\mathbb{E}_\infty$-rings is a map $f: A \to B$ of $\mathbb{E}_\infty$-rings which induces a continuous map on $\pi_0$. Any finitely generated ideal $I$ generating the topology of $\pi_0 A$ is called an **ideal of definition** for the topology; note that the ideal of definition is not itself part of the data of an adic $\mathbb{E}_\infty$-ring, only the topology it generates.

*Remark 8.16.1.* The **vanishing locus** of an adic $\mathbb{E}_\infty$-ring $A$ is the set $X_A \subseteq |\mathrm{Spec}\,A|$ of prime ideals which are open neighborhoods of $0$ in $\pi_0 A$; equivalently, primes which contain some (hence any) ideal of definition $I \subseteq \pi_0 A$. A map $\phi: A \to B$ of $\mathbb{E}_\infty$-rings is an adic map if and only if it sends $X_B$ into $X_A$; equivalently, if $\phi(I^n) \subseteq J$ for some $n$ where $I$ and $J$ are ideals of definition for $A$ and $B$ respectively [170, 8.1.1.3–4].

In particular, the topology on $\pi_0 A$ of an adic $\mathbb{E}_\infty$-ring $A$ is entirely determined by the vanishing locus.

### Completion at finitely generated ideals

Let $A \in \mathrm{CAlg}$ be an $\mathbb{E}_\infty$-ring (not necessarily connective). For every finitely generated ideal $I \subseteq \pi_0 A$ there is a notion of $I$-**complete** $A$-module. An $A$-algebra is called $I$-complete if its underlying module is so. There are adjoint pairs

$$M \mapsto M_I^\wedge: \mathrm{Mod}_A \rightleftarrows \mathrm{Mod}_A^{\mathrm{Cpt}(I)}\,, \qquad B \mapsto B_I^\wedge: \mathrm{CAlg}_A \rightleftarrows \mathrm{CAlg}_A^{\mathrm{Cpt}(I)}$$

whose right adjoint is the fully faithful inclusion of the category of $I$-complete objects, and whose left adjoint, called $I$-**completion**, is left exact. Furthermore, the notion of $I$-completeness and its associated completion functors depend only on the radical of $I$; hence, all ideals of definition of an adic $\mathbb{E}_\infty$-ring provide equivalent completion functors. See [170, 7.3] for more details.

*Remark 8.16.2.* Here is an explicit formula for $I$-completion on the level of modules. Given $a \in \pi_0 A$ let $\Sigma^{-1}(A/a^\infty) \in \mathrm{Mod}_A$ denote the homotopy fiber of the evident map $A \to A[a^{-1}]$. Then

$$M_I^\wedge \approx \underline{\mathrm{Hom}}_A(\Sigma^{-1}(A/a_1^\infty) \otimes_A \cdots \otimes_A \Sigma^{-1}(A/a_r^\infty), \, M)$$

where $(a_1, \ldots, a_r)$ is any finite sequence which generates the ideal $I$. The unit $M \to M_I^\wedge$ of the adjunction is induced by restriction along the evident map $\Sigma^{-1}(A/a_1^\infty) \otimes_A \cdots \otimes_A \Sigma^{-1}(A/a_r^\infty) \to A \otimes_A \cdots \otimes_A A \approx A$.

**Example 8.16.3.** If the vanishing ideal is $0 \subseteq \pi_0 A$, so that $\pi_0 A$ is equipped with the discrete topology, then every $A$-module is $I$-complete.

**Example 8.16.4.** If the vanishing ideal is $I = \pi_0 A$, so that $\pi_0 A$ is equipped with the trivial topology, then only the trivial $A$-module is $I$-complete.

**Example 8.16.5.** For a prime $p \in \mathbb{Z} = \pi_0 \mathbb{S}$, an $\mathbb{S}$-module is $(p)$-complete in the above sense if and only if it is a $p$-complete spectrum in the conventional sense, and $(p)$-completion coincides with the usual $p$-completion of spectra.

**Example 8.16.6** (Completion and $K(n)$-localization). Suppose $A$ is an $\mathbb{E}_\infty$-ring which $p$-local for some prime $p$, and is weakly 2-periodic and complex orientable (see (8.17) below). The complex orientation gives rise to a sequence of ideals $I_n = (p, u_1, \ldots, u_{n-1}) \subseteq \pi_0 A$; the ideal $I_n$ is called the *$n$th Landweber ideal*. It turns out that the underlying spectrum of $A$ is $K(n)$-local if and only if (i) $A$ is $I_n$-complete and (ii) $I_{n+1}(\pi_0 A) = \pi_0 A$ [167, 4.5.2].

## The formal spectrum of an adic $\mathbb{E}_\infty$-ring

Recall the $\infty$-topos $\mathrm{Shv}_A^{\text{ét}}$ of sheaves on the étale site of an $\mathbb{E}_\infty$-ring $A$. Given an adic $\mathbb{E}_\infty$-ring $A$, say that $F \in \mathrm{Shv}_A^{\text{ét}}$ is an **adic sheaf** if $F(A \to B) \approx *$ for étale morphisms $A \to B$ such that the image of $|\mathrm{Spec}\,\pi_0 B| \to |\mathrm{Spec}\,\pi_0 A|$ is disjoint from the vanishing locus $X_A$; i.e., if $I(\pi_0 B) = \pi_0 B$ for some (hence any) ideal of definition $I \subseteq \pi_0 B$. We thus obtain a full subcategory $\mathrm{Shv}_A^{\text{ad}} \subseteq \mathrm{Shv}_A^{\text{ét}}$ of adic sheaves, which in fact is an $\infty$-topos, and this inclusion is the right-adjoint of a geometric morphism $\mathrm{Shv}_A^{\text{ad}} \to \mathrm{Shv}_A^{\text{ét}}$.

*Remark 8.16.7.* That $\mathrm{Shv}_A^{\text{ad}}$ is an $\infty$-topos follows from the observation that $\mathrm{Shv}_A^{\text{ad}} \approx \mathrm{Shv}_{\pi_0 A/I}^{\text{ét}}$, where $I$ is an ideal of definition for $A$. See [170, 3.1.4].

We can now define the **formal spectrum** of an adic $\mathbb{E}_\infty$-ring $A$ to be the spectrally ringed $\infty$-topos $\mathrm{Spf}\,A := (\mathrm{Shv}_A^{\text{ad}}, \mathcal{O}_{\mathrm{Spf}\,A})$, where $\mathcal{O}_{\mathrm{Spf}\,A}$ is the composite functor

$$\mathrm{CAlg}_A^{\text{ét}} \rightarrowtail \mathrm{CAlg} \xrightarrow{(-)_I^\wedge} \mathrm{CAlg}.$$

Note that $\mathcal{O}_{\mathrm{Spf}\,A}$ is an adic sheaf because $B_I^\wedge \approx 0$ if $I(\pi_0 B) = \pi_0 B$ and because $I$-completion is limit preserving. It can be shown that $\mathrm{Spf}\,A$ is strictly Henselian and its structure sheaf is connective [170, 8.1.1.13].

## Formal spectral DM stacks

A **formal spectral Deligne–Mumford stack** is a spectrally ringed $\infty$-topos $X = (\mathcal{X}, \mathcal{O}_X)$ which admits a cover $\{U_i\} \subseteq \mathcal{X}$ such that each $X_{U_i} = (\mathcal{X}_{/U_i}, \mathcal{O}_X|_{U_i})$ is equivalent to $\mathrm{Spf}\, A_i$ for some adic $\mathbb{E}_\infty$-ring $A_i$. There is a full subcategory

$$\mathrm{fSpDM} \subseteq \infty\mathcal{T}\mathrm{op}^{\mathrm{sHen}}_{\mathrm{CAlg}}$$

of formal spectral Deligne–Mumford stacks and local maps between them.

**Example 8.16.8** (Spectral DM stacks are formal spectral DM stacks). If $A \in \mathrm{CAlg}^{\mathrm{ad}}$ is an adic $\mathbb{E}_\infty$-ring equipped with the discrete topology, then $\mathrm{Spf}\, A \approx \mathrm{Sp\acute{e}t}\, A$. In particular, any spectral DM stack is automatically a formal spectral DM stack, and $\mathrm{SpDM} \rightarrowtail \mathrm{fSpDM}$.

**Example 8.16.9** (Formal functor of points). There is a fully faithful embedding $\mathrm{fSpDM} \rightarrowtail \mathrm{Fun}(\mathrm{CAlg}^{\mathrm{cn}}, \mathcal{S})$ defined by sending $X$ to its functor of points $h_X(R) = \mathrm{Map}_{\mathrm{fSpDM}}(\mathrm{Sp\acute{e}t}\, R, X)$ on affine (not adic) spectral DM stacks [170, 8.1.5].

Furthermore, there is an explicit description of the functor of points of $\mathrm{Spf}\, A$:

$$h_{\mathrm{Spf}\, A}(R) = \mathrm{Map}_{\mathrm{fSpDM}}(\mathrm{Sp\acute{e}t}\, R, \mathrm{Spf}\, A) \approx \mathrm{Map}_{\mathrm{CAlg}^{\mathrm{ad}}}(A, R) \subseteq \mathrm{Map}_{\mathrm{CAlg}}(A, R).$$

Here $R$ is regarded as an adic $\mathbb{E}_\infty$-ring equipped with the discrete topology, so that $\phi\colon A \to R$ is a map of adic $\mathbb{E}_\infty$-rings if and only if $\phi(I^n) = 0$ for some $n$ and ideal of definition $I \subseteq \pi_0 A$ [170, 8.1.5].

*Remark 8.16.10.* The formal spectrum functor $\mathrm{Spf}\colon (\mathrm{CAlg}^{\mathrm{ad}})^{\mathrm{op}} \to \mathrm{fSpDM}$ is not fully faithful, or even conservative. However, we have the following. Say that $B \in \mathrm{CAlg}^{\mathrm{ad}}$ is **complete** if $B \xrightarrow{\sim} B_I^\wedge$ for some (and hence any) ideal of definition $I \subseteq \pi_0 B$. For complete adic $\mathbb{E}_\infty$-rings $B$ the evident map $\mathrm{Map}_{\mathrm{CAlg}^{\mathrm{ad}}}(B, R) \xrightarrow{\sim} \mathrm{Map}_{\mathrm{fSpDM}}(\mathrm{Spf}\, R, \mathrm{Spf}\, B)$ is always an equivalence [170, 8.1.5.4]. From this and the formal functor of points we see that the full subcategory of formal spectral DM stacks which are equivalent to $\mathrm{Spf}\, A$ for some adic ring $A$ is equivalent to opposite of the full subcategory of complete objects in $\mathrm{CAlg}^{\mathrm{ad}}$.

## Formal completion

Given a spectral DM stack $X$, one may form the **formal completion** $X_K^\wedge$ of $X$ with respect to a "cocompact closed subset $K \subseteq |X|$", which is a formal spectral DM stack equipped with a map $X_K^\wedge \to X$. We refer to [170, 8.1.6] for details, but note that in the case $X = \mathrm{Sp\acute{e}t}\, A$ for $A \in \mathrm{CAlg}^{\mathrm{cn}}$ we have that $|X|$ is precisely the prime ideal spectrum $|\mathrm{Spec}\, A|$, while $X_K^\wedge = \mathrm{Spf}\, A$, where $A$ is given the evident adic structure.

## 8.17    Formal groups in spectral geometry

Fix a connective $\mathbb{E}_\infty$-ring $R$. An *$n$-dimensional formal group* over $R$ is, roughly speaking, a formal spectral DM stack $\widehat{G}$ over $\mathrm{Sp\acute{e}t}\, R$ which (i) is an abelian group object in formal spectral DM stacks, and (ii) as a formal spectral DM stack is equivalent to $\mathrm{Spf}(A)$ where $A$ is an adic $\mathbb{E}_\infty$-ring which "looks like a ring of power series in $n$ variables over $R$".

## Smooth coalgebras

To make this precise, we need the notion of a **smooth commutative coalgebra**. Any symmetric monoidal $\infty$-category admits a notion of **commutative coalgebra** objects [166, 3.1]. If $C$ is a commutative coalgebra object in $\mathrm{Mod}_R$, then its $R$-linear dual $C^\vee := \underline{\mathrm{Hom}}_R(C, R)$ comes with the structure of a adic commutative $R$-algebra [167, 1.3.2].

We say that a commutative $R$-coalgebra $C$ is **smooth** if (i) $C$ is flat as an $R$-module and if (ii) there is an isomorphism of $\pi_0 R$-coalgebras

$$\pi_0 C \approx \bigoplus_{k \geq 0} \Gamma^k_{\pi_0 R}(M),$$

where the right-hand side is the divided polynomial coalgebra on some finitely generated projective $\pi_0 R$-module $M$; the rank of $M$ (if defined) is also called the dimension of $C$ [167, 1.2]. There is an associated $\infty$-category $\mathrm{cCAlg}^{\mathrm{sm}}_R$ of smooth commutative $R$-coalgebras.

*Remark 8.17.1.* The $R$-linear dual $C^\vee$ of $C$ as above satisfies

$$\pi_0 C^\vee \approx \prod_{k \geq 0} \mathrm{Sym}^k_{\pi_0 R}(M^\vee), \qquad M^\vee = \mathrm{Hom}_R(M, R).$$

In particular, if $M$ is free of rank $n$ then $\pi_* C^\vee \approx \pi_* R[[t_1, \ldots, t_n]]$ [167, 1.3.8].

For a *connective* $\mathbb{E}_\infty$-ring $R$, a functor $\mathrm{CAlg}^{\mathrm{cn}}_R \to \mathcal{S}$ represented by $\mathrm{Spf}(C^\vee)$ for some smooth commutative $R$-coalgebra is called a **formal hyperplane** over $R$ [167, 1.5.3]. It is said to be $n$-dimensional if $C$ is $n$-dimensional in the sense above. (The "hyperplane" terminology arises because our $\mathrm{Spf}(C^\vee)$ does not come equipped with a "base-point", i.e., there is no distinguished $R$-algebra map $C^\vee \to R$, despite the fact that $\pi_0 C^\vee$ is equipped with an adic topology.)

## Formal groups

An $n$-**dimensional formal group** over a *connective* $\mathbb{E}_\infty$-ring $R$ is a functor

$$\widehat{G} \colon \mathrm{CAlg}^{\mathrm{cn}}_R \to \mathrm{Mod}^{\mathrm{cn}}_{\mathbb{Z}}$$

such that the composite

$$\mathrm{CAlg}^{\mathrm{cn}}_R \xrightarrow{\widehat{G}} \mathrm{Mod}^{\mathrm{cn}}_{\mathbb{Z}} \xrightarrow{\Omega^\infty} \mathcal{S}$$

is represented by $\mathrm{Spf}(C^\vee)$ for some smooth commutative $R$-coalgebra $C$ of dimension $n$.

*Remark 8.17.2.* The definition of formal group I have given here is different than, but equivalent to, the one given in [167, 1.6]; see [167, 1.6.7]. In particular, the basic definitions given there are expressed more directly in terms of commutative coalgebras.

In particular, the functor $\mathrm{CAlg}_R^{\mathrm{cn}} \to \mathcal{S}$ represented by $\mathrm{Spf}(C^\vee)$, where $C$ is a commutative $R$-coalgebra, is equivalent to the **cospectrum** of $C$. The cospectrum is a functor sending an $R' \in \mathrm{CAlg}_R^{\mathrm{cn}}$ to a suitable space of "grouplike elements" in $R' \otimes_R C$ [167, 1.51].

*Remark 8.17.3.* What about the nonconnective case? Although smooth commutative coalgebras may be defined over any $\mathbb{E}_\infty$-ring, formal hyperplanes and formal groups have only been defined (following Lurie) over *connective* $\mathbb{E}_\infty$-rings.

This is awkward but it's okay! For instance, because smooth commutative $R$-coalgebras are *flat* over $R$, taking 0-connective covers gives an equivalence

$$\tau_{\geq 0} \colon \mathrm{cCAlg}_R^{\mathrm{sm}} \xrightarrow{\sim} \mathrm{cCAlg}_{\tau_{\geq 0}R}^{\mathrm{sm}}$$

between the $\infty$-categories of smooth commutative coalgebras over $R$ and over its connective cover $\tau_{\geq 0}R$ [167, 1.2.8]; compare (8.13.6).

So you can extend the notions of formal hyperplane and formal group to nonconnective ground rings, so that a formal hyperplane or formal group over $R$ is *defined* to be one over $\tau_{\geq 0}R$. In particular, for any $\mathbb{E}_\infty$-ring $R$ you get an $\infty$-category $\mathrm{FGroup}(R)$ of formal groups over $R$, which *by definition* satisfies $\mathrm{FGroup}(R) = \mathrm{FGroup}(\tau_{\geq 0}R)$.

## The Quillen formal group in spectral geometry

A complex oriented cohomology theory $R$ gives rise to a 1-dimensional formal group over $\pi_* R$, whose function ring is $R^* \mathbb{CP}^\infty$. When the theory is represented by an $\mathbb{E}_\infty$-ring which is suitably periodic, then we can upgrade this formal group to an object in spectral geometry.

Given $R \in \mathrm{CAlg}$ and $X \in \mathcal{S}$, write $C_*(X; R) := R \otimes_{\mathbb{S}} \Sigma_+^\infty X \in \mathrm{Mod}_R$ for the "$R$-module of $R$-chains on $X$". This object is in fact a commutative $R$-coalgebra, via the diagonal map on $X$ [168, 2.4.3.10].

An $\mathbb{E}_\infty$-ring $R$ is **weakly 2-periodic** if $\pi_2 R \otimes_{\pi_0 R} \pi_n R \to \pi_{n+2} R$ is an isomorphism for all $n \in \mathbb{Z}$. If $R$ is both weakly 2-periodic and complex orientable, then one can show that $C_*(\mathbb{CP}^\infty; R)$ is a *smooth* commutative $R$-coalgebra. Furthermore, it is a commutative group object in $\mathrm{cCAlg}_R$ (via the abelian group structure on $\mathbb{CP}^\infty$), and hence it gives rise to a 1-dimensional formal group $\widehat{G}_R^Q$, called the **Quillen formal group** of $R$ [167, 4.1.3].

*Remark 8.17.4.* In view of what I said about connectivity in relation to formal groups (8.17.3), the formal spectral DM stack associated to the Quillen formal group of $R$ is $\mathrm{Spf}((\tau_{\geq 0}C_*(\mathbb{CP}^\infty; R))^\vee)$. Note that $\tau_{\geq 0}C_*(\mathbb{CP}^\infty; R)$ is not at all the same as $C_*(\mathbb{CP}^\infty; \tau_{\geq 0}R)$, and that the latter does not give rise to a formal group in the sense defined above.

*Remark 8.17.5.* Let $R$ be an $\mathbb{E}_\infty$-ring which is weakly 2-periodic and complex orientable, with Quillen formal group $\widehat{G}_R^Q$. Then every commutative $R$-algebra $R \to R'$

is *also* weakly 2-periodic and complex orientable, and so also has a Quillen formal group, and in fact $\widehat{G}_{R'}^Q \approx \widehat{G}_R^Q \times_{\mathrm{Sp\acute{e}t}\, \tau_{\geq 0}R} \mathrm{Sp\acute{e}t}\, \tau_{\geq 0}R'$.

## Preorientations and orientations

Let $R$ be an $\mathbb{E}_\infty$-ring, not necessarily assumed to be connective, and $\widehat{G} \in \mathrm{FGroup}(R)$ a 1-dimensional formal group over it. We ask the question: What additional data do we need to identify $\widehat{G}$ with the Quillen formal group over $R$? Note that I don't want to presuppose that the Quillen formal group actually exists in this case, i.e., I don't assume that $R$ is weakly 2-periodic or complex orientable.

A **preorientation** of a 1-dimensional formal group $\widehat{G}$ over a (possibly nonconnective) $\mathbb{E}_\infty$-ring $R$ is a map

$$e \colon S^2 \to \widehat{G}(\tau_{\geq 0}R)$$

of based spaces, where the base point goes to the identity of the group structure. We write $\mathrm{Pre}(\widehat{G}) = \mathrm{Map}_{\mathcal{S}_*}(S^2, \widehat{G}(\tau_{\geq 0}R))$ for the space of preorientations.

**Proposition 8.17.6.** *Suppose $R$ is weakly 2-periodic and complex orientable. Then there is an equivalence*

$$\mathrm{Pre}(\widehat{G}) \approx \mathrm{Map}_{\mathrm{FGroup}(R)}(\widehat{G}_R^Q, \widehat{G})$$

*between the space of preorientations and the space of maps from the Quillen formal group.*

*Proof.* See [167, 4.3]. This is basically a formal consequence of the observation that the free abelian group on the based space $S^2$ is equivalent to $\mathbb{CP}^\infty$. □

Note that $\mathrm{Pre}(\widehat{G})$ is defined even when $R$ does not admit a Quillen formal group. We will now describe a condition on a *preorientation* $e \in \mathrm{Pre}(\widehat{G})$ which implies simultaneously (i) that $R$ is weakly 2-periodic and complex orientable, and (ii) that the map $\widehat{G}_R^Q \to \widehat{G}$ induced by $e$ is an isomorphism in $\mathrm{FGroup}(R)$.

Given $\widehat{G} \in \mathrm{FGroup}(R)$, let $\mathcal{O}_{\widehat{G}}$ denote its ring of functions, so that $\widehat{G} \approx \mathrm{Spf}(\mathcal{O}_{\widehat{G}})$. Note that by our definitions (8.17.3) the ring $\mathcal{O}_{\widehat{G}}$ is a connective $\tau_{\geq 0}R$-algebra, even if $R$ is not connective.

The **dualizing line** of a 1-dimensional formal group $\widehat{G}$ is an $R$-module defined by

$$\omega_{\widehat{G}} := R \otimes_{\mathcal{O}_{\widehat{G}}} \mathcal{O}_{\widehat{G}}(-\eta), \quad \text{where} \quad \mathcal{O}_{\widehat{G}}(-\eta) := \text{fiber of } (\mathcal{O}_{\widehat{G}} \xrightarrow{\eta} \tau_{\geq 0}R \to R),$$

where $\eta \in \widehat{G}(\tau_{\geq 0}R)$ is the identity element of the group structure. The $R$-module $\omega_{\widehat{G}}$ is in fact an $R$-module which is locally free of rank 1, and its construction is functorial with respect to isomorphisms of 1-dimensional formal groups [167, 4.1 and 4.2].

**Example 8.17.7.** Let $R$ be weakly 2-periodic and complex orientable, and $\widehat{G}_R^Q$ its Quillen formal group. Then there is a canonical equivalence of $R$-modules

$$\omega_{\widehat{G}_R^Q} \approx \Sigma^{-2}R.$$

This object is also canonically identified with $C^*_{\mathrm{red}}(\mathbb{CP}^1; R)$, the function spectrum representing the reduced $R$-cohomology of $\mathbb{CP}^1 \approx S^2$ as a $C^*(S^2; R)$-module.

For a 1-dimensional formal group $\widehat{G}$ over an $\mathbb{E}_\infty$-ring $R$, any preorientation $e \in$ $\text{Pre}(\widehat{G})$ determines a map

$$\beta_e \colon \omega_{\widehat{G}} \to \Sigma^{-2} R$$

of $R$-modules, called the **Bott map** associated to $e$. This map is constructed in [167, 4.2-3].

*Remark 8.17.8.* Here is one way to describe the construction of the Bott map [167, 4.2.10].

For any suspension $X = \Sigma Y$ of a based space, the object $C^*_{\text{red}}(X; R)$ is equivalent as a $C^*(X; R)$-module to the restriction of an $R$-module along the augmentation $\pi \colon C^*(X; R) \to R$ corresponding to the basepoint of $X$. ("The cup product is trivial on a suspension.") For instance if $X = S^2 = \Sigma S^1$ we have $C^*_{\text{red}}(X; R) \approx \pi^*(\Sigma^{-2} R)$.

A preorientation $e \colon S^2 \to \widehat{G}(\tau_{\geq 0} R)$ corresponds exactly to a map of $\mathbb{E}_\infty$-rings, $\tilde{e} \colon \mathcal{O}_{\widehat{G}} \to C^*(S^2; \tau_{\geq 0} R)$, compatible with augmentations to $\tau_{\geq 0} R$, and in turn induces a map

$$\mathcal{O}_{\widehat{G}}(-\eta) \to C^*_{\text{red}}(S^2; R) \approx \pi^*(\Sigma^{-2} R)$$

of $\mathcal{O}_{\widehat{G}}$-modules, which by the previous paragraph is adjoint to a map $\omega_{\widehat{G}} = R \otimes_{\mathcal{O}_{\widehat{G}}} \mathcal{O}_{\widehat{G}}(-\eta) \to \Sigma^{-2} R$ of $R$-modules, which is the Bott map of $e$.

An **orientation** of $\widehat{G}$ is a preorientation $e$ whose Bott map $\beta_e \colon \omega_{\widehat{G}} \to \Sigma^{-1} R$ is an equivalence. We write $\text{OrDat}(\widehat{G}) \subseteq \text{Pre}(\widehat{G})$ for the full subgroupoid consisting of orientations.

Now we can state the criterion for a preoriented 1-dimensional formal group to be isomorphic to the Quillen formal group.

*Proposition 8.17.9.* *A preorientation $e \in \text{Pre}(\widehat{G})$ of a formal group $\widehat{G}$ over an $\mathbb{E}_\infty$-ring $R$ is an orientation if and only if (i) $R$ is weakly 2-periodic and complex orientable, and (ii) the map $\widehat{G}^Q_R \to \widehat{G}$ of formal groups corresponding to $e$ is an isomorphism.*

*Proof.* See [167, 4.3.23]. That $R$ is weakly 2-periodic and complex orientable given the existence of an orientation is immediate from the fact that $\omega_{\widehat{G}}$ is locally free of rank 1, and also equivalent to $\Sigma^{-2} R$. □

## 8.18     Quasicoherent sheaves

Recall that we have defined a sheaf of $\mathbb{E}_\infty$-rings $\mathcal{O} \in \text{Shv}_{\text{CAlg}}(\mathcal{X})$ on an $\infty$-topos to be a limit preserving functor $\mathcal{X}^{\text{op}} \to \text{CAlg}$ (8.3). There is an alternate description: $\text{Shv}_{\text{CAlg}}(\mathcal{X})$ is equivalent to the $\infty$-category of *commutative monoid objects* in the symmetric monoidal $\infty$-category $(\text{Shv}_{\text{Sp}}(\mathcal{X}), \otimes)$ of sheaves of spectra, using a symmetric monoidal structure inherited from the usual one on spectra [163, VII 1.15].

This leads to notions of *sheaves of $\mathcal{O}$-modules* on a spectrally ringed $\infty$-topos, and eventually to *quasicoherent sheaves* on a nonconnective spectral DM stack.

## Sheaves of modules

To each spectrally ringed $\infty$-topos $X = (\mathcal{X}, \mathcal{O})$, there is an associated $\infty$-category $\mathrm{Mod}_{\mathcal{O}}$ of **sheaves of $\mathcal{O}$-modules** on $\mathcal{X}$, whose objects are sheaves of spectra which are modules over $\mathcal{O}$. (A precise description of this category requires the theory of $\infty$-operads; see [168, 3.3].)

The $\infty$-category $\mathrm{Mod}_{\mathcal{O}}$ is presentable (so is complete and cocomplete), stable, and symmetric monoidal, and the monoidal structure $\otimes_{\mathcal{O}}$ preserves colimits and finite limits in each variable [170, 2.1].

**Example 8.18.1.** Given an $\mathbb{E}_{\infty}$-ring $A$, any $A$-module $M \in \mathrm{Mod}_A$ can be promoted to a sheaf $\mathcal{M} \in \mathrm{Mod}_{\mathcal{O}}$ of $\mathcal{O}$-modules on $\mathrm{Sp\acute{e}t}\, A = (\mathrm{Shv}_A^{\mathrm{\acute{e}t}}, \mathcal{O})$, so that the underlying sheaf of spectra of $\mathcal{M}$ is

$$(A \to B) \mapsto B \otimes_A M : \mathrm{CAlg}_A^{\mathrm{\acute{e}t}} \to \mathrm{Sp}.$$

The resulting tuple $(\mathrm{Shv}_A^{\mathrm{\acute{e}t}}, \mathcal{O}, \mathcal{M})$ of $\infty$-topos, sheaf of rings, and sheaf of modules, is denoted $\mathrm{Sp\acute{e}t}(A, M)$; see [170, 2.2.1] for details.

## Quasicoherent sheaves

Now let $X = (\mathcal{X}, \mathcal{O}_X)$ be a nonconnective spectral DM stack. A sheaf of $\mathcal{O}_X$-modules $\mathcal{F} \in \mathrm{Mod}_{\mathcal{O}_X}$ is **quasicoherent** if there exists a set $\{U_i\}$ of objects in $\mathcal{X}$ which cover it (i.e., such that $\coprod_i U_i \to 1$ is effective epi), and there exist pairs $(A_i, M_i)$, $A_i \in \mathrm{CAlg}$, $M_i \in \mathrm{Mod}_{A_i}$, and equivalences

$$(\mathcal{X}_{/U_i}, \mathcal{O}_X|_{U_i}, \mathcal{F}|_{U_i}) \approx \mathrm{Sp\acute{e}t}(A_i, M_i)$$

of data consisting of (strictly Henselian spectrally ringed $\infty$-topos and sheaf of modules), where $\mathrm{Sp\acute{e}t}(A_i, M_i)$ is as in (8.18.1).

The $\infty$-category

$$\mathrm{QCoh}(X) \subseteq \mathrm{Mod}_{\mathcal{O}_X}$$

of quasicoherent sheaves on $X$ is defined to be the full subcategory of modules spanned by quasicoherent objects. It is presentable, stable, and symmetric monoidal (see [170, 2.2.4]).

For affine $X$, quasicoherent modules are just modules over the evident $\mathbb{E}_{\infty}$-ring.

**Proposition 8.18.2.** *If $X \approx \mathrm{Sp\acute{e}t}\, A$ for some $A \in \mathrm{CAlg}$, then there is an equivalence*

$$\mathrm{QCoh}(X) \approx \mathrm{Mod}_A$$

*of symmetric monoidal $\infty$-categories. The functor $\mathrm{QCoh}(X) \to \mathrm{Mod}_A$ sends a sheaf to its global sections; the functor $\mathrm{Mod}_A \to \mathrm{QCoh}(X)$ is $M \mapsto \mathrm{Sp\acute{e}t}(A, M)$.*

*Remark 8.18.3.* If $A \in \mathrm{CAlg}^{\heartsuit}$ is an ordinary ring, then

$$\mathrm{QCoh}(\mathrm{Sp\acute{e}t}\, A) \approx \mathrm{Mod}_A \approx \mathrm{Ch}(\mathrm{Mod}_A^{\heartsuit})[(\text{quasi-isos})^{-1}],$$

where $\mathrm{Mod}_A^{\heartsuit} \subseteq \mathrm{Mod}_A$ is the ordinary 1-category of $A$-modules.

There are other characterizations of quasicoherence. For instance, $\mathcal{F} \in \mathrm{Mod}_{\mathcal{O}_X}$ is quasicoherent if and only if the evident map

$$\mathcal{F}(V) \otimes_{\mathcal{O}_X(V)} \mathcal{O}_X(U) \to \mathcal{F}(U)$$

is an isomorphism for all maps $U \to V$ between affine objects in $\mathcal{X}$ [170, 2.2.4.3].

There are pairs of adjoint functors

$$\mathrm{QCoh}(X) \overset{\longrightarrow}{\underset{\longleftarrow}{\phantom{xx}}} \mathrm{Mod}_{\mathcal{O}_X} \overset{\mathcal{O}_X \otimes -}{\underset{\mathrm{forget}}{\rightleftarrows}} \mathrm{Shv}_{\mathrm{Sp}}(\mathcal{X}).$$

The left adjoints of these pairs are symmetric monoidal, and preserve finite limits but not arbitrary limits in general.

## Pullbacks and pushforwards of quasicoherent sheaves

Given a map $f : X \to Y$ of nonconnective spectral DM stacks, we have pairs of adjoint functors

$$\mathrm{QCoh}(X) \overset{\longrightarrow}{\underset{\longleftarrow}{\phantom{xx}}} \mathrm{Mod}_{\mathcal{O}_X} \overset{\longleftarrow}{\underset{\longrightarrow}{\phantom{xx}}} \mathrm{Shv}_{\mathrm{Sp}}(\mathcal{X})$$
$$f^* \Big\Uparrow\Big\Downarrow f_* \qquad f^* \Big\Uparrow\Big\Downarrow f_* \qquad f^* \Big\Uparrow\Big\Downarrow f_*$$
$$\mathrm{QCoh}(Y) \overset{\longrightarrow}{\underset{\longleftarrow}{\phantom{xx}}} \mathrm{Mod}_{\mathcal{O}_Y} \overset{\longleftarrow}{\underset{\longrightarrow}{\phantom{xx}}} \mathrm{Shv}_{\mathrm{Sp}}(\mathcal{Y})$$

so that each functor labeled $f^*$ is (strongly) symmetric monoidal, and such that the squares of *left adjoints* commute up to natural isomorphism, and the squares of *right adjoints* commute up to natural isomorphism. See [170, 2.5].

## Descent for modules and quasicoherent sheaves

It turns out that the formation of categories of either modules or quasicoherent sheaves satisfies a version of descent. Given a nonconnective spectral DM stack $X = (\mathcal{X}, \mathcal{O}_X)$, we have a functor

$$U \mapsto X_U = (\mathcal{X}_{/U}, \mathcal{O}_X|_U) : \mathcal{X} \to \mathrm{SpDM},$$

whose colimit exists and is equivalent to $X$ (8.8). For each $f : U \to V$ in $\mathcal{X}$ we have induced functors

$$f^* : \mathrm{Mod}_{\mathcal{O}_{X_V}} \to \mathrm{Mod}_{\mathcal{O}_{X_U}}, \qquad f^* : \mathrm{QCoh}(X_V) \to \mathrm{QCoh}(X_U),$$

which fit together to give functors $\mathcal{X}^{\mathrm{op}} \to \widehat{\mathrm{Cat}}_\infty$.

**Proposition 8.18.4.** *The functors $\mathcal{X}^{\mathrm{op}} \to \widehat{\mathrm{Cat}}_\infty$ defined by $U \mapsto \mathrm{Mod}_{\mathcal{O}_{X_U}}$ and $U \mapsto \mathrm{QCoh}(X_U)$ are limit preserving.*

*Proof.* See [170, 2.1.0.5] and [170, proof of 2.2.4.1]. $\qquad\qquad\square$

Thus, we may regard these constructions as defining sheaves of (presentable, stable, symmetric monoidal) $\infty$-categories on $\mathcal{X}$.

## Quasicoherent sheaves on quasiaffine spectral DM stacks

We have seen that $\mathrm{QCoh}(X) \approx \mathrm{Mod}_A$ if $X = \mathrm{Sp\acute{e}t}\,A$. This generalizes to $X$ which are *quasiaffine*.

A nonconnective spectral DM stack $X = (\mathcal{X}, \mathcal{O}_X)$ is **quasiaffine** if

1. the $\infty$-topos $\mathcal{X}$ is **quasicompact**, i.e., for any set $\{U_i\}$ of objects of $\mathcal{X}$ which is a cover, there is a finite subset $\{U_{i_k},\ k = 1,\dots,r\}$ which is a cover, and
2. it admits an *open immersion* into an affine, i.e., if there exists $A \in \mathrm{CAlg}$ and a $(-1)$-truncated object $U \in \mathrm{Shv}_A^{\text{ét}}$ such that $X \approx (\mathrm{Sp\acute{e}t}\,A)_U$.

**Theorem 8.18.5.** *If $X$ is quasiaffine, then taking global sections defines an equivalence of categories* $\mathrm{QCoh}(X) \xrightarrow{\sim} \mathrm{Mod}_A$ *where* $A = \Gamma(\mathcal{X}, \mathcal{O}_X)$.

*Proof.* See [170, 2.4]. □

**Example 8.18.6.** Here is an example which illustrates both the theorem and its proof. Let $R = \mathbb{S}[x,y]$, and $X = \mathbf{A}^2 = \mathrm{Sp\acute{e}t}\,R = (\mathrm{Shv}_R^{\text{ét}}, \mathcal{O})$. Define $U \in \mathrm{Shv}_R^{\text{ét}} \subseteq \mathrm{Fun}(\mathrm{CAlg}_R^{\text{ét}}, \mathcal{S})$ by

$$U(\mathbb{S}[x,y] \to B) := \begin{cases} * & \text{if } (x,y)\pi_0 B = \pi_0 B, \\ \varnothing & \text{if } (x,y)\pi_0 B \neq \pi_0 B. \end{cases}$$

Let $Y := X_U = \text{``}\mathbf{A}^2 \setminus \{0\}\text{''}$. Clearly $Y$ is quasiaffine.

We can write $U$ as a colimit in $\mathrm{Shv}_R^{\text{ét}}$ of a diagram $U_x \leftarrow U_{xy} \to U_y$, where $U_x, U_y \subseteq U$ are the subobjects which are "inhabited" exactly at those $\mathbb{S}[x,y] \to B$ such that $x \in (\pi_0 B)^\times$ or $y \in (\pi_0 B)^\times$ respectively, and $U_{xy} = U_x \times_U U_y$. There is an equivalence of commutative squares

$$
\begin{array}{ccc}
X_{U_{xy}} & \longrightarrow & X_{U_y} \\
\downarrow & & \downarrow \\
X_{U_x} & \longrightarrow & X_U
\end{array}
\quad \approx \quad
\begin{array}{ccc}
\mathrm{Sp\acute{e}t}\,\mathbb{S}[x^\pm, y^\pm] & \longrightarrow & \mathrm{Sp\acute{e}t}\,\mathbb{S}[x, y^\pm] \\
\downarrow & & \downarrow \\
\mathrm{Sp\acute{e}t}\,\mathbb{S}[x^\pm, y] & \longrightarrow & Y
\end{array}
$$

which are pushout squares in SpDM by (8.8). Taking quasicoherent sheaves, we obtain a commutative square of $\infty$-categories

$$
\begin{array}{ccc}
\mathrm{Mod}_{\mathbb{S}[x^\pm, y^\pm]} & \longleftarrow & \mathrm{Mod}_{\mathbb{S}[x, y^\pm]} \\
\uparrow & & \uparrow \\
\mathrm{Mod}_{\mathbb{S}[x^\pm, y]} & \longleftarrow & \mathrm{QCoh}(Y)
\end{array}
$$

which is a pullback by descent.

On the other hand, consider the ring of global sections

$$\Gamma := \Gamma(\mathcal{X}_{/U}, \mathcal{O}_X|_U) \approx \lim\big(\mathbb{S}[x^\pm, y] \to \mathbb{S}[x^\pm, y^\pm] \leftarrow \mathbb{S}[x, y^\pm]\big).$$

We have a commutative diagram

$$
\begin{array}{ccc}
\mathrm{Mod}_{\mathbb{S}[x^{\pm},y^{\pm}]} & \longleftarrow & \mathrm{Mod}_{\mathbb{S}[x,y^{\pm}]} \\
\uparrow & & \uparrow \\
\mathrm{Mod}_{\mathbb{S}[x^{\pm},y]} & \longleftarrow & \mathrm{Mod}_{\Gamma}
\end{array}
$$

which is also seen to be a pullback of $\infty$-categories. The equivalence

$$
\mathrm{Mod}_{\mathbb{S}[x^{\pm},y]} \times_{\mathrm{Mod}_{\mathbb{S}[x^{\pm},y^{\pm}]}} \mathrm{Mod}_{\mathbb{S}[x,y^{\pm}]} \to \mathrm{Mod}_{\Gamma}
$$

is realized by a functor which sends "descent data"

$$
\left( M_x \in \mathrm{Mod}_{\mathbb{S}[x^{\pm},y]},\ M_y \in \mathrm{Mod}_{\mathbb{S}[x,y^{\pm}]},\ \psi \colon M_x[y^{-1}] \xrightarrow{\sim} M_y[x^{-1}] \in \mathrm{Mod}_{\mathbb{S}[x^{\pm},y^{\pm}]} \right)
$$

to the limit $\lim(M_x \to M_x[y^{-1}] \xrightarrow{\sim} M_y[x^{-1}] \leftarrow M_y)$ in $\mathrm{Mod}_{\Gamma}$, while the inverse equivalence sends $N \in \mathrm{Mod}_{\Gamma}$ to $(\mathbb{S}[x^{\pm},y] \otimes_{\Gamma} N,\ \mathbb{S}[x,y^{\pm}] \otimes_{\Gamma} N,\ \mathrm{id})$. The key observation for proving the equivalence is that both these functors preserve arbitrary colimits and finite limits, and are easy to evaluate on the "generating" objects $\Gamma \in \mathrm{Mod}_{\Gamma}$ and $(\mathbb{S}[x^{\pm},y], \mathbb{S}[x,y^{\pm}], \mathrm{id})$ in the limit.

## 8.19    Elliptic cohomology and topological modular forms

I return to our motivating example of elliptic cohomology.

First, let us consider the moduli stack of (smooth) elliptic curves. This is an example of a "classical" Deligne–Mumford stack. However, according to (8.12) we can regard classical Deligne–Mumford stacks as a particular type of 0-truncated spectral DM stack, and since that is the language I have introduced in this paper, that is how I will generally talk about it.

### The moduli stack of elliptic curves

The moduli stack of elliptic curves is a (classical) DM stack $\mathcal{M}_{\mathrm{Ell}} = (\mathcal{X}_{\mathrm{Ell}}, \mathcal{O})$ such that, for ordinary ring $A \in \mathrm{CAlg}^{\heartsuit}$, we have

$$
\mathrm{Map}_{\mathrm{SpDM}}(\mathrm{Spét}\, A, \mathcal{M}_{\mathrm{Ell}}) \approx \{\text{elliptic curves over } \mathrm{Spét}\, A\}. \tag{8.19.1}
$$

The right-hand side of (8.19.1) represents the 1-groupoid of elliptic curves over $\mathrm{Spét}\, A$ and isomorphisms between them. (Note that an isomorphism of elliptic curves is necessarily compatible with the distinguished sections $e$; we usually omit $e$ from the notation.)

*Remark 8.19.1.*   Here "elliptic curve" means a classical smooth elliptic curve, i.e., a proper and smooth morphism $\pi \colon C \to \mathrm{Spét}\, A$ of schemes (i.e., of schematic DM stacks) whose geometric fibers are curves of genus 1, and which is equipped (as part of the data), with a section $e \colon \mathrm{Spét}\, A \to C$ of $\pi$.

I will not review the theory of elliptic curves here. However, we should note that every elliptic curve is an *abelian group scheme*; i.e., an elliptic curve $C \to \mathrm{Sp\acute{e}t}\, A$ is an abelian group object in the category of schemes over $A$. Furthermore, as it is 1-dimensional and smooth, the formal completion $C_e^\wedge$ at the identity section exists, and is an example of a 1-dimensional formal group over $A$.

That there exists such an object $\mathcal{M}_{\mathrm{Ell}}$ is a theorem, which we will take as given.

*Remark 8.19.2* (The étale site of $\mathcal{M}_{\mathrm{Ell}}$). As a DM stack, and hence as a spectral DM stack, $\mathcal{M}_{\mathrm{Ell}}$ is the colimit of a diagram whose objects are étale morphisms $\mathrm{Sp\acute{e}t}\, A \to \mathcal{M}_{\mathrm{Ell}}$, and since $\mathcal{M}_{\mathrm{Ell}}$ is 0-truncated the rings $A$ which appear in this diagram will be ordinary rings. Thus, $\mathcal{M}_{\mathrm{Ell}}$ can be reconstructed from the "étale site" of $\mathcal{M}_{\mathrm{Ell}}$, i.e., the category $\mathcal{U}$ whose objects are elliptic curves $C \to \mathrm{Sp\acute{e}t}\, A$ represented by an étale map $\mathrm{Sp\acute{e}t}\, A \to \mathcal{M}_{\mathrm{Ell}}$, and whose morphisms are commutative squares

$$
\begin{array}{ccc}
C & \longrightarrow & C' \\
\downarrow & & \downarrow \\
\mathrm{Sp\acute{e}t}\, A & \longrightarrow & \mathrm{Sp\acute{e}t}\, A'
\end{array}
$$

such that $C \to C' \times_{\mathrm{Sp\acute{e}t}\, A'} \mathrm{Sp\acute{e}t}\, A$ is an isomorphism of elliptic curves over $\mathrm{Sp\acute{e}t}\, A$.

It remains to characterize the objects of $\mathcal{U}$. Given elliptic curves $C \to S$ and $C' \to S'$, consider the functor

$$
T \mapsto \mathrm{Iso}_{C/S,C'/S'}(T) := \left\{ (f\colon T \to S, \ f'\colon T \to S', \ \alpha\colon f^*C \xrightarrow{\sim} f'^*C') \right\}
$$

which sends a scheme $T$ to the set of tuples consisting of maps of schemes $f$ and $f'$, and a choice of isomorphism $\alpha$ of elliptic curves over $T$. It turns out that this functor is itself representable by a *scheme* $I_{C/S,C'/S'}$:

$$
\mathrm{Iso}_{C/S,C'/S'}(T) \approx \mathrm{Map}_{\mathrm{Sch}}(\mathrm{Spec}\, T, I_{C/S,C'/S'}).
$$

An elliptic curve $C \to S$ is represented by an étale morphism $S \to \mathcal{M}_{\mathrm{Ell}}$ if and only if for every elliptic curve $C' \to S'$ the evident map $I_{C/S,C'/S'} \to S$ of schemes is étale.

See [144] for much more on the moduli stack of elliptic curves (although the word "stack" is rarely used there).

## The theorem of Goerss–Hopkins–Miller

Let $\mathcal{U}$ denote the étale site of $\mathcal{M}_{\mathrm{Ell}}$ as in (8.19.2). We note the functor

$$
\mathcal{O}\colon \mathcal{U}^{\mathrm{op}} \to \mathrm{CAlg}^\heartsuit.
$$

defined by $(C \to \mathrm{Sp\acute{e}t}\, A) \mapsto A$. Also recall that for each object $(C \to \mathrm{Sp\acute{e}t}\, A) \in \mathcal{U}$ we have 1-dimensional formal group law over $A$.

Question 8.19.3. Does there exist a functor $\mathcal{O}^{\mathrm{top}}\colon \mathcal{U}^{\mathrm{op}} \to \mathrm{CAlg}$ $\mathcal{O}^{\mathrm{top}}\colon \mathrm{CAlg}$ sitting in a commutative diagram

$$\begin{array}{ccc} & & \mathrm{CAlg} \\ & \nearrow^{\mathcal{O}^{\mathrm{top}}} & \downarrow \pi_0 \\ \mathcal{U}^{\mathrm{op}} & \xrightarrow[\mathcal{O}]{} & \mathrm{CAlg}^{\heartsuit} \end{array}$$ such that

1. for each object $C \to \mathrm{Sp\acute{e}t}\,A$, the corresponding ring $R = \mathcal{O}^{\mathrm{top}}(C \to \mathrm{Sp\acute{e}t}\,A)$ is weakly 2-periodic and has homotopy concentrated in even degrees, and hence is complex orientable; and

2. is equipped with natural isomorphisms $\mathrm{Sp\acute{e}t}(R^0(\mathbb{CP}^\infty)) \approx C_e^\wedge$ of formal groups between the formal groups of $R = \mathcal{O}^{\mathrm{top}}(C \to \mathrm{Sp\acute{e}t}\,A)$ and the formal completions $C_e^\wedge$ of elliptic curves

*Remark 8.19.4.* The formal groups $C_e^\wedge$ of elliptic curves $C \to \mathrm{Sp\acute{e}t}\,A$ in the étale site $\mathcal{U}$ satisfy the Landweber condition (see [74, Ch. 4]), and thus for each such curve there we can certainly construct a homotopy-commutative ring spectrum $R$ satisfying conditions (1) and (2). The point of the theorem is to rigidify this construction to an honest functor of $\infty$-categories, and while doing so lift it to a functor to structured commutative rings.

**Theorem 8.19.5** (Goerss–Hopkins–Miller). *The answer to (8.19.3) is yes. Furthermore, the resulting functor $\mathcal{O}^{\mathrm{top}}$ defines a sheaf of $\mathbb{E}_\infty$-rings on the étale site of $\mathcal{M}_{\mathrm{Ell}}$.*

The pair $(\mathcal{X}_{\mathrm{Ell}}, \mathcal{O}^{\mathrm{top}})$ is an example of a nonconnective spectral DM stack, whose 0-truncation is the classical DM stack $\mathcal{M}_{\mathrm{Ell}}$. (That this is the case is because $\pi_0\mathcal{O}^{\mathrm{top}} \approx \mathcal{O}$, the structure sheaf on $\mathcal{M}_{\mathrm{Ell}}$.)

Given (8.19.5), we can now define

$$\mathrm{TMF} := \Gamma(\mathcal{X}_{\mathrm{Ell}}, \mathcal{O}^{\mathrm{top}}) \approx \lim_{(C \to \mathrm{Sp\acute{e}t}\,A) \in \mathcal{U}} \mathcal{O}^{\mathrm{top}}(C \to \mathrm{Sp\acute{e}t}\,A),$$

the periodic $\mathbb{E}_\infty$-ring of **topological modular forms**.

*Remark 8.19.6.* See [74] for more on (8.19.5), including details about the original proof, as well as more information on TMF.

## 8.20    The classifying stack for oriented elliptic curves

It turns out that the nonconnective spectral Deligne–Mumford stack $(\mathcal{X}_{\mathrm{Ell}}, \mathcal{O}^{\mathrm{top}})$ admits a modular interpretation in spectral algebraic geometry: it is the classifying object for *oriented elliptic curves.*

### Elliptic curves in spectral geometry

A **variety** over an $\mathbb{E}_\infty$-ring $R$ is a flat morphism $X \to \mathrm{Sp\acute{e}t}\,R$ of nonconnective spectral DM stacks, such that the induced map $\tau_{\geq 0}X \to \mathrm{Sp\acute{e}t}\,\tau_{\geq 0}R$ of spectral DM stacks is: proper, locally almost of finite presentation, geometrically reduced, and geometrically connected [166, 1.1], [170, 19.4.5].

*Remark 8.20.1.* We have not and will not describe all the adjectives in the above definition. See [170, 5.1] for proper, [170, 4.2] for locally almost of finite presentation, [170, 8.6] for geometrically reduced and geometrically connected.

An **abelian variety** over an $\mathbb{E}_\infty$-ring $R$ is a variety $X$ over $R$ which is a commutative monoid object in $\mathrm{SpDM}_R^{\mathrm{nc}}$. It is an **elliptic curve** if it is of dimension 1.

*Remark 8.20.2.* "Commutative monoid object" is here taken in the sense of [168, 2.4.2]. In this case, it means that an abelian variety $X$ over $R$ represents a functor on $\mathrm{SpDM}_R^{\mathrm{nc}}$ which takes values in $\mathbb{E}_\infty$-spaces. In fact, one can show that every abelian variety in this sense is "grouplike", i.e., it actually represents a functor to grouplike $\mathbb{E}_\infty$-spaces [166, 1.4.4].

A **strict** abelian variety or elliptic curve is one in which the commutative monoid structure is equipped with a refinement to an abelian group structure; i.e., $X$ represents a functor to $\mathrm{Mod}_{\mathbb{Z}}^{\mathrm{cn}}$ (8.8.4).

*Remark 8.20.3.* Over an ordinary ring $R$, either notion of abelian variety reduces to the classical notion. In either case, the commutative monoid/abelian group structure coincides with the unique abelian group structure which exists on a classical abelian variety.

In the classical case, the underlying variety of an abelian variety admits a unique group structure compatible with a given identity section. In the spectral setting, this is no longer the case, and a group structure of some sort needs to be imposed.

There are $\infty$-categories $\mathrm{AbVar}(R)$ and $\mathrm{AbVar}^s(R)$ of abelian varieties and strict abelian varieties; morphisms are maps of nonconnective spectral DM stacks over $R$ which preserve the commutative monoid structure or abelian group structure as the case may be. We are going to be interested in $\mathrm{Ell}^s(R) \subseteq \mathrm{AbVar}^s(R)$, the full subcategory of strict elliptic curves.

*Remark 8.20.4.* Since abelian varieties over $R$ are in particular flat morphisms, we see that $\mathrm{AbVar}(R) \approx \mathrm{AbVar}(\tau_{\geq 0}R)$ and $\mathrm{AbVar}^s(R) \approx \mathrm{AbVar}^s(\tau_{\geq 0}R)$ by (8.13.6).

There is a moduli stack of strict elliptic curves.

**Theorem 8.20.5 (Lurie).** *There exists a spectral DM stack $\mathcal{M}_{\mathrm{Ell}}^s$ such that*

$$\mathrm{Map}_{\mathrm{SpDM}^{\mathrm{nc}}}(\mathrm{Sp\acute{e}t}\, R, \mathcal{M}_{\mathrm{Ell}}^s) \approx \mathrm{Ell}^s(R)^\simeq;$$

*the right-hand side is the maximal $\infty$-groupoid inside $\mathrm{Ell}^s(R)$. The underlying 0-truncated spectral DM stack of $\mathcal{M}_{\mathrm{Ell}}^s$ is equivalent to the classical moduli stack $\mathcal{M}_{\mathrm{Ell}}$.*

This is proved in [166, 2], using the spectral version of the Artin Representability Theorem [170, 18.3]. That $\mathcal{M}_{\mathrm{Ell}}^s$ is a connective object (i.e., not nonconnective) is immediate from the fact that $\mathrm{Ell}^s(R) \approx \mathrm{Ell}^s(\tau_{\geq 0}R)$. That the underlying 0-truncated stack of $\mathcal{M}_{\mathrm{Ell}}^s$ is the classical one is a consequence of the fact that strict elliptic curves over ordinary rings are just classical elliptic curves.

## Oriented elliptic curves

For any strict elliptic curve $C \to \operatorname{Sp\acute{e}t} R$, we may consider the formal completion $C_e^{\wedge}$ along the identity section. It turns out that $C_e^{\wedge}$ is a 1-dimensional formal group over $R$ [167, 7.1].

Thus, we define an **oriented** elliptic curve over $R$ to be a pair $(C, e)$ consisting of a strict elliptic curve $C \to \operatorname{Sp\acute{e}t} R$ together with an orientation $e \in \operatorname{OrDat}(\widehat{C_e})$ of its formal completion $\widehat{C}$ in the sense of (8.17). There is a corresponding $\infty$-category $\operatorname{Ell}^{\operatorname{or}}(R)$ of oriented elliptic curves: morphisms must preserve the orientation.

**Theorem 8.20.6 (Lurie).** *There exists a nonconnective spectral DM stack $\mathcal{M}_{\operatorname{Ell}}^{\operatorname{or}}$ such that*

$$\operatorname{Map}_{\operatorname{SpDM}^{\operatorname{nc}}}(\operatorname{Sp\acute{e}t} R, \mathcal{M}_{\operatorname{Ell}}^{\operatorname{or}}) \approx \operatorname{Ell}^{\operatorname{or}}(R)^{\simeq}.$$

*The map $\mathcal{M}_{\operatorname{Ell}}^{\operatorname{or}} \to \mathcal{M}_{\operatorname{Ell}}^{\operatorname{s}}$ classifying the strict elliptic curve induces an equivalence of underlying classical DM stacks.*

This is proved in [167, 7].

*Remark 8.20.7.* Taken together, we have maps of nonconnective spectral DM stacks

$$\mathcal{M}_{\operatorname{Ell}} \xrightarrow{i} \mathcal{M}_{\operatorname{Ell}}^{\operatorname{s}} \xleftarrow{p} \mathcal{M}_{\operatorname{Ell}}^{\operatorname{or}}$$

in which is $\mathcal{M}_{\operatorname{Ell}}^{\operatorname{s}}$ is a spectral DM stack (i.e., is connective), and $\mathcal{M}_{\operatorname{Ell}}$ is a 0-truncated spectral DM stack (and in fact is a DM stack). The map $i$ witnesses the fact that every classical elliptic curve is a strict elliptic curve, while the map $p$ forgets about orientation. All of these objects have the same underlying DM stack (i.e., they have equivalent $\infty$-topoi and $\pi_0$ of their structure sheaves coincide); in the case of $\mathcal{M}_{\operatorname{Ell}}^{\operatorname{or}}$ this is a non-trivial observation.

*Remark 8.20.8.* Note that if $\operatorname{Sp\acute{e}t} A \to \mathcal{M}_{\operatorname{Ell}}^{\operatorname{or}}$ is any map of nonconnective spectral DM stacks, then the theorem produces an oriented elliptic curve over $A$, and hence an oriented formal group over $A$. Thus (8.17.9) implies that the $\mathbb{E}_{\infty}$-ring $A$ must be weakly 2-periodic and complex orientable.

In fact, the proof of the theorem shows a little more in the case that $\operatorname{Sp\acute{e}t} A \to \mathcal{M}_{\operatorname{Ell}}^{\operatorname{or}}$ is étale. In this case, $A$ is not merely weakly 2-periodic; it also has the property that $\pi_{\operatorname{odd}}(A) \approx 0$.

As the underlying classical DM stack of $\mathcal{M}_{\operatorname{Ell}}^{\operatorname{or}}$ is $\mathcal{M}_{\operatorname{Ell}}$, we have that the full subcategory $\mathcal{U}' \subseteq \operatorname{SpDM}_{/\mathcal{M}_{\operatorname{Ell}}^{\operatorname{or}}}^{\operatorname{nc}}$ spanned by étale morphisms $\operatorname{Sp\acute{e}t} A \to \mathcal{M}_{\operatorname{Ell}}^{\operatorname{or}}$ is equivalent to the étale site of the classical stack $\mathcal{M}_{\operatorname{Ell}}$, which we called $\mathcal{U}$ in (8.19.2). Putting all this together, we see that we have functors

$$\mathcal{U}^{\operatorname{op}} \xleftarrow{\sim} \mathcal{U}'^{\operatorname{op}} \to \operatorname{CAlg}$$

given by

$$(\operatorname{Sp\acute{e}t} \pi_0 A \to \mathcal{M}_{\operatorname{Ell}}) \leftmapsto (\operatorname{Sp\acute{e}t} A \to \mathcal{M}_{\operatorname{Ell}}^{\operatorname{or}}) \mapsto A.$$

We see that the resulting functor $\mathcal{U}^{\operatorname{op}} \to \operatorname{CAlg}$ is precisely of the sort demanded by (8.19.3).

# Bibliography

[1]   J. F. Adams. "Lectures on generalised cohomology". In: *Category Theory, Homology Theory and their Applications, III (Battelle Institute Conference, Seattle, Wash., 1968, Vol. Three)*. Springer, Berlin, 1969, pp. 1–138.

[2]   J. F. Adams. *Stable homotopy and generalised homology*. Chicago Lectures in Mathematics. Reprint of the 1974 edition. University of Chicago Press, Chicago, IL, 1995, pp. x+373. isbn: 0-226-00524-0.

[3]   D. W. Anderson. "Convergent functors and spectra". In: *Localization in group theory and homotopy theory, and related topics (Sympos., Battelle Seattle Res. Center, Seattle, Wash., 1974)*. 1974, 1–5. Lecture Notes in Math., Vol. 418.

[4]   M. Ando, M. J. Hopkins, and N. P. Strickland. "Elliptic spectra, the Witten genus and the theorem of the cube". In: *Invent. Math.* 146.3 (2001), pp. 595–687. issn: 0020-9910.

[5]   Matthew Ando, Andrew J. Blumberg, David Gepner, Michael J. Hopkins, and Charles Rezk. "An $\infty$-categorical approach to $R$-line bundles, $R$-module Thom spectra, and twisted $R$-homology". In: *J. Topol.* 7.3 (2014), pp. 869–893. issn: 1753-8416.

[6]   Matthew Ando, Andrew J. Blumberg, David Gepner, Michael J. Hopkins, and Charles Rezk. "Units of ring spectra, orientations and Thom spectra via rigid infinite loop space theory". In: *J. Topol.* 7.4 (2014), pp. 1077–1117. issn: 1753-8416.

[7]   Matthew Ando, Michael J. Hopkins, and Charles Rezk. *Multiplicative orientations of $KO$-theory and the spectrum of topological modular forms*. url: http://www.math.uiuc.edu/~rezk/papers.html.

[8]   Vigleik Angeltveit. "Topological Hochschild homology and cohomology of $A_\infty$ ring spectra". In: *Geom. Topol.* 12.2 (2008), pp. 987–1032. issn: 1465-3060.

[9]   Vigleik Angeltveit, Michael A. Hill, and Tyler Lawson. "Topological Hochschild homology of $\ell$ and $ko$". In: *Amer. J. Math.* 132.2 (2010), pp. 297–330. issn: 0002-9327.

[10]  Benjamin Antieau and David Gepner. "Brauer groups and étale cohomology in derived algebraic geometry". In: *Geom. Topol.* 18.2 (2014), pp. 1149–1244. issn: 1465-3060.

[11]  Omar Antolín-Camarena and Tobias Barthel. "A simple universal property of Thom ring spectra". In: *J. Topol.* 12.1 (2019), pp. 56–78. issn: 1753-8416.

[12]  Maurice Auslander and Oscar Goldman. "The Brauer group of a commutative ring". In: *Trans. Amer. Math. Soc.* 97 (1960), pp. 367–409. issn: 0002-9947.

[13]    Christian Ausoni. "Topological Hochschild homology of connective complex $K$-theory". In: *Amer. J. Math.* 127.6 (2005), pp. 1261–1313. issn: 0002-9327.

[14]    Gorô Azumaya. "On maximally central algebras". In: *Nagoya Math. J.* 2 (1951), pp. 119–150. issn: 0027-7630.

[15]    Andrew Baker and Andrej Lazarev. "On the Adams spectral sequence for $R$-modules". In: *Algebr. Geom. Topol.* 1 (2001), pp. 173–199. issn: 1472-2747.

[16]    Andrew Baker and Andrey Lazarev. "Topological Hochschild cohomology and generalized Morita equivalence". In: *Algebr. Geom. Topol.* 4 (2004), pp. 623–645. issn: 1472-2747.

[17]    Andrew Baker and Birgit Richter. "Invertible modules for commutative $S$-algebras with residue fields". In: *Manuscripta Math.* 118.1 (2005), pp. 99–119. issn: 0025-2611.

[18]    Andrew Baker and Birgit Richter. "On the $\Gamma$-cohomology of rings of numerical polynomials and $E_\infty$ structures on $K$-theory". In: *Comment. Math. Helv.* 80.4 (2005), pp. 691–723. issn: 0010-2571.

[19]    Andrew Baker and Birgit Richter. "Uniqueness of $E_\infty$ structures for connective covers". In: *Proc. Amer. Math. Soc.* 136.2 (2008), pp. 707–714. issn: 0002-9939.

[20]    Andrew Baker, Birgit Richter, and Markus Szymik. "Brauer groups for commutative $S$-algebras". In: *J. Pure Appl. Algebra* 216.11 (2012), pp. 2361–2376. issn: 0022-4049.

[21]    David Barnes and Constanze Roitzheim. "Homological localisation of model categories". In: *Appl. Categ. Structures* 23.3 (2015), pp. 487–505. issn: 0927-2852.

[22]    David Barnes and Constanze Roitzheim. "Stable left and right Bousfield localisations". In: *Glasg. Math. J.* 56.1 (2014), pp. 13–42. issn: 0017-0895.

[23]    C. Barwick and D. M. Kan. "Relative categories: another model for the homotopy theory of homotopy theories". In: *Indag. Math. (N.S.)* 23.1-2 (2012), pp. 42–68. issn: 0019-3577.

[24]    Clark Barwick. "On exact $\infty$-categories and the theorem of the heart". In: *Compos. Math.* 151.11 (2015), pp. 2160–2186. issn: 0010-437X.

[25]    Clark Barwick. "On left and right model categories and left and right Bousfield localizations". In: *Homology Homotopy Appl.* 12.2 (2010), pp. 245–320. issn: 1532-0073.

[26]    Clark Barwick and Christopher Schommer-Pries. "On the unicity of the theory of higher categories". In: *J. Amer. Math. Soc.* 34.4 (2021), pp. 1011–1058. issn: 0894-0347.

[27]    M. Basterra. "André-Quillen cohomology of commutative $S$-algebras". In: *J. Pure Appl. Algebra* 144.2 (1999), pp. 111–143. issn: 0022-4049.

[28]    Maria Basterra and Michael A. Mandell. "Homology and cohomology of $E_\infty$ ring spectra". In: *Math. Z.* 249.4 (2005), pp. 903–944. issn: 0025-5874.

[29]    Maria Basterra and Michael A. Mandell. "The multiplication on BP". In: *J. Topol.* 6.2 (2013), pp. 285–310. issn: 1753-8416.

[30]    Maria Basterra and Birgit Richter. "(Co-)homology theories for commutative $(S$-)algebras". In: *Structured ring spectra.* Vol. 315. London Math. Soc. Lecture Note Ser. Cambridge Univ. Press, Cambridge, 2004, pp. 115–131.

[31]  Samik Basu, Steffen Sagave, and Christian Schlichtkrull. "Generalized Thom spectra and their topological Hochschild homology". In: *J. Inst. Math. Jussieu* 19.1 (2020), pp. 21–64. issn: 1474-7480.

[32]  Gilbert Baumslag. "Some aspects of groups with unique roots". In: *Acta Math.* 104 (1960), pp. 217–303. issn: 0001-5962.

[33]  Haldun Özgür Bayındır. "Topological equivalences of *E*-infinity differential graded algebras". In: *Algebr. Geom. Topol.* 18.2 (2018), pp. 1115–1146. issn: 1472-2747.

[34]  A. A. Beilinson, J. Bernstein, and P. Deligne. "Faisceaux pervers". In: *Analysis and topology on singular spaces, I (Luminy, 1981)*. Vol. 100. Astérisque. Soc. Math. France, Paris, 1982, pp. 5–171.

[35]  Tibor Beke. "Sheafifiable homotopy model categories". In: *Math. Proc. Cambridge Philos. Soc.* 129.3 (2000), pp. 447–475. issn: 0305-0041.

[36]  Clemens Berger. "Combinatorial models for real configuration spaces and $E_n$-operads". In: *Operads: Proceedings of Renaissance Conferences (Hartford, CT/Luminy, 1995)*. Vol. 202. Contemp. Math. Amer. Math. Soc., Providence, RI, 1997, pp. 37–52.

[37]  Clemens Berger and Benoit Fresse. "Combinatorial operad actions on cochains". In: *Math. Proc. Cambridge Philos. Soc.* 137.1 (2004), pp. 135–174. issn: 0305-0041.

[38]  Clemens Berger and Ieke Moerdijk. "On an extension of the notion of Reedy category". In: *Math. Z.* 269.3-4 (2011), pp. 977–1004. issn: 0025-5874.

[39]  Clemens Berger and Ieke Moerdijk. "On the homotopy theory of enriched categories". In: *Q. J. Math.* 64.3 (2013), pp. 805–846. issn: 0033-5606.

[40]  Julia E. Bergner. "A characterization of fibrant Segal categories". In: *Proc. Amer. Math. Soc.* 135.12 (2007), pp. 4031–4037. issn: 0002-9939.

[41]  Julia E. Bergner. "A model category structure on the category of simplicial categories". In: *Trans. Amer. Math. Soc.* 359.5 (2007), pp. 2043–2058. issn: 0002-9947.

[42]  Julia E. Bergner. "Three models for the homotopy theory of homotopy theories". In: *Topology* 46.4 (2007), pp. 397–436. issn: 0040-9383.

[43]  Andrew J. Blumberg. "Continuous functors as a model for the equivariant stable homotopy category". In: *Algebr. Geom. Topol.* 6 (2006), pp. 2257–2295. issn: 1472-2747.

[44]  Andrew J. Blumberg. *Progress towards the calculation of the K-theory of Thom spectra*. Thesis (Ph.D.)–The University of Chicago. ProQuest LLC, Ann Arbor, MI, 2005, p. 121. isbn: 978-0542-04081-8.

[45]  Andrew J. Blumberg, Ralph L. Cohen, and Christian Schlichtkrull. "Topological Hochschild homology of Thom spectra and the free loop space". In: *Geom. Topol.* 14.2 (2010), pp. 1165–1242. issn: 1465-3060.

[46]  Andrew J Blumberg, David Gepner, and Gonçalo Tabuada. "A universal characterization of higher algebraic K-theory". In: *Geometry & Topology* 17.2 (2013), pp. 733–838. doi: 10.2140/gt.2013.17.733.

[47]  Andrew J. Blumberg and Michael A. Mandell. "The strong Künneth theorem for topological periodic cyclic homology". arXiv:1706.06846. 2017.

[48]  J. M. Boardman and R. M. Vogt. *Homotopy invariant algebraic structures on topological spaces*. Lecture Notes in Mathematics, Vol. 347. Springer-Verlag, Berlin-New York, 1973, pp. x+257.

[49]  J. M. Boardman and R. M. Vogt. "Homotopy-everything $H$-spaces". In: *Bull. Amer. Math. Soc.* 74 (1968), pp. 1117–1122. issn: 0002-9904.

[50]  M. Bökstedt and I. Madsen. "Topological cyclic homology of the integers". In: 226. $K$-theory (Strasbourg, 1992). 1994, pp. 7–8, 57–143.

[51]  Marcel Bökstedt. *The topological Hochschild homology of $\mathbb{Z}$ and of $\mathbb{Z}/p\mathbb{Z}$*. preprint.

[52]  Marcel Bökstedt. *Topological Hochschild homology*. preprint.

[53]  Francis Borceux. *Handbook of categorical algebra. 2*. Vol. 51. Encyclopedia of Mathematics and its Applications. Categories and structures. Cambridge University Press, Cambridge, 1994, pp. xviii+443. isbn: 0-521-44179-X.

[54]  A. K. Bousfield. "The localization of spaces with respect to homology". In: *Topology* 14 (1975), pp. 133–150. issn: 0040-9383.

[55]  A. K. Bousfield. "The localization of spectra with respect to homology". In: *Topology* 18.4 (1979), pp. 257–281. issn: 0040-9383.

[56]  A. K. Bousfield and E. M. Friedlander. "Homotopy theory of $\Gamma$-spaces, spectra, and bisimplicial sets". In: *Geometric applications of homotopy theory (Proc. Conf., Evanston, Ill., 1977), II*. Vol. 658. Lecture Notes in Math. Springer, Berlin, 1978, pp. 80–130.

[57]  A. K. Bousfield and V. K. A. M. Gugenheim. "On PL de Rham theory and rational homotopy type". In: *Mem. Amer. Math. Soc.* 8.179 (1976), pp. ix+94. issn: 0065-9266.

[58]  A. K. Bousfield and D. M. Kan. *Homotopy limits, completions and localizations*. Lecture Notes in Mathematics, Vol. 304. Springer-Verlag, Berlin-New York, 1972, pp. v+348.

[59]  C. Braun, J. Chuang, and A. Lazarev. "Derived localisation of algebras and modules". In: *Adv. Math.* 328 (2018), pp. 555–622. issn: 0001-8708.

[60]  Edgar H. Brown Jr. and Franklin P. Peterson. "A spectrum whose $Z_p$ cohomology is the algebra of reduced $p^{th}$ powers". In: *Topology* 5 (1966), pp. 149–154. issn: 0040-9383.

[61]  Kenneth S. Brown. "Abstract homotopy theory and generalized sheaf cohomology". In: *Trans. Amer. Math. Soc.* 186 (1973), pp. 419–458. issn: 0002-9947.

[62]  Morten Brun, Zbigniew Fiedorowicz, and Rainer M. Vogt. "On the multiplicative structure of topological Hochschild homology". In: *Algebr. Geom. Topol.* 7 (2007), pp. 1633–1650. issn: 1472-2747.

[63]  R. R. Bruner, J. P. May, J. E. McClure, and M. Steinberger. $H_\infty$ *ring spectra and their applications*. Vol. 1176. Lecture Notes in Mathematics. Springer-Verlag, Berlin, 1986, pp. viii+388. isbn: 3-540-16434-0.

[64]  Gunnar Carlsson. "Derived completions in stable homotopy theory". In: *J. Pure Appl. Algebra* 212.3 (2008), pp. 550–577. issn: 0022-4049.

[65]   Carles Casacuberta. "Anderson localization from a modern point of view". In:
       *The Cech centennial (Boston, MA, 1993)*. Vol. 181. Contemp. Math. Amer. Math.
       Soc., Providence, RI, 1995, pp. 35–44.

[66]   Carles Casacuberta, Javier J. Gutiérrez, Ieke Moerdijk, and Rainer M. Vogt.
       "Localization of algebras over coloured operads". In: *Proc. Lond. Math. Soc. (3)*
       101.1 (2010), pp. 105–136. issn: 0024-6115.

[67]   Carles Casacuberta and Georg Peschke. "Localizing with respect to self-maps of
       the circle". In: *Trans. Amer. Math. Soc.* 339.1 (1993), pp. 117–140. issn: 0002-9947.

[68]   Steven Greg Chadwick and Michael A. Mandell. "$E_n$ genera". In: *Geom. Topol.*
       19.6 (2015), pp. 3193–3232. issn: 1465-3060.

[69]   Dustin Clausen, Akhil Mathew, Niko Naumann, and Justin Noel. "Descent in
       algebraic $K$-theory and a conjecture of Ausoni–Rognes". In: *J. Eur. Math. Soc.
       (JEMS)* 22.4 (2020), pp. 1149–1200. issn: 1435-9855.

[70]   Fred Cohen. "Homology of $\Omega^{(n+1)}\Sigma^{(n+1)}X$ and $C_{(n+1)}X$, $n > 0$". In: *Bull. Amer.
       Math. Soc.* 79 (1973), 1236–1241 (1974). issn: 0002-9904.

[71]   Frederick R. Cohen, Thomas J. Lada, and J. Peter May. *The homology of iterated
       loop spaces.* Lecture Notes in Mathematics, Vol. 533. Springer-Verlag, Berlin-New
       York, 1976, pp. vii+490.

[72]   Jean-Marc Cordier and Timothy Porter. "Vogt's theorem on categories of
       homotopy coherent diagrams". In: *Math. Proc. Cambridge Philos. Soc.* 100.1
       (1986), pp. 65–90. issn: 0305-0041.

[73]   Ethan S. Devinatz, Michael J. Hopkins, and Jeffrey H. Smith. "Nilpotence and
       stable homotopy theory. I". In: *Ann. of Math. (2)* 128.2 (1988), pp. 207–241. issn:
       0003-486X.

[74]   Christopher L. Douglas, John Francis, André G. Henriques, and Michael A. Hill,
       eds. *Topological modular forms.* Vol. 201. Mathematical Surveys and Monographs.
       American Mathematical Society, Providence, RI, 2014, pp. xxxii+318. isbn: 978-
       1-4704-1884-7.

[75]   Daniel Dugger. *A primer on homotopy colimits.* 2017. url: http://pages.
       uoregon.edu/ddugger/hocolim.pdf.

[76]   Daniel Dugger. "Coherence for invertible objects and multigraded homotopy
       rings". In: *Algebr. Geom. Topol.* 14.2 (2014), pp. 1055–1106. issn: 1472-2747.

[77]   Daniel Dugger. "Combinatorial model categories have presentations". In: *Adv.
       Math.* 164.1 (2001), pp. 177–201. issn: 0001-8708.

[78]   Daniel Dugger. "Replacing model categories with simplicial ones". In: *Trans.
       Amer. Math. Soc.* 353.12 (2001), pp. 5003–5027. issn: 0002-9947.

[79]   Daniel Dugger. "Universal homotopy theories". In: *Adv. Math.* 164.1 (2001),
       pp. 144–176. issn: 0001-8708.

[80]   Daniel Dugger and Brooke Shipley. "Topological equivalences for differential
       graded algebras". In: *Adv. Math.* 212.1 (2007), pp. 37–61. issn: 0001-8708.

[81]   Daniel Dugger and David I. Spivak. "Mapping spaces in quasi-categories". In:
       *Algebr. Geom. Topol.* 11.1 (2011), pp. 263–325. issn: 1472-2747.

[82]  Bjørn Ian Dundas, Ayelet Lindenstrauss, and Birgit Richter. "On higher topological Hochschild homology of rings of integers". In: *Math. Res. Lett.* 25.2 (2018), pp. 489–507. issn: 1073-2780.

[83]  Bjørn ian Dundas, Ayelet Lindenstrauss, and Birgit Richter. "Towards an understanding of ramified extensions of structured ring spectra". In: *Math. Proc. Cambridge Philos. Soc.* 168.3 (2020), pp. 435–454. issn: 0305-0041.

[84]  Bjørn Ian Dundas, Oliver Röndigs, and Paul Arne Østvær. "Enriched functors and stable homotopy theory". In: *Doc. Math.* 8 (2003), pp. 409–488. issn: 1431-0635.

[85]  Gerald Dunn. "Tensor product of operads and iterated loop spaces". In: *J. Pure Appl. Algebra* 50.3 (1988), pp. 237–258. issn: 0022-4049.

[86]  Gerald Dunn. "Uniqueness of $n$-fold delooping machines". In: *J. Pure Appl. Algebra* 113.2 (1996), pp. 159–193. issn: 0022-4049.

[87]  W. G. Dwyer and D. M. Kan. "Calculating simplicial localizations". In: *J. Pure Appl. Algebra* 18.1 (1980), pp. 17–35. issn: 0022-4049.

[88]  W. G. Dwyer and D. M. Kan. "Function complexes in homotopical algebra". In: *Topology* 19.4 (1980), pp. 427–440. issn: 0040-9383.

[89]  W. G. Dwyer and D. M. Kan. "Simplicial localizations of categories". In: *J. Pure Appl. Algebra* 17.3 (1980), pp. 267–284. issn: 0022-4049.

[90]  W. G. Dwyer, D. M. Kan, and J. H. Smith. "Homotopy commutative diagrams and their realizations". In: *J. Pure Appl. Algebra* 57.1 (1989), pp. 5–24. issn: 0022-4049.

[91]  W. G. Dwyer and J. Spaliński. "Homotopy theories and model categories". In: *Handbook of algebraic topology*. North-Holland, Amsterdam, 1995, pp. 73–126.

[92]  William G. Dwyer, Philip S. Hirschhorn, Daniel M. Kan, and Jeffrey H. Smith. *Homotopy limit functors on model categories and homotopical categories*. Vol. 113. Mathematical Surveys and Monographs. American Mathematical Society, Providence, RI, 2004, pp. viii+181. isbn: 0-8218-3703-6.

[93]  A. D. Elmendorf. "The Grassmannian geometry of spectra". In: *J. Pure Appl. Algebra* 54.1 (1988), pp. 37–94. issn: 0022-4049.

[94]  A. D. Elmendorf, I. Kriz, M. A. Mandell, and J. P. May. *Rings, modules, and algebras in stable homotopy theory*. Vol. 47. Mathematical Surveys and Monographs. With an appendix by M. Cole. American Mathematical Society, Providence, RI, 1997, pp. xii+249. isbn: 0-8218-0638-6.

[95]  A. D. Elmendorf and M. A. Mandell. "Rings, modules, and algebras in infinite loop space theory". In: *Adv. Math.* 205.1 (2006), pp. 163–228. issn: 0001-8708.

[96]  Emmanuel Dror Farjoun. *Cellular spaces, null spaces and homotopy localization*. Vol. 1622. Lecture Notes in Mathematics. Springer-Verlag, Berlin, 1996, pp. xiv+199. isbn: 3-540-60604-1.

[97]  Z. Fiedorowicz and R. M. Vogt. "An additivity theorem for the interchange of $E_n$ structures". In: *Adv. Math.* 273 (2015), pp. 421–484. issn: 0001-8708.

[98]  Zbigniew Fiedorowicz. "Symmetric Bar Construction". Preprint. 1996. url: https://people.math.osu.edu/fiedorowicz.1/symbar.ps.gz.

[99]    Ralph H. Fox. "Free differential calculus. I. Derivation in the free group ring". In: *Ann. of Math. (2)* 57 (1953), pp. 547–560. issn: 0003-486X.

[100]   P. Gabriel and M. Zisman. *Calculus of fractions and homotopy theory*. Ergebnisse der Mathematik und ihrer Grenzgebiete, Band 35. Springer-Verlag New York, Inc., New York, 1967, pp. x+168.

[101]   Nicola Gambino. "Weighted limits in simplicial homotopy theory". In: *J. Pure Appl. Algebra* 214.7 (2010), pp. 1193–1199. issn: 0022-4049.

[102]   Thomas Geisser and Lars Hesselholt. "Topological cyclic homology of schemes". In: *Algebraic K-theory (Seattle, WA, 1997)*. Vol. 67. Proc. Sympos. Pure Math. Amer. Math. Soc., Providence, RI, 1999, pp. 41–87.

[103]   David Gepner and Tyler Lawson. "Brauer groups and Galois cohomology of commutative ring spectra". In: *Compos. Math.* 157.6 (2021), pp. 1211–1264. issn: 0010-437X.

[104]   Victor Ginzburg and Mikhail Kapranov. "Koszul duality for operads". In: *Duke Math. J.* 76.1 (1994), pp. 203–272. issn: 0012-7094.

[105]   Saul Glasman. "Day convolution for $\infty$-categories". In: *Math. Res. Lett.* 23.5 (2016), pp. 1369–1385. issn: 1073-2780.

[106]   P. G. Goerss and M. J. Hopkins. *Moduli problems for structured ring spectra*. url: http://www.math.northwestern.edu/~pgoerss/.

[107]   P. G. Goerss and M. J. Hopkins. "Moduli spaces of commutative ring spectra". In: *Structured ring spectra*. Vol. 315. London Math. Soc. Lecture Note Ser. Cambridge Univ. Press, Cambridge, 2004, pp. 151–200.

[108]   Paul Goerss, Hans-Werner Henn, Mark Mahowald, and Charles Rezk. "On Hopkins' Picard groups for the prime 3 and chromatic level 2". In: *J. Topol.* 8.1 (2015), pp. 267–294. issn: 1753-8416.

[109]   Thomas G. Goodwillie. "Cyclic homology, derivations, and the free loopspace". In: *Topology* 24.2 (1985), pp. 187–215. issn: 0040-9383.

[110]   J. P. C. Greenlees. "Ausoni-Bökstedt duality for topological Hochschild homology". In: *J. Pure Appl. Algebra* 220.4 (2016), pp. 1382–1402. issn: 0022-4049.

[111]   J. P. C. Greenlees. "Spectra for commutative algebraists". In: *Interactions between homotopy theory and algebra*. Vol. 436. Contemp. Math. Amer. Math. Soc., Providence, RI, 2007, pp. 149–173.

[112]   J. P. C. Greenlees and J. P. May. "Completions in algebra and topology". In: *Handbook of algebraic topology*. North-Holland, Amsterdam, 1995, pp. 255–276.

[113]   A. Grothendieck. "À la Poursuite des Champs". In: (1983).

[114]   Alexander Grothendieck. "Le groupe de Brauer. I-III". In: *Dix exposés sur la cohomologie des schémas*. Vol. 3. Adv. Stud. Pure Math. North-Holland, Amsterdam, 1968, pp. 46–188.

[115]   Markus Hausmann. "$G$-symmetric spectra, semistability and the multiplicative norm". In: *J. Pure Appl. Algebra* 221.10 (2017), pp. 2582–2632. issn: 0022-4049.

[116]   Drew Heard, Akhil Mathew, and Vesna Stojanoska. "Picard groups of higher real $K$-theory spectra at height $p-1$". In: *Compos. Math.* 153.9 (2017), pp. 1820–1854. issn: 0010-437X.

[117]  Alex Heller. "Homotopy Theories". In: *Mem. Amer. Math. Soc.* 71.383 (1988), pp. vi+78. issn: 0065-9266.

[118]  Kathryn Hess, Magdalena Kędziorek, Emily Riehl, and Brooke Shipley. "A necessary and sufficient condition for induced model structures". In: *J. Topol.* 10.2 (2017), pp. 324–369. issn: 1753-8416.

[119]  Lars Hesselholt and Ib Madsen. "On the $K$-theory of finite algebras over Witt vectors of perfect fields". In: *Topology* 36.1 (1997), pp. 29–101. issn: 0040-9383.

[120]  M. A. Hill, M. J. Hopkins, and D. C. Ravenel. "On the nonexistence of elements of Kervaire invariant one". In: *Ann. of Math. (2)* 184.1 (2016), pp. 1–262. issn: 0003-486X.

[121]  Michael A. Hill and Lennart Meier. "The $C_2$-spectrum $\mathrm{Tmf}_1(3)$ and its invertible modules". In: *Algebr. Geom. Topol.* 17.4 (2017), pp. 1953–2011. issn: 1472-2747.

[122]  Michael Hill and Tyler Lawson. "Automorphic forms and cohomology theories on Shimura curves of small discriminant". In: *Adv. Math.* 225.2 (2010), pp. 1013–1045. issn: 0001-8708.

[123]  V. A. Hinich and V. V. Schechtman. "On homotopy limit of homotopy algebras". In: *K-theory, arithmetic and geometry (Moscow, 1984–1986)*. Vol. 1289. Lecture Notes in Math. Springer, Berlin, 1987, pp. 240–264.

[124]  Philip S. Hirschhorn. *Model categories and their localizations.* Vol. 99. Mathematical Surveys and Monographs. American Mathematical Society, Providence, RI, 2003, pp. xvi+457. isbn: 0-8218-3279-4.

[125]  A. Hirschowitz and C. Simpson. *Descente pour les n-champs.* arXiv:9807049. 2001. url: https://arxiv.org/abs/math/9807049.

[126]  M. J. Hopkins. "Algebraic topology and modular forms". In: *Proceedings of the International Congress of Mathematicians, Vol. I (Beijing, 2002)*. Higher Ed. Press, Beijing, 2002, pp. 291–317.

[127]  Michael J. Hopkins and Tyler Lawson. "Strictly commutative complex orientation theory". In: *Math. Z.* 290.1-2 (2018), pp. 83–101. issn: 0025-5874.

[128]  Michael J. Hopkins and Jacob Lurie. *On Brauer Groups of Lubin-Tate Spectra I.* url: http://https://www.math.ias.edu/~lurie/.

[129]  Michael J. Hopkins, Mark Mahowald, and Hal Sadofsky. "Constructions of elements in Picard groups". In: *Topology and representation theory (Evanston, IL, 1992)*. Vol. 158. Contemp. Math. Amer. Math. Soc., Providence, RI, 1994, pp. 89–126.

[130]  Mark Hovey. *Model categories.* Vol. 63. Mathematical Surveys and Monographs. American Mathematical Society, Providence, RI, 1999, pp. xii+209. isbn: 0-8218-1359-5.

[131]  Mark Hovey. *Smith ideals of structured ring spectra.* arXiv:1401.2850.

[132]  Mark Hovey. "Spectra and symmetric spectra in general model categories". In: *J. Pure Appl. Algebra* 165.1 (2001), pp. 63–127. issn: 0022-4049.

[133]  Mark Hovey, Brooke Shipley, and Jeff Smith. "Symmetric spectra". In: *J. Amer. Math. Soc.* 13.1 (2000), pp. 149–208. issn: 0894-0347.

[134]   Srikanth Iyengar. "André-Quillen homology of commutative algebras". In: *Interactions between homotopy theory and algebra*. Vol. 436. Contemp. Math. Amer. Math. Soc., Providence, RI, 2007, pp. 203–234.

[135]   J. F. Jardine. "Simplicial presheaves". In: *J. Pure Appl. Algebra* 47.1 (1987), pp. 35–87. issn: 0022-4049.

[136]   Michael Joachim. "A symmetric ring spectrum representing $KO$-theory". In: *Topology* 40.2 (2001), pp. 299–308. issn: 0040-9383.

[137]   Michael Joachim. "Higher coherences for equivariant $K$-theory". In: *Structured ring spectra*. Vol. 315. London Math. Soc. Lecture Note Ser. Cambridge Univ. Press, Cambridge, 2004, pp. 87–114.

[138]   Niles Johnson. "Azumaya objects in triangulated bicategories". In: *J. Homotopy Relat. Struct.* 9.2 (2014), pp. 465–493. issn: 2193-8407.

[139]   Niles Johnson and Justin Noel. "Lifting homotopy $T$-algebra maps to strict maps". In: *Adv. Math.* 264 (2014), pp. 593–645. issn: 0001-8708.

[140]   A. Joyal. "Quasi-categories and Kan complexes". In: *J. Pure Appl. Algebra* 175.1-3 (2002). Special volume celebrating the 70th birthday of Professor Max Kelly, pp. 207–222. issn: 0022-4049.

[141]   A. Joyal. *The theory of quasi-categories and its applications*. 2008. url: http://mat.uab.cat/~kock/crm/hocat/advanced-course/Quadern45-2.pdf.

[142]   André Joyal and Myles Tierney. "Quasi-categories vs Segal spaces". In: *Categories in algebra, geometry and mathematical physics*. Vol. 431. Contemp. Math. Amer. Math. Soc., Providence, RI, 2007, pp. 277–326.

[143]   Daniel M. Kan. "On homotopy theory and c.s.s. groups". In: *Ann. of Math. (2)* 68 (1958), pp. 38–53. issn: 0003-486X.

[144]   Nicholas M. Katz and Barry Mazur. *Arithmetic moduli of elliptic curves*. Vol. 108. Annals of Mathematics Studies. Princeton University Press, Princeton, NJ, 1985, pp. xiv+514.

[145]   G. M. Kelly. "Basic concepts of enriched category theory". In: *Repr. Theory Appl. Categ.* 10 (2005). Reprint of the 1982 original [Cambridge Univ. Press, Cambridge; MR0651714], pp. vi+137.

[146]   G. M. Kelly. "Elementary observations on 2-categorical limits". In: *Bull. Austral. Math. Soc.* 39.2 (1989), pp. 301–317. issn: 0004-9727.

[147]   G. M. Kelly and Ross Street. "Review of the elements of 2-categories". In: (1974), 75–103. Lecture Notes in Math., Vol. 420.

[148]   Gregory Maxwell Kelly. *Basic concepts of enriched category theory*. Vol. 64. London Mathematical Society Lecture Note Series. Cambridge University Press, Cambridge-New York, 1982, p. 245. isbn: 0-521-28702-2.

[149]   Inbar Klang. "The factorization theory of Thom spectra and twisted nonabelian Poincaré duality". In: *Algebr. Geom. Topol.* 18.5 (2018), pp. 2541–2592. issn: 1472-2747.

[150]   Nicholas J. Kuhn. "The McCord model for the tensor product of a space and a commutative ring spectrum". In: *Categorical decomposition techniques in*

*algebraic topology (Isle of Skye, 2001)*. Vol. 215. Progr. Math. Birkhäuser, Basel, 2004, pp. 213–236.

[151]  Tyler Lawson. "Commutative Γ-rings do not model all commutative ring spectra". In: *Homology Homotopy Appl.* 11.2 (2009), pp. 189–194. issn: 1532-0073.

[152]  Tyler Lawson. "Secondary power operations and the Brown-Peterson spectrum at the prime 2". In: *Ann. of Math. (2)* 188.2 (2018), pp. 513–576. issn: 0003-486X.

[153]  Tyler Lawson and Niko Naumann. "Commutativity conditions for truncated Brown-Peterson spectra of height 2". In: *J. Topol.* 5.1 (2012), pp. 137–168. issn: 1753-8416.

[154]  F. William Lawvere. "Functorial semantics of algebraic theories and some algebraic problems in the context of functorial semantics of algebraic theories". In: *Repr. Theory Appl. Categ.* 5 (2004). Reprinted from Proc. Nat. Acad. Sci. U.S.A. 50 (1963), 869–872 [MR0158921] and *Reports of the Midwest Category Seminar, II,* 41–61, Springer, Berlin, 1968 [MR0231882], pp. 1–121.

[155]  L. G. Lewis Jr., J. P. May, M. Steinberger, and J. E. McClure. *Equivariant stable homotopy theory.* Vol. 1213. Lecture Notes in Mathematics. With contributions by J. E. McClure. Springer-Verlag, Berlin, 1986, pp. x+538. isbn: 3-540-16820-6.

[156]  L. Gaunce Lewis Jr. "Is there a convenient category of spectra?" In: *J. Pure Appl. Algebra* 73.3 (1991), pp. 233–246. issn: 0022-4049.

[157]  L. Gaunce Lewis Jr. and Michael A. Mandell. "Modules in monoidal model categories". In: *J. Pure Appl. Algebra* 210.2 (2007), pp. 395–421. issn: 0022-4049.

[158]  Elon L. Lima. "The Spanier-Whitehead duality in new homotopy categories". In: *Summa Brasil. Math.* 4 (1959), 91–148 (1959). issn: 0039-498X.

[159]  John A. Lind. "Diagram spaces, diagram spectra and spectra of units". In: *Algebr. Geom. Topol.* 13.4 (2013), pp. 1857–1935. issn: 1472-2747.

[160]  Ayelet Lindenstrauss and Ib Madsen. "Topological Hochschild homology of number rings". In: *Trans. Amer. Math. Soc.* 352.5 (2000), pp. 2179–2204. issn: 0002-9947.

[161]  Jean-Louis Loday. *Cyclic homology.* Second. Vol. 301. Grundlehren der Mathematischen Wissenschaften [Fundamental Principles of Mathematical Sciences]. Appendix E by María O. Ronco, Chapter 13 by the author in collaboration with Teimuraz Pirashvili. Springer-Verlag, Berlin, 1998, pp. xx+513. isbn: 3-540-63074-0.

[162]  J. Lurie. "A survey of elliptic cohomology". In: *Algebraic topology.* Vol. 4. Abel Symp. Springer, Berlin, 2009, pp. 219–277.

[163]  Jacob Lurie. *Derived Algebraic Geometry I–XIV.* 2012. url: http://https://www.math.ias.edu/~lurie/.

[164]  Jacob Lurie. "Derived Algebraic Geometry VI: $E_k$ Algebras". arXiv:0911.0018.

[165]  Jacob Lurie. *Derived Algebraic Geometry VII: Spectral Schemes.* url: http://https://www.math.ias.edu/~lurie/.

[166]  Jacob Lurie. *Elliptic Cohomology I: Spectral Abelian Varieties.* 2018. url: http://https://www.math.ias.edu/~lurie/.

[167]  Jacob Lurie. *Elliptic Cohomology II: Orientations.* 2018. url: http://https://www.math.ias.edu/~lurie/.

[168]   Jacob Lurie. *Higher algebra*. 2017. url: http://https://www.math.ias.edu/~lurie/.

[169]   Jacob Lurie. *Higher topos theory*. Vol. 170. Annals of Mathematics Studies. Princeton University Press, Princeton, NJ, 2009, pp. xviii+925.

[170]   Jacob Lurie. *Spectral Algebraic Geometry*. 2018. url: http://https://www.math.ias.edu/~lurie/.

[171]   Manos Lydakis. *Simplicial functors and stable homotopy theory*. Preprint. 1998. url: http://hopf.math.purdue.edu/Lydakis/s_functors.pdf.

[172]   Manos Lydakis. "Smash products and Γ-spaces". In: *Math. Proc. Cambridge Philos. Soc.* 126.2 (1999), pp. 311–328. issn: 0305-0041.

[173]   Saunders Mac Lane and Ieke Moerdijk. *Sheaves in geometry and logic*. Universitext. A first introduction to topos theory, Corrected reprint of the 1992 edition. Springer-Verlag, New York, 1994, pp. xii+629.

[174]   Saunders MacLane. *Categories for the working mathematician*. Graduate Texts in Mathematics, Vol. 5. Springer-Verlag, New York-Berlin, 1971, pp. ix+262.

[175]   Mark Mahowald. "Ring spectra which are Thom complexes". In: *Duke Math. J.* 46.3 (1979), pp. 549–559. issn: 0012-7094.

[176]   Georges Maltsiniotis. "Le théorème de Quillen, d'adjonction des foncteurs dérivés, revisité". In: *C. R. Math. Acad. Sci. Paris* 344.9 (2007), pp. 549–552. issn: 1631-073X.

[177]   M. A. Mandell and J. P. May. "Equivariant orthogonal spectra and $S$-modules". In: *Mem. Amer. Math. Soc.* 159.755 (2002), pp. x+108. issn: 0065-9266.

[178]   M. A. Mandell, J. P. May, S. Schwede, and B. Shipley. "Model categories of diagram spectra". In: *Proc. London Math. Soc. (3)* 82.2 (2001), pp. 441–512. issn: 0024-6115.

[179]   Michael A. Mandell. "Cochain multiplications". In: *Amer. J. Math.* 124.3 (2002), pp. 547–566. issn: 0002-9327.

[180]   Michael A. Mandell. "Cochains and homotopy type". In: *Publ. Math. Inst. Hautes Études Sci.* 103 (2006), pp. 213–246. issn: 0073-8301.

[181]   Michael A. Mandell. "$E_\infty$ algebras and $p$-adic homotopy theory". In: *Topology* 40.1 (2001). A non-mangled version of this paper is available at https://pages.iu.edu/~mmandell/papers/einffinal.pdf, pp. 43–94. issn: 0040-9383.

[182]   Michael A. Mandell. "Equivariant $p$-adic homotopy theory". In: *Topology Appl.* 122.3 (2002), pp. 637–651. issn: 0166-8641.

[183]   Michael A. Mandell. "Equivariant symmetric spectra". In: *Homotopy theory: relations with algebraic geometry, group cohomology, and algebraic $K$-theory*. Vol. 346. Contemp. Math. Amer. Math. Soc., Providence, RI, 2004, pp. 399–452.

[184]   Michael A. Mandell. "The smash product for derived categories in stable homotopy theory". In: *Adv. Math.* 230.4-6 (2012), pp. 1531–1556. issn: 0001-8708.

[185]   Michael A. Mandell. "Topological André-Quillen cohomology and $E_\infty$ André-Quillen cohomology". In: *Adv. Math.* 177.2 (2003), pp. 227–279. issn: 0001-8708.

[186]   Akhil Mathew. "THH and base-change for Galois extensions of ring spectra". In: *Algebr. Geom. Topol.* 17.2 (2017), pp. 693–704. issn: 1472-2747.

[187]   Akhil Mathew and Lennart Meier. "Affineness and chromatic homotopy theory". In: *J. Topol.* 8.2 (2015), pp. 476–528. issn: 1753-8416.

[188]   Akhil Mathew, Niko Naumann, and Justin Noel. "On a nilpotence conjecture of J. P. May". In: *J. Topol.* 8.4 (2015), pp. 917–932. issn: 1753-8416.

[189]   Akhil Mathew and Vesna Stojanoska. "The Picard group of topological modular forms via descent theory". In: *Geom. Topol.* 20.6 (2016), pp. 3133–3217. issn: 1465-3060.

[190]   J. P. May. *Equivariant homotopy and cohomology theory.* Vol. 91. CBMS Regional Conference Series in Mathematics. With contributions by M. Cole, G. Comezaña, S. Costenoble, A. D. Elmendorf, J. P. C. Greenlees, L. G. Lewis, Jr., R. J. Piacenza, G. Triantafillou, and S. Waner. Published for the Conference Board of the Mathematical Sciences, Washington, DC; by the American Mathematical Society, Providence, RI, 1996, pp. xiv+366. isbn: 0-8218-0319-0.

[191]   J. P. May. "Memorial talk". In: *J. Homotopy Relat. Struct.* 2.2 (2007), pp. 11–12.

[192]   J. P. May. "Multiplicative infinite loop space theory". In: *J. Pure Appl. Algebra* 26.1 (1982), pp. 1–69. issn: 0022-4049.

[193]   J. P. May. "The construction of $E_\infty$ ring spaces from bipermutative categories". In: *New topological contexts for Galois theory and algebraic geometry (BIRS 2008).* Vol. 16. Geom. Topol. Monogr. Geom. Topol. Publ., Coventry, 2009, pp. 283–330.

[194]   J. P. May. *The geometry of iterated loop spaces.* Lectures Notes in Mathematics, Vol. 271. Springer-Verlag, Berlin-New York, 1972, pp. viii+175.

[195]   J. P. May. "The spectra associated to permutative categories". In: *Topology* 17.3 (1978), pp. 225–228. issn: 0040-9383.

[196]   J. P. May. "What are $E_\infty$ ring spaces good for?" In: *New topological contexts for Galois theory and algebraic geometry (BIRS 2008).* Vol. 16. Geom. Topol. Monogr. Geom. Topol. Publ., Coventry, 2009, pp. 331–365.

[197]   J. P. May. "What precisely are $E_\infty$ ring spaces and $E_\infty$ ring spectra?" In: *New topological contexts for Galois theory and algebraic geometry (BIRS 2008).* Vol. 16. Geom. Topol. Monogr. Geom. Topol. Publ., Coventry, 2009, pp. 215–282.

[198]   J. Peter May. "A general algebraic approach to Steenrod operations". In: *The Steenrod Algebra and its Applications (Proc. Conf. to Celebrate N. E. Steenrod's Sixtieth Birthday, Battelle Memorial Inst., Columbus, Ohio, 1970).* Lecture Notes in Mathematics, Vol. 168. Springer, Berlin, 1970, pp. 153–231.

[199]   J. Peter May. $E_\infty$ *ring spaces and* $E_\infty$ *ring spectra.* Lecture Notes in Mathematics, Vol. 577. With contributions by Frank Quinn, Nigel Ray, and Jørgen Tornehave. Springer-Verlag, Berlin-New York, 1977, p. 268.

[200]   J. Peter May. *Simplicial objects in algebraic topology.* Chicago Lectures in Mathematics. Reprint of the 1967 original. University of Chicago Press, Chicago, IL, 1992, pp. viii+161. isbn: 0-226-51181-2.

[201]   Randy McCarthy and Vahagn Minasian. "HKR theorem for smooth $S$-algebras". In: *J. Pure Appl. Algebra* 185.1-3 (2003), pp. 239–258. issn: 0022-4049.

[202]  J. E. McClure and R. E. Staffeldt. "On the topological Hochschild homology of $bu$. I". In: *Amer. J. Math.* 115.1 (1993), pp. 1–45. issn: 0002-9327.

[203]  J. McClure, R. Schwänzl, and R. Vogt. "$THH(R) \cong R \otimes S^1$ for $E_\infty$ ring spectra". In: *J. Pure Appl. Algebra* 121.2 (1997), pp. 137–159. issn: 0022-4049.

[204]  James E. McClure and Jeffrey H. Smith. "A solution of Deligne's Hochschild cohomology conjecture". In: *Recent progress in homotopy theory (Baltimore, MD, 2000)*. Vol. 293. Contemp. Math. Amer. Math. Soc., Providence, RI, 2002, pp. 153–193.

[205]  James E. McClure and Jeffrey H. Smith. "Multivariable cochain operations and little $n$-cubes". In: *J. Amer. Math. Soc.* 16.3 (2003), pp. 681–704. issn: 0894-0347.

[206]  M. C. McCord. "Classifying spaces and infinite symmetric products". In: *Trans. Amer. Math. Soc.* 146 (1969), pp. 273–298. issn: 0002-9947.

[207]  D. McDuff and G. Segal. "Homology fibrations and the "group-completion" theorem". In: *Invent. Math.* 31.3 (1975), pp. 279–284. issn: 0020-9910.

[208]  Dusa McDuff. "On the classifying spaces of discrete monoids". In: *Topology* 18.4 (1979), pp. 313–320. issn: 0040-9383.

[209]  C. A. McGibbon and J. A. Neisendorfer. "On the homotopy groups of a finite-dimensional space". In: *Comment. Math. Helv.* 59.2 (1984), pp. 253–257. issn: 0010-2571.

[210]  Haynes Miller. "The Sullivan conjecture on maps from classifying spaces". In: *Ann. of Math. (2)* 120.1 (1984), pp. 39–87. issn: 0003-486X.

[211]  Vahagn Minasian. "André-Quillen spectral sequence for $THH$". In: *Topology Appl.* 129.3 (2003), pp. 273–280. issn: 0166-8641.

[212]  John C. Moore. "Semi-simplicial complexes and Postnikov systems". In: *Symposium internacional de topología algebraica International symposium on algebraic topology.* Universidad Nacional Autónoma de México and UNESCO, Mexico City, 1958, pp. 232–247.

[213]  Jack Morava. "Noetherian localisations of categories of cobordism comodules". In: *Ann. of Math. (2)* 121.1 (1985), pp. 1–39. issn: 0003-486X.

[214]  Amnon Neeman. *Triangulated Categories.* Vol. 148. Annals of Mathematics Studies. Princeton University Press, Princeton, NJ, 2001, pp. viii+449.

[215]  Joseph Neisendorfer. *Algebraic methods in unstable homotopy theory.* Vol. 12. New Mathematical Monographs. Cambridge University Press, Cambridge, 2010, pp. xx+554. isbn: 978-0-521-76037-9.

[216]  Thomas Nikolaus and Peter Scholze. "On topological cyclic homology". In: *Acta Math.* 221.2 (2018), pp. 203–409. issn: 0001-5962.

[217]  Justin Noel. "The $T$-algebra spectral sequence: comparisons and applications". In: *Algebr. Geom. Topol.* 14.6 (2014), pp. 3395–3417. issn: 1472-2747.

[218]  Dominik Ostermayr. "Equivariant $\Gamma$-spaces". In: *Homology Homotopy Appl.* 18.1 (2016), pp. 295–324. issn: 1532-0073.

[219]  A. V. Pazhitnov and Yu. B. Rudyak. "Commutative ring spectra of characteristic 2". In: *Mat. Sb. (N.S.)* 124(166).4 (1984), pp. 486–494. issn: 0368-8666.

[220]  R. Pellissier. *Catégories enrichies faibles.* arXiv:math/0308246. 2002. url: https://arxiv.org/abs/math/0308246.

[221]   Maximilien Péroux and Brooke Shipley. "Coalgebras in symmetric monoidal categories of spectra". In: *Homology Homotopy Appl.* 21.1 (2019), pp. 1–18. issn: 1532-0073.

[222]   T. Pirashvili and B. Richter. "Robinson-Whitehouse complex and stable homotopy". In: *Topology* 39.3 (2000), pp. 525–530. issn: 0040-9383.

[223]   Teimuraz Pirashvili. "Hodge decomposition for higher order Hochschild homology". In: *Ann. Sci. École Norm. Sup. (4)* 33.2 (2000), pp. 151–179. issn: 0012-9593.

[224]   Daniel Quillen. "Higher algebraic $K$-theory. I". In: *Algebraic K-theory, I: Higher K-theories (Proc. Conf., Battelle Memorial Inst., Seattle, Wash., 1972)*. 1973, 85–147. Lecture Notes in Math., Vol. 341.

[225]   Daniel Quillen. "On the (co-) homology of commutative rings". In: *Applications of Categorical Algebra (Proc. Sympos. Pure Math., Vol. XVII, New York, 1968)*. Amer. Math. Soc., Providence, R.I., 1970, pp. 65–87.

[226]   Daniel Quillen. "On the formal group laws of unoriented and complex cobordism theory". In: *Bull. Amer. Math. Soc.* 75 (1969), pp. 1293–1298. issn: 0002-9904.

[227]   Daniel Quillen. "On the group completion of a simplicial monoid". In: *Filtrations on the homology of algebraic varieties*. Vol. 529. Mem. Amer. Math. Soc. Amer. Math. Soc., Providence, RI, 1994, pp. 89–105.

[228]   Daniel Quillen. "Rational homotopy theory". In: *Ann. of Math. (2)* 90 (1969), pp. 205–295. issn: 0003-486X.

[229]   Daniel G. Quillen. *Homotopical algebra*. Lecture Notes in Mathematics, No. 43. Springer-Verlag, Berlin-New York, 1967, iv+156 pp. (not consecutively paged).

[230]   Douglas C. Ravenel. "Localization with respect to certain periodic homology theories". In: *Amer. J. Math.* 106.2 (1984), pp. 351–414. issn: 0002-9327.

[231]   Douglas C. Ravenel. *Nilpotence and periodicity in stable homotopy theory*. Vol. 128. Annals of Mathematics Studies. Appendix C by Jeff Smith. Princeton University Press, Princeton, NJ, 1992, pp. xiv+209. isbn: 0-691-02572-X.

[232]   C. L. Reedy. *Homotopy theory of model categories*. 1974. url: ftp://hopf.math.purdue.edu/pub/Reedy/reedy.dvi.

[233]   Charles Rezk. "A model for the homotopy theory of homotopy theory". In: *Trans. Amer. Math. Soc.* 353.3 (2001), pp. 973–1007. issn: 0002-9947.

[234]   Charles Rezk. "Notes on the Hopkins-Miller theorem". In: *Homotopy theory via algebraic geometry and group representations (Evanston, IL, 1997)*. Vol. 220. Contemp. Math. Amer. Math. Soc., Providence, RI, 1998, pp. 313–366.

[235]   Birgit Richter and Steffen Sagave. "A strictly commutative model for the cochain algebra of a space". In: *Compos. Math.* 156.8 (2020), pp. 1718–1743.

[236]   Birgit Richter and Brooke Shipley. "An algebraic model for commutative $H\mathbb{Z}$-algebras". In: *Algebr. Geom. Topol.* 17.4 (2017), pp. 2013–2038. issn: 1472-2747.

[237]   Emily Riehl. *Categorical homotopy theory*. Vol. 24. New Mathematical Monographs. Cambridge University Press, Cambridge, 2014, pp. xviii+352. isbn: 978-1-107-04845-4.

[238] Emily Riehl. *Category theory in context.* Aurora: Modern Math Originals. Dover Publications, 2016.

[239] Emily Riehl. *Inductive presentations of generalized Reedy categories.* 2017. url: http://www.math.jhu.edu/~eriehl/generalized-reedy.pdf.

[240] Emily Riehl and Dominic Verity. "Fibrations and Yoneda's lemma in an ∞-cosmos". In: *J. Pure Appl. Algebra* 221.3 (2017), pp. 499–564. issn: 0022-4049.

[241] Emily Riehl and Dominic Verity. "Homotopy coherent adjunctions and the formal theory of monads". In: *Adv. Math.* 286 (2016), pp. 802–888. issn: 0001-8708.

[242] Emily Riehl and Dominic Verity. "Infinity category theory from scratch". In: *High. Struct.* 4.1 (2020), pp. 115–167.

[243] Emily Riehl and Dominic Verity. "The 2-category theory of quasi-categories". In: *Adv. Math.* 280 (2015), pp. 549–642. issn: 0001-8708.

[244] Emily Riehl and Dominic Verity. "The comprehension construction". In: *High. Struct.* 2.1 (2018), pp. 116–190. issn: 2209-0606.

[245] Emily Riehl and Dominic Verity. "The theory and practice of Reedy categories". In: *Theory Appl. Categ.* 29 (2014), pp. 256–301. issn: 1201-561X.

[246] Alan Robinson. "Gamma homology, Lie representations and $E_\infty$ multiplications". In: *Invent. Math.* 152.2 (2003), pp. 331–348. issn: 0020-9910.

[247] Alan Robinson. "Obstruction theory and the strict associativity of Morava $K$-theories". In: *Advances in homotopy theory (Cortona, 1988).* Vol. 139. London Math. Soc. Lecture Note Ser. Cambridge Univ. Press, Cambridge, 1989, pp. 143–152.

[248] Alan Robinson. "Spectral sheaves: a model category for stable homotopy theory". In: *J. Pure Appl. Algebra* 45.2 (1987), pp. 171–200. issn: 0022-4049.

[249] Alan Robinson. "The extraordinary derived category". In: *Math. Z.* 196.2 (1987), pp. 231–238. issn: 0025-5874.

[250] Alan Robinson and Sarah Whitehouse. "Operads and Γ-homology of commutative rings". In: *Math. Proc. Cambridge Philos. Soc.* 132.2 (2002), pp. 197–234. issn: 0305-0041.

[251] John Rognes. "Galois extensions of structured ring spectra. Stably dualizable groups". In: *Mem. Amer. Math. Soc.* 192.898 (2008), pp. viii+137. issn: 0065-9266.

[252] John Rognes. "Trace maps from the algebraic $K$-theory of the integers (after Marcel Bökstedt)". In: *J. Pure Appl. Algebra* 125.1-3 (1998), pp. 277–286. issn: 0022-4049.

[253] John Rognes, Steffen Sagave, and Christian Schlichtkrull. "Localization sequences for logarithmic topological Hochschild homology". In: *Math. Ann.* 363.3-4 (2015), pp. 1349–1398. issn: 0025-5831.

[254] J. Rosicky. "Are all cofibrantly generated model categories combinatorial?" In: *Cah. Topol. Géom. Différ. Catég.* 50.3 (2009), pp. 233–238. issn: 1245-530X.

[255] Steffen Sagave. "Logarithmic structures on topological $K$-theory spectra". In: *Geom. Topol.* 18.1 (2014), pp. 447–490. issn: 1465-3060.

[256]  Steffen Sagave and Christian Schlichtkrull. "Diagram spaces and symmetric spectra". In: *Adv. Math.* 231.3-4 (2012), pp. 2116–2193. issn: 0001-8708.

[257]  Steffen Sagave and Christian Schlichtkrull. "Virtual vector bundles and graded Thom spectra". In: *Math. Z.* 292.3-4 (2019), pp. 975–1016. issn: 0025-5874.

[258]  Marco Schlichting. "A note on $K$-theory and triangulated categories". In: *Invent. Math.* 150.1 (2002), pp. 111–116. issn: 0020-9910.

[259]  Christian Schlichtkrull. "Higher topological Hochschild homology of Thom spectra". In: *J. Topol.* 4.1 (2011), pp. 161–189. issn: 1753-8416.

[260]  R. Schwänzl, R. M. Vogt, and F. Waldhausen. "Adjoining roots of unity to $E_\infty$ ring spectra in good cases—a remark". In: *Homotopy invariant algebraic structures (Baltimore, MD, 1998)*. Vol. 239. Contemp. Math. Amer. Math. Soc., Providence, RI, 1999, pp. 245–249.

[261]  Stefan Schwede. "On the homotopy groups of symmetric spectra". In: *Geom. Topol.* 12.3 (2008), pp. 1313–1344. issn: 1465-3060.

[262]  Stefan Schwede. "$S$-modules and symmetric spectra". In: *Math. Ann.* 319.3 (2001), pp. 517–532. issn: 0025-5831.

[263]  Stefan Schwede. "Spectra in model categories and applications to the algebraic cotangent complex". In: *J. Pure Appl. Algebra* 120.1 (1997), pp. 77–104. issn: 0022-4049.

[264]  Stefan Schwede. "Stable homotopical algebra and $\Gamma$-spaces". In: *Math. Proc. Cambridge Philos. Soc.* 126.2 (1999), pp. 329–356. issn: 0305-0041.

[265]  Stefan Schwede and Brooke Shipley. "Equivalences of monoidal model categories". In: *Algebr. Geom. Topol.* 3 (2003), pp. 287–334. issn: 1472-2747.

[266]  Stefan Schwede and Brooke Shipley. "Stable model categories are categories of modules". In: *Topology* 42.1 (2003), pp. 103–153. issn: 0040-9383.

[267]  Stefan Schwede and Brooke E. Shipley. "Algebras and modules in monoidal model categories". In: *Proc. London Math. Soc. (3)* 80.2 (2000), pp. 491–511. issn: 0024-6115.

[268]  Graeme Segal. "Categories and cohomology theories". In: *Topology* 13 (1974), pp. 293–312. issn: 0040-9383.

[269]  Graeme Segal. "Classifying spaces and spectral sequences". In: *Inst. Hautes Études Sci. Publ. Math.* 34 (1968), pp. 105–112. issn: 0073-8301.

[270]  Graeme Segal. "Configuration-spaces and iterated loop-spaces". In: *Invent. Math.* 21 (1973), pp. 213–221. issn: 0020-9910.

[271]  Andrew Senger. *The Brown-Peterson spectrum is not* $E_{2(p^2+2)}$ *at odd primes.* arXiv:1710.09822. url: http://arxiv.org/abs/1710.09822.

[272]  Jean-Pierre Serre. "Groupes d'homotopie et classes de groupes abéliens". In: *Ann. of Math. (2)* 58 (1953), pp. 258–294. issn: 0003-486X.

[273]  Nobuo Shimada and Kazuhisa Shimakawa. "Delooping symmetric monoidal categories". In: *Hiroshima Math. J.* 9.3 (1979), pp. 627–645. issn: 0018-2079.

[274]  Brooke Shipley. "A convenient model category for commutative ring spectra". In: *Homotopy theory: relations with algebraic geometry, group cohomology, and algebraic K-theory.* Vol. 346. Contemp. Math. Amer. Math. Soc., Providence, RI, 2004, pp. 473–483.

[275]   Brooke Shipley. "$H\mathbb{Z}$-algebra spectra are differential graded algebras". In: *Amer. J. Math.* 129.2 (2007), pp. 351–379. issn: 0002-9327.

[276]   M. Shulman. *Homotopy limits and colimits and enriched homotopy theory.* arXiv 0610194. 2009.

[277]   Michael Shulman. "Comparing composites of left and right derived functors". In: *New York J. Math.* 17 (2011), pp. 75–125. issn: 1076-9803.

[278]   Carlos Simpson. *Homotopy theory of higher categories.* Vol. 19. New Mathematical Monographs. Cambridge University Press, Cambridge, 2012, pp. xviii+634. isbn: 978-0-521-51695-2.

[279]   V. A. Smirnov. "Homotopy theory of coalgebras". In: *Izv. Akad. Nauk SSSR Ser. Mat.* 49.6 (1985), pp. 1302–1321, 1343. issn: 0373-2436.

[280]   Markus Spitzweck. "Operads, algebras, and modules in general model categories". arXiv:0101102. 2001. url: https://arxiv.org/abs/math/0101102.

[281]   The Stacks Project Authors. *Stacks Project.* url: http://stacks.math.columbia.edu.

[282]   James Dillon Stasheff. "Homotopy associativity of $H$-spaces. I, II". In: *Trans. Amer. Math. Soc. 108 (1963), 275-292; ibid.* 108 (1963), pp. 293–312. issn: 0002-9947.

[283]   N. E. Steenrod. "Products of cocycles and extensions of mappings". In: *Ann. of Math. (2)* 48 (1947), pp. 290–320. issn: 0003-486X.

[284]   D. Sullivan. "Differential forms and the topology of manifolds". In: *Manifolds— Tokyo 1973 (Proc. Internat. Conf., Tokyo, 1973).* Univ. Tokyo Press, Tokyo, 1975, pp. 37–49.

[285]   Dennis Sullivan. "Genetics of homotopy theory and the Adams conjecture". In: *Ann. of Math. (2)* 100 (1974), pp. 1–79.

[286]   Dennis Sullivan. "Infinitesimal computations in topology". In: *Inst. Hautes Études Sci. Publ. Math.* 47 (1977), 269–331 (1978).

[287]   Dennis P. Sullivan. *Geometric topology: localization, periodicity and Galois symmetry.* Vol. 8. $K$-Monographs in Mathematics. The 1970 MIT notes, Edited and with a preface by Andrew Ranicki. Springer, Dordrecht, 2005, pp. xiv+283.

[288]   Robert M. Switzer. *Algebraic topology—homotopy and homology.* Classics in Mathematics. Reprint of the 1975 original [Springer, New York; MR0385836 (52 #6695)]. Springer-Verlag, Berlin, 2002, pp. xiv+526. isbn: 3-540-42750-3.

[289]   Markus Szymik. "Brauer spaces for commutative rings and structured ring spectra". In: *Manifolds and K-theory.* Vol. 682. Contemp. Math. Amer. Math. Soc., Providence, RI, 2017, pp. 189–208.

[290]   Bertrand Toën. *Derived Algebraic Geometry.* arXiv:1404.1044. 2014. url: http://arxiv.org/abs/1404.1044.

[291]   Bertrand Toën. "Derived Azumaya algebras and generators for twisted derived categories". In: *Invent. Math.* 189.3 (2012), pp. 581–652.

[292]   Bertrand Toën. "Vers une axiomatisation de la théorie des catégories supérieures". In: *K-Theory* 34.3 (2005), pp. 233–263.

[293]   D. R. B. Verity. "Weak complicial sets. I. Basic homotopy theory". In: *Adv. Math.* 219.4 (2008), pp. 1081–1149. issn: 0001-8708.

[294]   Vladimir Voevodsky. "$\mathbb{A}^1$-homotopy theory". In: *Proceedings of the International Congress of Mathematicians, Vol. I (Berlin, 1998)*. Extra Vol. I. 1998, pp. 579–604.

[295]   Rainer M. Vogt. "Homotopy limits and colimits". In: *Math. Z.* 134 (1973), pp. 11–52. issn: 0025-5874.

[296]   Friedhelm Waldhausen. "Algebraic $K$-theory of topological spaces. I". In: *Algebraic and geometric topology (Proc. Sympos. Pure Math., Stanford Univ., Stanford, Calif., 1976), Part 1*. Proc. Sympos. Pure Math., XXXII. Amer. Math. Soc., Providence, R.I., 1978, pp. 35–60.

[297]   Charles A. Weibel. *An Introduction to Homological Algebra*. Vol. 38. Cambridge Studies in Advanced Mathematics. Cambridge University Press, Cambridge, 1994, pp. xiv+450.

[298]   Charles A. Weibel and Susan C. Geller. "Étale descent for Hochschild and cyclic homology". In: *Comment. Math. Helv.* 66.3 (1991), pp. 368–388. issn: 0010-2571.

[299]   David White. "Model structures on commutative monoids in general model categories". In: *J. Pure Appl. Algebra* 221.12 (2017), pp. 3124–3168. issn: 0022-4049.

[300]   David White. *Monoidal Bousefield Localizations and Algebras over Operads*. Thesis (Ph.D.)–Wesleyan University. ProQuest LLC, Ann Arbor, MI, 2014, p. 191. isbn: 978-1369-19411-1.

[301]   David White and Donald Yau. "Bousfield localization and algebras over colored operads". In: *Appl. Categ. Structures* 26.1 (2018), pp. 153–203. issn: 0927-2852.

[302]   Richard Woolfson. "Hyper-$\Gamma$-spaces and hyperspectra". In: *Quart. J. Math. Oxford Ser. (2)* 30.118 (1979), pp. 229–255. issn: 0033-5606.

[303]   Urs Würgler. "Commutative ring-spectra of characteristic 2". In: *Comment. Math. Helv.* 61.1 (1986), pp. 33–45. issn: 0010-2571.

# Index

Printed in the United States
by Baker & Taylor Publisher Services